Algorithms and Combinatorics 19

Springer
Berlin
Heidelberg
New York
Barcelona
Hong Kong
London
Milan
Paris
Singapore
Tokyo

Adalbert Kerber

Applied Finite Group Actions

2nd, Revised and Expanded Edition

Springer

Adalbert Kerber
University of Bayreuth
Department of Mathematics
95440 Bayreuth
Germany
kerber@uni-bayreuth.de

Cataloging-in-Publication Data applied for

Die Deutsche Bibliothek – CIP-Einheitsaufnahme

Kerber, Adalbert:
Applied finite group actions / Adalbert Kerber. – 2., rev. and expanded ed. – Berlin; Heidelberg; New York; Barcelona; Hong Kong; London; Milan; Paris; Singapore; Tokyo: Springer, 1999
(Algorithms and combinatorics; 19)
1. Aufl. im BI-Wiss.-Verl. u.d.T.: Kerber, Adalbert: Algebraic combinatorics via finite group actions
ISBN 3-540-65941-2

The first edition of this book was published by BI Wissenschaftsverlag in 1991

Mathematics Subject Classification (1991):
05Exx, 20-01, 20-02, 20B30, 20B35, 20C15, 20C30

ISSN 0937-5511
ISBN 3-540-65941-2 Springer-Verlag Berlin Heidelberg New York

Computer to film: Mercedesdruck, Berlin
Binding: Buchbinderei Lüderitz & Bauer, Berlin
Typesetting: Typeset in LaTeX by the author.
SPIN 10654673 41/3143 - 5 4 3 2 1 0 - Printed on acid-free paper

To Dieta

Preface to the Second Edition

Also the present second edition of this book is an *introduction* to the theory of classification, enumeration, construction and generation of finite unlabeled structures in mathematics and sciences.

Since the publication of the first edition in 1991 the constructive theory of unlabeled finite structures has made remarkable progress. For example, the first 7-designs with moderate parameters were constructed, in Bayreuth, by the end of 1994 ([9]). The crucial steps were

- the prescription of a suitable group of automorphisms, i. e. a stabilizer, and the corresponding use of Kramer–Mesner matrices, together with
- an implementation of an improved version of the LLL-algorithm that allowed to find 0-1-solutions of a system of linear equations with the Kramer–Mesner matrix as its matrix of coefficients.

The Kramer–Mesner matrices can be considered as submatrices of matrices of the form $A^{\wedge}$ (see the chapter on group actions on posets, semigroups and lattices). They are associated with the action of the prescribed group G which is a permutation group on a set X of points induced on the power set of X. *Hence the discovery of the first 7-designs with small parameters is due to an application of finite group actions.* This method used by A. Betten, R. Laue, A. Wassermann and the present author is described in a section that was added to the manuscript of the first edition. In the meantime this team found millions of new 7-designs and the existence of $t-(v,k,\lambda)$-designs has been proved for many new quadruples (t,v,k,λ) of parameters. The first 8-designs with small parameters were constructed in 1997 by the same people ([10]).

Another interesting development is the enumeration and the construction of transversals of isometry classes of indecomposable error-correcting linear codes. A. Betten, H. Fripertinger, A. Wasserman and K.-H. Zimmermann should be mentioned here. The indecomposable codes can be applied since each linear code is an essentially unique sum of such indecomposable ones, and its minimal distance is the minimum of the minimal distances of these summands. Here again an intensive use has been made of actions of finite groups. Meanwhile a book containing most of the details appeared ([8]).

Concept analysis appears to be another interesting field of applications since the stabilizer of the context acts on the lattice of concepts. I would like to thank W. Lex in particular for a very interesting and stimulating cooperation.

I would also like to mention the continual effort put in the research on the generation of molecular graphs corresponding to a given chemical formula and optional further conditions. This solution of the isomerism problem has already been touched in the first edition. It is one of the basic tasks of graph theory to provide these graphs. Even the name *graph* comes from chemistry. J. J. Sylvester introduced it in his paper *Chemistry and Algebra*, Nature **17** (1877/8), as an abbreviation of the notion *chemicograph*. Moreover, it gave rise to algebraic combinatorics and finite group actions — Pólya's masterly paper *Anzahlbestimmungen für Gruppen, Graphen und Chemische Verbindungen*, Acta Sci. Math. **68** (1937), 145-254, shows this in a brilliant way. A very recent application of these methods it the mathematical simulation of *combinatorial chemistry,* the enumeration and construction of molecular *libraries.* It is described in this second edition, too. This problem, the corresponding research and the implementation of software that constructs molecular graphs efficiently, has been very helpful again in order to develop both the theory and its applications. But it also has shown how difficult it is, really to carry through a concrete and sophisticated application, to implement a *product* that can be used efficiently in research, education and industry. Therefore many thanks are due again to the DFG (Deutsche Forschungsgemeinschaft) and the BMBF (Bundesministerium für Bildung und Forschung) for very helpful financial support of several accompanying research projects mostly run together with R. Laue, special thanks are due to him. At present T. Grüner, M. Meringer and A. Ruckdeschel continue the successful work of their predecessors Ch. Benecke, R. Grund, D. Moser and T. Wieland.

All that — together with encouraging remarks of several colleagues — stimulated me to prepare a second edition. In particular further applications of finite group actions to the constructive theory of finite structures were added, so that a corresponding change of title may be justified. The former title was *Algebraic Combinatorics via Finite Group Actions*. Misprints and some errors were corrected, the notation changed (which may have caused new misprints, hopefully not too many), sections were rearranged, a chapter on labeled structures put in front etc.

I should like to express my cordial thanks to H. Fripertinger and M. Hofmeister who carefully read the manuscript and gave me very many useful hints. Many thanks are also due to an anonymous referee who made many helpful remarks which were gratefully acknowledged.

Last but not least I would also like to thank Dr. Peters, his crew and the Springer-Verlag for very good and efficient cooperation concerning the publication and the typesetting of this second edition.

Bayreuth, May 12, 1999 Adalbert Kerber

Preface to the First Edition

This book is an introduction to the theory of classification, enumeration, construction and generation of certain discrete structures in mathematics and the sciences. The structures in question are those which can be defined as equivalence classes on finite sets and in particular on finite sets of mappings. Prominent examples are graphs, switching functions, physical states and chemical isomers. Since powerful computers are now available for a cheap price, this theory has gained a rapidly increasing interest. The method used is to replace the equivalence relation by a finite group action and to apply algebraic tools like the Cauchy–Frobenius Lemma and its refinements. This will be worked out in full detail, starting with mere enumeration, refining it to enumeration by weight, counting by stabilizer class, by weight and stabilizer class etc. Finally we shall reach the point where we can describe several algorithmic methods which allow to construct the structures in question or to generate them uniformly at random.

In order to describe all this I assume on the side of the reader that he knows the basic concepts of algebra, but not at more than the usual undergraduate level.

I have tried to give a survey of the present situation of this theory, but in view of the flood of publications, many results had to be excluded, for example asymptotic methods. Other parts – like the theory of q-analogues, the theory of species or the theory of Schubert polynomials – are hardly touched, they deserve separate monographs. On the other hand the Pólya theory of enumeration is described in full detail, and most of the examples are taken from there. This theory is easy to absorb and it provides many beautiful examples, in particular in graph theory, but also in physics and chemistry. It was in fact the problem of chemical isomerism which led to its early development.

For a deeper insight I shall refine the basically permutation theoretical arguments to considerations of the corresponding linear representations of the groups in question and in particular of symmetric groups. Therefore, I did not hesitate to include linear representation theory giving a selfcontained and problem oriented introduction based mainly on finite group actions and set theoretic arguments.

The manuscript is the result of various lectures and seminars at Aachen and Bayreuth, of theses of my students, of research projects on the algebraic and combinatorial description of molecules, of helpful discussions, of many talks I heard or gave at the various meetings of the Lotharingian Seminar of Combinatorics and in particular of joint efforts of K.–J. Thürlings and myself which led to an earlier German version already published in the *Bayreuther Mathematische Schriften* (vols.

12 (1983), **15** (1983) and **21** (1986)). Hence I would like to express my sincerest thanks in particular to K.–J. Thürlings and also to M. Clausen, A. Dress, N. Esper, D. Foata, A. Golembiowski, J. Grabmeier, W. Hässelbarth, D. Jungnickel, A. Kohnert, A. Lascoux, R. Laue, A. O. Morris, J. Neubüser, W. Oberschelp, P. Paule, E. Ruch, F. Sänger, Th. Scharf, D. Stockhofe, F. Stötzer, V. Strehl, J. Tappe, B. Wagner, W. Lehmann and all the other people I had and have the pleasure to work with.

Moreover thanks are due to the Stiftung Volkswagenwerk and the Deutsche Forschungsgemeinschaft, which have supported research on molecular structure elucidation. It produced several results on combinatorial enumeration and a satisfactory solution of the basic problem of this theory, namely the program system MOLGRAPH that provides the molecular graphs corresponding to a chemical formula, the so–called connectivity isomers. This program system was implemented by D. Moser, and it was supported by many students among which I would like to mention in particular F. Bauer, W. Decker, R. Grund, R. Hager, B. Schmalz and W. Weber.

Furthermore I am indebted to H. Engesser for friendly, patient and efficient cooperation with the publishing company.

Last not least thanks are due to D.E. Knuth for providing both his generalization of the Robinson–Schensted construction and TEX, to L. Lamport for the development of LATEX, and to Stürtz, the printing company, for making the best of it.

Bayreuth, December 5, 1990 Adalbert Kerber

Table of Contents

List of Symbols

$\mathcal{S}$	a species, 2
$\mathcal{S}[M]$	the set of $\mathcal{S}$-structures on M, 2
$\mathcal{P}$	the species power set, 2
$\mathcal{P}^{[k]}$	the species k-subsets, 3
$\binom{M}{k}$	the set of all subsets of order k in M, 3
Par	the species set partitions, 3
$\mathbb{N}$	the set $\{0, 1, 2, \ldots\}$ of natural numbers, 3
$\mathbb{N}^*$	the set $\{1, 2, \ldots\}$ of nonzero natural numbers, 3
$\mathcal{G}$	the species simple graphs, 3
$\mathcal{G}^c$	the species connected graphs, 3
$\mathcal{G}^d$	the species directed graphs, 4
$\mathcal{T}$	the species trees, 4
$\mathcal{T}^r$	the species rooted trees, 4
Per	the species permutations, 4
$\mathcal{C}$	the species oriented cycles, 5
$\mathcal{L}$	the species linear orders, 5
End	the species endofunctions, 5
$\mathcal{X}$	the species singleton, 7
0	the empty species, 7
1	the species empty set, 7
$\mathcal{M}_k$	the species k-sets, 7
$\|M\|$	the cardinality of the set M, 7
n	the set $\{0, \ldots, n-1\}$ or its cardinality, 7
$\underline{n}$	the set $\{1, \ldots, n\}$, 7
s_n	$\|\mathcal{S}[n]\|$, 7
$\mathcal{S}(x)$	the generating function of the cardinal numbers s_n, 7
$\mathcal{S}_+[M]$	$\mathcal{S}[M]$, if $M \neq \emptyset$, and $\emptyset$, otherwise, 8
$\mathcal{R} \equiv \mathcal{S}$	equipotency of two species, 8

$(038)(124)(57)(6)$	a permutation in cycle notation, 8
$[324817650]$	a permutation in in list notation, 8
$\mathcal{R} \simeq \mathcal{S}$	isomorphy of two species, 8
$\mathcal{R} + \mathcal{S}$	the sum of the two species $\mathcal{R}$ and $\mathcal{S}$, 10
$\sum_{i \in I} \mathcal{S}_i$	the sum of a family of species, 10
$(\mathcal{S}_n)_{n \geq 0}$	the canonic decomposition of $\mathcal{S}$, 10
$\mathcal{R} \cdot \mathcal{S}$	the product of the two species $\mathcal{R}$ and $\mathcal{S}$, 11
Der	the species derangements, 11
$(\mathcal{R} \cdot \mathcal{S})\,(x)$	the cardinality of the product, 11
der_n	the number of derangements of a set of order n, 12
$\mathcal{R}(\mathcal{S})$	the partitional composition of $\mathcal{R}$ and $\mathcal{S}$, 13
c_n	the number of oriented cycles on n vertices, 15
$\mathcal{S}'$	the derivative of $\mathcal{S}$, 16
M^+	arises from M by adding a further element, 16
$\mathcal{S}^\bullet$	the pointing of $\mathcal{S}$, 16
$\mathcal{R} \times \mathcal{S}$	the cartesian product of two species $\mathcal{R}$ and $\mathcal{S}$, 18
$\mathcal{R} \circ \mathcal{S}$	the (functorial) composition of $\mathcal{R}$ and $\mathcal{S}$, 18
$\mathfrak{S}$	the class of isomorphism classes of species, 19
$\mathfrak{H}$	the ring $(\mathfrak{S} \times \mathfrak{S})/\sim$ of equivalence classes, 19
${}_GX$	an action of G on the set X from the left, 22
δ	a permutation representation, 23
$G_X := \ker(\delta)$	the kernel of ${}_GX$ and δ, 23
$N \trianglelefteq G$	N is a normal subgroup in G, 23
$x \sim_G x'$	the equivalence relation induced by ${}_GX$ on X, 24
$G(x)$	the orbit of x under the action of G, 24
$G \backslash\backslash X$	the set of all orbits of G on X, 24
$\mathcal{T}(G \backslash\backslash X)$	the set of all the transversals of $G \backslash\backslash X$, 24
$\oplus_i S_{X_i}$	a Young subgroup, 24
G_x	the stabilizer of x in G, 25
G_Δ	the pointwise stabilizer or centralizer of Δ, 25
$G_{\{\Delta\}}$	the setwise stabilizer or normalizer of Δ, 25
X_g	the set of fixed points of g on X, 25
X_S	the set of fixed points of $S \subseteq G$ on X, 25
X_G	the set of invariants of G on X, 25
$\widetilde{\mathcal{S}}(x)$	the type series of the species $\mathcal{S}$, 25
Ug	a right coset of U in G, 26

gU	a left coset of U in G, 26
$C^G(x)$	the conjugacy class of x in G, 27
$C_G(x)$	the centralizer of x in G, 27
G/U	the set of left cosets of U in G, 27
$L(G)$	the set (lattice) of subgroups of G, 27
$\widetilde{U}$	the conjugacy class of the subgroup U in G, 27
$N_G(U)$	the normalizer of U in G, 27
${}_GX \simeq {}_HY$	the isomorphy of two actions, 31
${}_GX \approx {}_HY$	the similarity of two actions, 31
${}_GX \sim {}_HY$	the homomorphy of two actions, 32
AgB	the (A, B)-double coset containing $g \in G$, 33
$\Delta(G \times G)$	the diagonal subgroup of $G \times G$, 33
$A\backslash G/B$	the set of all (A, B)-double cosets in G, 33
$X\dot{\cup}Y$	the disjoint union of X and Y, 34
$X \times Y$	the cartesian product of X and Y, 35
Y^X	the set of all the mappings from X into Y, 36
$E^{\bar{G}}$	the permutation group induced by G on Y^X, 37
$\bar{H}^E$	the permutation group induced by H on Y^X, 37
$\bar{H}^{\bar{G}}$	the permutation group induced by $H \times G$ on Y^X, 37
$H \wr_X G$	a wreath product, 37
$[\bar{H}]^{\bar{G}}$	the group induced by $H \wr_X G$ on Y^X, 38
$H \wr G$	a wreath product, 39
$\binom{v}{2}$	the set of 2-subsets of of the set v, 40
$GF(q)^n$	the n-dimensional vector space over $GF(q)$, 40
$GF(q)$	the Galois field of q elements, 40
$GL_k(q)$	a general linear group over $GF(q)$, 42
$N(n, k, q)$	$\lvert GL_k(q) \times GF(q)^* \wr S_n \backslash\backslash \left(GF(q)^k \backslash \{0\}\right)^n \rvert$, 43
$T(n, k, q)$	$N(n, k, q) - N(n, k-1, q)$, 43
$P_{k-1}(q)$	a projective space, 43
$H \sim_k K$	$H \backslash\backslash n^k = K \backslash\backslash n^k$, 46
$\bar{H}_{(k)}$	the k-closure of $\bar{H}$, 47
$\omega_{(x)}$	a particular orbit, 48
I_M	the indicator function of the set M, 49
$Hom_H(\mathbb{F}^{(n^k)}, \mathbb{F}^{(n^l)})$	an intertwining space, 49
$\mathrm{hol}(G)$	the holomorph of G, 51
χ	the character of an action, 54

$\bar{G}^+$	the kernel of the sign on $\bar{G}$, 57
G^+	the inverse image of $\bar{G}^+$, 57
δ	an embedding of $S_m \wr S_n$ into S_{mn}, 59
$\bar{H} \odot \bar{G}$	the image of $H \wr G$ under δ, the plethysm, 60
$h_\nu(\psi, \pi)$	a cycle product, 60
$a(\psi, \pi)$	the type of (ψ, π), 61
$l_\nu \cdot \alpha$	the partition $(l_\nu \cdot \alpha_0, l_\nu \cdot \alpha_1, \ldots)$, 61
$S_n[S_m]$	a composition of symmetric groups, 62
$\bar{G}[\bar{H}]$	a composition of groups, 63
$M^\pm$	two subsets of M such that $M = M^+ \dot{\cup} M^-$, 74
$\mathrm{sign}(m)$	a value of a sign function, 74
A^*	$\{a \mid$ there exists no i such that $P_i(a)\}$, 74
$\phi(n)$	a value of the Euler function, 75
Y^X_{inj}	the set of injective elements of Y^X, 77
Y^X_{sur}	the set of surjective elements of Y^X, 77
$S(n, m)$	a Stirling number of the second kind, 81
B_n	a Bell number, 81
$r(n, k)$	a Stirling number of the first kind, 82
$t(n, k)$	$\|\{\pi \in S_n \mid a_1(\pi) = k\}\|$, 82
$[x]_k$	$x(x-1) \cdot \ldots \cdot (x-k+1)$, 84
w	a weight function, 86
$c(f, -)$	the content of the mapping f, 87
$\left[{n \atop k} \right]$	a certain rational function, 89
$\left[{n \atop k} \right]_q$	a q-binomial number, 89
$C(G, X)$	the cycle indicator polynomial or cycle index, 91
$C(G, X; z_1, \ldots, z_{\|X\|})$	the cycle index with indeterminates displayed, 91
$C(G, X \mid p(u_0, \ldots))$	the cycle index with the polynomial p inserted, 92
$C(H, Y) \odot C(G, X)$	the plethysm of cycle indices, 94
$C(G, X, \chi)$	a generalized cycle index, 96
$C_{\mathcal{S}}$	the cycle series of $\mathcal{S}$-structures, 97
$[y]^n$	the rising factorial $y(y+1)\cdots(y+n-1)$, 99
$\mathbb{Q}[[z_1, z_2, \ldots]]$	a ring of formal power series, 101
$(a_{ik}) \sim_c (a'_{ik})$	column equivalence of matrices, 110
$(a_{ik})_c$	the equivalence class of (a_{ik}), 110
M_c	the set of matrix equivalence classes, 110
$C(H, n) \cap C(G, n)$	the cap product of two cycle indices, 115

$Y^c \cup \ldots \cup Y^d$	the cup product of monomials, 116
$G \backslash\!\backslash_{\widetilde{U}} X$	the $\widetilde{U}$–stratum of G on X, 121
$\mu(-,-)$	the Möbius function on $L(G)$, 122
U^i	the i-th subgroup of G, 122
$\zeta(G)$	the zeta-matrix of G, 122
$\mu(G)$	the Möbius-matrix of G, 122
$\widetilde{U}_i$	the i-th conjugacy class of subgroups of G, 123
U_i	a representative of $\widetilde{U}_i$, 123
$\widetilde{L}(G)$	the set of conjugacy classes of subgroups of G, 123
$\mu(U_i, \widetilde{U}_k)$	$\sum_{V \in \widetilde{U}_k} \mu(U_i, V)$, 123
$B(G) = (b_{ik})$	the Burnside matrix of G, 123
$\zeta(U_i, \widetilde{U}_k)$	$\sum_{V \in \widetilde{U}_k} \zeta(U_i, V)$, 123
$M(G)$	$B(G)^{-1}$, the table of marks, 123
σ	the Frobenius isomorphism, 126
$G \backslash\!\backslash_{\lvert G\rvert} X$	the set of orbits of length $\lvert G\rvert$, 127
$\mu(-)$	the number theoretic Möbius function, 127
l_{mn}	$\lvert C_n \backslash\!\backslash_n m^n\rvert$, a Dedekind number, 127
$L_n(m)$	the Lyndon words of length n over m, 128
$L(m)$	the Lyndon set over the alphabet m, 128
$l_0(f)\ldots l_{\lambda(f)-1}(f)$	a decomposition of f into Lyndon words, 128
$l_n(x,m)$	a certain generating function, 128
$L(x,m)$	a particular formal power series, 130
$\overline{{}_G X}$	the set of asymmetric elements, 130
$\overline{\mathcal{S}}$	the species asymmetric $\mathcal{S}$-structures, 130
$\overline{\mathcal{S}}[n]$	the asymmetric $\mathcal{S}$-structures on n, 130
$\zeta(\widetilde{U}, V)$	$\sum_{W \in \widetilde{U}} \zeta(W, V)$, 131
$\preceq$	order on the conjugacy classes of subgroups, 131
χ_i	the permutation character of G on G/U_i, 133
$A(G,X)$	the asymmetry indicator, 139
$(X, \leq)$	a poset, 142
${}_G(X, \leq)$	a poset action, 142
$(L, \wedge, \vee)$	a lattice, 145
${}_G(L, \wedge, \vee)$	a lattice action, 146
$\mathcal{O}$	a set of objects, 146
$\mathcal{P}$	a set of properties, 146
C	a context, 146

O', P'	the derivative of a set of objects, properties, 146
$\mathcal{C}$	a concept, 146
$\mathcal{CL}(C)$	the set of all the concepts of C, 146
$a_{ik}^{\wedge}$	$\vert\{x' \in \omega_k \mid x \leq x'\}\vert$, for an $x \in \omega_i$, 147
$a_{ik}^{\vee}$	$\vert\{x' \in \omega_k \mid x \geq x'\}\vert$, for an $x \in \omega_i$, 147
${}_G(X, \cdot)$	a semigroup action, 148
$a_{ijk}^{\cdot}$	$\vert\{(x, x') \in \omega_i \times \omega_j \mid x \cdot x' = z\}\vert$, for a $z \in \omega_k$, 148
$\mathbb{Z}_G^X$	the G–invariants, 148
$\mathbb{Z}_G^{X,\cdot}$	the ring of G–invariants, a subring of $\mathbb{Z}^X$, 149
$\underline{\omega_i}$	an orbit sum, 149
$\mathbb{Z}_G^{L,\wedge}$	a particular ring of invariants, 149
$\mathbb{Z}_G^{L,\vee}$	another ring of invariants, 149
$SP(n)$	the set of set partitions of n, 151
$\alpha(p)$	the type of the set partition p, 151
$(SP(n), \leq)$	the poset of set partitions of n, 152
$r(p)$	the rank of the set partition p, 152
$\mathcal{L}(d, q)$	the lattice of subspaces of $GF(q)^d$, 153
$G_{m,n}$	a Gaussian polynomial, 153
$(V, \mathcal{B})$	a $t - (v, k, \lambda)$-design , 157
$M_{t,k}^v = (m_{TB}^V)$	the matrix defining $t - (v, k, \lambda)$-designs, 158
$Aut(V, \mathcal{B})$	the full automorphism group of the design, 159
$M_{t,k}^G := (m_{T,K}^G)$	the Kramer–Mesner matrix for $t - (v, k, \lambda)$-designs with G as a group of automorphisms, 159
$\overline{G/U_i}$	a similarity class of transitive group actions, 162
n_t^G	the set of mappings f from G to n of weight t, 165
$\widetilde{\Omega}(G) := \mathbb{Z}_{\sim}^{L(G)}$	a ghost ring, 166
$\lambda \models n$	λ is an improper partition of n, 169
n^λ	a sequence of subsets of n, 169
n_i^λ	a subset of order λ_i in n, 169
$S(n^\lambda)$	the stabilizer of the sequence n^λ, 170
S_λ	the canonic Young subgroup associated with λ, 170
$\lambda(n)$	the set of λ-flags on n, 170
ξ^λ	the Young character associated with λ, 171
$IS_\lambda \uparrow S_n$	the representation of S_n induced by the identity representation of S_λ, 171
ι	the identity character, 171

$\mathcal{HA}_n[Y]$	the subspace of $\mathbb{Q}[Y]$ consisting of the *homogeneous* alternating polynomials of degree $n+\binom{m}{2}$, 237
$\mathcal{HA}[Y]$	the vector space of the alternating polynomials, 237
Δ_0	the Vandermonde determinant or the left multiplication by this polynomial, 238
$g^{\gamma}_{\alpha\beta}$	the multiplicity $([\alpha][\beta],[\gamma])$, 244
$\alpha^{i\pm}$	$(\alpha_0,\ldots,\alpha_{i-1},\alpha_i\pm 1,\alpha_{i+1},\ldots)$, 244
$[\alpha]\downarrow S_{n-1}$	the restriction of $[\alpha]$ of S_n to the subgroup S_{n-1}, 244
$[\alpha]\uparrow S_{n+1}$	the representation induced by $[\alpha]$ in S_{n+1}, 244
$P(\mathbb{N})$	the union of the sets of proper partitions $P(n)$, 245
$\alpha\wedge\beta$	the infimum of two partitions, 245
$\alpha\vee\beta$	the supremum of two partitions, 245
$(P(\mathbb{N}),\wedge,\vee)$	Young's Lattice, 245
$\gamma\backslash R^{\gamma}_{ij}$	the diagram after subtracting R^{γ}_{ij} from $[\gamma]$, 248
λ	a composition of a natural number, 250
$\mathcal{S}$	the restricted symmetric group on $\mathbb{N}$, 251
$\widetilde{\mathbf{D}}$	a representation of $H\wr_X G$ on $\mathbb{C}^{(Y^X)}$, 255
$D\boxdot D_i$	the $\lvert X\rvert$–fold symmetrization of D by D_i, 258
P^X_Y	the permutation representation on $\mathbb{C}^{(Y^X)}$, 258
$D\mathbin{\triangle_{\lvert X\rvert}} D_i$	the $\lvert X\rvert$–fold permutrization of D by D_i, 259
$[\alpha]\odot[\beta]$	the plethysm of $[\alpha]$ and $[\beta]$, 261
$\mathrm{Ch}(X)$	the set of all the chains on X, 267
$\mathrm{Ch}(X,l)$	the set of all the chains of length l on X, 267
$\mathrm{Ch}(X,l,\mathbb{C})$	the complex vector space with basis $\mathrm{Ch}(X,l)$, 267
$g^{(l)}$	the extension of the action of g to $\mathrm{Ch}(X,l,\mathbb{C})$, 267
d_l	the differential on $Ch(X,l,\mathbb{C})$, 267
$H_l(X,\mathbb{C})$	the l-th homology group of the chain complex, 268
$\Lambda_X(g)$	a Lefschetz number, 268
$\chi(X)$	the Euler characteristic of X, 268
$R(C)$	the rank set of C, 270
X_R	the subset of X consisting of the elements of rank in R, together with $\hat{0}$ and $\hat{1}$, 270
$\mathrm{MCh}(X,R)$	a set of rank selected maximal chains, 270
$\kappa_R(g)$	the trace of g on $\mathrm{MCh}(X,R,\mathbb{C})$, 270
$\gamma_{R,i}(g)$	the trace of g on $H_i(X_R,\mathbb{C})$, 270
ν_R	the Lefschetz character of G on X_R, 270

$P^{[k]}$, $P^{(k)}$, $P^{\langle k\rangle}$	the permutation groups induced by $P \leq S_n$ on the sets of k–subsets, k–tuples, injective k–tuples, 275
$\pi^{[k]}$, $\pi^{(k)}$, $\pi^{\langle k\rangle}$	the permutations induced by $\pi \in S_n$ on the sets of k–subsets, k–tuples and injective k–tuples, 276
d_α	the depth of the partition $\alpha \vdash n$, 283
$R_k(g)$	the number of k-th roots of $g \in G$, 284
p_k	the k-th power map on $G : g \mapsto g^k$, 284
r_k	the k-th root number function, 284
$c_{i,k}$	generalizes the Frobenius-Schur invariant $c_{i,2}$, 284
χ_n^k	a certain character, 286
C_n^k	a union of conjugacy classes, 286
$[\pi 0 \ldots \pi(n-1)]$	the list notation of $\pi \in S_n$, 297
$U[02137654]$	the up–down sequence of [02137654], 297
$E(n,k)$	an Eulerian number, 297
$\alpha(U)$	shape of the skew diagram corresponding to U, 297
$R(U)$	the rim hook of $[\alpha(U)]$, 297
$[\widetilde{R}(U)]$	the sum of the representations obtained from U, 300
t_n	a number of certain up–down sequences, 302
$\chi^{n,k}$	a Foulkes character of S_n, 303
F_n	the Foulkes table of S_n, 305
$\chi^{(m)}$	the character of the natural action of S_n on m^n, 306
∂_i	a linear operator on $\mathbb{Z}[x_0, \ldots, x_{n-1}]$, 307
X^E	the monomial $x_0^{n-1} x_1^{n-2} \ldots x_{n-2}^1$, 310
X_π	the Schubert polynomial corresponding to π, 310
Ω_i	an orbit, 318
$C_{\bar{G}}(k)$	the centralizer of the subset $k \subseteq n$, 319
$N_{\bar{G}}(k)$	the normalizer of the subset $k \subseteq n$, 319
$\pi_j^{(i)}$	a left coset representative of $C_{\bar{G}}(i)$ in $C_{\bar{G}}(i-1)$, 319
$r(i)$	the index of $C_{\bar{G}}(i)$ in $C_{\bar{G}}(i-1)$, 319
$S_p^{[2]}$	the group induced by S_p on the pairs, 320
m_λ^n	the set of elements of content λ in m^n, 322
$\omega_k^{(i)}$	a particular orbit, 327
$f <_X g$	a particular partial order, 328
$T_>(G \backslash\backslash Y^X)$	the canonic transversal consisting of the biggest elements in the orbits, 334
$p(-)$	a probability distribution, 337
$\hat{X}_U$	a subset of X_U, 340

V^i	a homogeneous component of V, 341
V^i_j	a subspace of V^i, 341
$b^{i,j}_k$	an element in a symmetry adapted basis, 341
$d^i_{kl}(g)$	an entry of the matrix representing g, 342
W^i_k	the subspace generated by $\{b^{i,j}_k \mid j \in n_i\}$, 343
S_X	the symmetric group on the set X, 398
$H \leq G$	H is a subgroup of G, 398
gH	a left coset of H, 398
Hg	a right coset of H, 398
$N \trianglelefteq G$	N is a normal subgroup of G, 398
$\ker(\varphi)$	the kernel of the homomorphism φ, 398
G/N	the factor group G modulo N, 398
$\alpha \vdash n$	α is a (proper) partition of n, 401
$a \models n$	a is a cycle type of n, 402
Σ_n, $\Sigma_{\underline{n}}$	the set of elementary transposition in S_n, $S_{\underline{n}}$, 403
$I(\pi)$	the set of inversions of π, 407
$L(\pi)$	the Lehmer code of π, 407
$L(\pi)^+$	the reduced Lehmer code of π, 407
$\pi \prec\!\cdot \rho$	the weak Bruhat order, 409
$RS(\pi)$	the set of reduced sequences for π, 411
$I_{\mathbb{F}}(P)$	the $\mathbb{F}$–algebra of incidence functions, 424
$\varphi \star \psi$	the convolution product of incidence functions, 425
$\delta(p,q)$	a value of the Kronecker δ-function, 425
$\zeta(p,q)$	a value of the zeta-function, 425
$\mu(-.-)$	a Möbius function, 426

0. Labeled Structures

Well-known discrete structures are graphs, linear codes, molecular graphs, designs. They usually occur in two forms, in a *labeled* form and in an *unlabeled* form. The difference between these forms is easily illustrated by graphs. Here is a *labeled graph:*

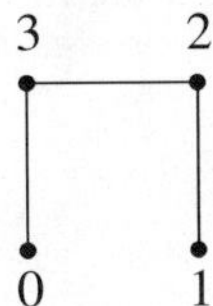

and here comes an *unlabeled graph:*

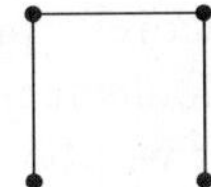

The difference is obvious, graphs — like many other incidence structures, for example, molecular graphs and designs — "live" on a set of points. In the labeled form these points carry labels, numbers or letters or other names. But these labels very often do not really matter, and so the corresponding unlabeled structure is defined to be the equivalence class of labeled structures, arising from each other by relabeling (renumbering, renaming etc.).

In other cases it is slightly more complicated but essentially the same situation. For example, the linear codes are finite vector spaces consisting of vectors or sequences of elements of a finite field. Their error correcting quality is not changed if we simultaneously permute the coordinates, and it does not change either if we simultaneously multiply a particular coordinate with a nonzero number. So we may consider the equivalence classes of linear codes with respect to these operations, the *isometry classes* of linear codes, and we may call the isometry classes of finite vector spaces the *unlabeled linear codes.*

We are mainly interested in unlabeled structures. The reason is clear in the case of graphs. They can be considered as *interaction models* for which the names of the objects that interact, and which are indicated by the vertices in the graph, do not matter at all. What matters is the interactions, and they are indicated by the edges. But we should keep in mind that a computer cannot directly deal with unlabeled objects like incidence structures unless the vertices carry numbers or names, and hence it is not very easy to enable a machine to handle unlabeled structures in a reasonable and efficient way.

In order to introduce unlabeled structures so that they can be handled properly we have therefore to begin with an introduction to labeled structures. The chosen way of dealing with them is the theory of *species,* which does not cover everything we want to demonstrate, but it is a modern and interesting way of formalizing at least the evaluation and examination of generating functions for many labeled structures.

Later on we shall introduce unlabeled structures as equivalence classes of labeled ones and we shall identify them with orbits of groups. This will bring us in a position where we can apply a rich set of algebraic results on finite group actions which will allow us to *count,* to *construct* and to *generate* many unlabeled structures uniformly at random. But this will be done in later chapters.

0.1 Species of Structures

Many finite structures consist of a uniquely defined collection of substructures. For example, a permutation consists of its cyclic factors, a graph consists of its connected components. Hence it is useful to look for an approach to structures that allows to formalize canonic constructions using disjoint union, cartesian product formation, recursion and other set theoretic procedures. A broad survey of many of these methods gives the book by Goulden and Jackson ([61]). Another approach makes use of the notion of species, it will be described now (more details on this theory can be found in [7], the standard reference for that theory).

To begin with we introduce the notion of *species.* It is defined to be a mapping $\mathcal{S}$ that associates to each finite set M a finite set

$$\mathcal{S}[M],$$

consisting of elements $\sigma \in \mathcal{S}[M]$ that can be expressed in terms of the labels m of the elements of M only. (This means in particular that for each bijection $\beta\colon M \to N$, say, we obtain the set $\mathcal{S}[N]$ from $\mathcal{S}[M]$ by simply replacing in each $\sigma \in \mathcal{S}[M]$ the labels m by the labels $\beta(m)$. We shall return to that later when we shall call this relabeling the *transport* of structure.)

The $\sigma \in \mathcal{S}[M]$ are called *the $\mathcal{S}$-structures on M*. In order to make that clear we collect a few easy set theoretic examples that are well known to the reader:

0.1.1 Basic set theoretic species and their notations:

- The species $\mathcal{P}$, associates to M its *power set,* the set of all subsets of M,

$$\mathcal{P}[M] := \{N \mid N \subseteq M\}.$$

It can be identified with the set

$$2^M := \{f : M \to \{0, 1\}\},$$

consisting of all the mappings f from M to the set $2 := \{0, 1\}$. In order to see this we only need to identify a subset $N \subseteq M$ with the mapping f that has N as inverse image of 1. The following mapping is a bijection:

$$2^M \to \mathcal{P}[M] : f \mapsto f^{-1}(1).$$

- By $\mathcal{P}^{[k]}$ we indicate the species *k-subsets* which maps M onto the set of all subsets of order k in M,

$$\mathcal{P}^{[k]}[M] := \binom{M}{k} := \{N \mid N \subseteq M, |N| = k\}.$$

- *Par* denotes the species *set partitions* of M, where $Par[M]$ is defined to be the set

$$\left\{\{N_0, \ldots, N_{n-1}\} \mid n \in \mathbb{N}^*, N_i \neq \emptyset, N_i \cap N_j = \emptyset, \text{ if } i \neq j, \cup_i N_i = M\right\}.$$

◇

Using these basic set theoretic species we can describe graph theoretic examples:

0.1.2 Basic graph theoretic species:

- The species *(simple) graphs,* which we denote by $\mathcal{G}$, associates to M the set of graphs on M. A graph on M has the elements of M as *vertices* or *points* and some of them are connected by an edge, but neither loops nor multiple edges are allowed to occur. Here are two elements of the set $\mathcal{G}[\{0, 1, 2, 3\}]$, i. e. graphs on $M := \{0, 1, 2, 3\}$:

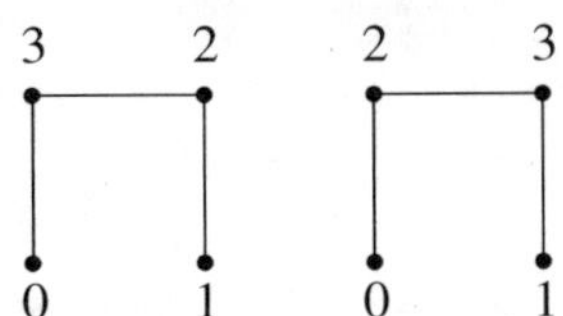

It is clear that a graph on M can be identified with the set of connected pairs of elements, and so the set $\mathcal{G}[M]$ of graphs on M can be identified with the power set

$$2^{\binom{M}{2}}$$

of the set of pairs of elements of M.

- The species *connected* graphs $\mathcal{G}^c$ associates to M the graphs on M, where we can reach each point (or vertex) from any other one by following edges. Both graphs in the preceding item are connected, here is a graph that is *not* connected:

3 2

$\in \mathcal{G}\,[\{0, 1, 2, 3\}]$

$\notin \mathcal{G}^c\,[\{0, 1, 2, 3\}]$

0 1

- The species *directed* graphs $\mathcal{G}^d$ yields the graphs on M, where each edge has a direction, but no parallel edges are allowed, no loops, but double edges which are not parallel are legitimate. Here is an example:

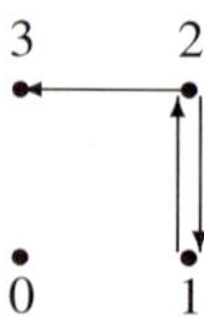

- The species $\mathcal{T}$, the *trees*, gives the connected simple graphs without cycles, which means that we cannot go from one point of the graph to another one and then return without using at least one edge twice. The examples of simple graphs shown in the corresponding item are trees, here is a graph that is *not* a tree:

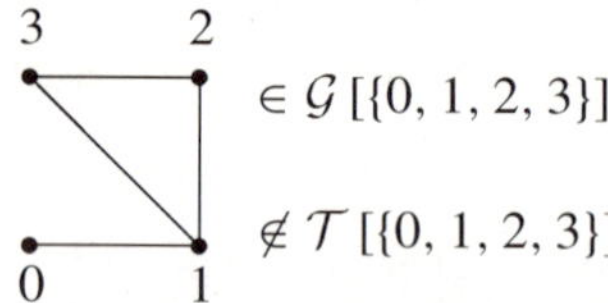

- $\mathcal{T}^r$, the *rooted* trees, which are trees with a distinguished point, the root. Here is a rooted tree with 4 vertices, the *root* (indicated by two small circles) is the vertex with number 0:

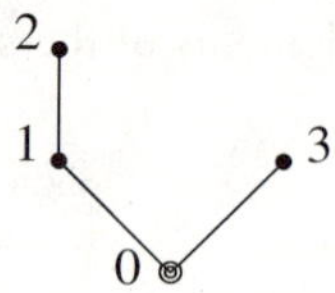

◇

Further important species associate to M particular sets of mappings:

0.1.3 Basic species of mappings:

- Per associates to M the set of *permutations* on M,

$$Per[M] := \{\pi \mid \pi : M \to M, \text{ bijectively}\}.$$

Here is a permutation π of the set $M := \{0, 1, 2, 3, 4, 5, 6, 7\}$:

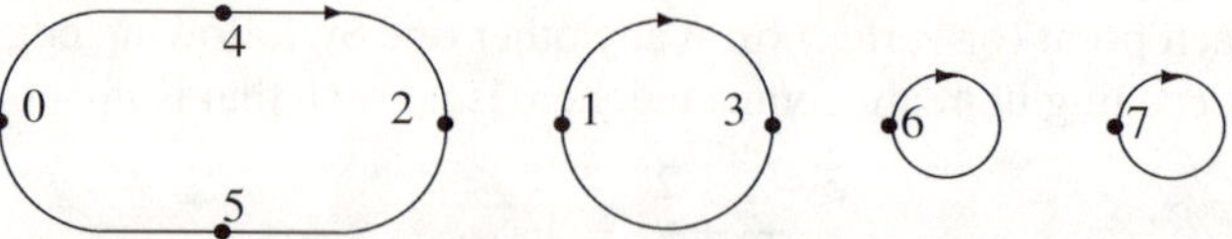

The arrows indicate the directions of the cyclic shifts in order to make clear that the permutation in question has the following decomposition into cyclic factors:

$$\pi = (0425)(13)(6)(7).$$

Nevertheless the reader should carefully note that the arrow in the cycle containing the point 0 does *not mean that the point 4 is replaced by the point 0.* In fact it is the other way round: In this book the mappings are written on the left hand side, $\pi(0) = 4$, i. e. the point 0 is replaced by the point 4 if π is applied to 0:

$$(0425)(13)(6)(7) = (0, \pi(0), \pi^2(0), \pi^3(0))(1, \pi(1))(6)(7).$$

In order to repeat this in other terms: In this book the identity $f(x) = y$ expresses two things. It indicates that x is mapped onto y, by f, and it means that x is replaced by y in the case when we apply f to x.

- $\mathcal{C}$ denotes the species *oriented cycles.* Each permutation is a collection of oriented cycles, the above example of a permutation consists of 4 such oriented cycles.
- $\mathcal{L}$ indicates the species *linear orders* or *total orders* as they are sometimes called. Here is one on $M = \{0, 1, 2, 3, 4, 5, 6, 7\}$:

$$2 < 4 < 3 < 6 < 7 < 5 < 1 < 0 \in \mathcal{L}[\{0, \dots, 7\}].$$

The usual order $0 < 1 < 2 < 3 < 4 < 5 < 6 < 7$ is called the *natural* order of this set of integers.
- *End*, the species of *endofunctions,* is defined by

$$End[M] := M^M := \{f \mid f : M \to M\}.$$

M is supposed to be a finite set. It is easy to check that an endofunction ε separates the elements of M into two classes, namely into the class of elements which are mapped onto itself by a suitable power of ε, $m = \varepsilon^n(m)$, for a suitable n in $\mathbb{N}^*$, which therefore lie on oriented cycles, and the rest of M. The following example shows that clearly: $M := \{4, 5, 6, 8, 9, a, b, c, d, e\}$ is separated into the set $\{a, b, c, d, e\}$ of elements that lie on cycles of ε and the rest $\{4, 5, 6, 8, 9\}$ of M, consisting of the elements that lie on rooted trees above the root (which lies on a cycle, of course):

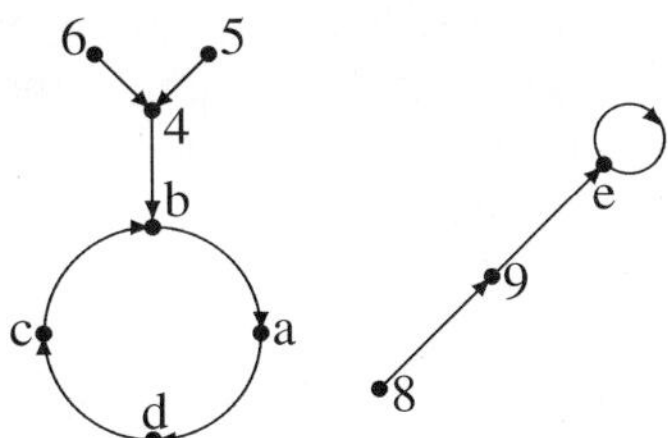

Hence an endofunction on a finite set is, in a certain sense, a permutation of rooted trees. Our present example can be considered as the permutation $(adcb)(e)$ of the trees with the roots a, b, c, d, e. We shall return to that later.

◇

It is clear from the definition of species that we can *transport* structure via bijections $\beta: M \to N$ from a set M to another set N by simply *renumbering* or *relabeling* the elements in the ground set of the structure in question. From the structure $\sigma \in \mathcal{S}[M]$ we obtain this way the structure $\beta(\sigma) = \tau \in \mathcal{S}[N]$. Here is an example from graph theory. A graph structure is transported from the ground set $M = \{0, 1, 2, 3\}$ to the set $N := \{a, b, c, d\}$ with the aid of the bijection

$$\beta : 0 \mapsto a, 1 \mapsto b, 2 \mapsto c, 3 \mapsto d,$$

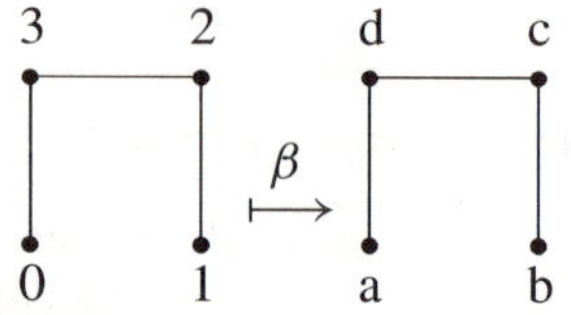

3 2 d c

β

0 1 a b

More generally, from the graph $\gamma \in \mathcal{G}[M]$ we obtain by an application of β the graph $\beta(\gamma) \in \mathcal{G}[N]$, where

$$\beta(\gamma) = \big\{\{\beta(m), \beta(m')\} \mid \{m, m'\} \in \gamma\big\}.$$

(Recall that we had identified a graph on M with a set of 2-subsets of M.) Slightly different but also easy is the transport of endofunctions with the aid of $\beta: M \to N$. We need only consider the endofunction $\varepsilon \in End[M]$ as the set of pairs $(m, \varepsilon(m))$, $m \in M$, and we have to apply to it the transport β:

$$\begin{aligned} \beta\big(\{(m, \varepsilon(m)) \mid m \in M\}\big) &= \big\{(\beta(m), \beta(\varepsilon(m))) \mid m \in M\big\} \\ &= \big\{(\beta(m), \beta\varepsilon\beta^{-1}(\beta(m))) \mid m \in M\big\} \\ &= \big\{(n, \beta\varepsilon\beta^{-1}(n)) \mid n \in N\big\}. \end{aligned}$$

In this way we obtain from ε the endofunction $\beta(\varepsilon) := \beta\varepsilon\beta^{-1}$ on N :

$$\beta(\varepsilon) = \beta\varepsilon\beta^{-1}.$$

We should now like to introduce further examples of species. They will help to prepare the introduction of various compositions of species as well as decompositions of species into constituents. For example, the reader will know about the decomposition of graphs into connected components as well as the decomposition of permutations into their cyclic factors.

0.1.4 Further examples of species

- The species *set*, $\mathcal{M}$, associates to M the one element set $\{M\}$,

$$\mathcal{M}[M] := \{M\}.$$

- The species *elements*, $\in$, gives the set M of its elements,

$$\in [M] := M.$$

- The species *singleton* is defined by:

$$\mathcal{X}[M] := \begin{cases} \{M\}, & \text{if } |M| = 1, \\ \emptyset, & \text{otherwise.} \end{cases}$$

- The *empty species,* denoted by 0, is obtained by putting $0[M] := \emptyset$.
- In contrast to the empty species the species *empty set* is introduced by putting

$$1[M] := \begin{cases} \{M\}, & \text{if } M = \emptyset, \\ \emptyset, & \text{otherwise.} \end{cases}$$

- Finally we set

$$\mathcal{M}_k[M] := \begin{cases} \{M\}, & \text{if } |M| = k, \\ \emptyset, & \text{otherwise,} \end{cases}$$

  which defines the species *k-set.*

(The notations 0 for the empty species and 1 for the species empty set will become clear as soon as we shall have introduced addition and multiplication of species, since they form neutral elements with respect to these compositions.) ◇

As every bijection $\beta: M \to N$ induces a bijection $\mathcal{S}[\beta]$ between $\mathcal{S}[M]$ and $\mathcal{S}[N]$, the "nature" of the elements of M does not really matter, what matters is its cardinality $|M|$ only. In particular the cardinal number $|\mathcal{S}[M]|$ does only depend on the cardinal number $|M|$. Therefore we may very well restrict attention to the standard sets

$$n := \{0, \ldots, n-1\}, \text{ or } \underline{n} := \{1, \ldots, n\}$$

of cardinality n and the structures living on these sets. Moreover, we put

$$s_n := |\mathcal{S}[n]| = |\mathcal{S}[\underline{n}]|.$$

The generating function of the cardinal numbers s_n, written *in exponential form,* i. e. the formal power series

0.1.5
$$\mathcal{S}(x) := \sum_{n \geq 0} s_n \frac{x^n}{n!} \in \mathbb{Q}[\![x]\!],$$

is called the *cardinality* of $\mathcal{S}$. Here are a few obvious examples:

0.1.6 Examples

$$\begin{aligned} \mathcal{L}(x) &= 1 + x + x^2 + x^3 + \ldots = \frac{1}{1-x} = Per(x), \\ \mathcal{M}(x) &= \sum_{n \geq 0} \frac{x^n}{n!} = e^x, \\ \in(x) &= \sum_{n > 0} n \cdot \frac{x^n}{n!} = x \cdot e^x, \end{aligned}$$

$$\begin{aligned}
\mathcal{P}(x) &= \sum_{n\geq 0} 2^n \cdot \frac{x^n}{n!} = e^{2x}, \\
\mathcal{X}(x) &= x, \\
1(x) &= 1, \\
0(x) &= 0, \\
\mathcal{G}(x) &= \sum_{n>0} 2^{\binom{n}{2}} \cdot \frac{x^n}{n!}, \\
End(x) &= \sum_{n\geq 0} n^n \cdot \frac{x^n}{n!}.
\end{aligned}$$

◇

Please note that, in accordance with these series, the following sets of structures on $\emptyset$ are *not* empty:

$$\mathcal{L}[\emptyset] = \mathcal{M}[\emptyset] = \mathcal{P}[\emptyset] = 1[\emptyset] = End[\emptyset] = \{\emptyset\} \neq \emptyset,$$

while

$$\in [\emptyset] = \mathcal{X}[\emptyset] = 0[\emptyset] = \mathcal{G}[\emptyset] = \emptyset.$$

Sometimes we shall prefer to avoid long winded explanations on the number of structures living on the particular ground set $M := \emptyset$. For this purpose we introduce the following notation:

0.1.7 $$\mathcal{S}_+[M] := \begin{cases} \mathcal{S}[M], & \text{if } M \neq \emptyset, \\ \emptyset, & \text{otherwise.} \end{cases}$$

Two species $\mathcal{R}$ and $\mathcal{S}$ with the same cardinality will be called *equipotent.* We shall indicate this by writing

$$\mathcal{R} \equiv \mathcal{S},$$

an abbreviaton of $\mathcal{R}(x) = \mathcal{S}(x)$. An obvious example is provided by

0.1.8 $$Per \equiv \mathcal{L}.$$

A canonic bijection between $Per[M]$ and $\mathcal{L}[M]$ uses the *list notation.*, For example, the permutation (038)(124)(57)(6) (see the appendix for this cycle notation, if necessary) has the list notation [324817650], which is defined to be the sequence of the images of the points 0, 1, ... It is mapped onto the linear order

$$3 < 2 < 4 < 8 < 1 < 7 < 6 < 5 < 0.$$

A more restrictive condition is that of *isomorphy*, indicated by

$$\mathcal{R} \simeq \mathcal{S},$$

where we assume the existence of canonic bijections $\Theta_M : \mathcal{R}[M] \to \mathcal{S}[M]$ and $\Theta_N : \mathcal{R}[N] \to \mathcal{S}[N]$, such that the following diagrams are commutative, for finite sets M, N and all bijections $\beta : M \to N$:

$$\begin{array}{ccc} \mathcal{R}[M] & \xrightarrow{\Theta_M} & \mathcal{S}[M] \\ \downarrow{\scriptstyle \mathcal{R}[\beta]} & & \downarrow{\scriptstyle \mathcal{S}[\beta]} \\ \mathcal{R}[N] & \xrightarrow{\Theta_N} & \mathcal{S}[N] \end{array}$$

It is clear that equipotency does *not* imply isomorphy, a standard example which shows that is the following:

0.1.9 Example We already know about the equipotency $Per \equiv \mathcal{L}$, and we will show now that these two species are *not isomorphic:*

$$Per \not\simeq \mathcal{L}.$$

In order to prove this we note that isomorphy $Per \simeq \mathcal{L}$ would imply the existence of a bijection $\Theta_M \colon Per[M] \to \mathcal{L}[M]$ (we chose $N := M$!) with

$$\mathcal{L}[\beta] \circ \Theta_M = \Theta_M \circ Per[\beta],$$

for every bijection $\beta \colon M \to M$. We apply this to a ground set M with $|M| > 1$, a bijection $\beta \neq \mathrm{id}_M$, and the permutation $\pi := \mathrm{id}_M \in Per[M]$. Consider the linear order

$$(m_0 < \ldots < m_{|M|-1}) := \Theta_M(\pi).$$

An application of the left hand side of the above identity to π gives

$$\begin{aligned} (\mathcal{L}[\beta] \circ \Theta_M)(\pi) &= \mathcal{L}[\beta](m_0 < \ldots < m_{|M|-1}) \\ &= \left(\beta(m_0) < \ldots < \beta(m_{|M|-1})\right), \end{aligned}$$

while an application of the right hand side yields

$$\begin{aligned} (\Theta_M \circ Per[\beta])(\pi) &= \Theta_M(\beta \circ \pi \circ \beta^{-1}) \\ &= \Theta_M(1) \\ &= (m_0 < \ldots < m_{|M|-1}) \\ &\neq \left(\beta(m_0) < \ldots < \beta(m_{|M|-1})\right), \end{aligned}$$

since β was assumed to differ from id_M. ◇

Hence equipotency does not imply isomorphy:

0.1.10 $$\mathcal{R} \equiv \mathcal{S} \not\Rightarrow \mathcal{R} \simeq \mathcal{S}.$$

Instead of $\mathcal{R} \simeq \mathcal{S}$ we shall also write

$$\mathcal{R} = \mathcal{S},$$

for sake of simplicity.

Exercises

Exercise 0.1.1 Prove that $\mathcal{S}[\beta]$ is in fact a bijection.

0.2 Sum and Product of Species

There are various compositions of species which allow to build new species from old ones, for example graphs from connected graphs, permutations from oriented cycles, and so on. The most easy one is the *sum* $\mathcal{R}+\mathcal{S}$ of two given species $\mathcal{R}$ and $\mathcal{S}$, where we take both the sets $\mathcal{R}[M]$ and $\mathcal{S}[M]$ of $\mathcal{R}$-structures and of $\mathcal{S}$-structures on M and form their disjoint union:

$$\mathcal{R}[M]+\mathcal{S}[M] := \mathcal{R}[M] \cup \mathcal{S}[M] = \mathcal{R}[M] \,\dot{\cup}\, \mathcal{S}[M],$$

in the case when $\mathcal{R}[M]\cap\mathcal{S}[M]=\emptyset$, otherwise we form by brute force the disjoint sets $\mathcal{R}[M]\times\{1\}$ and $\mathcal{S}[M]\times\{2\}$ and put

$$\begin{aligned}\mathcal{R}[M]+\mathcal{S}[M] &:= (\mathcal{R}[M]\times\{1\}) \cup (\mathcal{S}[M]\times\{2\})\\ &= (\mathcal{R}[M]\times\{1\}) \,\dot{\cup}\, (\mathcal{S}[M]\times\{2\}).\end{aligned}$$

This leads to the definition of the *sum* of two species $\mathcal{R}$ and $\mathcal{S}$,

$$(\mathcal{R}+\mathcal{S})[M] := \mathcal{R}[M]+\mathcal{S}[M].$$

The cardinality of this sum satisfies

0.2.1 $$(\mathcal{R}+\mathcal{S})(x) = \mathcal{R}(x)+\mathcal{S}(x).$$

Trivial examples are formed by structures for which we have a notion of connectivity. The graphs on a set form the sum of the connected and the not connected ones:

$$\mathcal{G} = \mathcal{G}^c + \mathcal{G}^{nc}.$$

The summation can be extended to families of structures: The family $(\mathcal{S}_i)_{i\in I}$ of species is called *summable,* if for each finite M set only finitely many $\mathcal{S}_i[M]$ are not empty. In this case we define as the sum of that family

$$\left(\sum_{i\in I}\mathcal{S}_i\right)[M] := \sum_{i\in I}\mathcal{S}_i[M] := \bigcup_{i\in I}\mathcal{S}_i[M]\times\{i\}.$$

This is compatible with the following notion of *decomposition:* The *canonic decomposition* of $\mathcal{S}$ is defined to be the summable family $(\mathcal{S}_n)_{n\geq 0}$, where

$$\mathcal{S}_n[M] := \begin{cases}\mathcal{S}[M], & \text{if } |M|=n,\\ \emptyset, & \text{otherwise.}\end{cases}$$

We abbreviate this by writing

$$\mathcal{S} = \mathcal{S}_0+\mathcal{S}_1+\mathcal{S}_2+\dots.$$

In the case when $\mathcal{S}_n=\emptyset$, for $n\neq k$, we say that $\mathcal{S}$ *concentrates* on k. A trivial example is the decomposition of $\mathcal{M}$ into k-sets:

$$\mathcal{M} = \sum_k \mathcal{M}_k.$$

In contrast to this the decompositions of permutations into permutations with a given number of cyclic factors and also the decompositions of set partitions into set partitions with prescribed number of blocks are *not canonic:*

$$Per = \sum_k Per^{[k]}, \ Par = \sum_k Par^{[k]}.$$

Besides addition there is also a *multiplication* of species. It uses the cartesian product formation, but it is not the same, in fact later on we shall introduce also a cartesian product of species. The *product* $\mathcal{R}\cdot\mathcal{S}$ of the two species $\mathcal{R}$ and $\mathcal{S}$ is defined as follows:

$$(\mathcal{R}\cdot\mathcal{S})\,[M] := \sum_{(M_0,M_1):\ M=M_0\dot{\cup}M_1} \mathcal{R}[M_0]\times\mathcal{S}[M_1],$$

where the summation is taken over all pairs of subsets which form a disjoint decomposition of M.

0.2.2 Examples

- An example is the (unique) decomposition of permutations into a set of fixed points together with a fixed point free permutation, a *derangement:*

$$Per = \mathcal{M}\cdot Der.$$

- Another example is the representation of the power set as a set of pairs, consisting of a subset and its complement:

$$\mathcal{P} = \mathcal{M}\cdot\mathcal{M}.$$

- The same holds for the set of k-subsets:

$$\mathcal{P}^{[k]} = \mathcal{M}_k\cdot\mathcal{M}.$$

- Finally we should like to mention that the product formation can be iterated in order to obtain *powers* of species:

$$Par^{[k]} = (\mathcal{M}_+)^k\,.$$

◇

This together with the following more or less obvious result (exercise 0.2.1)

0.2.3 $$(\mathcal{R}\cdot\mathcal{S})\,(x) = \mathcal{R}(x)\cdot\mathcal{S}(x)$$

on the cardinality of a product allows non-trivial applications. An application to the above mentioned identity $Per = \mathcal{M}\cdot Der$ yields

$$\frac{1}{1-x} = e^x \cdot Der(x),$$

from which we can obtain the following equation for the cardinality of the species derangements (see exercise 0.2.2):

$$Der(x) = \frac{e^{-x}}{1-x} = e^{-x} \cdot (1 + x + x^2 + x^3 + \ldots).$$

From this result we can obtain an explicit expression for the number of derangements of a set of order n :

0.2.4 $$der_n = n! \cdot \sum_{k=0}^{n} \frac{(-1)^k}{k!}.$$

Here is a table of the smallest numbers of derangements:

n	0	1	2	3	4	5	6	...
der_n	1	0	1	2	9	44	265	...

There is also a nice way to *visualize* compositions of species, it was developed by the Canadian group of people around A. Joyal who introduced, carefully elaborated and applied that theory. A structure σ is sketched as follows:

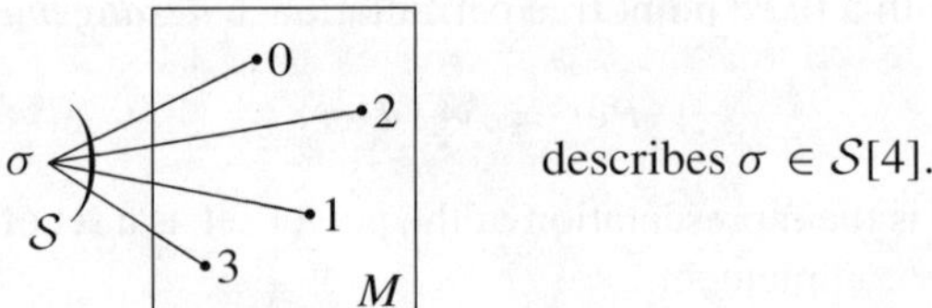

The frame around the set M together with the indexing of the structure with the name σ can be left out, for sake of simplicity. An element of the sum $(\mathcal{R} + \mathcal{S})[4]$ can be visualized by

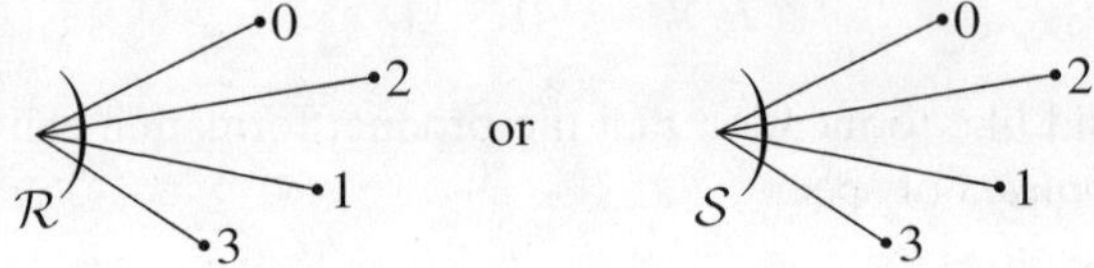

A structure of the product species may be indicated as follows:

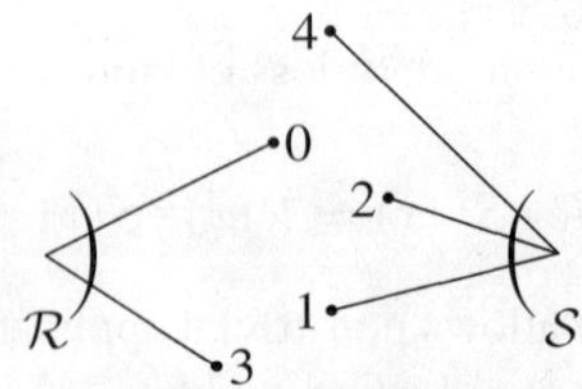

Since a set R together with operations $+$ and $\cdot$ is called a *semiring* if and only if both $(R,+)$ and $(R,\cdot)$ are commutative semigroups, 0 is neutral with respect to $+$, 1 is neutral with respect to $\cdot$, $0r = 0$, for each $r \in R$, and the distributivity law $(r+s)t = rt + st$ holds, we obtain (exercise 0.2.4):

0.2.5 Corollary *The isomorphism classes of species form a semiring with respect to addition and multiplication. The neutral element with respect to addition is the empty species* 0 *while the neutral element with respect to multiplication is the species empty set* 1.

Exercises

Exercise 0.2.1 Prove 0.2.3.

Exercise 0.2.2 Show that a formal power series with coefficients in a commutative ring with a multiplicative unity is invertible if and only if the constant term is invertible in the ground ring.

Exercise 0.2.3 Deduce a recursion formula for the numbers of derangements.

Exercise 0.2.4 Check 0.2.5.

0.3 Partitional Composition

Another slightly more complicated and very important composition of species is the *partitional composition* of two species, which was introduced by Foata and Schützenberger (see [48], [47]). Among other reasons it was introduced in order to define a species $\mathcal{R}(\mathcal{S})$ associated with $\mathcal{R}$ and $\mathcal{S}$, the cardinality of which is simply obtained by substitution: $\mathcal{R}(\mathcal{S})(x) = \mathcal{R}(\mathcal{S}(x))$.

Before this composition is introduced let me mention a few important examples: Permutations are obtained from oriented cycles and graphs are obtained from connected graphs, and so we should like to rephrase these species as partitional compositions. The same holds for endofunctions, they are obtained from rooted trees and oriented cycles.

Now we assume two species $\mathcal{R}$ and $\mathcal{S}$ where $\mathcal{S}[\emptyset] = \emptyset$. The *partitional composition* of $\mathcal{R}$ and $\mathcal{S}$ is defined by

$$\mathcal{R}(\mathcal{S})[M] := \bigcup_{p=\{p_0,\ldots,p_{r-1}\}\in Par[M]} \mathcal{R}[p] \times (\times_i \mathcal{S}[p_i]) \,.$$

This means that in order to obtain $\mathcal{R}(\mathcal{S})[M]$ we first of all form the set partitions p of M. If such a p consists of the (non-empty!) blocks p_i, $i = 0, \ldots, r-1$, we form on each of these blocks p_i the set of $\mathcal{S}$-structures, i. e. we form the sets $\mathcal{S}[p_i]$, and then the cartesian product over $i = 0, \ldots, r-1$ of all these sets of $\mathcal{S}$-structures together with $\mathcal{R}[p]$ (the set of $\mathcal{R}$-structures on the *set* p which is of order r). Finally the desired set $\mathcal{R}(\mathcal{S})[M]$ of composed structures is the union over all these cartesian products, the union being built over the set partitions p of M. Hence the elements of $\mathcal{R}(\mathcal{S})[M]$ are pairs of the form

$$\rho = (\sigma; \tau_0, \ldots, \tau_{r-1}) := (\rho, M) = ((\sigma, p); (\tau_0, p_0), \ldots, (\tau_{r-1}, p_{r-1})).$$

Here comes a sketch of a structure $\sigma \in \mathcal{R}(\mathcal{S})[6]$:

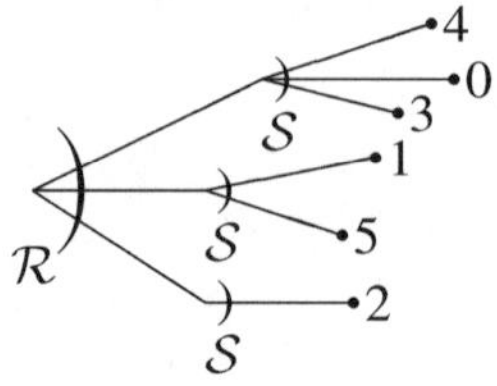

Let us consider a concrete application:

0.3.1 Example Let $M := 3 = \{0, 1, 2\}$. Here are the elements of the set $Par[3]$, consisting of all the set partitions of M :

$$\{\{0, 1, 2\}\}, \{\{0, 1\}, \{2\}\}, \{\{0, 2\}, \{1\}\}, \{\{1, 2\}, \{0\}\}, \{\{0\}, \{1\}, \{2\}\}.$$

We should like to evaluate $\mathcal{M}(\mathcal{C})[3]$. The set partition $\{\{0, 1, 2\}\}$ contributes

$$\mathcal{M}\left[\{\{0, 1, 2\}\}\right] \times \mathcal{C}[\{0, 1, 2\}] = \Big\{\{\{0, 1, 2\}\}\Big\} \times \{(012), (021)\}$$

$$= \big\{(\{0, 1, 2\}; (012)), (\{0, 1, 2\}; (021))\big\}.$$

The right factor in the middle expression is a two element set, consisting of the cycles (012) and (021), while the left factor is the one element set consisting of the one element partition $\{\{0, 1, 2\}\}$. The next summand of the set we want to evaluate belongs to the partition $\{\{0, 1\}, \{2\}\}$, it is equal to

$$\Big\{\{\{0, 1\}, \{2\}\}\Big\} \times \Big\{((01), (2))\Big\}.$$

Please note that the right factor means the cartesian product $\{(01)\} \times \{(2)\}$. The next summand arises from exchanging 1 and 2, which yields

$$\Big\{\{\{0, 2\}, \{1\}\}\Big\} \times \Big\{((02), (1))\Big\}.$$

The summand corresponding to $\{\{1, 2\}, \{0\}\}$ equals

$$\Big\{\{\{1, 2\}, \{0\}\}\Big\} \times \Big\{((12), (0))\Big\}.$$

Finally we have to build the summand corresponding to the partition $\{\{0\}, \{1\}, \{2\}\}$, it is

$$\Big\{\{\{0\}, \{1\}, \{2\}\}\Big\} \times \Big\{((0), (1), (2))\Big\}.$$

Forming the union of these five sets we obtain a set of order 6, and it is clear that it can be identified with the symmetric group (for its definition and its properties see the appendix) S_3. The left factors $\mathcal{M}[p]$ in the summands are redundant, and the union of the right factors, namely the set

$$\bigcup_p (\times_i \mathcal{C}[p_i]),$$

consists exactly of the different permutations of the symmetric group, since we can identify them with the elements of

$$\times_i \mathcal{C}[p_i].$$

More generally we have the isomorphy $\mathcal{M}(\mathcal{C}) \simeq Per$. We agreed about writing that as an identity, obtaining the equation

0.3.2
$$Per = \mathcal{M}(\mathcal{C}).$$

◇

After this example it should no longer be surprising that the cardinality of the partitional composition is the substitution of the cardinality $\mathcal{S}(x)$ for the variable in the cardinality of $\mathcal{R}$ (exercise 0.3.1):

0.3.3
$$\mathcal{R}(\mathcal{S})(x) = \mathcal{R}(\mathcal{S}(x)).$$

The above example shows that

$$\frac{1}{1-x} = Per(x) = \mathcal{M}(\mathcal{C}(x)) = e^{\mathcal{C}(x)},$$

from which we obtain the cardinality of the species oriented cycles:

0.3.4
$$\mathcal{C}(x) = \log(1-x)^{-1} = \sum_{n>0} \frac{x^n}{n}.$$

The number of oriented cycles on a set of order n is therefore equal to

$$c_n = (n-1)!,$$

a result which we can obtain, of course, also differently. It is the well known order of the conjugacy class of elements of a particular cycle type in the symmetric group. Quite analogously we can derive the following identities or identifications:

0.3.5
$$\mathcal{G} = \mathcal{M}(\mathcal{G}^c), \ Par = \mathcal{M}(\mathcal{M}_+),$$

which imply

$$\mathcal{G}^c(x) = \log(\mathcal{G}(x)), \ Par(x) = e^{e^x - 1}.$$

Exercises

Exercise 0.3.1 Verify 0.3.3.

Exercise 0.3.2 Prove formally that $End = Per(\mathcal{T}^r)$.

0.4 Derivation, Pointing, Functorial Composition

The derivative of the cardinality of a species should be the cardinality of the derived species, hence we introduce the derivative $\mathcal{S}'$ of a species $\mathcal{S}$ by putting

$$\mathcal{S}'[M] := \mathcal{S}[M^+],$$

where M^+ denotes a set obtained from M by adding a further element. We denote this element by $*$, assuming that M does not yet contain an element with this label (in which case we should use a different label for the new element). For short:

$$M^+ := M \cup \{*\}.$$

The transport is as follows (if β denotes a bijection between M and N):

$$\mathcal{S}'[\beta](\sigma) := \mathcal{S}[\beta^+](\sigma),$$

where

$$\beta^+ \colon M^+ \to N^+ \colon m \mapsto \begin{cases} \beta(m), & \text{if } m \in M, \\ *, & \text{if } m = *. \end{cases}$$

It is clear, for example, that from an oriented cycle $\zeta \in \mathcal{C}'[M]$, which contains, by definition of $\mathcal{C}'$, the point $*$, we obtain, by removing the element $*$, a linear order on M, and hence the following is true:

0.4.1 $$\mathcal{C}' = \mathcal{L}.$$

For the cardinality of a species and its derivative we have, as it was announced already, the identity

0.4.2 $$\mathcal{S}'(x) = \frac{d}{dx}\mathcal{S}(x).$$

(We are working over a field of formal power series, and the derivative is defined algebraically, by linear extension from the action on x^n.) An application to 0.4.1 gives

0.4.3 $$\mathcal{C}(x) = \int_0^x \frac{dx}{1-x} = \log\frac{1}{1-x} = \sum_{n>0} \frac{x^n}{n},$$

which again shows that there are exactly $c_n = (n-1)!$ oriented cycles on n points.

Instead of deriving a cardinality of a species we can multiply it by x, obtaining the cardinality of the species $\mathcal{S}^\bullet$ that arises by *pointing* $\mathcal{S}$, which means by distinguishing a single point of M,

$$\mathcal{S}^\bullet[M] := \mathcal{S}[M] \times M.$$

The transport of that structure along a bijection $\beta \colon M \to N$ is

$$\mathcal{S}^\bullet[\beta]((\sigma, m)) := (\mathcal{S}[\beta](\sigma), \beta(m)).$$

The cardinality of the pointed species is obtained as announced:

0.4.4 $$\mathcal{S}^\bullet(x) = x \cdot \mathcal{S}(x).$$

It is obvious that in this way we obtain rooted trees from trees:

0.4.5 $$\mathcal{T}^\bullet = \mathcal{T}^r.$$

0.4.6 Application (Cayley's Theorem) A very elegant application of the formalism of species theory is A. Joyal's proof ([73]) of the famous result of Cayley, that there are exactly n^{n-2} trees on n vertices. Joyal uses the cardinalities of the species End_+, Per_+, $\mathcal{L}_+$, obtained from End, Per, $\mathcal{L}$ as indicated in 0.1.7. The basic idea of this proof is the fact that an endofunction can be considered as a permutation of rooted trees. This becomes clear when we look again at the above example of an endofunction:

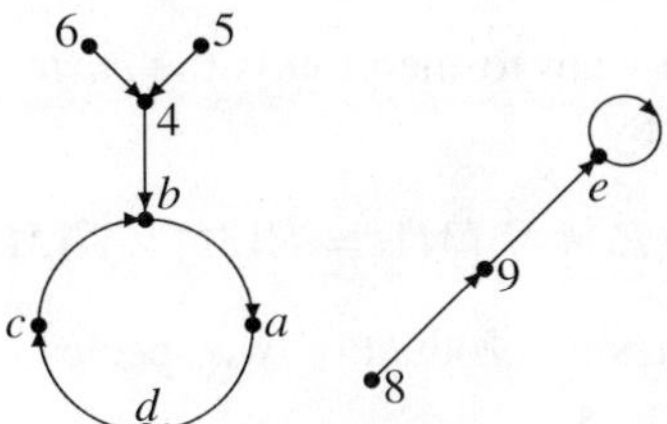

This particular endofunction can be identified with $(adcb)(e)$, a permutation of rooted trees, the letters are the roots. Moreover, as we are only interested in *numbers* of structures, we can replace permutations by linear orders, and so permutations of rooted trees are equipotent to linear orders of rooted trees. A canonic mapping from the permutations to the linear orders is obtained — as it was already mentioned — via the list notation (see appendix) of the permutation in question, in our example it is

$$(adcb)(e) = [d, a, b, c, e],$$

since we mean by the list notation the sequence of images $\pi(i)$. (Recall the remarks on the permutation notation used in this book and on the interpretation of $f(x) = y$ made in 0.1.1.) Hence this particular permutation is mapped onto the linear order

$$d < a < b < c < e.$$

Thus the endofunction given is mapped onto the following linear order of rooted trees:

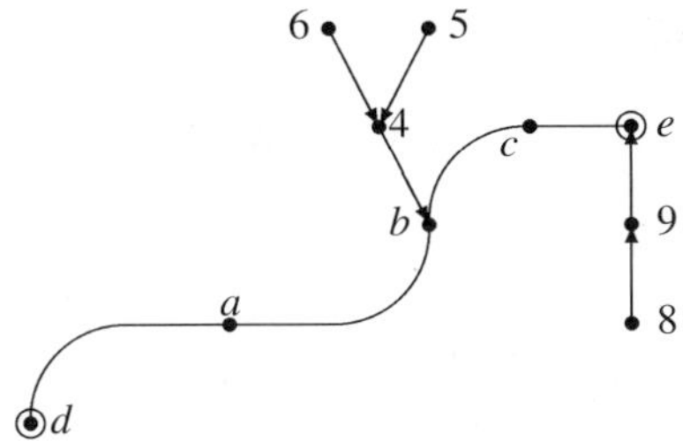

Finally we note that a linear order of rooted trees is a tree which is pointed twice, and that there are n^2 ways of pointing a structure on n vertices *twice*. This gives the elegant proof of the following

0.4.7 Theorem (Cayley) *The number of labeled trees on $n \geq 2$ vertices is*

$$t_n = n^{n-2}.$$

Proof: From

$$End_+ = Per_+(\mathcal{T}^r) \equiv \mathcal{L}_+(\mathcal{T}^r) = \mathcal{T}^{\bullet\bullet}.$$

we obtain by comparing coefficients of the left and of the right hand side:

$$n^n = n^2 \cdot t_n.$$

◇

The next construction we want to mention is the *cartesian product* $\mathcal{R} \times \mathcal{S}$ of two species $\mathcal{R}$ and $\mathcal{S}$, defined by

$$(\mathcal{R} \times \mathcal{S})[M] := \mathcal{R}[M] \times \mathcal{S}[M].$$

The cardinality of the cartesian product of two species is the *Hadamard product* of the cardinalities of the factors:

0.4.8 $$(\mathcal{R} \times \mathcal{S})(x) = \sum_n r_n s_n \frac{x^n}{n!}.$$

We may also want to form the set of $\mathcal{R}$-structures on a set of $\mathcal{S}$-structures, which leads us to the following definition: The *(functorial) composition* $\mathcal{R} \circ \mathcal{S}$ of $\mathcal{R}$ and $\mathcal{S}$ is defined by

$$(\mathcal{R} \circ \mathcal{S})[M] := \mathcal{R}(\mathcal{S}[M]),$$

together with the transport along a bijection $\beta \colon M \to N$,

$$(\mathcal{R} \circ \mathcal{S})[\beta] := \mathcal{R}\left[\mathcal{S}[\beta]\right].$$

A lucid example is formed by the graphs: Recall that a graph on n vertices can be considered as a subset of the set of 2-element subsets of the set of vertices, namely the set of connected pairs of vertices. Hence

0.4.9 $$\mathcal{G} = \mathcal{P} \circ \mathcal{P}^{[2]}.$$

We can further decompose this using the product formation, obtaining

$$\mathcal{G} = (\mathcal{M} \cdot \mathcal{M}) \circ (\mathcal{M}_2 \cdot \mathcal{M}).$$

The cardinality of the composition is

0.4.10 $$(\mathcal{R} \circ \mathcal{S})(x) = \sum_n (r \circ s)_n \frac{x^n}{n!} = \sum_n r_{s_n} \frac{x^n}{n!}.$$

Exercises

Exercise 0.4.1 Describe the derivative $\mathcal{L}'$.

Exercise 0.4.2 What is the pointing of $\mathcal{T}^r$ (The elements of this species are sometimes called *vertebrates,* for more or less obvious reasons.)

0.5 The Ring of Isomorphism Classes of Species

Having a semiring of isomorphism classes of species at hand (recall 0.2.5), we can construct a commutative ring with these classes as a $\mathbb{Z}$-basis. We use the same procedure as in the construction of $\mathbb{Z}$ from the semiring $\mathbb{N}$ of natural numbers, i. e. we form the cartesian square of the class of $\mathfrak{S}$ of isomorphism classes of species and introduce the following equivalence relation on pairs $(\mathcal{R}, \mathcal{S})$ of isomorphism classes of species:

$$(\mathcal{R}_1, \mathcal{S}_1) \sim (\mathcal{R}_2, \mathcal{S}_2) \iff \mathcal{R}_1 + \mathcal{S}_2 = \mathcal{R}_2 + \mathcal{S}_1.$$

For sake of simplicity of notation we shall replace the pair $(\mathcal{R}, \mathcal{S})$ of isomorphism classes by the expression

$$\mathcal{R} - \mathcal{S}.$$

An easy example is obtained from the fact that the set of permutations of a set M is the disjoint union of the set of fixed point free permutations on M, which is $Der[M]$, and the set of permutations *with* fixed points, which we indicate by $\mathcal{F}[M]$, and so

$$Per = \mathcal{F} + Der,$$

which gives that

$$\mathcal{F} = Per - Der.$$

The set of equivalence classes in the cartesian square of $\mathfrak{S}$ will be indicated as follows:

$$\mathfrak{H} := (\mathfrak{S} \times \mathfrak{S})/\sim.$$

It is a commutative ring, its elements $(\mathcal{S}, 0)$ will be denoted by $\mathcal{S}$, for sake of simplicity, while $(0, \mathcal{S})$ will be indicated as $-\mathcal{S}$, it is called a *virtual* (isomorphism class of) species.

0.5.1 Application (trees and rooted trees) A nice example is a description of trees in terms of rooted trees. Of course, it is easy to express rooted trees in terms of trees: $\mathcal{T}^r = \mathcal{T}^\bullet$, but this needs pointing, and, moreover, we want to do it the other way round, we want to express trees by rooted ones.

To begin with, we introduce the notion of *excentricity* of a vertex v in a tree τ, which is defined to be the maximal distance between v and other nodes of the tree:

$$e(v) := \max\{d(u, v) \mid u \text{ a vertex of the tree } \tau\}.$$

A vertex is now called *central,* if its excentricity is minimal, i. e. if

$$e(v) = \min\{e(u) \mid u \in \tau\}.$$

It is not difficult to see (exercise 0.5.1) that there are either one or two central vertices in each tree. For short: The center of a tree either consists of a vertex or of an edge. Hence each tree can be *canonically* colored by marking its center. Moreover, the partitional composition

$$\mathcal{M}_2(\mathcal{T}^r)[M] = \bigcup_{(M_0,M_1):\ M=M_0\dot{\cup}M_1} \mathcal{M}_2[\{M_0, M_1\}] \times \mathcal{T}^r[M_0] \times \mathcal{T}^r[M_1]$$

can be identified with the trees on M with a distinguished edge, since each element of that set is of the form (ρ, σ), where $\rho \in \mathcal{T}^r[M_0]$ and $\sigma \in \mathcal{T}^r[M_1]$ so that we simply need to join the roots of ρ and σ by an edge. In addition to that we note that the set $(\mathcal{T}^r)^2[M]$ and the set of *not canonically* colored trees on M are bijective (exercise 0.5.2). Putting things together we obtain the following expression for the species trees in terms of rooted trees:

0.5.2 $$\mathcal{T} = \mathcal{T}^r + \mathcal{M}_2(\mathcal{T}^r) - (\mathcal{T}^r)^2.$$

This is called the *dissymmetry theorem for trees.* An immediate consequence is the following identity for the cardinalities (due to Otter):

$$\mathcal{T}(x) = \mathcal{T}^r(x) + \mathcal{M}_2(\mathcal{T}^r)(x) - (\mathcal{T}^r(x))^2 = \mathcal{T}^r(x) - \frac{x}{2}(\mathcal{T}^r(x)^2 - \mathcal{T}^r(x^2)).$$

◇

Exercises

Exercise 0.5.1 Show that a tree contains either one or two central vertices.

Exercise 0.5.2 Prove that $(\mathcal{T}^r)^2[M]$ and the set of not canonically colored trees on M are bijective.

1. Unlabeled Structures

The main aim of this book is an introduction to the theory of classification, enumeration, construction and generation of *unlabeled* structures which means equivalence classes of *labeled* structures. A typical example is formed by incidence structures, where the names of the points (the labels) do not really matter, and so two incidence structures that differ only by a renumbering of the points are considered as *essentially the same*, as *equivalent* or as *isomorphic*. Prominent examples are unlabeled graphs, isometry classes of linear codes, isomorphism classes of designs, equivalence classes of switching functions, physical states and chemical isomers. We shall discuss these cases in detail.

In order to prepare this, the present chapter contains the basic notions: actions of groups, orbits (which will replace the equivalence classes), stabilizers, fixed points, and so on. The close relationship between orbits, cosets and double cosets is emphasized, since it is very important for enumerative and for constructive purposes. Various examples are given.

The paradigmatic actions which we shall discuss in full detail here and later on are several natural actions on the set Y^X, consisting of all the mappings from X into Y. These actions on Y^X are induced in a natural way by actions of groups on X or on Y. The corresponding orbits are called symmetry classes of mappings and there are many structures in mathematics and sciences which can be defined as symmetry classes of this kind.

1.1 Group Actions

It was already mentioned (see the definition of the cardinality series) that the labels of the elements of the ground set M, on which we form the set $\mathcal{S}[M]$ of $\mathcal{S}$-structures, do not really matter. The reason is that we can transport the structure from $\mathcal{S}[M]$ to $\mathcal{S}[N]$ via bijections $\beta\colon M \to N$, which means relabeling. It therefore suffices to consider, for each $n \in \mathbb{N}$, the set of structures $\mathcal{S}[n]$ on the set n. Moreover, there is also the transport of structure *within the ground set n,* i. e. with the aid of bijections β on n,

$$S_n \times \mathcal{S}[n] \to \mathcal{S}[n]\colon (\beta, \sigma) \mapsto \mathcal{S}[\beta](\sigma).$$

These mappings collect such $\mathcal{S}$–structures into classes, the *types of structures,* which arise from each other by an application of a suitable bijection β on n. We shall discuss this in a more general context and in more detail in a minute. To begin with, we should only like to mention examples which make clear what is meant by this. Here are two graphs on the set $4 = \{0, 1, 2, 3\}$ which are obviously of the same type, they were already shown:

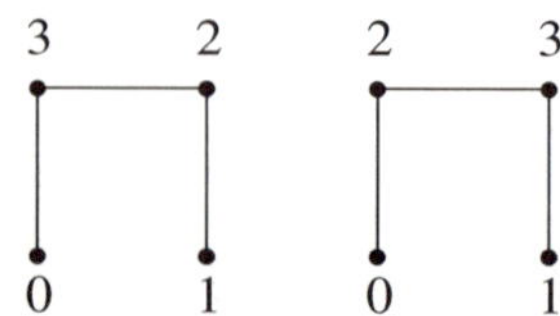

They are of the same type since they can be mapped onto each other by an application of the bijection $\beta: 4 \to 4$, defined by

$$0 \mapsto 0, 1 \mapsto 1, 2 \mapsto 3, 3 \mapsto 2,$$

whereas the graph

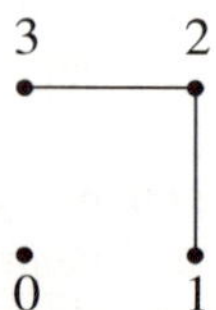

cannot be obtained from either of these two via transport of structure, and hence it is of a *different* type.

Now we note in passing that the definition of species implies, for the transport of $\mathcal{S}$-structure along bijections α and β on the given ground set,

$$\mathcal{S}[\alpha](\mathcal{S}[\beta](\sigma)) = \mathcal{S}[\alpha] \circ \mathcal{S}[\beta](\sigma) = \mathcal{S}[\alpha \circ \beta](\sigma), \text{ and } \mathcal{S}[\mathrm{id}_n](\sigma) = \mathrm{id}_{\mathcal{S}[n]}(\sigma) = \sigma.$$

This shows that we are faced with an *action* of the symmetric group. This notion — when generalized to arbitrary finite groups G — *is the basic concept of the present book,* and we are going to introduce it right now.

Let G denote a multiplicative group (the basic definitions and facts from group theory can be found in the appendix, if necessary) and X a nonempty set. An *action* of G on X (from the left) is described by a mapping

$$G \times X \to X: (g, x) \mapsto gx,$$

such that, for each $x \in X$ and any $g, g' \in G$, the following holds:

1.1.1 $$g(g'x) = (gg')x, \text{ and } 1x = x.$$

We abbreviate this by saying that G *acts* on X or simply by calling X a *G–set* or by writing

$$_GX,$$

in short, since G acts from the *left* on X. It is clear that analogously actions of groups on sets *from the right* can be defined and applied. In fact we shall use such actions, too, sometimes even in connection with left actions.

Before we provide examples, we mention a second formulation of the conditions 1.1.1 which is easy to check (exercise 1.1.1):

1.1.2 Lemma *The mapping*

$$\delta: G \to S_X: g \mapsto \bar{g},$$

where $\bar{g}: x \mapsto gx$, *is a homomorphism, a* permutation representation *of* G *on* X.

We shall call $\bar{g}$ the permutation *induced by* g *on* X and $\bar{G} := \delta(G)$ the permutation group *induced by* G *on* X. The *kernel* of δ, i. e. the set of elements which are mapped onto the identity mapping (the unit element of S_X), will be denoted as follows:

$$G_X := \ker(\delta) := \{g \mid \bar{g} = \delta(g) = \mathrm{id}_X\} = \{g \mid \forall\, x \in X: \; gx = x\}.$$

Being the kernel of a homomorphism, it is a *normal* subgroup of G, $G_X \trianglelefteq G$ for short, and the homomorphism theorem yields the *isomorphism*

1.1.3 $$\varphi: G/G_X \simeq \bar{G}: g \cdot G_X \mapsto \bar{g}.$$

In the case when $G_X = \{1\}$, the action is said to be *faithful.*

1.1.4 Examples

- A very trivial example is the *natural action* of S_X on X itself, where the corresponding permutation representation $\delta: \pi \mapsto \bar{\pi}$ is the identity mapping: $\pi = \bar{\pi}$.

- Another easy example is the following action of a group $G := \{1, g\}$ of order 2 (i. e. $g^2 = 1$) on the set

  $$D(n) := \{d \mid d \in \mathbb{N}^*, d \text{ divides } n\}$$

  of positive divisors of a given positive natural number n: The action of g is defined to be the mapping

  $$\bar{g}: D(n) \to D(n): d \mapsto n/d.$$

  The action of the identity element 1 of G is, by definition (see 1.1.1) of action, trivial:

  $$\bar{1}: D(n) \to D(n): d \mapsto d.$$

◇

An action of G on X has first of all the following property which is immediate from the two conditions 1.1.1 mentioned in its definition:

1.1.5 $$gx = x' \iff x = g^{-1}x'.$$

This is the reason for the fact that ${}_GX$ induces several structures on X and G, and *it is the close arithmetic and algebraic connection between these structures which makes the concept of group action so efficient.*

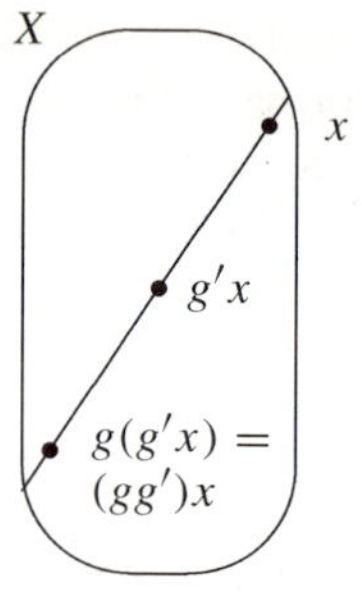

To begin with, the action induces the following equivalence relation on X (exercise 1.1.2):

$$x \sim_G x' :\Longleftrightarrow \exists\, g \in G\colon x' = gx.$$

The equivalence classes $G(x) := \{gx \mid g \in G\}$ are called *orbits*. Here is a sketch of the orbit of $x \in X$. Please note that each of its elements can be reached from every other element of this orbit, and so this orbit is in fact the orbit of every element of it under the action of G. As $\sim_G$ is an equivalence relation on X, a *transversal* T of the orbits, which means a complete set of representatives of the equivalence classes, yields a *set partition* of X, i. e. a complete dissection of X into the pairwise disjoint and nonempty subsets $G(t)$:

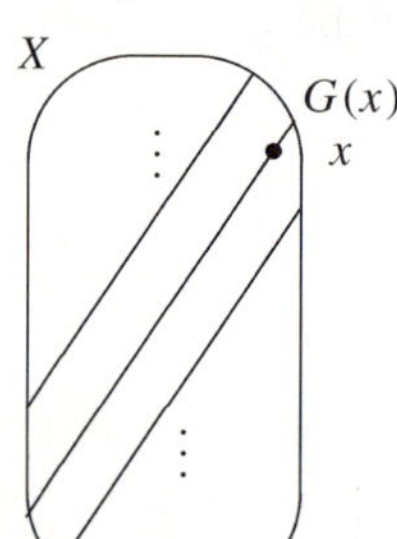

$$X = \dot{\bigcup}_{t \in T} G(t).$$

This may be visualized as indicated in the picture. The *set of all orbits* will be denoted by

$$G \backslash\!\backslash X := \{G(t) \mid t \in T\} = \{G(x) \mid x \in X\},$$

and we shall denote the *set* of all their transversals by

$$\mathcal{T}(G \backslash\!\backslash X) := \{T \mid T \text{ is a transversal of } G \backslash\!\backslash X\}.$$

In the case when both G and X are finite, we call the action a *finite action*. We notice that, according to 1.1.3, for each finite G–set X, we may also assume without loss of generality that G is finite. If G has exactly one orbit on X, i. e. if and only if $G \backslash\!\backslash X = \{X\}$, then we say that the action is *transitive*, or that G acts *transitively* on the set X.

As it was mentioned above, an action of G on X yields a partition of X. It is trivial but very important to notice that also the converse is true: Each set partition of X (and therefore each equivalence relation on X, too) gives rise to an action of a certain group G on X as follows. Let, for an arbitrary index set I, X_i, $i \in I$, denote the blocks of the set partition (the classes of the equivalence relation) in question, i. e. the X_i are nonempty, pairwise disjoint, and their union is equal to X. Then the following subgroup of the symmetric group S_X acts in a natural way on X and it has the X_i as its orbits:

1.1.6 $$\bigoplus_i S_{X_i} := \{\pi \in S_X \mid \forall\, i \in I\colon \pi X_i = X_i\},$$

where $\pi X_i := \{\pi x \mid x \in X_i\}$. Summarizing our considerations in two sentences, we have obtained:

1.1.7 Corollary *An action of a group G on a set X is equivalent to a permutation representation of G on X and it yields a set partition of X into orbits. Conversely, each set partition of (or equivalence relation on) X corresponds in a natural way to an action of a certain subgroup of the symmetric group S_X which has the blocks of the partition (or the equivalence classes) as its orbits.*

These two facts are very important since we are aiming at a constructive theory of finite unlabeled structures. The preceding result in fact shows that

> *all the structures in mathematics and sciences that can be defined as equivalence classes on sets can be described as orbits of groups.*

Prominent examples are linear codes, designs, graphs, switching functions, physical states and chemical isomers. We shall give details in due course.

To the orbits $G(x)$, which are *subsets* of X, there correspond certain *subgroups* of G. For each $x \in X$ we introduce its *stabilizer*:

$$G_x := \{g \in G \mid gx = x\}.$$

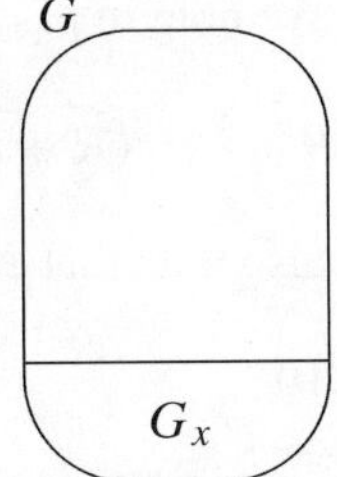

This is the subgroup (exercise 1.1.3) of elements of G that *stabilize* the *point* x. There is also the subgroup of elements which pointwise stabilize the elements of a *subset* Δ of X :

$$G_\Delta := \{g \in G \mid \forall\, x \in \Delta\colon\ gx = x\},$$

and which is called the *pointwise stabilizer* or the *centralizer* of Δ, in contrast to the *setwise stabilizer* or *normalizer* of Δ :

$$G_{\{\Delta\}} := \{g \in G \mid \forall\, x \in \Delta :\ gx \in \Delta\}.$$

We note in passing that $G_{\{x\}} = G_x$, that $G_\Delta \subseteq G_{\{\Delta\}}$ and that this notation is compatible with the notation G_X for the kernel of the permutation representation δ corresponding to ${}_GX$ which is in fact the pointwise stabilizer of X, by definition.

The last one of the fundamental concepts induced by an action of G on X is that of *fixed points*. A point $x \in X$ is said to be *fixed* under $g \in G$ if and only if $gx = x$, and the set of all the fixed points of g is indicated by

$$X_g := \{x \in X \mid gx = x\}.$$

More generally, for any subset $S \subseteq G$, we put

$$X_S := \{x \in X \mid \forall\, g \in S\colon\ gx = x\}.$$

The particular case X_G is called the set of *invariants*.

1.1.8 Application (type series of structures) For a given species $\mathcal{S}$ we introduced the action

$$S_n \times \mathcal{S}[n] \to \mathcal{S}[n]\colon (\beta, \sigma) \mapsto \mathcal{S}[\beta](\sigma).$$

Its orbits are called *types of structures*. The corresponding generating function of the numbers of types is

$$\widetilde{\mathcal{S}}(x) := \sum_n \widetilde{s}_n x^n, \ \text{ where } \widetilde{s}_n := |\, S_n \backslash\!\backslash \mathcal{S}[n]\,|\,.$$

It is called the *type series* of $\mathcal{S}$. Here are a few easy examples for which we do not need the general methods that will be introduced in the later sections:

- $\widetilde{\mathcal{L}}(x) = \widetilde{\mathcal{M}}(x) = \frac{1}{1-x} = 1 + x + x^2 + x^3 + \ldots,$
- $\widetilde{\mathcal{C}}(x) = \widetilde{\mathbb{E}}(x) = \frac{x}{1-x} = x + x^2 + x^3 + \ldots,$
- $\widetilde{\mathcal{P}}(x) = \left(\frac{1}{1-x}\right)^2 = 1 + x + 2x^2 + 3x^3 + \ldots,$
- $\widetilde{\mathcal{X}}(x) = x, \widetilde{1}(x) = 1, \widetilde{0}(x) = 0.$

We note in passing that for the type series of sum and product we have

1.1.9 $$\widetilde{(\mathcal{R} + \mathcal{S})}(x) = \widetilde{\mathcal{R}}(x) + \widetilde{\mathcal{S}}(x), \ \widetilde{(\mathcal{R} \cdot \mathcal{S})}(x) = \widetilde{\mathcal{R}}(x) \cdot \widetilde{\mathcal{S}}(x).$$

We also note that equipotency does *not* imply the equality of the type series:

1.1.10 $$\mathcal{R} \equiv \mathcal{S} \not\Rightarrow \widetilde{\mathcal{R}}(x) = \widetilde{\mathcal{S}}(x).$$

◇

The next bunch of examples will show that various important group theoretical structures can be considered as orbits or stabilizers:

1.1.11 Application (conjugacy classes, centralizers, cosets in groups) If G denotes a group, then

- G acts on itself by *left multiplication*:

$$G \times G \to G: (g, x) \mapsto g \cdot x.$$

  This action is called the *(left) regular representation* of G, it is obviously transitive, and all the stabilizers are equal to the identity subgroup $\{1\}$.

- If we restrict attention to the subgroup U, then we obtain the action

$$U \times G \to G: (u, x) \mapsto u \cdot x$$

  of U on G and the orbits of the elements are the *right cosets*

$$U(g) = Ug$$

  of the subgroup U. The stabilizers are trivial again:

$$U_g = \{1\}.$$

  Correspondingly, we obtain the *left cosets* gU of U as orbits, if we consider the *right* regular representation.

This shows that *different right cosets as well as different left cosets are disjoint* and that both *the set of right cosets and the set of left cosets of U is a set partition of G.*

– G acts on itself by *conjugation*:

$$G \times G \to G\colon (g, x) \mapsto g \cdot x \cdot g^{-1}.$$

The orbits of this action are the *conjugacy classes* of elements,

$$G(x) = C^G(x) := \{gxg^{-1} \mid g \in G\},$$

and the stabilizers are the *centralizers* of elements:

$$G_x = C_G(x) := \{g \in G \mid gxg^{-1} = x\}.$$

An immediate consequence is that *different conjugacy classes of elements are disjoint,* since they are orbits, *they also form a set partition of* G. Moreover, *centralizers of elements are subgroups,* since they are stabilizers.

– If U denotes a subgroup of G (in short: $U \leq G$), then G acts on the set $G/U := \{xU \mid x \in G\}$ of its *left cosets* as follows:

$$G \times G/U \to G/U\colon (g, xU) \mapsto gxU.$$

This action is transitive, and the stabilizer of xU is the subgroup xUx^{-1} which is conjugate to U.

Thus xUx^{-1} *is a subgroup,* too.

– G acts on the set $L(G) := \{U \mid U \leq G\}$ of all its subgroups by *conjugation*:

$$G \times L(G) \to L(G)\colon (g, U) \mapsto gUg^{-1}.$$

The orbits of this action are the *conjugacy classes of subgroups*, and the stabilizers are the *normalizers* :

$$G(U) = \widetilde{U} := \{\, gUg^{-1} \mid g \in G\},$$

and

$$G_U = N_G(U) := \{g \in G \mid gU = Ug\}.$$

Hence *different conjugacy classes of subgroups are disjoint, and normalizers of subgroups are also subgroups.* ◇

Returning to the general case we first state the main (and obvious) properties of the stabilizers of elements belonging to the same orbit:

1.1.12 $$G_{gx} = gG_xg^{-1}, \widetilde{G_x} = \{gG_xg^{-1} \mid g \in G\} = \{G_{x'} \mid x' \in G(x)\}.$$

It means that the total set of stabilizers G_{gx}, $g \in G$, of the elements in the orbit $G(x)$ form the conjugacy class $\widetilde{G_x}$ of G_x (recall the last item of the examples given in 1.1.11).

Another important fact is the close connection between orbit sets and stabilizers. Consider the set $L(G) = \{U \mid U \leq G\}$ of subgroups of G, ordered by inclusion, $(L(G), \leq)$, together with the set

$$Par[X] = \{p \mid p \text{ is a set partition of } X\}$$

of set partitions $p = \{p_0, \ldots\}$ of X, ordered by *refinement*:

$$p \leq q :\Longleftrightarrow \forall\, p_i \in p\, \exists\, q_j \in q\colon\ p_i \subseteq q_j.$$

We introduce the following two mappings between these sets which depend strongly on the action of G:

$$\mathrm{orb}_G\colon L(G) \to Par[X]\colon U \mapsto U \backslash\!\backslash X,\ \ \mathrm{stab}_G\colon Par[X] \to L(G)\colon p \mapsto G_p,$$

where $G_p := \{g \in G \mid \forall\, i\colon\ g[p_i] = p_i\}$ is the *stabilizer* of p in G. These mappings are monotone:

$$U \leq V \Rightarrow \mathrm{orb}_G(U) \leq \mathrm{orb}_G(V),\ \ p \leq q \Rightarrow \mathrm{stab}_G(p) \leq stab(q),$$

their composition $\mathrm{stab}_G \circ \mathrm{orb}_G$ is increasing, while $\mathrm{orb}_G \circ \mathrm{stab}_G$ is decreasing:

$$U \leq (\mathrm{stab}_G \circ \mathrm{orb}_G)(U),\ \ p \geq (\mathrm{orb}_G \circ \mathrm{stab}_G)(p).$$

This shows that orb_G is a mapping of the following particular form: Assume two posets $(M, \leq)$ and $(N, \leq)$, then $\alpha\colon M \to N$ is called a *Galois function* if there exists a mapping β from N to M such that both these mappings are monotone, while $\beta \circ \alpha$ is increasing and $\alpha \circ \beta$ is decreasing. Thus we obtain from exercise 11.6.6:

1.1.13 Corollary (Galois theory of a group action) *The mappings* orb_g *and* stab_G *have the following properties:*

- $\mathrm{orb}_G\colon L(G) \to Par[X]$ *is a* Galois function, *since* $\mathrm{stab}_G\colon Par[X] \to L(G)$ *is also monotone, while* $\mathrm{stab}_G \circ \mathrm{orb}_G$ *is increasing and* $\mathrm{orb}_G \circ \mathrm{stab}_G$ *is decreasing.*
- $stab_G \circ \mathrm{orb}_G\colon L(G) \to Par[X]$ *is a closure operator. This means that it is increasing, monotone and idempotent. The corresponding set of* closed elements *is*

$$\{U \mid \overline{U} := (\mathrm{stab}_G \circ \mathrm{orb}_G)(U) = U\} = \mathrm{stab}_G(Par[X]) = \{G_p \mid p \in Par[X]\}.$$

- $\mathrm{orb}_G \circ \mathrm{stab}_G\colon Par[X] \to L(G)$ *is a co-closure operator on* $Par[X]$, *which means that it is decreasing, monotone and idempotent. The set of* co-closed elements *is*

$$\{p \mid \overline{p} := (\mathrm{orb}_G \circ \mathrm{stab}_G)(p) = p\} = \mathrm{orb}_G(L(G)) = \{U \backslash\!\backslash X \mid U \leq G\}.$$

- *These sets of closed and of coclosed elements are order isomorphic,*

$$\mathrm{orb}_G\colon\ \big(\{G_p \mid p \in Par[X]\}, \leq\big) \simeq \big(\{U \backslash\!\backslash X \mid U \leq G\}, \leq\big).$$

Exercises

Exercise 1.1.1 Assume X to be a G–set and check carefully that $g \mapsto \overline{g}$ is in fact a permutation representation, i. e. that $\overline{g} \in S_X$ and that

$$\overline{g_1 \cdot g_2} = \overline{g_1} \cdot \overline{g_2}.$$

Exercise 1.1.2 Prove that $\sim_G$ is an equivalence relation.

Exercise 1.1.3 Prove that G_x is a subgroup of G.

1.2 Orbits, Cosets and Double Cosets

The crucial point is the indicated natural bijection between the orbit of x and the set of left cosets of its stabilizer G_x.

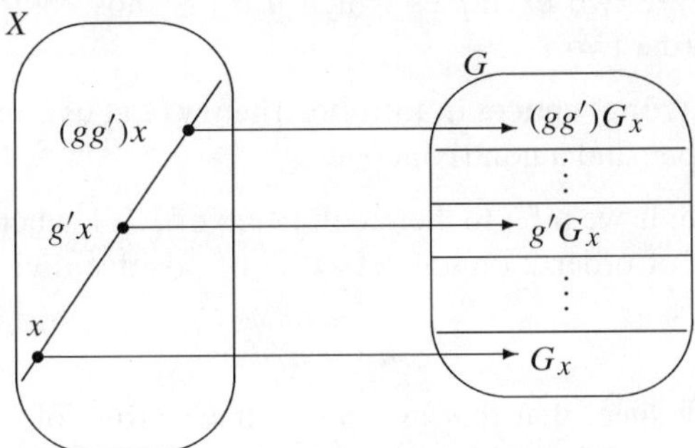

It allows to replace the elements of the orbit by subsets of the acting group. The group is, of course, usually much bigger, but it carries a very helpful algebraic structure that can be used both for enumerative and for constructive purposes.

1.2.1 The Fundamental Lemma *The mapping $G(x) \to G/G_x\colon gx \mapsto gG_x$ is a bijection between the orbit $G(x)$ and the set of left cosets G/G_x.*

Proof: It is clear from 1.1.5 that

$$gx = g'x \iff g^{-1}g' \in G_x \iff g'G_x = gG_x.$$

Reading this from left to right we see that $gx \mapsto gG_x$ defines a mapping, reading it from right to left we obtain that it is injective. Furthermore it is obvious that this mapping is also surjective. □

This result implies

1.2.2 Corollary *The length of the orbit is the index of the stabilizer:*

$$|G(x)| = |G/G_x|.$$

In particular, if $|G|$ is finite, then $|G(x)| = |G|/|G_x|$, and so the length of each orbit is a divisor of the group order in this finite case.

1.2.3 Application (divisibilities in groups) An application to the examples given in 1.1.11 yields: If G is finite, $g \in G$, and $U \leq G$, then the order of U divides the order of G :

$$|U| \text{ divides } |G|,$$

and the orders of the conjugacy classes of elements and of subgroups satisfy the following equations:

$$|C^G(g)| = |G|/|C_G(g)|\ , \text{ and } |\widetilde{U}| = |G|/|N_G(U)|.$$

◇

We saw that lengths of orbits are divisors of the group order. This suggests to try congruences. Here are two examples which use the most simple case where the acting group is of order two.

1.2.4 Applications (congruences in number theory) Let us consider two such applications, a trivial one and a nontrivial one:

- For the easy example we refer to the second item of 1.1.4, where we met an action of a group $\{1, g\}$ of order 2 on the set $D(n)$ of positive divisors of the positive natural number n :
$$\bar{g}: d \mapsto n/d.$$
From 1.2.2 we deduce, that this group can have orbits of length 1 or 2 only. Moreover we see that it has an orbit of order 1 if and only if there exists a divisor d with $d = n/d$, which means that $n = d^2$. This shows that *a positive natural number n has an odd number of divisors if and only if it is a square.*
- A less trivial example is the following proof (due to D. Zagier) of the fact that *every prime number p which is congruent 1 modulo 4 can be expressed as a sum of two squares of positive natural numbers.* The idea is to consider a suitable set and to show that it is of odd order (by showing that there exists an *involution,* i. e. a group element of order 2 which has exactly one fixed point on this set), and therefore each group element of order 2 must have a fixed point there.
Consider the set
$$S := \left\{(x, y, z) \in (\mathbb{N}^*)^3 \,\middle|\, x^2 + 4yz = p\right\}$$
and the following map defined on S :
$$\tau: (x, y, z) \mapsto \begin{cases} (x + 2z, z, y - x - z), & \text{if } x < y - z, \\ (2y - x, y, x - y + z), & \text{if } y - z < x < 2y, \\ (x - 2y, x - y + z, y), & \text{if } x > 2y. \end{cases}$$
We note that it has the following properties (exercise 1.2.2):

- τ is an involution: $\tau^2\colon (x, y, z) \mapsto (x, y, z)$.
- τ has exactly one fixed point, namely $(1, 1, k)$, if $p = 4k + 1$.

Therefore $|S|$ must be odd if $p = 4k + 1$, and consequently the involution

$$\sigma\colon (x, y, z) \mapsto (x, z, y)$$

possesses a fixed point, too, which shows that S contains an element (x, y, z) where $y = z$, and so we obtain that $p = x^2 + 4y^2$, a sum of two squares.

◇

Besides these divisibility results we note that the mapping

$$\varphi\colon G(x) \to G/G_x\colon gx \mapsto gG_x$$

commutes with the action of G, $\varphi(gx) = g\varphi(x)$. This shows that in fact the restricted action of G on the orbit $G(x)$,

$$G \times G(x) \to G(x)\colon (g, hx) \mapsto ghx,$$

is essentially the same — in a sense which we are going to describe next — as the action of G on G/G_x via left multiplication of the cosets:

1.2.5 $$G \times G/G_x \to G/G_x\colon (g, hG_x) \mapsto ghG_x.$$

Under enumerative aspects ${}_GX$ is essentially the same as ${}_{\bar{G}}X$. This leads to the question of a suitable concept of morphism between actions of groups. To begin with, two actions $G \times X \to X$ and $H \times Y \to Y$ will be called *isomorphic* iff they differ only by an isomorphism $\eta\colon G \simeq H$ of the groups and a bijection $\theta\colon X \to Y$ between the sets which satisfy $\eta(g)\theta(x) = \theta(gx)$. This means that the following diagram is commutative:

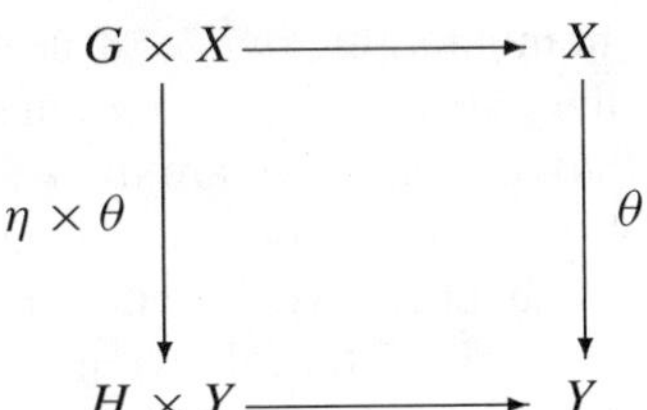

In this case we shall write

$${}_GX \simeq {}_HY,$$

in order to indicate the existence of such a pair of mappings. If $G = H$ we call ${}_GX$ and ${}_GY$ *similar* actions, if and only if they are isomorphic by (η, θ), where moreover $\eta = \mathrm{id}_G$, the identity mapping (cf. exercise 1.2.4). We indicate this in the following way:

$${}_GX \approx {}_GY.$$

Recall that this means the existence of a bijection $\theta\colon X \to Y$, such that, for each $x \in X$ and every $g \in G$ we have $g\theta(x) = \theta(gx)$. An important special case follows directly from the proof of 1.2.1:

1.2.6 Corollary *If ${}_GX$ is an action then, for any $x \in X$, the action of G on the orbit $G(x)$ is similar to the action of G on the set of left cosets of the stabilizer G_x,*

$${}_GG(x) \approx {}_G(G/G_x).$$

In particular, if ${}_GX$ is a transitive action, then for each $x \in X$ the action of G on X is similar to the action of G on G/G_x,

$${}_GX \approx {}_G(G/G_x).$$

A weaker condition than isomorphy is that of *homomorphy.* We shall write

$${}_GX \sim {}_HY$$

if and only if there exist a mapping $\theta: X \to Y$ and a homomorphism $\eta: G \to H$ such that $\theta(gx) = \eta(g)\theta(x)$. Later on we shall see that the use of homomorphisms is one of the most important tools in the constructive theory of finite discrete structures which can be defined as orbits of finite groups on finite sets. In the case of ${}_GX \sim {}_GY$ where $\eta = \text{id}$, the identity mapping, we say that ${}_GX$ and ${}_GY$ are *G–homomorph.* An example of homomorphy is

1.2.7
$${}_GX \sim {}_{\bar{G}}X.$$

Similarity is a special case of homomorphy, too. We see from these examples that in the case of homomorphy ${}_GX \sim {}_HY$ there *may* be a close connection between $G \backslash\backslash X$ and $H \backslash\backslash Y$, but that is not always the case.

From a given action we can derive various other actions in a natural way, e. g. ${}_GX$ yields ${}_{\bar{G}}X$, $\bar{G}$ being the homomorphic image of G in S_X, which was already mentioned. We also obtain the *subactions* ${}_GM$ on subsets $M \subseteq X$ which are nonempty unions of orbits. Furthermore there are the *restrictions* ${}_UX$ to the subgroups U of G. As the orbits of ${}_GX$ are unions of orbits of ${}_UX$, the comparison of subactions and restrictions is a suitable way of generalizing or specializing structures if they can be defined as orbits. The following example will show what is meant by this.

1.2.8 Applications (bilateral classes, cosets and double cosets) Let U denote a subgroup of the direct product $G \times G$. Then U acts on G as follows:

$$U \times G \to G: ((a, b), g) \mapsto agb^{-1}.$$

The orbits $U(g) = \{agb^{-1} \mid (a, b) \in U\}$ of this action are called the *bilateral classes* of G with respect to U (this notion was introduced by Hässelbarth, Klein, Richter, Ruch and Seligman, see [69]). We note that therefore *different bilateral classes, being orbits, are disjoint.*

By specializing U we obtain various interesting group theoretical structures some of which have already been mentioned:

- If A is a subgroup of G, then both $A \times \{1\}$ and $\{1\} \times A$ are subgroups of $G \times G$. Their orbits are the subsets

$$(A \times \{1\})(g) = Ag,$$

the *right cosets* of A in G, and

$$(\{1\} \times A)(g) = gA,$$

the *left cosets* of A in G.
- If B denotes a second subgroup of G, then we can put U equal to the subgroup $A \times B$, obtaining as orbits the (A, B)–*double cosets* of G:

$$(A \times B)(g) = AgB := \{agb \mid a \in A, b \in B\}.$$

- Another subgroup of $G \times G$ is its *diagonal* subgroup

$$\Delta(G \times G) := \{(g, g) \mid g \in G\}.$$

Its orbits are the conjugacy classes:

$$\Delta(G \times G)(g) = \{g'gg'^{-1} \mid g' \in G\} = C^G(g).$$

Hence left and right cosets, double cosets and conjugacy classes turn out to be special cases of bilateral classes. Being orbits, two left cosets, two right cosets, two double cosets and two conjugacy classes are either equal or disjoint and the stabilizer of an element is a subgroup of the group in question. Moreover, the order of each of these orbits is equal to the index of the stabilizer. We have mentioned this in connection with conjugacy classes and centralizers of elements already, here is the consequence for double cosets: Since the stabilizer of $g \in G$ in $A \times B$ is

$$(A \times B)_g = \{(gbg^{-1}, b) \mid b \in B, gbg^{-1} \in A\} = A \cap gBg^{-1},$$

we obtain, for finite groups G with subgroups A and B,

1.2.9
$$|AgB| = \frac{|A||B|}{|A \cap gBg^{-1}|},$$

and if D denotes a transversal of the set $A\backslash G/B$ of (A, B)-double cosets, then

1.2.10
$$|G| = \sum_{g \in D} |AgB| = \sum_{g \in D} \frac{|A||B|}{|A \cap gBg^{-1}|}.$$

◇

Having double cosets now at hand, we can formulate another very interesting and useful consequence of 1.2.1 and 1.2.6:

1.2.11 Breaking of Symmetry *If ${}_GX$ is an action and U a subgroup of G, then for each $x \in X$ we have the following bijection between the orbits of U on $G(x)$ and the set of (U, G_x)-double cosets:*

$$\psi_x : U \backslash\backslash G(x) \to U \backslash G / G_x : U(gx) \mapsto UgG_x,$$

with respect to the action

$$U \times G(x) \to G(x) : (u, gx) \mapsto ugx.$$

We shall return to that later, since it is of enormous importance, in particular for constructive purposes. It was intensively used for the construction of unlabeled graphs, representatives of isometry classes of linear codes, isomorphism types of designs, chemical isomers, physical states, and it can be applied to very many other cases, too. The method used is the following application of the Breaking of Symmetry 1.2.11. Detailed applications of this approach will be given later.

1.2.12 Application (a method for the construction of finite unlabeled structures)

- Define the desired set of finite unlabeled structures, e. g. the set of unlabeled graphs on a given number of points, as the set of classes of an equivalence relation on the set of labeled structures X.
- Replace the equivalence relation by a suitable action of a group U on X, so that the structures in question correspond to the set of orbits $U \backslash\backslash X$.
- In order to reduce complexity by restricting attention to a suitable subset of $U \backslash\backslash X$, introduce a suitable bigger group $G \supset U$ and an action ${}_GX$ for which ${}_UX$ is the restriction of ${}_GX$ to U.
- Restrict attention to the orbits $G(x)$, for example to the graphs with given numbers of points and edges.
- Evaluate a transversal T_G of $G \backslash\backslash X$, and for each $x \in T_G$ construct a transversal T_x of $U \backslash G / G_x$.
- Retranslate the elements of the T_x into elements of X and obtain this way *a transversal T_U of the equivalence classes $U \backslash\backslash X$, i. e. of the desired structures in question:*

$$T_U := \bigcup_{x \in T_G} \psi_x^{-1}(T_x).$$

◇

Now we take *two* actions into account, say ${}_GX$ and ${}_HY$, and we derive further actions from these. Without loss of generality we can assume $X \cap Y = \emptyset$ since otherwise we can rename the elements of X, in order to replace ${}_GX$ by a similar action ${}_GX'$, for which $X' \cap Y = \emptyset$. We form the (disjoint) union $X \dot{\cup} Y$ and let $G \times H$ act on this set as follows:

$$\text{1.2.13} \qquad (G \times H) \times (X \dot{\cup} Y) \to X \dot{\cup} Y : ((g, h), z) \mapsto \begin{cases} gz, & \text{if } z \in X, \\ hz, & \text{if } z \in Y. \end{cases}$$

The corresponding permutation group will be denoted by $\bar{G} \oplus \bar{H}$ (cf. 1.1.6) and called the *direct sum* of $\bar{G}$ and $\bar{H}$. Another canonical action of $G \times H$ is that on the cartesian product:

1.2.14 $$(G \times H) \times (X \times Y) \to X \times Y : \big((g,h),(x,y)\big) \mapsto (gx, hy).$$

The corresponding permutation group will be denoted by $\bar{G} \otimes \bar{H}$ and called the *cartesian product* of $\bar{G}$ and $\bar{H}$. An important particular case is

1.2.15 Application (Mackey's Theorem) Assume two finite and transitive actions of G on X and Y. They yield, as was just described, a canonical action of $G \times G$ on $X \times Y$ which has as one of its restrictions the action of $\Delta(G \times G)$, the diagonal, which is isomorphic to G, on $X \times Y$. We notice that, for fixed $x \in X$, $y \in Y$, the following holds (exercise 1.2.6):

- Each orbit of G on $X \times Y$ contains an element of the form (x, gy).
- The stabilizer of (x, gy) is $G_x \cap gG_yg^{-1}$, hence, according to 1.2.6, the action of G on the orbit of (x, gy) is similar to the action of G on $G/(G_x \cap gG_yg^{-1})$.
- (x, gy) lies in the orbit of $(x, g'y)$ if and only if

$$G_xgG_y = G_xg'G_y.$$

Thus the following is true:

1.2.16 Mackey's Theorem for group actions *If G acts transitively on both X and Y, then, for fixed $x \in X$, $y \in Y$, the mapping*

$$G\backslash\!\backslash(X \times Y) \to G_x\backslash G/G_y : G(x, gy) \mapsto G_xgG_y$$

is a bijection (note that $G\backslash\!\backslash(X \times Y)$ stands for $\Delta(G \times G)\backslash\!\backslash(X \times Y)$). Moreover, the action of G on the orbit $G(x, gy)$ and on the set of left cosets

$$G/(G_x \cap gG_yg^{-1})$$

are similar. Hence, if D denotes a transversal of the set of double cosets $G_x\backslash G/G_y$, for fixed $x \in X$, $y \in Y$, then we have the following similarity:

$$_G(X \times Y) \approx {}_G\Big(\dot{\bigcup}_{g \in D} G/(G_x \cap gG_yg^{-1})\Big).$$

(It is usually a generalization to linear group representations which carries this name, but the argument for its proof is essentially the same as the one that is used here.) ◇

Exercises

Exercise 1.2.1 Consider a G–set X, a normal subgroup $U \trianglelefteq G$, and the corresponding restriction $_UX$. Check the following facts:

- For each orbit $U(x)$ and any $g \in G$, the set $gU(x)$ is an orbit of U on X.
- The orbits of U on X form a G/U–set, in a natural way.
- The orbits of G/U on $U \backslash\backslash X$ are just the orbits of G on X.
- The U–orbits which belong to the same G–orbit are of the same order.

Exercise 1.2.2 Check all the details of the second item in 1.2.4.

Exercise 1.2.3 Check that the G-isomorphy $\simeq$ (and hence also the G-similarity $\approx$) is an equivalence relation on group actions.

Exercise 1.2.4 Consider the following definition: We call actions ${}_GX$ and ${}_GY$ *inner isomorphic* if and only if there exists a pair (η, θ) such that ${}_GX \simeq {}_GY$ and where η is an *inner* automorphism , which means that

$$\eta: G \to G: g \mapsto g'gg'^{-1},$$

for a suitable $g' \in G$. Show that this equivalence relation has the same classes as $\approx$.

Exercise 1.2.5 If the actions ${}_GX$ and ${}_HY$ are homomorphic by (η, θ), then for each $x \in X$ we have that

$$\eta(G_x) \subseteq H_{\theta(x)}.$$

Conversely, if we are given two actions ${}_GX$ and ${}_HY$ and a homomorphism η from G to H such that, for each $x \in X$, there exists an $y \in Y$ for which $\eta(G_x) \subseteq H_y$, then these actions are homomorphic by (η, θ), where

$$\theta: gt \mapsto \eta(g)y(t),$$

for any fixed transversal $T \in \mathcal{T}(G \backslash\backslash X)$ and chosen $y(t) \in Y$, for every $t \in T$.

Exercise 1.2.6 Prove the statements of 1.2.15.

1.3 Symmetry Classes of Mappings

Now we are going to introduce the actions derived from ${}_GX$ and ${}_HY$ which form our *paradigmatic examples*, and which will be discussed in full detail and applied to many cases in later sections. In order to prepare this, we form the set of all the mappings from X into Y:

$$Y^X := \{f \mid f : X \to Y\}.$$

Well-known examples are the finite-dimensional vector spaces $\mathbb{R}^n$, $GF(q)^n$ etc., where Y is the ground field and X the set $n := \{0, \ldots, n-1\}$, say, of numbers of coordinates. Other prominent examples are incidence structures. An incidence structure consisting of m points and n lines can be represented by 0-1-matrices with m rows and n columns, i. e. by elements of the set $\{0, 1\}^{m \times n} = 2^{m \times n}$. The rows (columns) correspond to the points (lines). It is clear that these matrices are uniquely determined up to permutations of rows and columns, which means that they form orbits under a suitable group action. Similar things hold for codes and graphs, we shall come to this quite soon. Let us introduce the corresponding actions on such sets:

1.3.1 Canonic actions on Y^X : *If G acts on X and H acts on Y, then G, H and $H \times G$ act on Y^X as follows:*

- *G acts on Y^X by*

$$G \times Y^X \to Y^X : (g, f) \mapsto f \circ \bar{g}^{-1},$$

i. e. (g, f) is mapped onto $\tilde{f}$, where

$$\tilde{f}(x) := \left(f \circ \bar{g}^{-1}\right)(x) = f(g^{-1}x).$$

The corresponding permutation group on Y^X will be denoted by $E^{\bar{G}}$.

- *H acts on Y^X via*

$$H \times Y^X \to Y^X : (h, f) \mapsto \bar{h} \circ f,$$

i. e. (h, f) is mapped onto $\tilde{f}$, where

$$\tilde{f}(x) := \left(\bar{h} \circ f\right)(x) = hf(x).$$

The corresponding permutation group will be denoted by $\bar{H}^E$.

- *The direct product $H \times G$ acts in the following way:*

$$(H \times G) \times Y^X \to Y^X : ((h, g), f) \mapsto \bar{h} \circ f \circ \bar{g}^{-1},$$

i. e. $((h, g), f)$ is mapped onto $\tilde{f}$, where

$$\tilde{f}(x) := \left(\bar{h} \circ f \circ \bar{g}^{-1}\right)(x) = hf(g^{-1}x).$$

The corresponding permutation group on Y^X will be denoted by $\bar{H}^{\bar{G}}$, and it will be called the power group *of $\bar{H}$ by $\bar{G}$.*

There is a fourth action which contains these three actions as subactions, but in order to describe it we first need to introduce the wreath product *$H \wr_X G$: Its underlying set is*

$$H \wr_X G := H^X \times G = \{(\psi, g) \mid \psi : X \to H, g \in G\}$$

(this does not *mean that $H \wr_X G$ is a direct product of subgroups, it only means that it is the* set theoretic cartesian product *of the subsets H^X and G), and the multiplication is defined by*

$$(\psi, g)(\psi', g') := (\psi\psi'_g, gg'),\ \psi\psi'_g(x) := \psi(x)\psi'_g(x) := \psi(x)\psi'(g^{-1}x).$$

We are now in a position to introduce the announced action of the wreath product $H \wr_X G$ on the set of mappings:

– *The actions $_GX$ and $_HY$ yield the following natural action of $H \wr_X G$ on Y^X:*

1.3.2 $$H \wr_X G \times Y^X \to Y^X : ((\psi, g), f) \mapsto \tilde{f},$$

where $\tilde{f}$ is defined by

$$\tilde{f}(x) := \psi(x) f(g^{-1}x).$$

The corresponding permutation group on Y^X will be denoted by $[\bar{H}]^{\bar{G}}$, and it will be called the exponentiation group *of $\bar{H}$ by $\bar{G}$.*

The orbits of G, H, $H \times G$ and $H \wr_X G$ on Y^X will be called *symmetry classes of mappings.* A few remarks concerning the wreath product $H \wr_X G$ are in order, they will in particular show that the actions of G, H, and $H \times G$ on Y^X are restrictions of the action of $H \wr_X G$ on Y^X defined above. The reader is kindly asked carefully to check the following statements on wreath products $H \wr_X G$:

1.3.3 Lemma *The wreath product $H \wr_X G$ has the following properties:*

– *The identity element of $H \wr_X G$ is $(\iota, 1)$, where $\iota : x \mapsto 1_H$.*

– *If we define $\psi^{-1} \in H^X$ by $\psi^{-1}(x) := \psi(x)^{-1}$, we get*

$$(\psi, g)^{-1} = (\psi^{-1}_{g^{-1}}, g^{-1}), \text{ where } \psi^{-1}_{g^{-1}} := (\psi^{-1})_{g^{-1}} = (\psi_{g^{-1}})^{-1}.$$

– *The normal subgroup*

$$H^* := \{(\psi, 1) \mid \psi \in H^X\} \trianglelefteq H \wr_X G,$$

is called the base group, *it is the direct product of $|X|$ copies H^x of H:*

$$H^x := \{(\psi, 1) \mid \forall\, x' \neq x : \psi(x') = 1_H\} \simeq H, \text{ for each } x \in X.$$

– *The subgroup $G' := \{(\iota, g) \mid g \in G\} \simeq G$ is a complement of H^*, so that we have*

$$H \wr_X G = H^* \cdot G', \; H^* \trianglelefteq H \wr_X G, \; H^* \cap G' = \{(\iota, 1)\}.$$

– *The diagonal*

$$\Delta(H^*) := \{(\psi, 1) \mid \psi \text{ constant}\} \simeq H,$$

satisfies

$$\Delta(H^*) \cdot G' = \{(\psi, g) \mid \psi \text{ constant}, \; g \in G\} \simeq H \times G.$$

□

This shows that the subgroups G', $\Delta(H^*)$ and $\Delta(H^*) \cdot G'$ are natural embeddings of G, H and $H \times G$ into $H \wr_X G$, in short:

1.3.4 $$G \hookrightarrow H \wr_X G, \; H \hookrightarrow H \wr_X G, \; H \times G \hookrightarrow H \wr_X G,$$

so that in fact the actions of G, H, and $H \times G$ on Y^X introduced above are restrictions of the action of $H \wr_X G$ on Y^X to subgroups. In the case when G is a permutation group, say $G \leq S_n$, and when we take for ${}_GX$ the natural action of G on n, we shorten the notation by putting

$$H \wr G := H \wr_n G.$$

A particular case is $H \wr S_n$, the *complete monomial group* of degree n over H. Several important groups are of this form, examples will be given later.

1.3.5 Applications (necklaces, graphs, linear codes) Here is a brief description of a series of interesting examples, they will be discussed later on in more detail. The crucial point here is just to note that these structures can be identified with orbits of suitable groups on suitable sets Y^X of mappings.

- *Necklaces* can be considered as colorings of the vertices of a regular n-gon in m colors, say. Thus the necklaces with n beads in up to m colors form the set of mappings $Y^X := m^n$. This means that we number the n vertices of the n-gon from 0 to $n-1$, i. e. we take the set $n := \{0, \ldots, n-1\}$ as the set of labels of the vertices of the n-gon, while we number the colors from 0 to $m-1$, thus taking the set $m := \{0, \ldots, m-1\}$ as set of colors, hoping that it will always become clear from the context if m or n means a cardinality or a set.
  We may consider two such colorings as equivalent if and only if one arises from the other by a cyclic shift. The equivalence classes, i. e. the sets of essentially different necklaces are formed by the set of orbits

  $$C_n \backslash\backslash m^n,$$

  where C_n means the cyclic group of order n, generated by a cyclic shift of the beads. (Of course, we may also allow reflections together with cyclic shifts, in which case we need only replace the cyclic group C_n by the dihedral group D_n, both these groups will be described in detail in the appendix.) This action which describes necklaces as orbits is an action of the form ${}_G(Y^X)$.

- *Incidence structures* consist of points and lines, and hence they can be described by 0-1-matrices where the rows correspond to the points, the columns to the lines, and where an entry is 1 or 0 depending if the corresponding point lies on the corresponding line or not. Thus the incidence structures with p points and l lines "are" the orbits of the cartesian product of symmetric groups $S_p \times S_l$ acting on the set $Y^X := 2^{p \times l}$. Recall that $2 = \{0, 1\}$, $p = \{0, 1, \ldots, p-1\}$, $l = \{0, \ldots, l-1\}$ and so

  $$p \times l = \{(i, k) \mid i \in p, k \in l\}$$

  stands for the set of pairs (i, k) consisting of a point i and a line k. An incidence structure can therefore be identified with its *characteristic function*

  $$f \in 2^{p \times l},$$

  which has the value 1 on the pair (i, k) — consisting of the point i and the line k — if and only if that pair is incident. Hence the equivalence classes of incidence structures, i. e. the sets of incidence structures which differ only by renumbering of points or lines, form the set of orbits

$$S_p \times S_l \,\backslash\!\!\backslash 2^{p\times l}.$$

This is again an action of the form ${}_G(Y^X)$.

- *Graphs* on the set $v = \{0, \ldots, v-1\}$ of vertices (or, to be exact, of *numbers* of vertices) can be considered as mappings from the set $\binom{v}{2}$ consisting of all the *2-element subsets* or (unordered) *pairs* of points into the set $\{0, 1\}$, where a pair of points is mapped onto 1 if and only if the two points are connected in the graph. This set of mappings $Y^X := 2^{\binom{v}{2}}$ is the set of *labeled* graphs on v vertices. The symmetric group S_v acts on the set v of vertices, and so it also acts on the set $\binom{v}{2}$ of pairs of vertices in the following way:

$$S_v \times \binom{v}{2} \to \binom{v}{2} \colon \left(\pi, \{i, j\}\right) \mapsto \{\pi i, \pi j\}.$$

  The corresponding set of orbits

$$S_v \,\backslash\!\!\backslash 2^{\binom{v}{2}}$$

  is the set of *unlabeled* graphs on v vertices. Hence, the set of unlabeled graphs on a given set M of vertices can be identified with the set of orbits of another action of the form ${}_G(Y^X)$.

- The *complementary* graph arises from a graph by putting edges where no edges were before and vice versa, i. e. — in terms of mappings $f \in Y^X = 2^{\binom{v}{2}}$ — by interchanging zeros and ones. This means to consider instead of the group $G = S_v$ the group $H \times G := S_2 \times S_v$. Its orbits are unions of one or two orbits of S_v. The union consists of an equivalence class of labeled graphs together with the class consisting of the complements. Hence the orbits of $S_2 \times S_v$ that coincide with orbits of S_v consist of *selfcomplementary* labeled graphs. The number of such classes is therefore twice the number of orbits of $S_2 \times S_v$ minus the number of orbits of S_v. Hence graphs together with their complements form the orbits of an action of the form ${}_{H\times G}(Y^X)$.

- *Linear codes* are subspaces of finite vector spaces, say of the n-dimensional vector space $Y^X := GF(q)^n$ over the Galois field $GF(q)$. Their error correcting properties are described in terms of the *Hamming distance* which counts the number of different coordinates of a given pair of vectors:

$$d(u, v) := |\{i \mid u_i \neq v_i\}|.$$

  The *minimum distance* of a linear code $C \leq GF(q)^n$ is defined to be the minimum

$$d := \min\{d(c, c') \mid c, c' \in C, c \neq c'\}.$$

It expresses the error correcting property of C to some extent, since the canonic way of decoding a received message into one of the nearest elements of C with respect to the Hamming distance allows to correct $\lfloor (d-1)/2 \rfloor$ transmission errors. Hence we can ask for the equivalence classes of linear codes with respect to linear mappings that keep the Hamming distance (which is in fact a metric) fixed.
The Hamming distance between any two different vectors, and therefore also the minimum distance of the code in question is neither changed by permuting the coordinates nor by simultaneous multiplication of coordinates by nonzero elements of the ground field. These permutations and multiplications form the set of *isometries* of $GF(q)^n$. The set of all isometries is the *isometry group*

$$H \wr_X G = GF(q)^* \wr S_n$$

($GF(q)^*$ means the multiplicative group of the Galois field). It acts according to 1.3.2 in the following way:

$$GF(q)^* \wr S_n \times GF(q)^n \to GF(q)^n : ((\psi, \pi), u) \mapsto \psi(i) \cdot u_{\pi^{-1}i}.$$

Hence the examination of necklaces, incidence structures or graphs amounts to a study of actions of the form ${}_G(Y^X)$, and we can cover the complementary structures by considering actions of the form ${}_{H\times G}(Y^X)$. The theory of linear codes leads to the consideration of an action of the form ${}_{H\wr_X G}(Y^X)$. ◇

In order to prepare further applications we mention that actions of direct products and of wreath products can be reformulated in terms of the respective factors of the direct and of the wreath product. Here is, to begin with, a lemma on actions of direct products, which is very easy to check (exercise 1.3.2):

1.3.6 Lemma *In each case when a direct product $H \times G$ acts on a set M, we obtain both a natural action of H on the set of orbits of G:*

$$H \times (G \backslash\backslash M) \to G \backslash\backslash M : (h, G(m)) \mapsto G(hm),$$

and a natural action of G on the set of orbits of H:

$$G \times (H \backslash\backslash M) \to H \backslash\backslash M : (g, H(m)) \mapsto H(gm).$$

Moreover the orbit of $G(m) \in G \backslash\backslash M$ under H is the set consisting of the orbits of G on M that form $(H \times G)(m)$, while the orbit of $H(m) \in H \backslash\backslash M$ under G is the set consisting of the orbits of H on M that form $(H \times G)(m)$, and therefore the following identity holds:

$$|H \backslash\backslash (G \backslash\backslash M)| = |G \backslash\backslash (H \backslash\backslash M)| = |(H \times G) \backslash\backslash M|.$$

In particular each action of the form ${}_{H\times G}Y^X$ can be considered as an action of H on $G \backslash\backslash Y^X$ or as an action of G on $H \backslash\backslash Y^X$. The corresponding result on wreath products is due to W. Lehmann ([97],[96]), and it reads as follows:

1.3.7 Lemma *The following mapping is a bijection:*

$$\Phi: H \wr_X G \backslash\!\backslash Y^X \to G \backslash\!\backslash (H \backslash\!\backslash Y)^X : H \wr_X G(f) \mapsto G(F),$$

if $F \in (H \backslash\!\backslash Y)^X$ *is defined by* $F(x) := H(f(x))$. *In particular,*

$$\left| H \wr_X G \backslash\!\backslash Y^X \right| = \left| G \backslash\!\backslash (H \backslash\!\backslash Y)^X \right|.$$

Proof: It is easy to see that Φ is well defined. In order to prove that Φ is injective assume $G(F) = G(F')$, so that there exist $g \in G$ such that $F' = F \circ \bar{g}^{-1}$. But this implies

$$H(f'(x)) = F'(x) = F(g^{-1}x) = H(f(g^{-1}x)),$$

for each $x \in X$. Therefore there must exist, for each $x \in X$, an element $\psi(x)$ in H such that $f'(x) = \psi(x) f(g^{-1}x)$. Summarizing, there exist $(\psi, g) \in H \wr_X G$, for which $f' = (\psi, g) f$, and so $H \wr_X G(f) = H \wr_X G(f')$.

In order to show that Φ is surjective assume that $G(F') \in G \backslash\!\backslash (H \backslash\!\backslash Y)^X$. Defining $f \in Y^X$ in such a way that $f(x) \in F'(x) = H(y_x)$, say $f(x) = y_x$, for all $x \in X$, then, for F defined as above, we obtain

$$F(x) = H(f(x)) = H(y_x) = F'(x),$$

which gives $\Phi(H \wr_X G(f)) = G(F) = G(F')$ and it completes the proof. □

1.3.8 Application (isometry classes of linear codes) It was mentioned that the isometry group on $GF(q)^n$ (with respect to the Hamming metric) is the wreath product $GF(q)^* \wr S_n$. It also acts on the power set of $GF(q)^n$, which is

$$Y^X := 2^{(GF(q)^n)}.$$

The set of orbits of the isometry group on this usually very big power set contains particular orbits that consist of subspaces only. These orbits are called the *isometry classes* of linear codes of length n. Hence *the isometry classes of linear codes of length n over* $GF(q)$ *are elements of the set of orbits*

$$GF(q)^* \wr S_n \backslash\!\backslash 2^{(GF(q)^n)}.$$

In the theory of linear codes we are interested in exactly these orbits. The problem is that the power set of $GF(q)^n$ is much too big to be handled by a computer, except for very small cases of n and q. In order to reduce the set which has to be examined, we can introduce bases and replace the sets of vectors on which the isometry group acts by generator matrices, i. e. by matrices the rows of which form a basis of a subspace, say of a subspace of dimension k. These codes will be called *(n, k)-codes.*

Each such matrix is a $k \times n$-matrix over $GF(q)$ and hence it can be considered as an element of the set

$$\left(GF(q)^k\right)^n.$$

But we have to take into account that each subspace has many bases and usually many more generator matrices. The general linear group $GL_k(q)$ is transitive on the set of bases of such a subspace, and therefore we are now dealing with orbits not just of the isometry group but of its direct product with $GL_k(q)$. The isometry classes are elements of the orbit set

$$GL_k(q) \times \left(GF(q)^* \wr S_n\right) \backslash\!\backslash \left(GF(q)^k\right)^n .$$

For obvious reasons, we can restrict attention to generator matrices with no column consisting of zeros only, so we may restrict attention to the following subset of orbits:

$$GL_k(q) \times \left(GF(q)^* \wr S_n\right) \backslash\!\backslash \left(GF(q)^k \backslash \{0\}\right)^n .$$

This set of *all* the $k \times n$-matrices over $GF(q)$ without colums of zeros is, of course, still much too big, but we can use it efficiently for enumerative purposes: Using the notation

$$N(n,k,q) := \left| GL_k(q) \times \left(GF(q)^* \wr S_n\right) \backslash\!\backslash \left(GF(q)^k \backslash \{0\}\right)^n \right|$$

for the number of these orbits, we get an expression *for the number* $T(n,k,q)$ *of* (n,k)*-codes, i. e. codes of length n and dimension* k*, without zero columns in their generator matrices:*

1.3.9 $$T(n,k,q) = N(n,k,q) - N(n,k-1,q).$$

For computing $N(n,k,q)$ we have to apply 1.3.6, obtaining the set

$$GL_k(q) \backslash\!\backslash GF(q)^* \wr S_n \backslash\!\backslash \left(GF(q)^k \backslash \{0\}\right)^n ,$$

and then to apply 1.3.7 which gives

$$GL_k(q) \backslash\!\backslash S_n \backslash\!\backslash \left(GF(q)^* \backslash\!\backslash GF(q)^k \backslash \{0\}\right)^n .$$

The inner set is a set of orbits of the multiplicative group $GF(q)^*$ of the ground field on the set of nonzero vectors of length k over $GF(q)$. The action is (recall 1.3.1, second item) the multiplication of the vectors with nonzero elements of the ground field:

$$GF(q)^* \times GF(q)^k \backslash \{0\} \to GF(q)^k \backslash \{0\} \colon (\kappa, v) \mapsto \kappa \cdot v.$$

Hence the orbit of $v \in GF(q)^k \backslash \{0\}$ is just the one-dimensional vector space generated by v. The set of these one-dimensional vector spaces is called the $(k-1)$-dimensional *projective space* over $GF(q)$, and it is indicated as follows:

$$P_{k-1}(q) := GF(q)^* \backslash\!\backslash GF(q)^k \backslash \{0\}.$$

Thus we finally end up with the following orbit set:

1.3.10 $$GL_k(q) \backslash\!\backslash \left(S_n \backslash\!\backslash P_{k-1}(q)^n\right).$$

This is the reason for the close connection between the theory of linear codes and projective geometry.

Particular orbits of S_n on $P_{k-1}(q)^n$ are those consisting of injective mappings into the projective space. If these mappings are also surjective, then these codes are the *simplex-codes*, the duals of *Hamming codes*. This means for the binary case ($q = 2$) that the generator matrix consists of columns which are the binary representations of the natural numbers from one to $2^k + 1$. ◇

The result 1.2.16 on the close relationship between double cosets and the orbits of a group on a cartesian product of sets on which it acts transitively suggests an application to symmetry classes of mappings:

1.3.11 Application (Mackey's theorem and symmetry classes of mappings) Consider a finite action $_GX$ and assume that $X = X_1 \dot{\cup} X_2$, where each one of the (disjoint) components X_i is supposed to be a union of orbits. Then

$$Y^{X_1 \dot{\cup} X_2} = Y^{X_1} \times Y^{X_2},$$

and each of the two cartesian factors is a union of orbits. Let

$$\{f_{ij}, j = 1, \ldots, |G \backslash\backslash Y^{X_i}|\}$$

denote a transversal of the orbits of G on Y^{X_i}, for $i = 1, 2$, so that we have the following decomposition into orbits

$$Y^{X_1 \dot{\cup} X_2} = \Big[\dot{\bigcup_j} G(f_{1j})\Big] \times \Big[\dot{\bigcup_k} G(f_{2k})\Big] = \dot{\bigcup_{j,k}} \big[G(f_{1j}) \times G(f_{2k})\big].$$

Thus we can rewrite the set of orbits in the following form:

$$G \backslash\backslash Y^{X_1 \dot{\cup} X_2} = G \backslash\backslash Y^{X_1} \times Y^{X_2} = G \backslash\backslash \dot{\bigcup_{j,k}} \big[G(f_{1j}) \times G(f_{2k})\big]$$

$$= \dot{\bigcup_{j,k}} \big[G \backslash\backslash G(f_{1j}) \times G(f_{2k})\big].$$

An application of Mackey's Theorem yields the bijection

$$G \backslash\backslash G(f_{1j}) \times G(f_{2k}) \to G_{f_{1j}} \backslash G / G_{f_{2k}} : G(f_{1j}, gf_{2k}) \mapsto G_{f_{1j}} g G_{f_{2k}}.$$

This finally yields a bijection between the set of symmetry classes of mappings and a set of double cosets:

$$G \backslash\backslash Y^X \longrightarrow \bigcup_{j,k} G_{f_{1j}} \backslash G / G_{f_{2k}}.$$

◇

Exercises

Exercise 1.3.1 Show that $E^{\bar{G}}$ is normal in $\bar{H}^{\bar{G}}$ and in $[\bar{H}]^{\bar{G}}$, and that $\bar{H}^{\bar{G}}$ is not in general normal in $[\bar{H}]^{\bar{G}}$. Check that the factor group $\bar{H}^{\bar{G}}/E^{\bar{G}}$ is isomorphic to $\bar{H}$, while $[\bar{H}]^{\bar{G}}/[\bar{H}]^{\bar{E}}$ is isomorphic to $\bar{G}$. What does this mean, in the light of exercise 1.2.1, for the enumeration of the orbits of $\bar{H}^{\bar{G}}$ and $[\bar{H}]^{\bar{G}}$?

Exercise 1.3.2 Check lemma 1.3.6.

1.4 Invariant Relations

Another interesting example of a symmetry class of mappings is an *invariant relation,* as we are going to show now. It is an orbit of the form ${}_H(Y^X)$. If H acts on the set n, then it also acts on the sets of mappings n^k, $k \in \mathbb{N}^*$ in a canonic way:

$$H \times n^k \to n^k : (h, f) \mapsto \bar{h} \circ f.$$

The subsets of n^k are the *k-ary* relations over the set n, and so the orbits of H on this set are the *minimal H-invariant k*-relations over n. The set of *all the H-invariant k-relations over n* is the power set of $H \backslash\!\backslash n^k$, which can be identified with the set of mappings

$$2^{H \backslash\!\backslash n^k}.$$

(To be exact: an H-invariant relation is a *union* of the elements — which are in fact orbits — contained in an element of this power set.) There are several operations on H-invariant relations that obviously keep the invariance:

1.4.1 Lemma *The following operations on H-invariant relations can be used in order to construct further H-invariant relations:*

- *The union $R \cup R'$ and the intersection $R \cap R'$ of two H-invariant k-relations $R, R' \subseteq n^k$ is an H-invariant k-relation.*
- *The* complement $R^c := n^k \backslash R$ *of an H-invariant k-relation is an H-invariant k-relation, too.*
- *The natural action of S_k on n^k maps an H-invariant k-relation R onto an H-invariant k-relation πR, for every $\pi \in S_k$.*
- *The* restriction $R \downarrow k-1$ *of an H-invariant k-relation R to $k-1$ is an H-invariant $(k-1)$-relation.*
- *The cartesian product $R \times R'$ of an H-invariant k-relation R and an H-invariant l-relation R' is an H-invariant $(k+l)$-relation.*

The consideration of invariant relations allows *certain classifications of the acting groups.* We are going to describe a few aspects of this. The interested reader who wants to get more details is referred to H. Wielandt's Ohio Lectures *Permutation groups through invariant relations and invariant functions* ([164]) and the booklet by M. Klin, R. Pöschel and Rosenbaum ([82]).

Using the notion of invariant relations we introduce the following equivalence relation on the set of all the groups that act on n:

$$H \sim_k K :\Longleftrightarrow H \backslash\backslash n^k = K \backslash\backslash n^k.$$

Moreover, we denote by n^k_{inj} the set of injective mappings:

$$n^k_{inj} := \{f \in n^k \mid f \text{ injective}\},$$

and we say that H is *k-transitive* or *k-fold transitive* on n iff H is transitive on n^k_{inj}. The following implications are helpful:

1.4.2 Lemma *If both H and K act on n, then the following is true:*

- *$H \sim_1 K$ is the same as $H \backslash\backslash n = K \backslash\backslash n$.*
- *$H \sim_n K$ is equivalent to $\bar{H} = \bar{K}$.*

Moreover, for $2 \leq k \leq n$, we have that

- *$H \sim_k K$ implies that $H \sim_{k-1} K$.*
- *$H \sim_k K$ is equivalent to $H \backslash\backslash n^k_{inj} = K \backslash\backslash n^k_{inj}$.*
- *If $H \sim_k K$ and H is k-transitive on n, then K is k-transitive, too.*

Proof: The characterization of $H \sim_1 K$ follows from $n = n^1$. The statement on $H \sim_n K$ is a consequence of the fact that the H-orbit of the identity mapping can be identified with $\bar{H}$, as it is easy to see:

$$H(\mathrm{id}_n) = \{(h0, h1, \ldots, h(n-1)) = \bar{h} \mid h \in H\},$$

while its K-orbit can be identified with $\bar{K}$. The next statement is clear from the fact that each H- or K-orbit on n^{k-1} can be obtained from an orbit on n^k by restricting its elements $f \in n^k$ to the subset $k-1$.

In order to prove the following statement, we note that the set n^k_{inj} of injective mappings is a union of orbits of H as well as of K, and so $H \sim_k K$ implies $H \backslash\backslash n^k_{inj} = K \backslash\backslash n^k_{inj}$. Conversely, if $H \backslash\backslash n^k_{inj} = K \backslash\backslash n^k_{inj}$, we use induction on k. $H \backslash\backslash n^k_{inj} = K \backslash\backslash n^k_{inj}$ gives, by restriction, that $H \backslash\backslash n^l_{inj} = K \backslash\backslash n^l_{inj}$, for $l < k$. Now we use that the orbit $H(f)$ of a noninjective mapping $f \in n^k$ can be obtained from the orbit $H(f')$ of an injective $f' \in n^l_{inj}$, for a suitable $l < k$, so that $H(f) = K(f)$ follows from $H \backslash\backslash n^l_{inj} = H \backslash\backslash n^l_{inj}$.

The final statement is immediate from the fourth, together with the definition of k-fold transitivity. □

Now we look for a distinguished transversal of the k-equivalence classes of the groups acting on n. As

$$\bar{H} \sim_k H,$$

each k-equivalence class contains a subgroup of S_n. The interesting ones are those which are in a certain sense maximal. We indicate the maximal subgroup of S_n which is k-equivalent to H as follows:

$$\bar{H}_{(k)} := \{\pi \in S_n \mid \pi(\omega) = \omega, \text{ for each } \omega \in H \backslash\backslash n^k\}.$$

This group is called the *k-closure* of $\bar{H}$. The following lemma lists extremal cases which are easy to check:

1.4.3 Lemma *If H acts on n, then we have:*

- *The group $\bar{H}_{(1)}$ is the Young subgroup (see appendix) of S_n, defined by the orbits of H :*

$$\bar{H}_{(1)} = \bigoplus_{\omega \in H \backslash\backslash n} S_\omega .$$

- *The n-closure of $\bar{H}$ is its image under the permutation representation:*

$$\bar{H}_{(n)} = \bar{H}.$$

- *The k-closures of $\bar{H}$ form the chain*

$$\bar{H} = \bar{H}_{(n)} \leq \bar{H}_{(n-1)} \leq \ldots \leq \bar{H}_{(1)} \leq S_n .$$

- *The equality $\bar{H}_{(k)} = S_n$ holds if and only if H is k-fold transitive on n.*

Another result proves the existence of certain cycles in $\bar{H}$:

1.4.4 Lemma *If H and K are k-equivalent groups acting on n then each m-cycle, $m \leq k$, of an element in $\bar{H}$ is also a cyclic factor of an element in $\bar{K}$.*

Proof: From $H \sim_m K$ we obtain that both H and K have the same orbits on n^m_{inj}. We consider a cyclic factor $(i_0, \ldots, i_{m-1})$, say, of an element $\pi \in \bar{H}$, and the orbit of the corresponding mapping $f := (i_0, \ldots, i_{m-1}) \in n^m_{inj}$. There exists an element $\rho \in \bar{K}$, such that $\pi f = \rho f$, i. e. $\pi i_\nu = \rho i_\nu$, $0 \leq \nu \leq m-1$. Hence,

$$i_1 = \pi i_0 = \rho i_0, \ldots, i_{m-1} = \pi i_{m-2} = \rho i_{m-2}, i_0 = \pi i_{m-1} = \rho i_{m-1}.$$

This shows that $(i_0, \ldots, i_{m-1}) = (i_0, \rho i_0, \rho^2 i_0, \ldots)$ is a cyclic factor of ρ, too.

□

Sylow's theorem gives that a prime number p divides the order of a finite group H if and only if H contains elements of that order. This, together with 11.2.7 yields

1.4.5 Corollary *Assume that $P, Q \leq S_n$, $k \in \mathbb{N}^*$, and that p is a prime number $\leq k$. Then, if $P \sim_k Q$, we have:*

$$p \text{ divides } |P| \iff p \text{ divides } |Q|.$$

Now we call $P \leq S_n$ a *k-closed* permutation group if and only if

$$P = P_{(k)}.$$

1.4.6 Theorem *A subgroup P of S_n is k-closed if and only if there exists a set of P-invariant k-relations over n such that P is the maximal subgroup of S_n which contains this set among the set of its invariant relations.*

Proof: In the case when $P = P_{(k)}$ we can simply choose the complete set of its invariant k-relations. In order to prove the converse, we assume that $M \subseteq 2^{(n^k)}$ is this set of invariant relations. $P_{(k)}$ is contained in the maximal subgroup of S_n the invariant k-relations set of which comprises M. Thus $P_{(k)} \leq P$ and therefore $P = P_{(k)}$, and so P is k-closed. □

For example, the symmetry groups of the regular tetrahedron, the cube, the octahedron, the icosahedron and the dodecahedron are 2-closed. Here is another example:

1.4.7 Example The regular representation of a finite group H, i. e. the permutation group $\bar{H}$ induced by the action ${}_H H$ of H on itself by left multiplication, is 2-closed. In order to check this, we note (recall 1.2.15) that each orbit of H on H^2 contains exactly one element of the form $(1, x)$, for a suitable element $x \in H$. We indicate this orbit by

$$\omega_{(x)} := \{(h, hx) \mid h \in H\}.$$

Thus all the orbits except $\omega_{(1)}$ consist of injective pairs only, and $\omega_{(1)}$ consists of all the noninjective ones. The 2-closure $\bar{H}_{(2)}$ of H consists of the $\pi \in S_H$ which fix each of these orbits setwise. We would like to show that it acts transitively on H (which is clear from its action on $\omega_{(1)}$) and that, moreover, the stabilizer of each $h \in H$ in the 2-closure is trivial. The latter statement immediately follows if we consider the stabilizer of $(1, x)$. But these two properties show that the order of the 2-closure is $|H|$, which is also the order of H, so that H must be 2-closed. ◇

Now we take a field $\mathbb{F}$ and consider the free vector space over $\mathbb{F}$ with n^k as basis:

$$\mathbb{F}^{(n^k)} := \{F : n^k \to \mathbb{F}\}.$$

The group H acts on n^k as was described at the beginning of this section. Hence it also acts in a natural way on $\mathbb{F}^{(n^k)}$:

1.4.8 $$H \times \mathbb{F}^{(n^k)} \to \mathbb{F}^{(n^k)} : (h, F) \mapsto hF,$$

where, for each $f \in n^k$, we have $hF(f) := F(h^{-1}f) = F(\bar{h}^{-1} \circ f)$. An important subspace is the subset of H-invariants

$$\mathbb{F}_H^{(n^k)} = \{F \in \mathbb{F}^{(n^k)} \mid \forall\, h \in H : hF = F\}.$$

From the foregoing we clearly get

1.4.9 Corollary *The following properties of $F \in \mathbb{F}^{(n^k)}$ are equivalent:*

- *F is H-invariant,*
- *F is constant on each orbit of H on n^k,*
- *each niveau set of F, i. e. each*

$$F_\kappa := \{f \in n^k \mid F(f) = \kappa\},\ \kappa \in \mathbb{F},$$

is an H-invariant k-relation of H over n.

This allows to characterize the invariant relations as niveau sets of spaces of invariant functions. It is obvious that $\mathbb{F}^{(n^k)}$ together with pointwise addition and multiplication is an $\mathbb{F}$-algebra of dimension n^k. The subspace of invariants $\mathbb{F}_H^{(n^k)}$ is clearly a subalgebra. If we decompose n^k into orbits, say

$$n^k = \omega_0 \,\dot{\cup} \ldots \dot{\cup}\, \omega_{r-1},$$

then we can rewrite $F \in \mathbb{F}_H^{(n^k)}$, using its values on the ω_i and the *indicator functions* I of the subsets M of n^k :

$$I_M(f) := \begin{cases} 1_{\mathbb{F}}, & f \in M, \\ 0, & \text{otherwise.} \end{cases}$$

In terms of the indicator functions of the orbits, we can rewrite F as follows:

$$\kappa_i := F(f),\ f \in \omega_i,$$

for which we have

$$F = \sum_i \kappa_i I_{\omega_i}.$$

Hence the I_{ω_i} span $\mathbb{F}_H^{(n^k)}$. Since they are linearly independent, we obtain for the dimension of that space the equation

1.4.10 $$\dim_{\mathbb{F}}(\mathbb{F}_H^{(n^k)}) = |H \backslash\backslash n^k| = r.$$

The *action* of H on $\mathbb{F}^{(n^k)}$ is described in 1.4.8. The corresponding *linear representation* (see the appendix) of H on $\mathbb{F}^{(n^k)}$ is the k-fold inner tensor power

$$\otimes^k \delta, \text{ where } \delta\colon h \mapsto \bar{h}$$

of the permutation representation δ of H on n. In the present context the main property of this representation is:

1.4.11 Lemma *The intertwining space $Hom_H(\mathbb{F}^{(n^k)}, \mathbb{F}^{(n^l)})$ of two such representations is isomorphic to the space of invariants*

$$\mathbb{F}_H^{(n^{(k+l)})}.$$

Hence, in particular

$$\dim\left(Hom_H(\mathbb{F}^{(n^k)}, \mathbb{F}^{(n^l)})\right) = \left| H \backslash\backslash n^{(k+l)} \right|.$$

Proof: The intertwining space consists, by definition, of all the matrices that have n^k rows, n^l columns, and which commute with the action of H. This means just H-invariance, since the action of H on $\mathbb{F}^{(n^k)}$ and on $\mathbb{F}^{(n^l)}$ consists of the representing matrices, which consist of n^k and n^l rows, respectively, and the elements of a matrix representation that corresponds to $\otimes^k \delta$ and $\otimes^l \delta$. □

1.4.12 Corollary *If both H and K act on n, then the equality of intertwining spaces*

$$Hom_H\left(\mathbb{F}^{(n^k)}, \mathbb{F}^{(n^l)}\right) = Hom_K\left(\mathbb{F}^{(n^k)}, \mathbb{F}^{(n^l)}\right)$$

holds if and only if

$$H \sim_{k+l} K.$$

Exercises

Exercise 1.4.1 Check that the symmetry groups of the regular tetrahedron, the cube, the octahedron and the dodecahedron are 2-closed.

Exercise 1.4.2 Describe *invariant functions,* i. e. elements of $\mathbb{F}_H^{(n^k)}$, for subgroups $H \leq S_n$, and develop methods to obtain such functions.

1.5 Hidden Symmetries

In the case when we are given an action ${}_GX$, where G is a *symmetry group* on X, the decomposition $G \backslash\backslash X$ of X into its orbits under G may be called the induced *symmetry.* For example, X may be the set of atoms in a crystal, or the set of vertices of an n-gon, or the set of carbon atoms in a benzene molecule, while G denotes the symmetry group of the crystal, of the n-gon or of the benzene molecule. The orbits form the sets of atoms or vertices which are equivalent with respect to physical or common-sense symmetry, and the number of orbits gives the number of such sets of equivalent atoms or nodes. This number quite often agrees, in the case of molecules, with the number of signals in a spectrum which come from the atoms that form the orbits. For example, in the benzene case, there is only one signal coming from carbon since, with respect to the symmetry group, all six carbon atoms fall in the same orbit.

But there may also be *relationships between different orbits,* for example, some of them may have the same order, they may even have the same conjugacy class of stabilizers, and so on. A certain kind of such further relations can be described in a natural way by a bigger group that also acts on X, as we are going to describe now. (Hermann Weyl pointed to this fact in a discussion of problems Immanuel Kant had when he discussed in his "Prolegomena" what the difference might be between the right and the left ear!)

The action of G on X gives the same orbits as the action of the image

$$\bar{G} = \{\bar{g} \mid \bar{g}: x \mapsto gx, g \in G\}$$

of G in the symmetric group S_X. Somewhere between $\bar{G}$ and the symmetric group S_X there is the *normalizer* of $\bar{G}$ (cf. example 1.1.11):

$$N_{S_X}(\bar{G}) := \{\pi \in S_X \mid \pi\bar{G} = \bar{G}\pi\}.$$

Since this group is a subgroup of the symmetric group, it consists of permutations of X, and so we have a natural action of the normalizer on X, too.

1.5.1 Lemma *Assume an action ${}_GX$. The orbits of $N_{S_X}(\bar{G})$ on X are unions of orbits of G. Moreover, the orbits of G on X are blocks of $N_{S_X}(\bar{G})$, i. e. each ω in $G\backslash\backslash X$ is mapped under any $\pi \in N_{S_X}(\bar{G})$ either onto itself or onto a subset of X that is disjoint to ω. In particular, $N_{S_X}(\bar{G})$ permutes the orbits of G on X.*

Proof: The statement on the orbits of $N_{S_X}(\bar{G})$ is clear since $N_{S_X}(\bar{G})$ lies above $\bar{G}$. The next statement can be obtained this way: Assume that $\omega \in G\backslash\backslash X$, and $\pi \in N_{S_X}(\bar{G})$. If $x \in \pi\omega \cap \omega$, then $x = \pi x'$, for a suitable $x' \in \omega$, and so $\omega = G(x) = G(\pi x') = \pi G(x') = \pi\omega$.

□

Thus there exists a natural symmetry on the set $G\backslash\backslash X$ of orbits of G which is induced by the normalizer of $\bar{G}$ in S_X : *The normalizer of $\bar{G}$ in S_X acts on the orbits of G and therefore we obtain a decomposition of the orbit set $G\backslash\backslash X$ in the following way:*

$$N_{S_X}(\bar{G})\backslash\backslash(G\backslash\backslash X)\,.$$

We call this symmetry the *hidden symmetry* of G on X, while the orbits of G on X exhibit the *obvious symmetry*. The hidden symmetry is in fact induced on the set of orbits of G by the factor group

$$N_{S_X}(\bar{G})/\bar{G}.$$

H. Weyl pointed to this fact also at the end of his famous booklet on symmetry, and therefore physicists call *Weyl's Recipe* the recommendation, to *consider both the obvious and the hidden symmetries.*

1.5.2 Application (the holomorph of a group) In the case when the action ${}_GX$ is *regular,* which means that it is transitive and that each stabilizer is trivial, i. e. $G_x = \{1\}$, the identity subgroup. The definition of hidden symmetry leads to a well known notion of group theory. Since in this particular case the set X can be identified with G and the action is similar to the regular action,

$${}_GX \approx {}_G(G/1) \approx {}_GG,$$

the hidden symmetry is induced by the normalizer of $\bar{G} \simeq G$ in the symmetric group S_G (the permutation $\bar{g}$ means left multiplication of G by g). This group is called the *holomorph* of G and we claim that the following ist true:

1.5.3 $$\mathrm{hol}(G) := N_{S_G}(\bar{G}) = \bar{G}\cdot\mathrm{Aut}(G).$$

Proof: Consider the subgroup of S_G generated by the complex product

$$\bar{G} \cdot \mathrm{Aut}(G) = \{\pi\rho \mid \pi \in \bar{G}, \rho \in \mathrm{Aut}(G)\}.$$

This product is a group, since $\bar{G}$ is a normal subgroup of the generated group: $\pi := \binom{g'}{gg'} \in \bar{G}$ and $\rho = \binom{g''}{\rho(g'')} \in \mathrm{Aut}(G)$ satisfy

$$\rho\pi\rho^{-1} = \begin{pmatrix} g'' \\ \rho(g'') \end{pmatrix}\begin{pmatrix} g' \\ gg' \end{pmatrix}\begin{pmatrix} \rho(g'') \\ g'' \end{pmatrix}$$

$$= \begin{pmatrix} gg'' \\ \rho(gg'') \end{pmatrix}\begin{pmatrix} g'' \\ gg'' \end{pmatrix}\begin{pmatrix} \rho(g'') \\ g'' \end{pmatrix} = \begin{pmatrix} \rho(g'') \\ \rho(gg'') \end{pmatrix} = \begin{pmatrix} \rho(g'') \\ \rho(g)\rho(g'') \end{pmatrix} \in \bar{G}.$$

Thus $\bar{G} \cdot \mathrm{Aut}(G)$ is identical with the group generated by this set. Moreover, $\bar{G}$ and $\mathrm{Aut}(G)$ have a trivial intersection since each element of $\mathrm{Aut}(G)$ leaves 1 fixed, while $\bar{1}$ is the only element of $\bar{G}$ that fixes 1. We have obtained that

$$\bar{G} \cap \mathrm{Aut}(G) = \{1\}.$$

This shows that each element π of $\bar{G} \cdot \mathrm{Aut}(G)$ has a unique decompostion $\pi = \rho\sigma$, where $\rho \in \bar{G}$ and $\sigma \in \mathrm{Aut}(G)$. In fact this product is *semidirect*. (Recall that a *semidirect product* of two groups A and B is any group Γ whose underlying set is the cartesian product $A \times B$ and the multiplication of which is of the form $(a, b)(a', b') = (a \cdot \hat{b}(a'), bb')$, for a suitable homomorphism $b \mapsto \hat{b}$ of B into the automorphism group of A. A nice example is the wreath product $H \wr_X G$, see exercise 1.5.1.)

We apply this in order to describe the stabilizer of $1 \in G$. Assume $\sigma = \pi\rho$ in $(\bar{G} \cdot \mathrm{Aut}(G))_1$. It implies $1 = \sigma 1 = \pi\rho 1 = \pi 1$, since each automorphism ρ fixes the identity element. Hence $\pi = 1$, and so $\sigma = \rho$, an element of the automorphism group. This shows that

$$(\bar{G} \cdot \mathrm{Aut}(G))_1 = \mathrm{Aut}(G).$$

It remains to prove that $\bar{G} \cdot \mathrm{Aut}(G)$ is the normalizer of $\bar{G}$ in the symmetric group S_G. Take an element ν in the normalizer. It maps 1 onto another element of G, say $\nu 1 = g$. If π is the permutation of G induced by the left multiplication of the elements of G by g, then $\pi^{-1}\nu 1 = \pi^{-1} g = 1$, and hence $\pi^{-1}\nu \in \mathrm{Aut}(G)$. This proves that

$$N_{S_G}(\bar{G}) \subseteq \bar{G} \cdot \mathrm{Aut}(G).$$

The reverse inclusion is obvious, and this completes the proof. □

◇

Exercises

Exercise 1.5.1 Show that $H \wr_X G$ is a semidirect product.

Exercise 1.5.2 Describe the holomorph of C_m, derive its order and show that it contains the dihedral group D_m.

2. Enumeration of Unlabeled Structures

We are now going to examine unlabeled structures. The first step is to compute their total number, for example, the number of graphs with prescribed number of vertices. Unlabeled structures will be described as orbits of groups on sets. For the purpose of their enumeration, the Cauchy–Frobenius Lemma is derived, it yields the number of orbits in the case when both the group and the set on which it acts are finite.

The enumeration of symmetry classes of mappings is displayed in full detail, in order to prepare refinements which are given in the following chapters.

A very simple case of a group action leads to another important enumerative concept, the Involution Principle. Finally we discuss the enumeration of symmetry classes which consist of injective or of surjective mappings.

2.1 The Number of Orbits

The equation $|G(x)| = |G/G_x|$, obtained in 1.2.2, was already seen to be of fundamental importance. It is also essential in the proof of the following counting lemma which, together with later refinements, forms *the basic tool of the theory of enumeration under finite group action*:

2.1.1 The Cauchy–Frobenius Lemma *The number of orbits of a finite group G acting on a finite set X is equal to the average number of fixed points:*

$$|G \backslash\backslash X| = \frac{1}{|G|} \sum_{g \in G} |X_g|.$$

Proof:

$$\sum_g |X_g| = \sum_g \sum_{x \in X_g} 1 = \sum_x \sum_{g \in G_x} 1 = \sum_x |G_x|$$

$$=_{1.2.2} |G| \sum_x |G(x)|^{-1} = |G| \cdot |G \backslash\backslash X|,$$

since each orbit contributes exactly 1 to the sum $\sum_x |G(x)|^{-1}$. □

The next remark helps considerably to shorten the calculations necessary for applications of this lemma. It shows that we can replace the summation over all $g \in G$ by a summation over a *transversal* of the conjugacy classes, as the number of fixed points turns out to be constant on each such class:

2.1.2 Lemma *For each finite group action, the mapping*

$$X_{g'} \to X_{gg'g^{-1}} : x \mapsto gx$$

is a bijection between these two sets of fixed points, and hence

$$\chi : G \to \mathbb{N} : g \mapsto |X_g|$$

is a class function, *i. e. it is constant on the conjugacy classes of* G. *More formally, for any* $g, g' \in G$, *we have that* $|X_{g'}| = |X_{gg'g^{-1}}|$.

Proof: That $x \mapsto gx$ establishes a bijection between $X_{g'}$ and $X_{gg'g^{-1}}$ is clear from the following equivalence:

$$g'x = x \iff gg'g^{-1}(gx) = gx.$$

□

The mapping χ is called the *character* of ${}_GX$.

2.1.3 Corollary *Let* ${}_GX$ *be a finite action and let* C *denote a transversal of the conjugacy classes of* G. *Then*

$$|G \backslash\backslash X| = \frac{1}{|G|} \sum_{g \in C} |C^G(g)||X_g| =_{1.2.3} \sum_{g \in C} |C_G(g)|^{-1}|X_g|.$$

Another formulation of the Cauchy–Frobenius Lemma makes use of the permutation representation $g \mapsto \bar{g}$ defined by the action in question. The permutation group $\bar{G}$ which is the image of G under this representation, yields an action ${}_{\bar{G}}X$ of $\bar{G}$ on X,which has the same orbits, and so we also have:

2.1.4 Corollary *If* X *denotes a finite* G*–set, then (for* any *group* G*) the following identity holds:*

$$|G \backslash\backslash X| = \frac{1}{|\bar{G}|} \sum_{\bar{g} \in \bar{G}} |X_{\bar{g}}| = \frac{1}{|\bar{G}|} \sum_{\bar{g} \in \bar{C}} |C^{\bar{G}}(\bar{g})||X_{\bar{g}}|,$$

where $\bar{C}$ *denotes a transversal of the conjugacy classes of* $\bar{G}$.

2.1.5 Application (the numbers of bilateral classes and of double cosets) The Cauchy–Frobenius Lemma yields among many other cardinalities the number of bilateral classes and therefore also the number of double cosets. Recall from 1.2.8 that bilateral classes are the orbits of the following action of a subgroup U of the direct product $G \times G$:

$$U \times G \to G : ((a, b), g) \mapsto agb^{-1}.$$

In order to apply the Lemma of Cauchy–Frobenius to that situation, we have to evaluate the number $|G_{(a,b)}|$ of fixed points of $(a, b) \in U$ on G which is

$$|\{g \mid a = gbg^{-1}\}| = \begin{cases} |C_G(a)| = |C_G(b)|, & \text{if } a, b \text{ are conjugates,} \\ 0, & \text{otherwise.} \end{cases}$$

Thus, by the Cauchy–Frobenius Lemma, we obtain the number of bilateral classes in the following way:

$$\begin{aligned} |U \backslash\backslash G| &= \frac{1}{|U|} \sum_{(a,b)} |G_{(a,b)}| \\ &= \frac{1}{|U|} \sum_{(a,b):a \sim b} |C_G(a)| \\ &= \frac{1}{|U|} \sum_{(a,b)} \frac{|C^G(a) \cap C^G(b)|}{|C^G(a)|} |C_G(a)| \\ &= \frac{|G|}{|U|} \sum_{(a,b) \in U} \frac{|C^G(a) \cap C^G(b)|}{|C^G(a)|^2}. \end{aligned}$$

It can be simplified since characters are constant on conjugacy classes of elements, and so *the number of bilateral classes* turns out to be

2.1.6
$$|U \backslash\backslash G| = \frac{|G|}{|U|} \sum_{g \in C} \frac{|(C^G(g) \times C^G(g)) \cap U|}{|C^G(g)|},$$

if C denotes a transversal of the conjugacy classes of elements in G. In particular, *the set*

$$A \backslash G / B := \{AgB \mid g \in G\} = (A \times B) \backslash\backslash G$$

of (A, B)*-double cosets has the order*

2.1.7
$$|A \backslash G / B| = \frac{|G|}{|A||B|} \sum_{g \in C} \frac{|C^G(g) \cap A||C^G(g) \cap B|}{|C^G(g)|}.$$

◇

2.1.8 Application (congruences) We already saw in 1.2.4 that group actions can also be useful in number theory, and it seems appropriate to emphasize the following immediate implication of the Cauchy–Frobenius Lemma. For any action of a finite group G on a finite set X we have the following congruence modulo the group order:

2.1.9
$$\sum_{g \in G} |X_g| \equiv 0 \ (|G|).$$

Here is an easy example. The symmetric group S_p, p a prime number, acts trivially on a one element set $X = \{x\}$:

$$S_p \times X \to X \colon (\pi, x) \mapsto x.$$

There is clearly one orbit only, and so, by the Cauchy–Frobenius Lemma in its form 2.1.3, together with 11.2.7,

$$p! = \sum_{a \vdash p} \frac{p!}{\prod_i i^{a_i} a_i!} |X_\pi|.$$

As each $|X_\pi| = 1$ and since the only class orders that are *not* divisible by p are the orders of the class containing the identity element and the class containing the cycles of full length p (this class has order $(p-1)!$, see the appendix on symmetric groups), this yields the congruence

$$0 \equiv 1 + (p-1)!\ (p),$$

a congruence that is called *Wilson's theorem.*

Later on we shall return to 2.1.9, we shall refine it considerably (numbers will be replaced by polynomials, and the congruence will be a congruence for the coefficients of the monomial summands).

◇

Further actions of G which can be derived from ${}_GX$ are the actions of G on the sets of *k–subsets* of X, $1 \leq k \leq |X|$,

2.1.10 $$G \times \binom{X}{k} \to \binom{X}{k} : (g, M) \mapsto \bar{g}M = \{gm \mid m \in M\}.$$

The action ${}_GX$ is called *k–homogeneous* if and only if the corresponding action of G on $\binom{X}{k}$ is transitive. An obvious example is the natural action of S_X on X, it is k–homogeneous for $k \leq |X|$. The following is a very important application of actions on k–subsets, it is in fact one of the highlights of applied finite group actions:

2.1.11 Application (existence and properties of Sylow subgroups) The regular representation of G yields, in accordance with 2.1.10, the G–sets $\binom{G}{k}$, for $1 \leq k \leq |G|$. If G is finite and p a prime dividing $|G|$, say $|G| = p^r \cdot q$, where $r \geq 1$, $q = p^s t$, and p does not divide t. Then we can put $k := p^r$ and we may consider the particular G–set $\binom{G}{p^r}$. H. Wielandt used this in his famous proof of *Sylow's Theorem* ([162]) in order to show that G possesses subgroups of order p^r. His argument runs as follows: p^s is the exact power of p dividing the order of $\binom{G}{p^r}$. This is clear from

$$\left|\binom{G}{p^r}\right| = \frac{p^r q}{p^r} \cdot \frac{p^r q - 1}{1} \cdots \frac{p^r q - (p^r - 1)}{p^r - 1},$$

as each power of p contained in the denominator cancels. Thus p^r–subsets M exist, the orbit length of which *is not divisible by* p^{s+1}. We consider such an M and show that its stabilizer G_M is of order p^r by proving that p^r is both an upper and lower bound:

1. For each $m \in M$ and $g \in G_M$ we have that $gm \in M$, hence

$$|G_M| \leq |M| = p^r.$$

2. On the other hand, p^{s+1} does not divide the length of the orbit, which is $|G(M)| = |G|/|G_M|$, and so

$$|G_M| \geq p^r.$$

This proves the opening item of

2.1.12 Sylow's Theorem *Assume G to be a finite group and p to be a prime divisor of its order. Then*

– *G contains subgroups of order p^r, for each power p^r dividing its order $|G|$.*

The subgroups $S \leq G$ of the maximal p-power order are called the Sylow p–subgroups *of G. They have the following properties:*

– *Each p-subgroup U of G is contained in a suitable Sylow p-subgroup S.*

– *Any two Sylow p-subgroups of G are conjugate subgroups.*

Proof: The proofs of the other two items follow from a consideration of double cosets. Assume a p-subgroup U of G and a Sylow p-subgroup S. Then we derive from 1.2.10 that

$$\frac{|G|}{|S|} = \sum_{g \in D} \frac{|U|}{|U \cap gSg^{-1}|}$$

where D denotes a transversal of $U\backslash G/S$. If all the intersections in the denominator on the right hand side were proper subgroups of U, then the right hand side were divisible by p, which contradicts the left hand side. Hence there must exist a $g \in D$, such that $U \leq gSg^{-1}$. Since gSg^{-1} is a Sylow p-subgroup, too, U is contained in a Sylow subgroup, which proves the second item.

The final item follows by taking for U a Sylow p-subgroup S' :

$$\frac{|G|}{|S|} = \sum_{g \in D'} \frac{|S'|}{|S' \cap gSg^{-1}|}$$

shows that $S' = gSg^{-1}$, for a suitable $g \in D'$, where D' denotes a transversal of $S'\backslash G/S$.

◇

In the case when ${}_GX$ is a finite action, we can apply the sign map ϵ (see the appendix) to $\bar{G}$, the permutation group induced by G on X. Its kernel

$$\bar{G}^+ := \{\bar{g} \in \bar{G} \mid \epsilon(\bar{g}) = 1\}$$

is either $\bar{G}$ itself or a subgroup of index 2, this is easy to see. Denoting its inverse image by

$$G^+ := \{g \in G \mid \epsilon(\bar{g}) = 1\},$$

we obtain a useful interpretation of the *alternating sum* of fixed point numbers:

2.1.13 Lemma *For any finite action $_GX$ such that $G \neq G^+$, the number of orbits of G on X which split over G^+ (i. e. which decompose into more than one — and hence into two — G^+–orbits) is equal to*

$$\frac{1}{|G|}\sum_{g\in G}\epsilon(\bar{g})|X_g| = \frac{1}{|\bar{G}|}\sum_{\bar{g}\in\bar{G}}\epsilon(\bar{g})|X_{\bar{g}}|.$$

Proof: As $G \neq G^+$, and hence $|G^+| = |G|/2$, we have

$$\frac{1}{|G|}\sum_{g\in G}\epsilon(\bar{g})|X_g| = \frac{2}{|G|}\sum_{g\in G^+}|X_g| - \frac{1}{|G|}\sum_{g\in G}|X_g|$$

$$= \frac{1}{|G^+|}\sum_{g\in G^+}|X_g| - \frac{1}{|G|}\sum_{g\in G}|X_g| = |G^+ \backslash\backslash X| - |G \backslash\backslash X|.$$

Each orbit of G on X is either a G^+–orbit or it splits into two orbits of G^+, since $|G^+| = |G|/2$. Hence $|G^+ \backslash\backslash X| - |G \backslash\backslash X|$ is the number of orbits which split over G^+. Finally the stated identity is obtained by an application of the homomorphism theorem. □

2.1.14 Corollary *In the case when $G \neq G^+$, the number of G–orbits on X which do not split over G^+ is equal to*

$$\frac{1}{|G|}\sum_{g\in G}(1 - \epsilon(\bar{g}))|X_g|.$$

Note what this means. If G acts on a finite set X in such a way that $G \neq G^+$, then we can group the orbits of G on X into a set of orbits which are also G^+–orbits. In figure 2.1 we denote these orbits by the symbol ⊗. The other G–orbits split into two G^+–orbits, we indicate one of them by ⊕, the other one by ⊖, and call the pair $\{\oplus, \ominus\}$ an *enantiomorphic pair* of G–orbits.

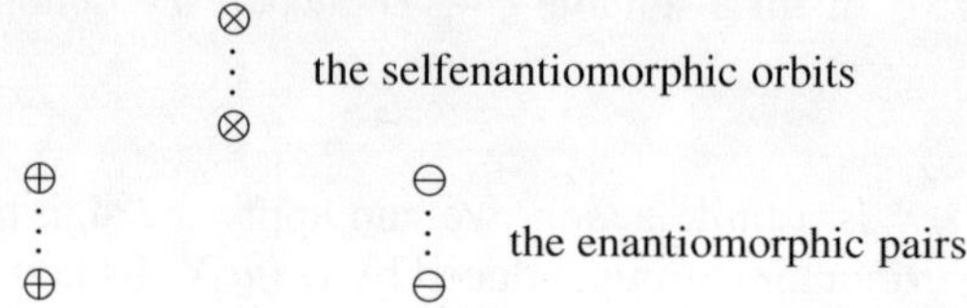

Fig. 2.1. Enantiomorphic pairs and selfenantiomorphic orbits

Hence 2.1.13 gives us the number of enantiomorphic pairs of orbits, while 2.1.14 yields the number of *selfenantiomorphic* orbits of G on X. The elements $x \in X$ belonging to selfenantiomorphic orbits are called *achiral* objects, while the others are called *chiral*. These notions of *enantiomerism* and *chirality* are taken from chemistry, where G is usually the symmetry group of the molecule while G^+ is its subgroup consisting of the proper rotations. We call ${}_GX$ a *chiral* action if and only if $G \neq G^+$. Using this notation we can now rephrase 2.1.13 and 2.1.14 in the following way:

2.1.15 Corollary *If ${}_GX$ is a finite chiral action, then the number of selfenantiomorphic orbits of G on X is equal to*

$$\frac{1}{|G|}\sum_{g\in G}(1-\epsilon(\bar{g}))|X_g| = 2|G\backslash\backslash X| - |G^+\backslash\backslash X|,$$

while the number of enantiomorphic pairs of orbits is

$$\frac{1}{|G|}\sum_{g\in G}\epsilon(\bar{g})|X_g| = |G^+\backslash\backslash X| - |G\backslash\backslash X|.$$

Exercises

Exercise 2.1.1 Let ${}_GX$ be finite and transitive. Consider an arbitrary $x \in X$ and prove that

$$|G_x\backslash\backslash X| = \frac{1}{|G|}\sum_{g\in G}|X_g|^2.$$

Exercise 2.1.2 Prove that the number of Sylow p-subgroups divides the order of G and is congruent 1 modulo p, for each prime divisor of the order of G.

2.2 Enumeration of Symmetry Classes

Now we are going to enumerate the symmetry classes of mappings introduced above. We shall do this by enumerating the set of orbits of $H \wr_X G$ on Y^X and then restricting the attention to subgroups. In order to enumerate the orbits of the wreath product in an economic way we shall use the fact that the number of fixed points is constant on the conjugacy classes of elements, so let us consider a special case of wreath product where the conjugacy classes are well known.

The complete monomial group $S_m \wr S_n$ has the following natural embedding (i. e. an injective homomorphism) into S_{mn}:

2.2.1 $$\delta\colon S_m \wr S_n \hookrightarrow S_{mn}\colon (\psi,\pi) \mapsto \begin{pmatrix} j\cdot m+i \\ \pi j\cdot m+\psi(\pi j)i \end{pmatrix}_{i\in m, j\in n}.$$

This can be seen as follows: Recall the direct factor S_m^j, for $j \in n$, of the base group S_m^* of $S_m \wr S_n$ (cf. the remark on H^x in 1.3.3). Its image $\delta\,[S_m^j]$ acts on the block $\{j\cdot m+0,\ldots, j\cdot m+m-1\}$ in a similar way as S_m does on m, while the image

$\delta\,[S_n{}']$ of the complement $S_n{}'$ of the base group acts on the set of these n subsets $\{j \cdot m, \ldots, jm+m-1\}$ of length m of mn in a similar way as S_n acts on n. For example the element

$$(\psi, \pi) := (\psi(0), \psi(1), \psi(2), \pi) := ((01), (012), 1, (12)) \in S_3 \wr S_3$$

is mapped under δ onto

$$\underbrace{(01)(345)}_{=\delta((\psi,1))}\,\underbrace{(36)(47)(58)}_{=\delta((\iota,\pi))} = (01)(364758) \in S_9.$$

The image of $H \wr G$ under δ will be denoted as follows:

2.2.2 $$\bar{H} \odot \bar{G} := \delta\,[H \wr G].$$

It is called the *plethysm* of $\bar{G}$ and $\bar{H}$, for reasons which will become clear later. As an application of this permutation representation we obtain a description of the centralizers of elements in finite symmetric groups. To see this note that $\delta[C_m \wr S_n]$, where $C_m := \langle(0, \ldots, m-1)\rangle$, is just the centralizer of

$$\sigma := (0, \ldots, m-1)(m, \ldots, 2m-1)\ldots((n-1)m, \ldots, nm-1) \in S_{mn}.$$

This follows from $\delta[C_m \wr S_n] \subseteq C_{S_{mn}}(\sigma)$, which is clear from 11.2.6 together with 2.2.1 and the fact that $|C_{S_{mn}}(\sigma)| = m^n n! = |C_m \wr S_n|$ (cf. 11.2.7). The general case is now easy:

2.2.3 Corollary *If $\sigma \in S_n$ is of type $a = (a_1, \ldots, a_n)$, then $C_{S_n}(\sigma)$ is a subgroup of S_n which is similar to the following direct sum of plethysms:*

$$\oplus_i (C_i \odot S_{a_i}).$$

Analogously we can show (recall 1.1.11)

2.2.4 Corollary *The normalizer of the n–fold direct sum*

$$\oplus^n S_m := S_m \oplus \ldots \oplus S_m,$$

is conjugate to the plethysm $S_m \odot S_n$.

Thus centralizers of elements and normalizers of specific subgroups of symmetric groups turn out to be direct sums of complete monomial groups. Since such groups will also occur as acting groups later on, we also describe their conjugacy classes. Consider an element (ψ, π) in $H \wr S_n$ and assume that $C^1, C^2, \ldots$ are the conjugacy classes of H. If

$$\pi = \prod_{\nu \in c(\pi)} (j_\nu \ldots \pi^{l_\nu - 1} j_\nu),$$

in standard cycle notation, then we associate with its ν-th cyclic factor the element

2.2.5 $$h_\nu(\psi, \pi) := \psi(j_\nu)\psi(\pi^{-1} j_\nu) \cdots \psi(\pi^{-l_\nu+1} j_\nu) = \psi \cdots \psi_{\pi^{l_\nu - 1}}(j_\nu)$$

of H and call it the *ν-th cycleproduct of* (ψ, π) or the cycleproduct *associated* to $(j_\nu \dots \pi^{l_\nu - 1} j_\nu)$ with respect to (ψ, π). In this way we obtain a total of $c(\pi)$ cycleproducts, $a_k(\pi)$ of them arising from the cyclic factors of π which are of length k. Now let $a_{ik}(\psi, \pi)$ be the number of these cycleproducts which are associated to a k-cycle of π and which belong to the conjugacy class C^i of H,

$$a_{ik}(\psi, \pi) = |\{\nu \mid \nu \in c(\pi),\ l_\nu = k,\ h_\nu(\psi, \pi) \in C^i\}|.$$

We put these natural numbers together into the matrix

$$a(\psi, \pi) := (a_{ik}(\psi, \pi)).$$

This matrix has n columns (k is the column index) and as many rows as there are conjugacy classes in H (i is the row index). Its entries satisfy the following conditions:

2.2.6 $$a_{ik}(\psi, \pi) \in \mathbb{N},\ \sum_i a_{ik}(\psi, \pi) = a_k(\pi),\ \sum_{i,k} k \cdot a_{ik}(\psi, \pi) = n.$$

We call this matrix $a(\psi, \pi)$ the *type* of (ψ, π) and we say that (ψ, π) is *of type* $a(\psi, \pi)$.

2.2.7 Lemma *The conjugacy classes of complete monomial groups $H \wr S_n$ have the following properties:*

- $C^{H \wr S_n}(\psi', \pi') = C^{H \wr S_n}(\psi, \pi)$ *if and only if* $a(\psi', \pi') = a(\psi, \pi)$.
- *The order of the conjugacy class of elements of type (a_{ik}) in $H \wr S_n$, H finite, is equal to*

$$\frac{|H|^n n!}{\prod_{i,k} a_{ik}!(k|H|/|C^i|)^{a_{ik}}}.$$

- *Each matrix (b_{ik}) with n columns and as many rows as H has conjugacy classes, the elements of which satisfy*

$$b_{ik} \in \mathbb{N},\ \sum_{i,k} k \cdot b_{ik} = n,$$

  occurs as the type of an element $(\psi, \pi) \in H \wr S_n$.

- *If H is a permutation group and $\alpha := \alpha(h_\nu(\psi, \pi))$ the cycle partition of $h_\nu(\psi, \pi)$ (see appendix), then the cycle partition $\alpha(\delta(\psi, \pi))$, where δ denotes the permutation representation of 2.2.1, is equal to*

$$\sum_\nu l_\nu \cdot \alpha(h_\nu(\psi, \pi)),$$

  where $l_\nu \cdot \alpha$, $\alpha := \alpha(h_\nu(\psi, \pi))$, is defined to be $(l_\nu \cdot \alpha_0, l_\nu \cdot \alpha_1, \dots)$, and where $\sum_\nu \dots$ means that the proper partition has to be formed that consists of all the parts of all the summands $l_\nu \cdot \alpha(h_\nu(\psi, \pi))$.

Proof: To begin with, we consider the cycleproducts introduced in 2.2.5. Since in each group G the products xy and yx of two elements are conjugate, we have that $h_\nu(\psi, \pi)$ is conjugate to

$$\psi\psi_\pi \dots \psi_{\pi^{l_\nu - 1}}(\pi^z j_\nu),$$

for any integer z.

The next remark is, that for each $\pi' \in S_n$ and every $\psi' \in H^n$,

$$a(\psi, \pi) = a((\iota, \pi')(\psi, \pi)(\iota, \pi')^{-1}) = a((\psi', 1)(\psi, \pi)(\psi', 1)^{-1}).$$

This follows from the fact that both $(\psi_{\pi'}, \pi'\pi\pi'^{-1})$ and $(\psi'\psi\psi'^{-1}_\pi, \pi)$ are of type $a(\psi, \pi)$.

Now we note that $a(\psi, \pi) = a(\psi', \pi')$ implies the existence of an element $\pi'' \in S_n$ which satisfies $\pi = \pi''\pi'\pi''^{-1}$, and for which the cycleproducts $h_\nu(\psi, \pi)$ and $h_\nu(\psi'_{\pi''}, \pi)$ are conjugate.

It is not difficult to check these remarks and then to derive the statement (exercise 2.2.2). □

A numerical example is provided by $S_3 \wr S_2$. The set of proper partitions characterizing the conjugacy classes of S_2 is

$$\{\alpha \mid \alpha \vdash 2\} = \{(2), (1^2)\},$$

the set of corresponding cycle types is

$$\{a \mid a \vdash\!\dashv 2\} = \{(0, 1), (2, 0)\}.$$

Thus the types of $S_3 \wr S_2$ turn out to be

$$\begin{pmatrix} 0 & 1 \\ 0 & 0 \\ 0 & 0 \end{pmatrix}, \begin{pmatrix} 0 & 0 \\ 0 & 1 \\ 0 & 0 \end{pmatrix}, \begin{pmatrix} 0 & 0 \\ 0 & 0 \\ 0 & 1 \end{pmatrix}, \begin{pmatrix} 2 & 0 \\ 0 & 0 \\ 0 & 0 \end{pmatrix}, \begin{pmatrix} 0 & 0 \\ 2 & 0 \\ 0 & 0 \end{pmatrix},$$

$$\begin{pmatrix} 0 & 0 \\ 0 & 0 \\ 2 & 0 \end{pmatrix}, \begin{pmatrix} 1 & 0 \\ 1 & 0 \\ 0 & 0 \end{pmatrix}, \begin{pmatrix} 1 & 0 \\ 0 & 0 \\ 1 & 0 \end{pmatrix}, \begin{pmatrix} 0 & 0 \\ 1 & 0 \\ 1 & 0 \end{pmatrix}.$$

The orders of the conjugacy classes are 6,18,12,1,9,4,6,4,12. We now describe an interesting action of $S_m \wr S_n$ which is in fact an action of the form ${}_G(Y^X)$.

2.2.8 Application (partitions of numbers with restricted length of parts and number of parts) The action of $S_m \wr S_n$ on mn is obviously similar to the following action of $S_m \wr S_n$ on the set $m \times n$:

$$S_m \wr S_n \times (m \times n) \to m \times n \colon ((\psi, \pi), (i, j)) \mapsto (\psi(\pi j)i, \pi j).$$

The corresponding permutation group on $m \times n$ will be denoted by

$$S_n[S_m]$$

and called the *composition* of S_n and S_m, while

$$\bar{G}[\bar{H}]$$

will be used for the permutation group on $Y \times X$, induced by the natural action of $H \wr_X G$ on $Y \times X$.

The action of the wreath product $S_m \wr S_n$ on $m \times n$ induces a natural action of $S_m \wr S_n$ on the set

$$Y^X := 2^{m \times n} = \{(a_{ij}) \mid a_{ij} \in \{0,1\}, i \in m, j \in n\},$$

i. e. on the set of 0-1-matrices consisting of m rows and n columns: As $(\psi, \pi)^{-1} = (\psi^{-1}_{\pi^{-1}}, \pi^{-1})$, the following mapping describes an action:

$$S_m \wr S_n \times 2^{m \times n} \to 2^{m \times n} \colon ((\psi, \pi), (a_{ij})) \mapsto (a_{(\psi,\pi)^{-1}(i,j)}) = (a_{\psi^{-1}(j)i, \pi^{-1}j}).$$

Since $(\psi, \pi)^{-1} = (\iota, \pi^{-1})(\psi^{-1}, 1)$, we can do this in two steps:

$$(\psi, \pi) \colon (a_{ij}) \longmapsto (a_{\psi(j)^{-1}i, j}) \longmapsto (a_{\psi^{-1}(j)i, \pi^{-1}j}).$$

Hence we can permute the rows of (a_{ij}) in such a way that the numbers of 1's in the rows of the resulting matrix are nonincreasing from top to bottom. And after having carried out this permutation with a suitable $\psi \in S_m^*$, we can find a $\pi \in S_n$ that moves the 1's of each row in flush left position. This proves that the orbit of (a_{ij}) under $S_m \wr S_n$ is characterized by an element of the form

$$\begin{pmatrix} 1 & \dots & \dots & \dots & 1 & \\ \vdots & & & \cdot^{\cdot^{\cdot}} & & \\ 1 & \dots & 1 & & & \\ & & & & & 0 \end{pmatrix} \in 2^{m \times n},$$

i. e. by a proper partition of $k := \sum_{i,j} a_{ij}$. Hence the orbits of $S_m \wr S_n$ on $2^{m \times n}$ are characterized by the proper partitions α, where each part α_i is $\leq n$ and where the total number of parts is $\leq m$:

2.2.9 Corollary *There exists a natural bijection*

$$S_m \wr S_n \backslash\backslash 2^{m \times n} \to \{\alpha \vdash k \mid k \leq mn,\ \alpha_0 \leq n,\ \alpha_0' \leq m\}.$$

Hence an application of the Cauchy–Frobenius Lemma yields the following formula for the number of partitions of this restricted form:

2.2.10 $$\left| S_m \wr S_n \backslash\backslash 2^{m \times n} \right| = \frac{1}{m!^n n!} \sum_{(\psi,\pi) \in S_m \wr S_n} 2^{\Sigma_\nu c(h_\nu(\psi,\pi))},$$

which can be made more explicit by an application of 2.2.7. ◇

Our paradigmatic examples are the actions of G, H, $H \times G$ and $H \wr_X G$ on Y^X, obtained from given actions ${}_GX$ and ${}_HY$. The orbits of these groups are called *symmetry classes of mappings*. If we want to be more explicit, we call them *G–classes, H–classes, H × G–classes and H ≀ₓ G–classes*, respectively. Their total number can be obtained by an application of the Cauchy–Frobenius Lemma as soon as we know the number of fixed points for each element of the respective group. In order to derive these numbers we characterize the fixed points of each $(\psi, g) \in H \wr_X G$ on Y^X and then we use the natural embedding of G, H and $H \times G$ in $H \wr_X G$ as described above. Thus the following lemma will turn out to be crucial:

2.2.11 Lemma *Consider an $f \in Y^X$, an element (ψ, g) of $H \wr_X G$ and assume that*

$$\bar{g} = \prod_{\nu \in c(\bar{g})} (x_\nu \, g x_\nu \dots g^{l_\nu - 1} x_\nu)$$

is the disjoint cycle decomposition of $\bar{g}$, the permutation of X which corresponds to g. Then f is a fixed point of (ψ, g) if and only if the following two conditions hold:

- *Each $f(x_\nu)$ is a fixed point of the cycleproduct $h_\nu(\psi, g)$:*

$$f(x_\nu) \in Y_{h_\nu(\psi,g)}.$$

- *The other values of f arise from the values $f(x_\nu)$ according to the following equations:*

$$f(x_\nu) = \psi(x_\nu) f(g^{-1}x_\nu) = \psi(x_\nu)\psi(g^{-1}x_\nu) f(g^{-2}x_\nu) = \dots .$$

Proof: 1.3.2 says that f is fixed under (ψ, g) if and only if its values $f(x)$ satisfy the equations

$$f(x) = \psi(x) f(g^{-1}x) = \psi(x)\psi(g^{-1}x) f(g^{-2}x) \dots$$
$$\dots = \psi(x)\psi(g^{-1}x) \dots \psi(g^{-l+1}x) f(x),$$

where l denotes the length of the cyclic factor of $\bar{g}$ containing the point $x \in X$. Hence in particular the following must be true:

$$f(x_\nu) = h_\nu(\psi, g) f(x_\nu),$$

which means that $f(x_\nu)$ is a fixed point of $h_\nu(\psi, g)$, as claimed. Thus any fixed $f \in Y^X$ clearly has the stated properties, and vice versa. □

This, together with the Cauchy–Frobenius Lemma, yields the number of $H \wr_X G$–classes on Y^X, and the restrictions to the subgroups G, H and $H \times G$ give the numbers of G–, H– and $H \times G$–classes on Y^X:

2.2.12 Theorem *If both ${}_GX$ and ${}_HY$ are finite actions, then we obtain the following expression for the number of orbits of the corresponding action of $H \wr_X G$ on Y^X:*

$$\left| H \wr_X G \backslash\backslash Y^X \right| = \frac{1}{|H|^{|X|}|G|} \sum_{(\psi,g) \in H \wr_X G} \prod_{\nu \in c(\bar{g})} \left| Y_{h_\nu(\psi,g)} \right| .$$

The restrictions to G, H and $H \times G$ according to 1.3.4 yield:

$$\left| G \backslash\!\backslash Y^X \right| = \frac{1}{|G|} \sum_{g \in G} |\, Y \,|^{c(\bar{g})}, \quad \left| H \backslash\!\backslash Y^X \right| = \frac{1}{|H|} \sum_{h \in H} |\, Y_h \,|^{|X|},$$

and

$$\left| (H \times G) \backslash\!\backslash Y^X \right| = \frac{1}{|H||G|} \sum_{(h,g) \in H \times G} \prod_{i=1}^{|X|} |\, Y_{h^i} \,|^{a_i(\bar{g})}.$$

2.2.13 Application (Fermat's congruence) According to 2.1.9 we obtain from theorem 2.2.12 the following congruences modulo group orders:

2.2.14 Corollary *For any subgroups $G \leq S_n$ and $H \leq S_m$, the following congruences hold:*

$$\sum_{\pi \in G} m^{c(\pi)} \equiv 0\ (|G|), \quad \sum_{\rho \in H} a_1(\rho)^n \equiv 0\ (|H|),$$

and also

$$\sum_{(\rho,\pi) \in H \times G} \prod_{i=1}^{n} a_1(\rho^i)^{a_i(\pi)} \equiv 0\ (|H||G|),$$

as well as

$$\sum_{(\psi,\pi) \in H \wr G} \prod_{\nu \in c(\pi)} a_1(h_\nu(\psi,\pi)) \equiv 0\ (|H|^n|G|).$$

Let us now consider a nice application of the first one of these congruences in order to derive a famous number theoretic result from a particular necklace enumeration. Let C_p denote the following cyclic subgroup of S_p:

$$C_p := \langle (0 \ldots p-1) \rangle \leq S_p.$$

We assume that p is a prime and we consider the action of C_p on m^p. We should like to count the number of orbits, which is in fact the number of necklaces. Here is an example: Three colourings of the regular 5–gon.

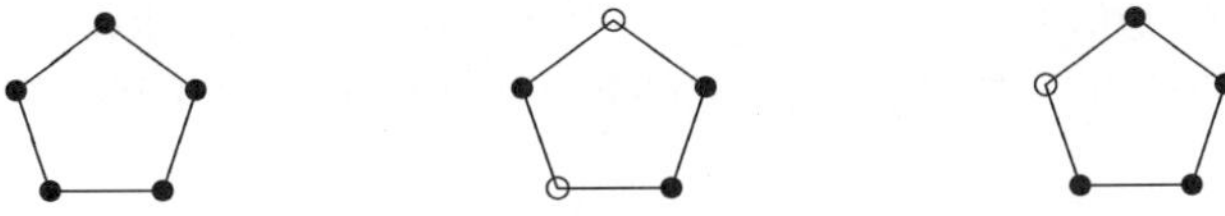

C_p contains, besides the identity element, p–cycles only. The identity element of C_p keeps each $f \in m^p$ fixed, while each p–cycle fixes the m monochromatic colourings only (notice that $(0 \ldots p-1)$ acts as a clockwise rotation of the p–gon after having numbered the vertices of the p-gon from 0 to $p-1$, counterclockwise). Hence we obtain from the Cauchy–Frobenius Lemma that

$$|\, C_p \backslash\!\backslash m^p \,| = \frac{1}{p}(m^p + (p-1)m),$$

provided that p is a prime number. (If $p = 5$ and $m = 2$ there are exactly eight necklaces, three of them are shown above, the other five are as easy to find.) This implies that $m^p + (p-1)m$, and hence also $m^p - m$, is congruent zero modulo p, for each positive integer m. It is clear that this is then also true for any integer z.

2.2.15 Fermat's Congruence *If p is a prime number and z an integer, then*

$$z^p \equiv z\ (p).$$

◇

Further congruences show up in the enumeration of group elements with prescribed properties. This theory of enumeration in finite groups is, besides the enumeration of chemical graphs, one of the main sources for the theory of enumeration which we are discussing here. A prominent example taken from this complex of problems is the following one due to Frobenius: The number of solutions of the equation $x^n = 1$ in a finite group G is divisible by n, if n divides the order of G. There are many proofs of this result and also many generalizations. Later on we return to this problem, at present we can only discuss a particular case which can be treated with the tools we already have at hand.

2.2.16 Application (number of p-th roots of group elements) Let g denote an element of a finite group which forms its own conjugacy class and consider a prime number p, which divides $|G|$. We want to show that the number of solutions $x \in G$ of the equation $x^p = g$ is divisible by p. In order to prove this we consider the action of C_p on the set $Y^X := G^p$. The orbits are of length 1 or p. An orbit is of length 1 if and only if it consists of a single and therefore of a constant mapping $(g', \ldots, g')$, say. We now restrict our attention to the following subset $M \subseteq G^p$:

$$M := \{(g_0, \ldots, g_{p-1}) \mid g_0 \cdots g_{p-1} = g\}.$$

As g forms its own conjugacy class, we obtain a subaction of C_p on M (for example $g_0 \cdots g_{p-1}$ is conjugate to $g_{p-1}g_0 \cdots g_{p-2}$). Hence the desired number k of solutions of $x^p = g$ is equal to the number of orbits of length 1 in M. Now we consider the number l of orbits of length p in M. It satisfies the equation $k + pl = |M|$. As each equation $g_0 g' = g$ has a unique solution g_0 in G, we moreover have that $|M| = |G|^{p-1}$. Thus

$$k \equiv k + pl = |M| \equiv 0\ (p),$$

which completes the proof. We note in passing that the *center* of G consists of the elements which form their own conjugacy class, so that we have proved the following:

2.2.17 Corollary *If the prime p divides the order of the group G, then the number of p-th roots of each element in the center of G is divisible by p. In particular the number of p-th roots of the unit element* 1 *of G has this property and it is nonzero, since* 1 *is a p-th root of* 1. *Thus G contains elements of order p.*

This result can be used in order to give an *inductive* proof of Sylow's Theorem which we proved *directly* in example 2.1.11. ◇

Exercises

Exercise 2.2.1 Prove that the conjugacy class (in S_n) of an even element $\pi \in S_n$ splits into two conjugacy classes with respect to the alternating group A_n (for its definition see the appendix) if and only if the lengths of the cyclic factors of π are pairwise different and odd (hint: use 2.2.3).

Exercise 2.2.2 Fill in the details of the proof of 2.2.7.

Exercise 2.2.3 Prove, by considering a suitable action, the following congruence due to Euler:

$$\forall z \in \mathbb{Z}, n \in \mathbb{N}^* \text{ such that } \gcd(z, n) = 1 : \ z^{\phi(n)} \equiv 1 \ (n),$$

where ϕ denotes the *Euler function,* its definition is given in the appendix.

Exercise 2.2.4 Give another proof of 2.2.12 using the obvious equations for $|G \backslash\backslash Y^X|$ and $|H \backslash\backslash Y|$ together with 1.3.7.

2.3 Application to Incidence Structures

We have already met 0-1-matrices with m rows and n columns, say, they form the set of mappings

$$Y^X = 2^{m \times n}.$$

There are many natural actions on this set, and various finite structures can be introduced as corresponding orbits. The example we met was the natural action of the wreath product $S_m \wr S_n$ on this set of mappings, the orbits turned out to be partitions of natural numbers, the number of parts being bounded by m, the lengths of the parts being bounded by n.

Another canonic action, in fact an even easier case, is the action of the direct product $S_m \times S_n$ on this set, a restriction of the action of the wreath product:

$$(S_m \times S_n) \times 2^{m \times n} \to 2^{m \times n} : \big((\pi, \rho), (a_{ij})\big) \mapsto (a_{\pi^{-1}i, \rho^{-1}j}).$$

The orbits are, of course, the 0-1-matrices *up to reordering rows and columns.* Therefore we may consider them as (unlabeled) *incidence structures,* where a 0-1-matrix describes — after numbering points and lines — which *point* (that corresponds to a row of the matrix) is incident with a given *line* (that corresponds to a column of the matrix). We can therefore identify the incidence structures with m points and n lines with the set of orbits

$$S_m \times S_n \backslash\backslash 2^{m \times n}.$$

The count of incidence structures with given parameters m and n is therefore an obvious application of the Cauchy–Frobenius Lemma: The number of (unlabeled) incidence structures on m points and n lines is equal to

$$\frac{1}{m!n!}\sum_{(\pi,\rho)\in S_m\times S_n} 2^{c(\overline{(\pi,\rho)})},$$

where $\overline{(\pi,\rho)}$ means the permutation induced by $(\pi,\rho)\in S_m\times S_n$ on the set $m\times n$. Hence it remains to deduce the number of cyclic factors of $\overline{(\pi,\rho)}$. But this is easy: An element $i\in m$ that lies in an r-cycle of π, together with an element $j\in n$ in an s-cycle of ρ forms a pair (i,j) that lies in a cycle of length $\mathrm{lcm}(r,s)$. Hence the elements in the $a_r(\pi)$ r-cycles of π together with the elements in the $a_s(\rho)$ s-cycles of ρ contribute to $\overline{(\pi,\rho)}$ exactly so many cycles of length $\mathrm{lcm}(r,s)$:

$$\gcd(r,s)\cdot a_r(\pi)\cdot a_s(\rho).$$

This yields the desired number of unlabeled incidence structures:

2.3.1 Corollary *The number of (unlabeled) incidence structures on m points and n lines is*

$$\left|\, S_m\times S_n \backslash\backslash 2^{m\times n}\,\right| = \sum_{a\dashv m,\, b\dashv n} \frac{2^{c(a,b)}}{\prod_i i^{a_i}a_i!\prod_j j^{b_j}b_j!},$$

where

$$c(a,b)=\sum_{k=1}^{m\cdot n}\sum_{i,j:\ lcm(i,j)=k}\gcd(i,j)\cdot a_i\cdot a_j.$$

M. Wiesend has calculated the following table of numbers of unlabeled incidence structures:

$m\backslash n$	1	2	3	4	5	6	7	8
1	2	3	4	5	6	7	8	9
2	3	7	13	22	34	50	70	95
3	4	13	36	87	190	386	734	1324
4	5	22	87	317	1053	3250	9343	25207
5	6	34	190	1053	5624	28576	136758	613894
6	7	50	386	3250	28576	251610	2141733	17256831

A nice and useful application of 0-1-matrices and their orbits under this direct product of symmetric groups is the *concept analysis,* invented by R. Wille. It is devoted to the analysis of *contexts* which are tables representing knowledge on *objects* (corresponding to the rows of the table) and *attributes* (corresponding to the columns). An entry 1 in the i-th row and j-th column of such a table means that the i-th object has the j-th attribute. Hence contexts — of this particular form — are 0-1-matrices, and the obvious equivalence is described by that action of the direct product of symmetric groups. Hence *the above table gives numbers of essentially different contexts* as well.

We would like also to derive from 2.2.12 a formula for the number of graphs, for the number of multigraphs with restricted multiplicities and for the number of selfcomplementary graphs on v vertices. Let us see how suitably chosen G, X and Y can be used in order to define and enumerate these graph structures on v vertices. (Later on we shall refine this method by counting these graphs according to the number of edges or according to their automorphism group. Finally we shall even use this *Ansatz* in order to construct such graphs and also to generate them uniformly at random.) The way of defining graphs as orbits of groups may still seem to be circumstantial, but we shall see that this definition is very flexible since it can easily be generalized considerably in order to cover also multigraphs, directed graphs and so on. We recall from 1.3.5 that the set of unlabeled graphs with v vertices can be identified with the following set of orbits:

$$S_v \backslash\backslash 2^{\binom{v}{2}}.$$

If we want to allow multiple edges, say up to the multiplicity k, then we again consider $X := \binom{v}{2}$, but we change Y into $Y := \{0, \ldots, k\} = k + 1$. The elements of

$$Y^X = (k+1)^{\binom{v}{2}}$$

are called *labeled k–graphs,* while the set of orbits

$$S_v \backslash\backslash (k+1)^{\binom{v}{2}}$$

can be identified with the set of *(unlabeled) k-graphs* on v vertices. If we want to allow loops or multiple loops, we replace X by the union $\binom{v}{2} \cup \binom{v}{1}$, and now $f(\{i\}) = j$, for $\{i\} \in \binom{v}{1}$, means that the vertex with the number i carries a j-fold loop. If we want to consider *digraphs* (i. e. the edges are directed and neither loops nor parallel edges are allowed), we simply put

$$X := v^2_{inj} := \{(i, j) \in v^2 \mid i \neq j\},$$

the set of *injective pairs* over v. Hence the set of orbits

$$S_v \backslash\backslash 2^{(v^2_{inj})}$$

is essentially the set of unlabeled digraphs on v vertices.

Thus graphs, k–graphs, k–graphs with loops and also digraphs can be considered as symmetry classes of mappings. Several other structures will later on be obtained in the same way. Having *defined* the graphs this way we want to *count* them by an application of 2.2.12 which means that we have to derive a formula for $c(\bar{\pi})$, the number of cyclic factors of $\bar{\pi}$, the permutation induced by $\pi \in S_v$ on the set $\binom{v}{2}$ of pairs of points, expressed in terms of the cycle structure of π. In fact we can do better, we can derive the cycle type of $\bar{\pi}$ from the cycle type of π.

2.3.2 Lemma *If $\pi \in S_v$, then*

- *Each i-cycle of π, i odd, contributes to $\bar{\pi}$ exactly $(i-1)/2$ cyclic factors, each of which is an i-cycle.*
- *Each i-cycle of π, i even, contributes to $\bar{\pi}$ exactly one cycle of length $i/2$ and $(i/2)-1$ further cycles, which are of length i.*
- *Each pair of cyclic factors of π, say an i-cycle and a j-cycle, contributes to $\bar{\pi}$ exactly $\gcd(i,j)$ cyclic factors, each of which has the length $\operatorname{lcm}(i,j)$.*
- *All the cyclic factors of $\bar{\pi}$ arise in this way.*

Proof: To begin with, let i be odd. Without loss of generality we may assume that the i-cycle in question is $(0,\ldots,i-1)$. Consider a $k<(i-1)/2$. Then $\bar{\pi}$ contains the following cyclic permutation of 2-subsets:

$$(\{0,k\},\{1,k+1\},\ldots,\{i-k-1,i-1\},\{0,i-k\},\ldots,\{k-1,i-1\}).$$

This cycle is of length i, and the cycles of this form arising from different $k<(i-1)/2$ are pairwise disjoint. Furthermore these are all the cycles arising from $(0,\ldots,i-1)$ since, for each such k, we have, as i is odd:

$$(\{0,k\},\{1,k+1\},\ldots)=(\{0,i-k\},\{1,i-k+1\},\ldots).$$

If i is even, say $i=2j$, then we may assume that the cyclic factor of π is $(0,\ldots,2j-1)$. It yields, for $2\le k\le j-1$, the $(i/2)-1$ different i-cycles

$$(\{0,k-1\},\{1,k\},\ldots,\{i-k,i-1\},\{0,i-k+1\},\ldots,\{k-2,i-1\}),$$

together with the $(i/2)$-cycle

$$(\{0,j\},\{1,j+1\},\ldots,\{j-1,2j-1\}).$$

In order to prove the third item we consider a pair of cyclic factors of π, say the pair $(0,\ldots,i-1)(i,\ldots,i+j-1)$. It contributes to $\bar{\pi}$ a product of disjoint cycles of the following form:

$$(\{0,i\},\{1,i+1\},\ldots,\{i-1,i+j-1\})$$

or

$$(\{0,i+k\},\{1,i+k+1\},\ldots,\{i-1,i+k-1\})\text{ for }k\ge 1.$$

The length of each of these cyclic factors is $\operatorname{lcm}(i,j)$, and therefore their number is equal to $\gcd(i,j)$.

The final item is clearly true. □

This lemma yields the cycle structure of $\bar{\pi}$ and the desired number $c(\bar{\pi})$ of cyclic factors which we need in order to evaluate the number of graphs on v vertices by an application of the Cauchy–Frobenius Lemma:

2.3.3 Corollary *For each $\bar{\pi}$ on $\binom{v}{2}$ we have:*

- *If i is odd, then*

$$a_i(\bar{\pi}) = \frac{a_i(\pi)}{2}(i \cdot a_i(\pi) - 1) + a_{2i}(\pi) + \sum_{r<s,\ lcm(r,s)=i} a_r(\pi)a_s(\pi)\gcd(r,s).$$

- *If i is even, then*

$$a_i(\bar{\pi}) = \frac{a_i(\pi)}{2}(i \cdot a_i(\pi) - 2) + a_{2i}(\pi) + \sum_{r<s,\ lcm(r,s)=i} a_r(\pi)a_s(\pi)\gcd(r,s).$$

- *The total number of cyclic factors is*

$$c(\bar{\pi}) = \frac{1}{2}\sum_i i \cdot a_i(\pi)^2 - \frac{1}{2}\sum_{i\ odd} a_i(\pi) + \sum_{r<s} a_r(\pi)a_s(\pi)\gcd(r,s).$$

Thus, by an application of the Cauchy–Frobenius Lemma, we obtain

2.3.4 Corollary *The number of k-graphs on v vertices is equal to*

$$\frac{1}{v!}\sum_{\pi\in S_v}(k+1)^{c(\bar{\pi})},$$

where $c(\bar{\pi})$ is as above. More explicitly and in terms of cycle types of v (see 11.2.7) this number is equal to

$$\sum_{a \vdash v}\frac{(k+1)^{c_{\bar{a}}}}{\prod_i i^{a_i}a_i!},\quad where\ c_{\bar{a}} := \frac{1}{2}\sum_i i \cdot a_i^2 - \frac{1}{2}\sum_{i\ odd} a_i + \sum_{r<s} a_r a_s \gcd(r,s).$$

The upper left hand corner of a table of these numbers looks as follows:

$v\backslash k$	0	1	2	3	4
1	1	1	1	1	1
2	1	2	3	4	5
3	1	4	10	20	35
4	1	11	66	276	900
5	1	34	792	10688	90005
6	1	156	25506	1601952	43571400

We have evaluated the number of graphs on v vertices by examining a certain action of the form ${}_G(Y^X)$. We shall now give an example of the form ${}_{H\times G}(Y^X)$, i. e. a power group action. Afterwards we shall see how these two examples can be combined in order to prove an existence theorem for a certain class of graphs.

While doing so we shall meet an interesting and useful counting principle which uses suitable actions of S_2, the smallest nontrivial group.

Two labeled graphs $f, \tilde{f} \in 2^{\binom{v}{2}}$ are called *complementary* if and only if

$$\tilde{f}(\{i, j\}) = 0 \iff f(\{i, j\}) = 1.$$

Correspondingly we say that two graphs, i. e. isomorphism classes of labeled ones, are *complementary* graphs if and only if one class arises from the other by forming the complements. The example

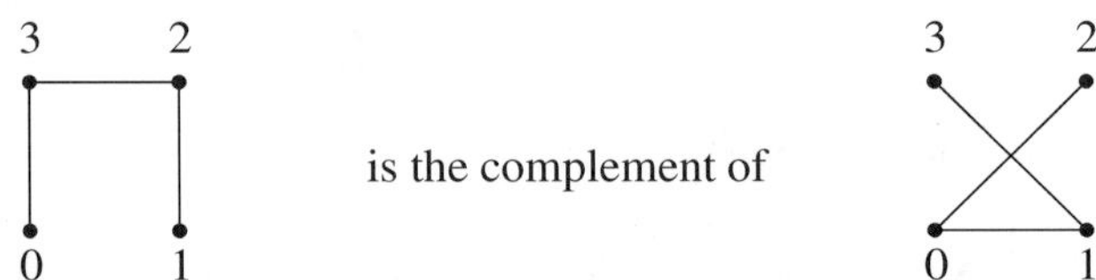

shows that graphs may very well be *selfcomplementary*, and hence we ask for the number of selfcomplementary graphs on v vertices. In order to prepare the derivation of this number, we notice that the classes which we obtain by putting a graph and its complement together in one class are just the orbits of the group $H \times G := S_2 \times S_v$ on the set $Y^X := 2^{\binom{v}{2}}$ (recall 1.3.1). According to 2.2.12 the number of these orbits is equal to

$$\frac{1}{2 \cdot v!} \sum_{(\rho,\pi)\in S_2\times S_v} \prod_{i=1}^{\binom{v}{2}} |2_{\rho^i}|^{a_i(\bar{\pi})}.$$

But $|2_{\rho^i}|$, the number of fixed points of ρ^i, $\rho \in S_2$, acting on the set 2, is either 2 or zero. Hence we separate the sum over the $(\rho, \pi) \in S_2 \times S_v$ into two sums depending on $\rho \in S_2$ and get the following expression for the number of these orbits:

2.3.5 $$\left| (S_2 \times S_v) \backslash\backslash 2^{\binom{v}{2}} \right| = \frac{1}{2 \cdot v!} \left(\sum_{\pi} 2^{c(\bar{\pi})} + \sum_{\pi}{}' 2^{c(\bar{\pi})} \right),$$

where $\sum_\pi$ means the sum over all the elements of S_v, while $\sum'_\pi$ means the sum over all the elements π of S_v such that $\bar{\pi}$ does not contain a cycle of odd length, i. e. $a_{2i+1}(\bar{\pi}) = 0$, for each i.

In order to derive from this equation the total number of selfcomplementary graphs on v vertices we use an easy argument which we have already met when we introduced the notions of enantiomorphic pairs and selfenantiomorphic orbits. A group of order 2 consisting of the identity map and the "complementation" acts on the set of graphs. This action is chiral if $v \geq 2$, and hence the desired number of selfcomplementary graphs is twice the number given in 2.3.5 minus the number in 2.3.4 (cf. 2.1.15).

We have obtained

2.3.6 Corollary *The number of selfcomplementary graphs on v vertices is equal to*

$$\frac{1}{v!}\sum_{\pi}{}' 2^{c(\bar{\pi})} = \sum_{a}{}' \frac{2^{c_{\bar{a}}}}{\prod_i i^{a_i} a_i!},$$

where $c_{\bar{a}}$ is as in 2.3.4, and where $\sum_a'$ denotes the sum over all the cycle types a such that for each element π with this cycle type, the corresponding $\bar{\pi}$ does not contain any cycle of odd length.

This leads to an existence theorem (cf. exercise 2.3.1):

2.3.7 Theorem *Selfcomplementary graphs on v vertices exist if and only if v is congruent to 0 or 1 modulo 4.*

Proof: By 2.3.6, a selfcomplementary graph exists if and only if there are $\pi \in S_v$ such that $\bar{\pi}$ does not contain any cycle of odd length.

Now, if neither $v \equiv 0\,(4)$ nor $v \equiv 1\,(4)$, then $\binom{v}{2}$ is odd so that each $\bar{\pi}, \pi \in S_v$, must contain a cyclic factor of odd length, and hence no selfcomplementary graph can exist in these cases.

On the other hand, if $v \equiv 0\,(4)$, then to the full cycle $\pi := (0, \dots, v-1) \in S_v$ there corresponds a permutation $\bar{\pi}$ which, by 2.3.2, does not contain a cyclic factor of odd length, and so a selfcomplementary graph exists. Finally, if $v \equiv 1\,(4)$, we consider the permutation $\pi := (0, \dots, v-2)(v-1)$. Again by 2.3.2, $\bar{\pi}$ does not contain any cyclic factor of odd length, and hence a selfcomplementary graph exists in this case, too. □

Exercises

Exercise 2.3.1 Prove 2.3.7 directly.

2.4 Special Symmetry Classes

Before we generalize the method used for complementation of graphs we should recall 1.3.6. It shows that each action of the form ${}_{H\times G}Y^X$ can be considered as an action of H on $G\,\backslash\!\backslash Y^X$ or as an action of G on $H\,\backslash\!\backslash Y^X$. Hence each such action of $S_2 \times G$ on 2^X gives rise to an action of S_2 on $G\,\backslash\!\backslash 2^X$ and leads us to the discussion of the Involution Principle. We call a group element $\tau \neq 1$ an *involution* if and only if $\tau^2 = 1$. The Involution Principle is a method of counting objects by simply defining an involution τ on a suitably chosen set M and using the fact that $S_2 \simeq \{1, \tau\}$ has orbits of length 1 (the selfenantiomorphic orbits) and of length 2 (which form the enantiomorphic pairs) only. A typical example is the complementation τ of graphs which is, for $v \geq 2$, an involution on the set of graphs on v vertices. An even easier case was described in 1.1.4. We should like now to look closer at actions of involutions. The following remark is trivial but very helpful: Let $\tau \in S_M$ be an involution with the following reversion property with respect to $T, U \subseteq M$:

2.4.1 $$m \in T \iff \tau m \in U.$$

Then the restriction of τ to T establishes a bijection between T and U. We shall apply this to disjoint decompositions $M = M^+ \dot{\cup} M^-$ of M into subsets $M^\pm$. Each such disjoint decomposition gives rise to a *sign function* on M:

$$\mathrm{sign}(m) := \begin{cases} 1, & m \in M^+, \\ -1, & m \in M^-. \end{cases}$$

2.4.2 The Involution Principle *Let $M = M^+ \dot{\cup} M^-$ be a disjoint decomposition of a finite set M and let $\tau \in S_M$ be a* sign reversing *involution:*

$$\forall\, m \notin M_\tau\colon\ \mathrm{sign}(\tau m) = -\mathrm{sign}(m).$$

Then the restriction of τ to $M^+\backslash M_\tau$ is a bijection onto $M^-\backslash M_\tau$. Moreover

$$\sum_{m \in M} \mathrm{sign}(m) = \sum_{m \in M_\tau} \mathrm{sign}(m).$$

If in addition $M_\tau \subseteq M^+$, then

$$\sum_{m \in M} \mathrm{sign}(m) = |M_\tau| = |M^+| - |M^-|.$$

Proof: $\sum_{m \in M} \mathrm{sign}(m)$ is equal to

$$\sum_{m \in M_\tau} \mathrm{sign}(m) + \underbrace{\sum_{m \in M^+\backslash M_\tau} \mathrm{sign}(m) + \sum_{m \in M^-\backslash M_\tau} \mathrm{sign}(m)}_{=0,\ by\ 2.4.1}.$$

□

A beautiful example is

2.4.3 Application (the Principle of Inclusion and Exclusion) Let A denote a finite set and consider a sequence $P_0, \ldots, P_{n-1}$ of properties. We want to express the number of elements of A which have *none* of these properties in terms of numbers of elements which have *some* of these properties. In order to do this we indicate by $P_i(a)$ the fact that $a \in A$ has the property P_i, and for an index set $I \subseteq n$ we put

$$A_I := \{a \mid \forall\, i \in I\colon P_i(a)\}, \text{ in particular } A_\emptyset = A.$$

Furthermore we put

$$A^* := \{a \mid \text{there exists no } i \text{ such that } P_i(a)\},$$

and it is our aim to express $|A^*|$ in terms of the $|A_I|$. The set on which we shall define an involution is

$$M := \{(a, I) \mid I \subseteq n, a \in A_I\}.$$

A disjoint decomposition of this set is $M = M^+ \cup M^-$, where

$$M^+ := \{(a, I) \mid |I| \text{ even}\}, \text{ and } M^- := M \backslash M^+.$$

Now we introduce, for $a \notin A^*$, the number $s(a) := \min\{i \mid P_i(a)\} \in n$, and define τ on M as follows:

$$\tau(a, I) := \begin{cases} (a, I \cup \{s(a)\}), & \text{if } a \notin A^*, s(a) \notin I, \\ (a, I \backslash \{s(a)\}), & \text{if } a \notin A^*, s(a) \in I, \\ (a, I) = (a, \emptyset), & \text{if } a \in A^*. \end{cases}$$

Obviously $\mathrm{id}_M \neq \tau \in S_M$ provided that $A^* \neq A$. Furthermore $\tau^2 = \mathrm{id}_M$ and τ is sign–reversing, so that from 2.4.2 we obtain

$$|A^*| = |M_\tau| = |M^+| - |M^-| = \sum_{(a,I)\in M^+} 1 - \sum_{(a,I)\in M^-} 1$$

$$= \sum_{|I| \text{ even}} |A_I| - \sum_{|I| \text{ odd}} |A_I| = \sum_{I \subseteq n} (-1)^{|I|} |A_I|.$$

Thus we have proved

2.4.4 The Principle of Inclusion and Exclusion *Let A be a finite set and assume a sequence $P_0, \ldots, P_{n-1}$ of properties, while A_I denotes the set of elements of A having each of the properties $P_i, i \in I \subseteq n$. Then the order of the subset A^* of elements having none of these properties is*

$$|A^*| = \sum_{I \subseteq n} (-1)^{|I|} |A_I|.$$

A nice application is the following derivation of a formula for the values of the Euler function ϕ : We would like to express the value $\phi(n) = |\{i \in n \mid \gcd(i, n) = 1\}|$, $n > 0$, of the Euler function in terms of the different prime divisors $p_0, \ldots, p_{m-1}$ of n. In order to do this we put $A := n$ and we define $P_i, i \in m$, to be the property "divisible by p_i". The Principle of Inclusion and Exclusion yields the following expression for the order of A^*, the set of elements of n which are relatively prime to n :

$$|A^*| = |n| - |\{n/p_i \mid i \in m\}| + |\{n/p_i p_j \mid i, j \in m\}| - + \ldots,$$

which gives the desired expression for the value of the Euler function at n :

2.4.5
$$\phi(n) = n \cdot \prod_{i \in m} \left(1 - \frac{1}{p_i}\right).$$

◇

In situations where *two involutions* act we can use the following result which allows to replace certain identities of orders by bijections:

2.4.6 The Garsia–Milne bijection *We assume that*

$$M = M^+\dot{\cup}M^-, \quad N = N^+\dot{\cup}N^-$$

are disjoint decompositions of the finite sets M *and* N, *that* $\phi\colon M \to N$ *is a sign-preserving bijection, and that* $\sigma \in S_M$, $\tau \in S_N$ *are sign-reversing involutions such that* $M_\sigma \subseteq M^+$, $N_\tau \subseteq N^+$. *Then the following mapping is a bijection:*

$$\gamma\colon M_\sigma \to N_\tau\colon m \mapsto \sigma^*(\tau^*\sigma^*)^{k(m)}(m),$$

where

$$k(m) := \min\{k \in \mathbb{N} \mid \sigma^*(\tau^*\sigma^*)^k(m) \in N_\tau\}, \ \sigma^* := \phi\sigma, \tau^* := \phi^{-1}\tau.$$

Proof: Easy checks give the following implications:

$$m \in M_\sigma \cup M^- \Rightarrow \sigma^*(m) \in N^+, n \in N^+\backslash N_\tau \Rightarrow \tau^*(n) \in M^-,$$

and as an immediate consequence, for each natural number k:

$$\sigma^*(\tau^*\sigma^*)^k(m) \in N^+\backslash N_\tau$$

implies that

$$(\tau^*\sigma^*)^{k+1}(m) \in M^-, \text{ and } \sigma^*(\tau^*\sigma^*)^{k+1}(m) \in N^+.$$

We now prove the existence of $k(m)$ indirectly. Consider $m \in M_\sigma$. If $k(m)$ does not exist, then the implications mentioned above yield that

$$\sigma^*(\tau^*\sigma^*)^k(m) \in N^+\backslash N_\tau,$$

for each natural k. But this set is supposed to be finite. Hence there would be i, j in $\mathbb{N}$, $i \neq j$, such that $\sigma^*(\tau^*\sigma^*)^j(m) = \sigma^*(\tau^*\sigma^*)^i(m)$, and so (assume $j > i$):

$$(\tau^*\sigma^*)^{j-i}(m) = m \in M_\sigma,$$

a contradiction to the earlier implications since $M_\sigma \subseteq M^+$.

Finally we mention that γ is injective for the following reason: Assume m, m' in M_σ, for which $\gamma(m) = \gamma(m')$. They satisfy

$$\sigma^*(\tau^*\sigma^*)^{k(m)}(m) = \sigma^*(\tau^*\sigma^*)^{k(m')}(m'),$$

and hence also (assuming $k(m') - k(m) \geq 0$)

$$(\tau^*\sigma^*)^{k(m')-k(m)}(m') = m \in M_\sigma.$$

Thus either $k(m') = k(m)$, which is equivalent to $m = m'$, or there must exist a $j < k(m') - k(m)$ such that

$$\sigma^*(\tau^*\sigma^*)^j(m') \in N_\tau,$$

since otherwise we had $(\tau^*\sigma^*)^{k(m')-k(m)}(m') \in M^-$. This contradicts the minimality of $k(m')$. □

An application to certain inclusion–exclusion situations runs as follows. Consider two families $\mathcal{A} := \{A_0, \dots, A_{n-1}\}$ and $\mathcal{B} := \{B_0, \dots, B_{n-1}\}$ of subsets of two finite sets A and B. For each $I \subseteq n$ we put

$$A_I := \bigcap_{i\in I} A_i,\ B_I := \bigcap_{i\in I} B_i,\ A^* := A \backslash \bigcup_{i\in n} A_i,\ B^* := B \backslash \bigcup_{i\in n} B_i.$$

The Principle of Inclusion and Exclusion yields

$$|A^*| = \sum_{I\subseteq n} (-1)^{|I|} |A_I|,\ |B^*| = \sum_{I\subseteq n} (-1)^{|I|} |B_I|.$$

In the case when $|A_I| = |B_I|$, for each $I \subseteq n$, these two families $\mathcal{A}$ and $\mathcal{B}$ are called *sieve–equivalent*, a property which implies $|A^*| = |B^*|$.

Now we assume that this holds and that furthermore we are given, for each $I \subseteq n$, a bijection

$$\phi_I : A_I \to B_I.$$

Then the Garsia–Milne construction in fact allows us to sharpen the identity $|A^*| = |B^*|$ by replacing it by a bijection

2.4.7 $$\gamma : A^* \to B^*,$$

since we need only put

$$M := \{(a, I) \mid a \in A, \forall i \in I : a \in A_i\},\ M^+ := \{(a, I) \mid |I|\ even\},$$

$$N := \{(b, I) \mid b \in B, \forall i \in I : b \in B_i\},\ N^+ := \{(b, I) \mid |I|\ even\}.$$

A sign preserving bijection is

$$\phi : M \to N : (a, I) \mapsto (\phi_I(a), I),$$

and the involutions σ, τ are as in the proof of the Principle of Inclusion and Exclusion. Another consequence of 2.4.6 is (exercise 2.4.3):

2.4.8 The Two Involutions Principle *If $M = M^+ \dot\cup M^-$ is a disjoint decomposition of the finite set M, and $\sigma, \tau \in S_M$ are sign-reversing involutions such that $M_\sigma, M_\tau \subseteq M^+$, then each orbit of $G := \langle\sigma, \tau\rangle$ either consists of a fixed point of G alone, or it contains exactly one fixed point of σ and exactly one fixed point of τ, or it contains neither a fixed point of σ nor a fixed point of τ. This means in particular that there is a canonical bijection between M_σ and M_τ, namely the γ that maps $m \in M_\sigma$ onto the fixed point of τ that lies in the orbit $G(m)$ of m.*

We now return to Y^X and consider its subsets consisting of the injective and the surjective maps f only:

$$Y^X_{inj} := \{f \in Y^X \mid f \text{ injective}\} \text{ and } Y^X_{sur} := \{f \in Y^X \mid f \text{ surjective}\}.$$

It is clear that each of these sets is both a G–set and an H–set and therefore it is also an $H \times G$–set, but it will not in general be an $H \wr_X G$-set. The corresponding orbits of G, H and $H \times G$ on Y^X_{inj} are called *injective* symmetry classes, while those on Y^X_{sur} will be called *surjective* symmetry classes. We should like to determine their number. In order to do this we describe the fixed points of $(h, g) \in H \times G$ on these sets to prepare an application of the Cauchy–Frobenius Lemma. An introductory remark shows how the fixed points of (h, g) on Y^X can be constructed with the aid of $\bar{h}$ and $\bar{g}$, the permutations induced by h on Y and by g on X (use 2.2.11):

2.4.9 Corollary *If $\bar{g} = \prod_\nu (x_\nu \dots g^{l_\nu - 1} x_\nu)$, then $f \in Y^X$ is fixed under (h, g) if and only if the following two conditions are satisfied:*

$$f(x_\nu) \in Y_{h^{l_\nu}},$$

and the other values of f arise from the values $f(x_\nu)$ according to

$$f(x_\nu) = hf(g^{-1}x_\nu) = h^2 f(g^{-2}x_\nu) = \dots .$$

This together with 11.2.12 yields:

2.4.10 Corollary *The fixed points of (h, g) are the $f \in Y^X$ which can be obtained in the following way:*

- *To each cyclic factor of $\bar{g}$, let l denote its length, we associate a cyclic factor of $\bar{h}$ of length d dividing l.*
- *If x is a point in this cyclic factor of $\bar{g}$ and y a point in the chosen cyclic factor of $\bar{h}$, then put*

$$f(x) := y,\ f(gx) := hy,\ f(g^2x) := h^2y, \dots .$$

Such an f is injective if and only if the mapping described in the upper item of 2.4.10 is injective and corresponding cyclic factors of $\bar{g}$ and $\bar{h}$ have the same length. The number of such mappings is

$$\prod_j \binom{a_j(\bar{h})}{a_j(\bar{g})} a_j(\bar{g})!,$$

while the second item of 2.4.10 says that we have to multiply this number by $\prod_j j^{a_j(\bar{g})}$ in order to get the number $|Y^X_{inj,(h,g)}|$ of fixed points of (h, g) on Y^X_{inj}. Thus we have proved

2.4.11 Corollary *The number of fixed points of (h, g) on Y^X_{inj} is*

$$\left|Y^X_{inj,(h,g)}\right| = \prod_j \binom{a_j(\bar{h})}{a_j(\bar{g})} j^{a_j(\bar{g})} a_j(\bar{g})!,$$

and hence, by restriction, the numbers of fixed points of g and of h are:

$$\left|Y^X_{inj,g}\right| = \begin{cases} \binom{|Y|}{|X|}|X|!, & \text{if } \bar{g} = 1, \\ 0, & \text{otherwise,} \end{cases} \quad \text{and} \quad \left|Y^X_{inj,h}\right| = \binom{a_1(\bar{h})}{|X|}|X|!.$$

An application of the Cauchy–Frobenius Lemma yields the desired number of injective symmetry classes:

2.4.12 Corollary *The number of injective $H \times G$–classes is*

$$\left|(H \times G)\backslash\!\backslash Y^X_{inj}\right| = \frac{1}{|H||G|}\sum_{(h,g)}\prod_j \binom{a_j(\bar{h})}{a_j(\bar{g})} j^{a_j(\bar{g})} a_j(\bar{g})!,$$

so that we obtain by restriction the number of injective G–classes

$$\left|G \backslash\!\backslash Y^X_{inj}\right| = \frac{|X|!}{|\bar{G}|}\binom{|Y|}{|X|} = \binom{|Y|}{|X|}|S_X/\bar{G}|,$$

and the number of injective H–classes

$$\left|H \backslash\!\backslash Y^X_{inj}\right| = \frac{|X|!}{|H|}\sum_{k=|X|}^{|Y|} |\{h \in H \mid a_1(\bar{h}) = k\}|\binom{k}{|X|}.$$

In order to derive the number of surjective fixed points of (h, g) we use the preceding corollaries together with an application of the Principle of Inclusion and Exclusion in order to get rid of the nonsurjective fixed points. We denote by Y_ν the set of points $y \in Y$ contained in the ν–th cyclic factor of $\bar{h}$ and put, for each index set $I \subseteq c(\bar{h})$:

$$Y^X_{(h,g),I} := \{f \in Y^X_{(h,g)} \mid \forall\, \nu \in I : f^{-1}[Y_\nu] = \emptyset\}.$$

Then, by the Principle of Inclusion and Exclusion, we obtain for the desired number of surjective fixed points of (h, g) the following expression:

$$\left|Y^X_{sur,(h,g)}\right| = \left|Y^{X*}_{(h,g)}\right| = \sum_{I \subseteq c(\bar{h})} (-1)^{|I|}\left|Y^X_{(h,g),I}\right|$$

$$= \sum_{I \subseteq c(\bar{h})} (-1)^{c(\bar{h})-|I|}\left|Y^X_{(h,g),c(\bar{h})\setminus I}\right|.$$

Now we recall that

$$Y^X_{(h,g),c(\bar{h})\setminus I} = \{f \in Y^X_{(h,g)} \mid \forall\, \nu \notin I : f^{-1}[Y_\nu] = \emptyset\}.$$

This set can be identified with $\widetilde{Y}^X_{(\tilde{h},g)}$, where $\tilde{h}$ denotes the product of the cyclic factors of $\bar{h}$ the numbers of which lie in I, and where $\widetilde{Y}$ is the set of points contained in these cyclic factors. Thus

$$\left|Y^X_{(h,g),c(\bar{h})\setminus I}\right| = \left|\widetilde{Y}^X_{(\tilde{h},g)}\right| = \prod_j |\widetilde{Y}_{\tilde{h}^j}|^{a_j(\bar{g})}.$$

We can make this more explicit by an application of 11.2.12 which yields:

2.4.13 $$|\widetilde{Y}_{\tilde{h}^j}| = a_1(\tilde{h}^j) = \sum_{d|j} d \cdot a_d(\tilde{h}).$$

Putting these things together we conclude

2.4.14 Corollary *The number of surjective fixed points of* (h, g) *is*

$$\left|Y^X_{sur,(h,g)}\right| = \sum_{k=1}^{c(\bar{h})}(-1)^{c(\bar{h})-k}\sum_a \prod_{i=1}^{|Y|}\binom{a_i(\bar{h})}{a_i}\prod_{j=1}^{|X|}\Big(\sum_{d|j} d\cdot a_d\Big)^{a_j(\bar{g})},$$

where the middle sum is taken over all the sequences $a = (a_1, \ldots, a_{|Y|})$ *of natural numbers* a_j *such that* $\sum a_j = k$ *(they correspond to all possible choices of* $\tilde{h}$ *out of* h*, where* a_i *of the chosen cyclic factors of* $\tilde{h}$ *are* i*–cycles). Hence the numbers of surjective fixed points of* g *and of* h *amount to:*

$$\left|Y^X_{sur,g}\right| = \sum_{k=1}^{|Y|}(-1)^{|Y|-k}\binom{|Y|}{k}k^{c(\bar{g})},$$

and

$$\left|Y^X_{sur,h}\right| = \sum_{k=1}^{c(\bar{h})}(-1)^{c(\bar{h})-k}\sum_a\Big(\prod_{i=1}^{|Y|}\binom{a_i(\bar{h})}{a_i}\Big)a_1^{|X|},$$

if the sum is taken over all the sequences $(a_1, \ldots, a_{|Y|})$*,* $a_i \in \mathbb{N}$ *and* $\sum a_i = k$*.*

An application of the Cauchy–Frobenius Lemma finally yields the desired numbers of surjective symmetry classes:

2.4.15 Theorem *The number* $\left|(H \times G)\backslash\backslash Y^X_{sur}\right|$ *of surjective* $H \times G$*–classes is*

$$\frac{1}{|H||G|}\sum_{(h,g)}\sum_{k=1}^{c(\bar{h})}(-1)^{c(\bar{h})-k}\sum_a\prod_{i=1}^{|Y|}\binom{a_i(\bar{h})}{a_i}\prod_{j=1}^{|X|}\Big(\sum_{d|j} d\cdot a_d\Big)^{a_j(\bar{g})},$$

where the inner sum is taken over the sequences $a = (a_1, \ldots a_{|Y|})$ *described in the corollary above. This implies, by restriction, the equations*

$$\left|G\backslash\backslash Y^X_{sur}\right| = \frac{1}{|G|}\sum_g\sum_{k=1}^{|Y|}(-1)^{|Y|-k}\binom{|Y|}{k}k^{c(\bar{g})},$$

and

$$\left|H\backslash\backslash Y^X_{sur}\right| = \frac{1}{|H|}\sum_h\sum_{k=1}^{c(\bar{h})}(-1)^{c(\bar{h})-k}\sum_a\Big(\prod_{i=1}^{|Y|}\binom{a_i(\bar{h})}{a_i}\Big)a_1^{|X|},$$

where the last sum is to be taken over all the sequences $a = (a_1, \ldots, a_{|Y|})$ *such that* $a_i \in \mathbb{N}$ *and* $\sum a_i = k$*.*

These considerations lead to various combinatorial numbers so that a few remarks concerning these are in order. For example, $Y^X_{sur,g}$ is the set of surjective mappings $f \in Y^X$ which are constant on the $c(\bar{g})$ cyclic factors of $\bar{g}$. Hence this set can be identified with the set $Y^{c(\bar{g})}_{sur}$. More generally we consider the set m^n_{sur} and define the numbers $S(n, m)$ by

$$m!S(n,m) := \left| m^n_{sur} \right|, \text{ where } m, n \in \mathbb{N}.$$

These $S(n,m)$ are called the *Stirling numbers* of the *second kind*. If n is used as row index and m as column index, then the upper left hand corner of the table of Stirling numbers of the second kind is as follows:

2.4.16
$$\big(S(n,m)\big) = \begin{pmatrix} 1 & & & & & \\ 0 & 1 & & & & \\ 0 & 1 & 1 & & & \\ 0 & 1 & 3 & 1 & & \\ 0 & 1 & 7 & 6 & 1 & \\ 0 & 1 & 15 & 25 & 10 & 1 \\ \vdots & & & & & & \ddots \end{pmatrix}.$$

By definition $m!S(n,m)$ is equal to the number of *ordered* set partitions

$$\big(n^{(0)}, \dots, n^{(m-1)}\big)$$

of n into m *blocks*, i. e. into m nonempty subsets $n^{(i)}$. This is clear since each such ordered partition can be identified with $f\colon n \to m$ where $f^{-1}(i) := n^{(i)}, i \in m$. Thus $S(n,m)$ is the number of set partitions $\{n^{(0)}, \dots, n^{(m-1)}\}$ of the set n into m blocks, and hence B_n, the number of *all* the set partitions of n satisfies the equation

2.4.17
$$B_n = \sum_{k=0}^{n} S(n,k).$$

These numbers B_n are called *Bell numbers*. Another consequence of the definition of Stirling numbers of the second kind and 2.4.14 is

$$|Y|! \cdot S\big(c(\bar{g}), |Y|\big) = \left| Y^X_{sur,g} \right| = \sum_{k=1}^{|Y|} (-1)^{|Y|-k} \binom{|Y|}{k} k^{c(\bar{g})},$$

which implies:

2.4.18 Stirling's Formula *For $m > 0$ and any natural number n we have:*

$$S(n,m) = \frac{1}{m!} \sum_{k=1}^{m} (-1)^{m-k} \binom{m}{k} k^n.$$

A further series of combinatorial numbers shows up if we rewrite 2.4.15 in the following form:

2.4.19
$$\left| G \backslash\backslash Y^X_{sur} \right| = \frac{|Y|!}{|G|} \sum_{k=|X|}^{|X|} S\big(k, |Y|\big) \cdot \left| \{ g \in G \mid c(\bar{g}) = k \} \right|.$$

We put

$$r(n,k) := |\{\pi \in S_n \mid c(\pi) = k\}|,$$

and call these the *signless Stirling numbers of the first kind*. They satisfy the following recursion formula, since in $\pi \in S_n$ the point $n-1$ either forms a 1–cycle or does not:

2.4.20 Lemma *For $n, k > 1$ we have*

$$r(n,k) = r(n-1,k-1) + (n-1)r(n-1,k)$$

while the initial values are $r(0,0) = 1$ and $r(n,0) = r(0,k) = 0$, for $n, k > 0$.

The upper left hand corner of a table containing these numbers $r(n,k)$, for $n, k \in \mathbb{N}$, is as follows

2.4.21
$$(r(n,k)) = \begin{pmatrix} 1 & & & & & & \\ 0 & 1 & & & & & \\ 0 & 1 & 1 & & & & \\ 0 & 2 & 3 & 1 & & & \\ 0 & 6 & 11 & 6 & 1 & & \\ 0 & 24 & 50 & 35 & 10 & 1 & \\ \vdots & & & & & & \ddots \end{pmatrix}.$$

We now return to the number $\left|S_X \backslash\backslash Y^X_{sur}\right|$. The exercise 2.4.4 together with the identity 2.4.19 yields

2.4.22
$$\frac{|Y|!}{|X|!}\sum_{k=|Y|}^{|X|} r\big(|X|,k\big)S\big(k,|Y|\big) = \binom{|X|-1}{|Y|-1}.$$

Another series of combinatorial numbers arises when we count certain injective symmetry classes, since 2.4.12 implies

2.4.23
$$\left| S_Y \backslash\backslash Y^X_{inj} \right| = \frac{|X|!}{|Y|!}\sum_{k=|X|}^{|Y|} t\big(|Y|,k\big)\cdot\binom{k}{|X|},$$

where $t(n,k) := |\{\pi \in S_n \mid a_1(\pi) = k\}|$. It is easy to derive these numbers from the *rencontre numbers* $R(n) := t(n,0)$, since obviously the following is true:

2.4.24
$$t(n,k) = \binom{n}{k}R(n-k).$$

This can be made more explicit by an application of exercise 2.4.5 which yields

2.4.25
$$R(n) = n!\sum_{k=0}^{n}\frac{(-1)^k}{k!}.$$

Now we use that, for $|Y| \geq |X|$, $|S_Y \backslash\backslash Y^X_{inj}| = 1$, so that by the last three equations:

2.4.26 $$1 = \sum_{k=|X|}^{|Y|} \frac{1}{(k-|X|)!} \sum_{j=0}^{|Y|-k} \frac{(-1)^j}{j!}, \text{ if } |Y| \geq |X|.$$

It is in fact an important and interesting task of enumeration theory to derive identities in this way since they are understood as soon as they are seen to describe a combinatorial situation. Another example is the identity

2.4.27 Lemma *For natural numbers n and k the following identities hold:*

$$\sum_{m=1}^{k} \binom{k}{m}\binom{n-1}{m-1} = \binom{n+k-1}{n} = \frac{1}{n!}\sum_{\pi \in S_n} k^{c(\pi)}.$$

Proof: Exercise 2.4.4 implies that, for $m \leq k$,

$$\binom{k}{m}\binom{n-1}{m-1}$$

is equal to the number of symmetry classes of S_n on k^n, the elements of which satisfy $|f[n]| = m$. Thus the left hand side is $|S_n \backslash\backslash k^n|$. But the orbit of $f \in k^n$ under S_n is characterized by the orders of the inverse images $|f^{-1}(i)|$, $i \in k$. Hence the number of these orbits is equal to the number of k–tuples $(n_0, \ldots, n_{k-1})$, $n_i \in \mathbb{N}$, and $\sum n_i = n$, therefore the left side of the identity follows from exercise 2.4.6. The right hand side is already clear from 2.2.12. □

Exercises

Exercise 2.4.1 Assume X to be a finite set with subsets $X_0, \ldots, X_{n-1}$. Use the Principle of Inclusion and Exclusion in order to derive the number of elements of X which lie in precisely m of these subsets X_i.

Exercise 2.4.2 Evaluate again (cf. 0.2.4) the *derangement number*

$$\mathrm{der}_n := |\{\pi \in S_n \mid \forall\, i \in n\colon \pi i \neq i\}|,$$

i. e. the number of fixed point free elementes in S_n. Derive a recursion for these numbers and show that $\mathrm{der}_n/n!$ tends to $1/e$.

Exercise 2.4.3 Prove 2.4.8.

Exercise 2.4.4 Prove that $|S_n \backslash\backslash m^n_{sur}| = \binom{n-1}{m-1}$.

Exercise 2.4.5 Use the Principle of Inclusion and Exclusion in order to prove 2.4.25.

Exercise 2.4.6 Show that the number of k–tuples $(n_0, \ldots, n_{k-1})$ such that $n_i \in \mathbb{N}$ and $\sum n_i = n$ is equal to

$$\binom{n+k-1}{n}.$$

Exercise 2.4.7 Prove that the Stirling numbers of the second kind satisfy the equation

$$x^n = \sum_{k=0}^{n} S(n,k)[x]_k,$$

where $[x]_k := x(x-1) \cdot \ldots \cdot (x-k+1)$, for $k > 0$, while $[x]_0 := 1$.

Exercise 2.4.8 Prove the following recursion:

$$S(n+1,m) = \sum_{l=0}^{n} \binom{n}{l} S(l, m-1).$$

Exercise 2.4.9 Prove that a transitive action ${}_G X$ is k–fold transitive if and only if, for each $x \in X$,

$${}_{G_x}\left(X \backslash \{x\}\right) \text{ is } (k-1)\text{–fold transitive.}$$

Exercise 2.4.10 Prove that $|G|$ is divisible by $[|X|]_k$ if ${}_G X$ is finite and k–fold transitive.

Exercise 2.4.11 Show that S_n is n–fold transitive on n while A_n is $(n-2)$–fold transitive on n, but not $(n-1)$–fold transitive, for $n \geq 3$.

3. Enumeration by Weight

Now we are going to refine our methods in order to enumerate *orbits with prescribed properties*. For this purpose we introduce weight functions on X, i. e. mappings from X into commutative rings which are constant on the orbits of G on X. The Cauchy–Frobenius Lemma will be refined in order to count orbits with prescribed *weight*. We shall use it, for example, in order to enumerate graphs with prescribed numbers of vertices *and* edges. This leads us to certain generating functions, the *cycle indicator polynomials*.

We shall see that the enumeration of rooted trees amounts to the consideration of *sums* of cycle indicator polynomials and leads to recursive methods. These recursions can be used even for constructive purposes, and they stimulate the formalization of the calculation of generating functions.

Afterwards we *generalize* the enumeration of symmetry classes by introducing a combinatorial situation which covers both the notion of symmetry class (which was in fact Pólya's approach to the enumeration of graphs) and the situation which J. H. Redfield studied in order to enumerate superpositions of graphs.

3.1 Weight Functions

In the preceding chapter group actions were introduced, the Cauchy–Frobenius Lemma was proved, and we studied certain actions of groups on sets of the form Y^X in some detail. We saw that various structures like graphs and partitions can be defined as orbits on such sets of mappings, and so we already have a method at hand to evaluate the *total number* of such structures. The question arises how these methods can be refined in such a way that we can also derive the *number of orbits with certain prescribed properties* like, for example, the number of graphs on v vertices which have e edges. The answer to many such questions can be given by introducing a *weight* which means a mapping, defined on the set on which the group is acting and which is supposed to be *constant on each orbit.* The range of the weights which we will consider is mostly a polynomial ring over $\mathbb{Q}$. The final result will be a *generating function* for the enumeration problem in question, i. e. we shall obtain a polynomial which has the desired numbers of orbits as coefficients of its different monomial summands. The basic tool is

3.1.1 The Cauchy–Frobenius Lemma, weighted form *Let* ${}_GX$ *denote a finite action and* $w\colon X \to R$ *a map from* X *into a commutative ring* R *containing* $\mathbb{Q}$ *as a*

subring. If w is constant on the orbits of G on X, then we have, for any transversal T of the orbits:

$$\sum_{t\in T} w(t) = \frac{1}{|G|}\sum_{g\in G}\sum_{x\in X_g} w(x) = \frac{1}{|\bar{G}|}\sum_{\bar{g}\in\bar{G}}\sum_{x\in X_{\bar{g}}} w(x).$$

Proof: The following equations are clear from the foregoing, except maybe the last one which uses the assumption that w is constant on the orbits:

$$\sum_{g\in G}\sum_{x\in X_g} w(x) = \sum_{x}\sum_{g\in G_x} w(x)$$

$$= \sum_{x} |G_x| w(x) = |G| \sum_{x} |G(x)|^{-1} w(x) = |G| \sum_{t\in T} w(t).$$

This proves the left one of the stated equations, the right one follows by an application of the homomorphism theorem. □

This result will be used in the following way for the enumeration of orbits with prescribed properties. We consider weights w with values in a *vector space*, for example in a polynomial ring. These weight functions will be chosen in such a way that the values are the same on orbits with the considered property, while their values on orbits which we should like to distinguish are *linearly independent*. Then we obviously can read off from $\sum w(t)$ the number of orbits with the property in question, it is just the coefficient of the vector $w(t_0)$, if t_0 has that property, in the vector $\sum w(t)$. For example, we can associate with a labeled graph with e edges the weight y^e. Then $\sum w(t)$ is a polynomial in y, where the coefficient of y^e is the desired number of unlabeled graphs on v vertices having exactly e edges.

The lemma 3.1.1 implies 2.1.1 (put $w: x \mapsto 1$), which we shall sometimes call the *constant form* of the Cauchy–Frobenius Lemma. In order to apply 3.1.1 to the enumeration of symmetry classes of mappings f in Y^X we introduce, for a given $W: Y \to R$, R a commutative ring with $\mathbb{Q}$ as a subring, the *multiplicative weight* w, defined by

3.1.2
$$w: Y^X \to R: f \mapsto \prod_{x\in X} W(f(x)),$$

and notice that for any finite actions ${}_GX$ and ${}_HY$ the following is easy to verify:

3.1.3 Lemma *If W is constant on the orbits of H on Y, then w is constant on the orbits of $H \wr_X G$, $H \times G$, H and G on Y^X. Moreover, for any W, the corresponding multiplicative weight w is constant on the orbits of G on Y^X.*

Thus 3.1.1 can be applied as soon as we have evaluated the sum of the weights of those f which are fixed under $(\psi, g) \in H \wr_X G$. But this sum of weights follows directly from the characterization of the fixed points of (ψ, g) given in 2.2.11: Using the same notation as in 2.2.11, for (ψ, g) in $H \wr_X G$, a function $W: Y \to R$ which is constant on the orbits of H, and the corresponding multiplicative weight w on Y^X, we obtain the equation

$$\sum_{f\in Y^X_{(\psi,g)}} w(f) = \prod_\nu \sum_{y\in Y_{h_\nu(\psi,g)}} W(y)^{l_\nu}.$$

Now an application of the weighted form of the Cauchy–Frobenius Lemma to 3.1.3 yields the desired generating function for the enumeration of symmetry classes by the multiplicative weight w:

3.1.4 Theorem *Let ${}_GX$ and ${}_HY$ be finite actions, $W: Y \to R$ a mapping into a commutative ring containing $\mathbb{Q}$ as a subring, and denote by w the corresponding multiplicative weight function on Y^X.*

- *If W is constant on the orbits of H on Y, then w is constant on the orbits of $H \wr_X G$ on Y^X, and, for each transversal T of these orbits, we have*

$$\sum_{t\in T} w(t) = \frac{1}{|H|^{|X|}|G|} \sum_{(\psi,g)\in H\wr_X G} \prod_{\nu\in c(\bar{g})} \sum_{y\in Y_{h_\nu(\psi,g)}} W(y)^{l_\nu}.$$

  Moreover w is also constant on the orbits of $H \times G$ and H, so that, by restriction, we obtain for the sum of the weights of the elements in a transversal the expressions

$$\frac{1}{|H||G|} \sum_{(h,g)\in H\times G} \prod_{i=1}^{|X|} \Big(\sum_{y\in Y_{h^i}} W(y)^i\Big)^{a_i(\bar{g})},$$

  and

$$\frac{1}{|H|} \sum_{h\in H} \Big(\sum_{y\in Y_h} W(y)\Big)^{|X|}.$$

- *For any $W: Y \to R$, the multiplicative weight $w: f \mapsto \prod_x W(f(x))$ is constant on the orbits of G on Y^X, and the sum of its values on a transversal of the orbits is equal to*

$$\frac{1}{|G|} \sum_{g\in G} \prod_{i=1}^{|X|} \Big(\sum_{y\in Y} W(y)^i\Big)^{a_i(\bar{g})}.$$

The most general weight function is obtained when we take for W a mapping which sends each $y \in Y$ to a separate indeterminate of a polynomial ring. For the sake of notational simplicity we can do this by taking the elements $y \in Y$ themselves as indeterminates and putting

$$W: Y \to \mathbb{Q}[Y]: y \mapsto y,$$

where $\mathbb{Q}[Y]$ denotes the polynomial ring over $\mathbb{Q}$ in the set Y of commuting indeterminates. This yields the multiplicative weight $w(f) = \prod_x f(x)$, a monomial in $\mathbb{Q}[Y]$. If we define the *content* of $f \in Y^X$ to be the mapping

3.1.5 $$c(f, -): Y \to \mathbb{N}: y \mapsto |f^{-1}(y)|,$$

i. e. $c(f, y)$ is the multiplicity with which f takes the value y, then we get

3.1.6 Corollary *The number of $G-$classes on Y^X, the elements of which have the same content as $f \in Y^X$, is equal to the coefficient of the monomial $\prod_y y^{c(f,y)}$ in the polynomial*

$$\frac{1}{|G|}\sum_{g\in G}\prod_{i=1}^{|X|}\Big(\sum_{y\in Y} y^i\Big)^{a_i(\overline{g})}.$$

A nice example is the solution of the *necklace problem*:

3.1.7 Application (necklaces by weight) We ask for the number of different necklaces with n beads in up to m colours and given content. In order to bring this problem within reach of unromantic mathematics, we consider such a necklace again as a colouring of the vertices of a regular $n-$gon, i. e. as an $f \in m^n$. Two such necklaces or colourings are different if and only if none of them can be obtained from the other one by a rotation. Hence we are faced with an action of the form $_G(Y^X)$, namely the natural action of the cyclic group $G := C_n$ on the set $Y^X := m^n$. (In case we want to allow reflections, we have to consider $G := D_n$, the dihedral group.) A particular case was already discussed in 2.2.13, where we obtained the total number of orbits in the special case when n is prime. Now we are in a position to count these orbits by content. In order to do this for general m and n we take from 11.2.14 the cycle structure of the elements of C_n, obtaining by an application of 3.1.6 the desired solution of the necklace problem:

3.1.8 Corollary *The number of different necklaces containing b_i beads of the i–th colour, $i \in m$, is the coefficient of $y_0^{b_0}\dots y_{m-1}^{b_{m-1}}$ in the polynomial*

$$\frac{1}{n}\sum_{d|n}\phi(d)\big(y_0^d+\ldots+y_{m-1}^d\big)^{n/d},$$

where ϕ denotes the Euler function (see 11.2.14).

For an example we take $m := 2$ and $n := 5$, obtaining the generating function

$$\frac{1}{5}\Big((y_0+y_1)^5+4(y_0^5+y_1^5)\Big)=y_0^5+y_0^4y_1+2y_0^3y_1^2+2y_0^2y_1^3+y_0y_1^4+y_1^5.$$

Recall that the monomial summand $2y_0^3y_1^2$ means that there are exactly two different necklaces consisting of 5 beads three of which are of the 0-th colour and two of which are of the colour with number 1. The drawing shows four of the eight different necklaces, the remaining ones are obtained by simply exchanging the two colours.

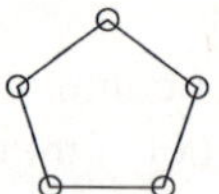
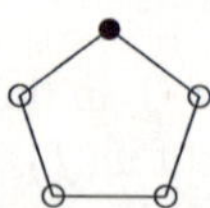
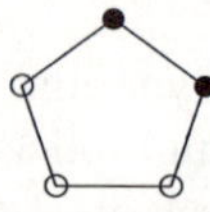
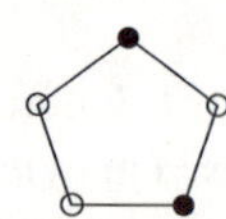

◇

The result 3.1.6 has an important consequence which can be used for a systematic study of certain congruences:

3.1.9 Corollary *For each finite action ${}_GX$, $m \in \mathbb{N}$, the coefficients of monomial summands in the generating function are divisible by $|G|$, in formal terms:*

$$\sum_{g\in G}\prod_{i=1}^{|X|}\Bigl(\sum_{v\in m-1} y_v^i\Bigr)^{a_i(\bar g)} \equiv 0 \ (|G|).$$

□

Besides the Cauchy–Frobenius Lemma also the Involution Principle admits a generalization to a weighted form:

3.1.10 The Involution Principle, weighted form *Assume that τ is a sign-reversing involution acting on the finite set $X = X^+\dot\cup X^-$ and that $w\colon X \to R$ is a weight function which is constant on the orbits of τ. Then*

$$\sum_{x\in X}\operatorname{sign}(x)w(x) = \sum_{x\in X_\tau}\operatorname{sign}(x)w(x).$$

The proof is trivial but the applications are not as is shown by

3.1.11 Application (the q-binomial identity) We would like to prove the q–*binomial theorem.* It says that, for any $q \in \mathbb{R}$ and $n \in \mathbb{N}^*$, the following polynomial identity holds:

3.1.12 $$(x-1)(x-q)\dots(x-q^{n-1}) = \sum_{k=0}^{n}\begin{bmatrix} n\\ k\end{bmatrix}_q q^{\binom{n-k}{2}}(-1)^{n-k}x^k,$$

where, for $n, k \in \mathbb{N}$, the rational function $\left[{n \atop k}\right]$ is defined by

$$\begin{bmatrix} n\\ k\end{bmatrix} = \begin{cases} [n]![k]!^{-1}[n-k]!^{-1}, & \text{if } 0 \le k \le n,\\ 0, & \text{otherwise,}\end{cases}$$

and

$$[0]! := 1,\ [n]! := [n][n-1]\dots[1],\ \text{for } n \ge 1,$$

while, for $n \ge 1$:

$$[n] := 1 + x + \dots + x^{n-1},$$

and finally the q–*binomial number* is defined to be the value of the *binomial function* $\left[{n \atop k}\right]$ at q:

$$\begin{bmatrix} n\\ k\end{bmatrix}_q := \begin{bmatrix} n\\ k\end{bmatrix}(q),\ \text{in particular}\ \begin{bmatrix} n\\ k\end{bmatrix}_1 = \begin{bmatrix} n\\ k\end{bmatrix}(1) = \binom{n}{k}.$$

We consider the identity we want to prove as an identity of polynomials in x over $\mathbb{R}$. This single identity is equivalent to the following system of identities:

$$\forall\, n > l \ge 0\colon \sum_{k=0}^{n}\begin{bmatrix} n\\ k\end{bmatrix}_q q^{\binom{n-k}{2}}(-1)^{n-k}(q^l)^k = 0.$$

We prove this system by an application of the weighted form of the Involution Principle. The set on which we shall define an involution is

$$X := \{(I, J) \mid n = I \dot{\cup} J\},$$

which we decompose into

$$X^+ := \{(I, J) \in X \mid |J| \text{ even}\}, \ X^- := X \backslash X^+.$$

The involution which we use is defined, for a fixed $l \in n$, by

$$\tau(I, J) := \begin{cases} (I \backslash \{n-l\}, J \cup \{n-l\}), & \text{if } n-l \in I, \\ (I \cup \{n-l\}, J \backslash \{n-l\}), & \text{if } n-l \notin I. \end{cases}$$

A weight function is introduced by

$$w(I, J) := q^{\binom{|J|}{2} + i\,(I,J) + l \cdot |I|},$$

where $i\,(I, J)$ means the number of *inversions between I and J*, i. e. the number of pairs $(i, j) \in I \times J$ such that $i > j$. It is clear that τ is a sign–reversing involution on X, and $X_\tau = \emptyset$. It remains to check that w is constant on the orbits of τ. Clearly $i\,(I \backslash \{n-l\}, J \cup \{n-l\})$ is equal to

$$i\,(I, J) - i\,(\{n-l\}, J) + i\,(I \backslash \{n-l\}, \{n-l\}) = i\,(I, J) + l - |J|.$$

This gives

$$w(I \backslash \{n-l\}, J \cup \{n-l\}) = q^{\binom{|J|+1}{2} + i\,(I,J) + l \cdot |I| - |J|} = w(I, J).$$

We are now in a position to apply 3.1.10 which yields, as $X_\tau = \emptyset$, that 0 is equal to

$$\sum_{(I,J)} (-1)^{|J|} q^{\binom{|J|}{2} + i\,(I,J) + l \cdot |I|} = \sum_{k=0}^{n} (-1)^{n-k} q^{\binom{n-k}{2} + k \cdot l} \sum_{(I,J):\, |J| = n-k} q^{i\,(I,J)},$$

so that it remains to prove

3.1.13 $$\begin{bmatrix} n \\ k \end{bmatrix}_q = \sum_{(I,J):\, |J| = n-k} q^{i\,(I,J)},$$

an identity that follows from exercise 3.1.4. ◇

Exercises

Exercise 3.1.1 Prove the following combinatorial principle: If X and Y are finite sets and R is a commutative ring, and $\varphi: Y \times X \to R$, then

$$\sum_{f \in Y^X} \prod_{x \in X} \varphi(f(x), x) = \prod_{x \in X} \sum_{y \in Y} \varphi(y, x).$$

Exercise 3.1.2 Derive 3.1.6 directly, using the fact that $f \in Y^X$ is fixed under $g \in G$ if and only if f is constant on the cyclic factors of $\bar{g}$.

Exercise 3.1.3 Prove by induction that

$$\sum_{\pi \in S_n} q^{l(\pi)} = [n]!.$$

Exercise 3.1.4 Derive 3.1.13 from exercise 3.1.3 by considering a transversal of the left cosets of $S_k \oplus S_{n \backslash k}$. (Hint: Show that the permutations π in S_n which are increasing both on k and $n \backslash k$ form such a transversal.)

3.2 Cycle Indicator Polynomials

We have seen in 3.1.6 that the generating function for the enumeration of the G–classes on Y^X by weight is equal to

$$\frac{1}{|G|} \sum_{g \in G} \prod_{i=1}^{|X|} \Big(\sum_{y \in Y} y^i\Big)^{a_i(\bar{g})}.$$

This polynomial can be obtained from the polynomial

$$C(G, X) := \frac{1}{|G|} \sum_{g \in G} \prod_{i=1}^{|X|} z_i^{a_i(\bar{g})} \in \mathbb{Q}[z_1, \dots, z_{|X|}]$$

by simply replacing the indeterminate z_i by the polynomial $\sum_y y^i$. It is therefore the polynomial $C(G, X)$ which really matters. We call this polynomial the *cycle indicator polynomial* or the *cycle index* of ${}_GX$, since it displays the cycle structure of the elements $\bar{g} \in \bar{G}$. But it should be mentioned that $C(G, X) = C(H, X)$ does *not* mean that $\bar{G}$ and $\bar{H}$ are isomorphic. There are counterexamples known: For odd primes p there exist exactly two nonabelian groups of order p^3, and the regular representation of one of them consists of the identity element together with $p^3 - 1$ elements of order p which consist of p-cycles only. Hence this nonabelian group has the same cycle indicator polynomial as the regular representation of the abelian group $C_p \times C_p \times C_p$ (exercise 3.2.3).

In the case when we wish to display the indeterminates in the cycle indicator, we shall write $C(G, X; z_1, \dots, z_{|X|})$, and if we replace z_i by $r_i \in \mathbb{Q}$, we simply write $C(G, X; r_1, \dots, r_{|X|})$. Hence we obtain, for example (cf. 2.2.12):

3.2.1 Corollary *The number of G–classes on Y^X is equal to*

$$C(G, X; |Y|, \dots, |Y|).$$

In order to generalize the substitution process mentioned above which yields the generating function from $C(G, X)$, we put, for any p in $\mathbb{Q}[u_0, \ldots, u_{m-1}]$,

$$C(G, X \mid p(u_0, \ldots, u_{m-1})) := C(G, X)|_{z_i := p(u_0^i, \ldots, u_{m-1}^i)}$$

$$= \frac{1}{|G|} \sum_{g \in G} \prod_{i=1}^{|X|} p(u_0^i, \ldots, u_{m-1}^i)^{a_i(\bar{g})} \in \mathbb{Q}[u_0, \ldots, u_{m-1}],$$

and call this process *Pólya–substitution*. Using this notation we can rephrase Corollary 3.1.6 as follows:

3.2.2 Pólya's Theorem *The generating function for the enumeration of G–classes on Y^X by content can be obtained from the cycle indicator of ${}_GX$ by Pólya–substituting $\sum_{y \in Y} y$ in the cycle indicator polynomial $C(G, X)$. Hence this generating function is equal to*

$$C\big(G, X \,\big|\, \sum_{y \in Y} y\big) = C(G, X)\Big|_{z_i := \Sigma y^i}.$$

In order to count G–classes by content it therefore remains to evaluate the cycle indicator of ${}_GX$. A few examples should be welcome:

3.2.3 Examples of cycle indicator polynomials

- The cycle indicator of the identity subgroup of S_n and its natural action on n is

$$C(\{1\}, n) = z_1^n.$$

- The cycle indicator of the natural action of the cyclic group C_n on n is (recall 3.1.8)

$$C(C_n, n) = \frac{1}{n} \sum_{d \mid n} \phi(d) z_d^{n/d}.$$

- The cycle indicator of the natural action of the dihedral group D_n of order $2n$ on the set n (which can be considered as being the set of vertices of the regular n–gon, of which D_n is the symmetry group) satisfies, if $n \geq 3$,

$$C(D_n, n) = \frac{1}{2} C(C_n, n) + \begin{cases} \frac{1}{2} z_1 z_2^{(n-1)/2}, & \text{if } n \text{ is odd}, \\ \frac{1}{4} z_2^{n/2} + \frac{1}{4} z_1^2 z_2^{(n-2)/2}, & \text{if } n \text{ is even}. \end{cases}$$

  This follows from the fact that the rotations form a cyclic subgroup of order n, while the remaining reflections either leave exactly one vertex fixed (in the case when n is odd) or two vertices or none (in the case when n is even), and group the remaining vertices into pairs of vertices that are mapped onto each other.

- Furthermore we have the following cycle indicators of the natural actions of S_n and A_n (cf. 11.2.7 and exercise 2.2.1):

$$C(S_n, n) = \sum_{a \dashv n} \prod_{k=1}^{n} \frac{1}{a_k!}\left(\frac{z_k}{k}\right)^{a_k},$$

$$C(A_n, n) = \sum_{a \dashv n} (1 + (-1)^{a_2 + a_4 + \ldots}) \prod_{k=1}^{n} \frac{1}{a_k!}\left(\frac{z_k}{k}\right)^{a_k}.$$

◇

3.2.4 Applications (graphs and selfcomplementary graphs) In order to enumerate the graphs on 4 vertices we use the cycle index (exercise 3.2.1)

3.2.5 $$C\left(S_4, \binom{4}{2}\right) = \frac{1}{24}(z_1^6 + 9z_1^2 z_2^2 + 8z_3^2 + 6z_2 z_4),$$

which is obtained from 2.3.3. It yields the generating function

$$C\left(S_4, \binom{4}{2} \,\Big|\, y_0 + y_1\right) = y_0^6 + y_0^5 y_1 + 2y_0^4 y_1^2 + 3y_0^3 y_1^3 + 2y_0^2 y_1^4 + y_0 y_1^5 + y_1^6,$$

in accordance with figure 3.1.

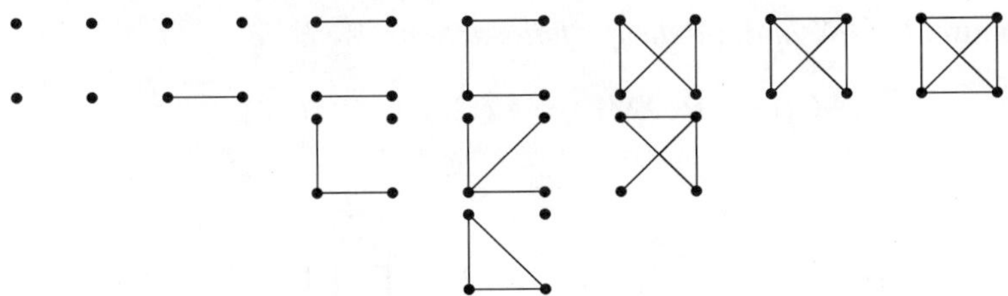

Fig. 3.1. The graphs on 4 points

In order to remove redundant information, we can replace $y_0 + \ldots + y_{m-1}$ by $1 + y_1 + \ldots + y_{m-1}$, and in the case when $|Y| = 2$ we can even take $1 + y$ instead of $y_0 + y_1$, so that, for example, the generating function for the graphs on 4 vertices by their number of edges takes the following form:

$$C\left(S_4, \binom{4}{2} \,\Big|\, 1 + y\right) = 1 + y + 2y^2 + 3y^3 + 2y^4 + y^5 + y^6.$$

Here the coefficient of y^e is equal to the number of graphs on 4 vertices which possess exactly e edges, so that we obtain (recall 2.3.4 and 2.3.6):

3.2.6 Corollary *The total number of graphs on v vertices is equal to*

$$C\left(S_v, \binom{v}{2}; 2, \ldots, 2\right),$$

while the number of selfcomplementary graphs on v vertices is equal to

$$C\left(S_v, \binom{v}{2}; 0, 2, 0, 2, \ldots\right).$$

The number of graphs on v vertices which contain e edges is the coefficient of y^e *in the polynomial*

$$C\left(S_v, \binom{v}{2} \,\Big|\, 1 + y\right).$$

◇

These examples have shown that the evaluation of the cycle indicator polynomial is an important step towards the solution of various enumeration problems, so that a few remarks concerning this are well in order. The cycle indicators of the natural actions of S_n, A_n, C_n, and D_n are at hand and we therefore put the question how we can obtain cycle indicators of groups, which are direct sums or certain products of symmetric, alternating, cyclic or dihedral groups. More generally we want to know how the cycle indicator of a sum or a product can be obtained from the cycle indicators of the summands or factors, depending of course on the action in question. The following remarks are clear from the definitions 1.2.13 and 1.2.14 of direct sum and cartesian product:

3.2.7 Lemma *Let* ${}_GX$ *and* ${}_HY$ *denote finite actions. We have*

$$C(G \times H, X \dot{\cup} Y) = C(G, X) \cdot C(H, Y),$$

and

$$C(G \times H, X \times Y) = \frac{1}{|G||H|} \sum_{(g,h) \in G \times H} \prod_{i=1}^{|X|} \prod_{k=1}^{|Y|} z_{lcm(i,k)}^{gcd(i,k) a_i(\bar{g}) a_k(\bar{h})}.$$

Further important constructions are the plethysm $H \odot G$ which we introduced in 2.2.2, and the composition which was introduced in 2.2.8. They are defined as permutation groups induced by *similar actions* of $H \wr_X G$, and so the corresponding cycle indicator polynomials are the same:

3.2.8 $$C(G[H], Y \times X) = C(H \odot G, |Y||X|).$$

From the description of the conjugacy classes of elements of complete monomial groups given in 2.2.7 we obtain for $C(H \odot G, |Y||X|)$ the following *plethysm of cycle indicators* (recall 2.2.7, in particular its last item):

3.2.9 $$C(H, Y) \odot C(G, X) := \frac{1}{|G|} \sum_{g \in G} \prod_{k=1}^{|X|} \left(\frac{1}{|H|} \sum_{h \in H} \prod_{i=1}^{|Y|} z_{i \cdot k}^{a_i(\bar{h})} \right)^{a_k(\bar{g})}.$$

Finally we show how the cycle indicator polynomial $C(G, X)$ can be obtained from the character $\chi: g \mapsto |X_g|$ of the action ${}_GX$ in question. In order to do this we have to evaluate the numbers $a_i(\bar{g}), i \in \mathbb{N}^*$, for each $g \in G$, and we shall do this by inverting the equations

3.2.10 $$\chi(g^k) = a_1(g^k) = a_1(\bar{g}^k) = \sum_{d|k} d \cdot a_d(\bar{g})$$

(cf. 11.2.12 for the last equation). The inversion procedure uses the number theoretic Möbius function and the corresponding inversion theorem (see appendix, 11.2.12, 11.6.3 and 3.2.10) which gives the following interesting formulation of the cycle index in terms of the character of the group action in question:

3.2.11 Corollary *For each finite action ${}_GX$ we can obtain the cycle structure of the permutation $\bar{g}$ from the numbers of fixed points of its powers:*

$$\forall\, k: \; a_k(\bar{g}) = \frac{1}{k} \sum_{d|k} \mu(k/d) a_1(\bar{g}^d),$$

where μ denotes the number theoretic Möbius function. (This system of equations is in fact equivalent to 3.2.10.) In particular we obtain the following expression of the cycle indicator of ${}_GX$ in terms of the character χ of the group action ${}_GX$:

$$C(G, X) = \frac{1}{|G|} \sum_{g \in G} \prod_{i=1}^{|X|} z_i^{\,i^{-1} \sum_{d|i} \mu(i/d) \chi(g^d)}.$$

3.2.12 Further examples

- The *exterior cycle index* of $\bar{G} \leq S_X$ is defined to be $C(S_X, S_X/\bar{G})$, the cycle indicator of the natural action of S_X on the set of left cosets $S_X/\bar{G}$. The permutation character of this action is (exercise 3.2.5)

$$\chi(\pi) = \frac{|X|!|C^{S_X}(\pi) \cap \bar{G}|}{|\bar{G}||C^{S_X}(\pi)|}.$$

- The character of S_X on the set $\binom{X}{k}$ of all the k–subsets of X is

$$\chi(\pi) = \sum_{b \dashv k} \prod_{i=1}^{k} \binom{a_i(\pi)}{b_i},$$

  since a k–subset is fixed under π if and only if it is the union of cyclically permuted points. Hence we can use 3.2.11 for an evaluation of $C(S_v, \binom{v}{2})$ which is obtained from 2.3.3:

$$\chi(\bar{\pi}) = \binom{a_1(\pi)}{2} + a_2(\pi).$$

◇

We should like to introduce now an interesting generalization of the cycle index polynomial. In order to define it we consider the cycle index again:

$$C(G,X) := \frac{1}{|G|}\sum_{g\in G}\prod_{i=1}^{|X|} z_i^{a_i(\bar{g})} = \frac{1}{|\bar{G}|}\sum_{\bar{g}\in\bar{G}}\prod_{i=1}^{|X|} z_i^{a_i(\bar{g})} = \frac{1}{|\bar{G}|}\sum_{a\dashv|X|}|C^a\cap\bar{G}|\prod_{i=1}^{|X|} z_i^{a_i}.$$

If we insert a class function $\chi\colon G\to\mathbb{C}$, we obtain a *generalized cycle index*

3.2.13 $$C(G,X,\chi) := \frac{1}{|G|}\sum_{g\in G}\chi(g)\prod_{i=1}^{|X|} z_i^{a_i(\bar{g})} = \frac{1}{|\bar{G}|}\sum_{a\dashv|X|}\chi_a|C^a\cap\bar{G}|\prod_{i=1}^{|X|} z_i^{a_i},$$

and the equation

$$C(G,X) = \frac{1}{|X|!}\sum_{a\dashv|X|}\frac{|X|!|C^a\cap\bar{G}|}{|\bar{G}||C^a|}|C^a|\prod_{i=1}^{|X|} z_i^{a_i}$$

shows that the following is true:

3.2.14 Corollary *Each cycle index can be considered as a generalized cycle index of the symmetric group:*

$$C(G,X) = C(S_X,X,\chi),\ \textit{where}\ \chi\colon S_X\to\mathbb{C}\colon\beta\mapsto\frac{|X|!|C^{S_n}(\beta)\cap\bar{G}|}{|\bar{G}||C^{S_n}(\beta)|}.$$

3.2.15 Application (the cycle series) Consider the following sum of generalized cycle indices:

$$C(S_n,n,\chi^{\mathcal{S}}) = \frac{1}{n!}\sum_{a\dashv n}\chi_a^{\mathcal{S}}|C^a|\prod_{i=1}^{n} z_i^{a_i},$$

where

$$\chi^{\mathcal{S}}\colon S_n\to\mathbb{C}\colon\beta\mapsto a_1(\mathcal{S}[\beta]) = |\,\mathcal{S}[n]_\beta\,|,$$

is the character of the action of S_n on the set of structures $\mathcal{S}[n]$. The value of this character is a class function. We denote by $\chi_a^{\mathcal{S}}$ its value on the conjugacy class of elements of cycle type a, recalling that $a_i(\beta)$ means the number of cycles of $\beta\in S_n$ but *not* numbers of cycles of the permutation induced by β on $\mathcal{S}[n]$, which was denoted by $\mathcal{S}[\beta]$. We are now going to apply suitable substitutions in order to obtain summands of cardinalities or of type series. An application of the above explicit form of cycle indices shows that the substitution $z_1 := x, z_2 := 0, z_3 := 0, \ldots$ yields

3.2.16 $$C(S_n,n,\chi^{\mathcal{S}};x,0,0,\ldots) = |\mathcal{S}[n]|\cdot\frac{x^n}{n!},$$

since the only remaining summand is the one corresponding to the identity element. And from the Lemma of Cauchy-Frobenius we get that the substitution $z_1 := x, z_2 := x^2, z_3 := x^3, \ldots$ gives

3.2.17 $$C(S_n, n, \chi^{\mathcal{S}}; x, x^2, x^3, \ldots) = \mid S_n \backslash\backslash \mathcal{S}[n] \mid \cdot x^n.$$

Defining the *cycle series* of $\mathcal{S}$-structures by

$$C_{\mathcal{S}} := C_{\mathcal{S}}(z_1, z_2, \ldots) := \sum_{n \geq 0} C(S_n, n, \chi^{\mathcal{S}}; z_1, z_2, \ldots)$$

we now obtain the announced result that this series is in fact an interesting generalization of both the cardinality and the type series (recall 0.1.5 and 1.1.8):

3.2.18 Corollary *The cycle series yields via suitable substitutions both the cardinality and the type series of structures:*

$$\mathcal{S}(x) = C_{\mathcal{S}}(x, 0, \ldots), \; \widetilde{\mathcal{S}}(x) = C_{\mathcal{S}}(x, x^2, \ldots).$$

Another interesting property of this series can be obtained from the following

3.2.19 Lemma *Assume a finite action $_GX$ with orbits ω_i, representatives of these orbits $x_i \in \omega_i$, stabilizers $G_i := G_{x_i}$, and a class function χ on G together with a weight function w on X, which has on ω_i the value w_i, then*

$$\sum_i w_i \frac{1}{|G_i|} \sum_{g \in G_i} \chi(g) = \frac{1}{|G|} \sum_{g \in G} \chi(g) \sum_{x \in X_g} w(x).$$

Proof:

$$\begin{aligned} \sum_{g \in G} \chi(g) \sum_{x \in X_g} w(x) &= \sum_{x \in X} w(x) \sum_{g \in G_x} \chi(g) \\ &= \sum_i |\omega_i| \cdot w_i \cdot |G_i| \cdot \frac{1}{|G_i|} \sum_{g \in G_i} \chi(g) \\ &= \sum_i w_i \cdot |G| \cdot \frac{1}{|G_i|} \sum_{g \in G_i} \chi(g). \end{aligned}$$

□

We should like to apply this result to species, putting $w: \sigma \mapsto 1, \; \chi: \beta \mapsto \prod_k z_k^{a_k(\beta)}$. This gives (the check is easy) for the $\mathcal{S}$-structures on n, their S_n-orbits ω_i, representatives $\sigma_i \in \omega_i$ and their stabilizers $(S_n)_i$:

$$\begin{aligned} \frac{1}{n!} \sum_{\beta \in S_n} | \, \mathcal{S}[n]_\beta \, | \prod_k z_k^{a_k(\beta)} &= \frac{1}{n!} \sum_{\beta \in S_n} a_1(\mathcal{S}[\beta]) \prod_k z_k^{a_k(\beta)} \\ &= \sum_i \frac{1}{|(S_n)_i|} \sum_{\beta \in (S_n)_i} \prod_k z_k^{a_k(\beta)}. \end{aligned}$$

Replacing in all these expressions the indeterminate z_k by x^k we get

$$|S_n \backslash\backslash \mathcal{S}[n]| \cdot x^n = C(S_n, n, \chi^{\mathcal{S}}; x, x^2, \ldots) = \sum_i C((S_n)_i, n; x, x^2, \ldots).$$

Summation over all n yields in particular the equation

3.2.20
$$\widetilde{\mathcal{S}}(x) = \sum_n \sum_i C\big((S_n)_i, n; x, x^2, \ldots\big).$$

Here are a few examples that are easy to check:

3.2.21
$$C_{\mathcal{X}} = z_1,\ C_{\mathcal{L}} = \frac{1}{1-z_1},\ C_{Per} = \prod_{i\geq 1} \frac{1}{1-z_i}.$$

$\diamond$

Comparing the cycle series of Per and $\mathcal{L}$ we see, that equipotency does *not* necessarily imply equality of the corresponding cycle series:

3.2.22
$$\mathcal{S} \equiv \mathcal{T} \not\Rightarrow C_{\mathcal{S}} = C_{\mathcal{T}}.$$

We can also generalize weight functions to species. Such a *weight function* is defined to be a mapping w, defined on each set $\mathcal{S}[M]$, into a commutative ring R, containing $\mathbb{Q}$ as subring, and being compatible with transport, in the following sense:

$$w\big(\mathcal{S}[\beta](\sigma)\big) = w(\sigma).$$

This immediately implies that w is constant on the types of $\mathcal{S}$-structures. The pair $(\mathcal{S}, w)$, i. e. the weighted species, will be indicated by $\mathcal{S}_w$. The cardinality of $\mathcal{S}_w$ is defined as

$$\mathcal{S}_w(x) := \sum_{n\geq 0}\Big(\sum_{\sigma\in\mathcal{S}[n]} w(\sigma)\Big)\frac{x^n}{n!},$$

and the type series by

$$\widetilde{\mathcal{S}_w}(x) := \sum_{n\geq 0}\Big(\sum_{t\in T_n} w(t)\Big)x^n$$

(where T_n means a transversal of $S_n \backslash\backslash \mathcal{S}[n]$), while the cycle series is

$$C_{\mathcal{S}_w}(z_1, z_2, \ldots) := \sum_n C\big(S_n, n, \chi_w^{\mathcal{S}}; z_1, z_2, \ldots\big),$$

where

$$\chi_w^{\mathcal{S}} \colon \beta \mapsto \sum_{\sigma\in\mathcal{S}[n]_\beta} w(\sigma).$$

3.2.23 Examples

– The following mapping w into $\mathbb{Q}[y]$ is a weight function on Per :

$$w\colon \beta \mapsto y^{c(\beta)},\ c(\beta) := \sum_i a_i(\beta).$$

The cardinality turns out to be

$$Per_w(x) = \sum_{n\geq 0}\Big(\sum_{a\dashv n}\frac{n!}{\prod_i i^{a_i}a_i!}\cdot y^{\Sigma_i a_i}\Big)\cdot\frac{x^n}{n!},$$

while the type series is

$$\widetilde{Per_w}(x) = \sum_{n\geq 0}\Big(\sum_{a\dashv n} y^{\Sigma_i a_i}\Big)\cdot x^n = \sum_n [y]^n x^n,$$

if $[y]^n := y(y+1)\cdots(y+n-1)$ is the rising factorial (exercise 3.2.7).

– A second weight function on Per (it generalizes the former) is

$$v\colon \beta \mapsto y_1^{a_1(\beta)} y_2^{a_2(\beta)} \cdots \in \mathbb{Q}[y_1, y_2, \ldots].$$

Here we obtain

$$Per_v(x) = \sum_{n\geq 0}\Big(\sum_{a\dashv n}\frac{n!}{\prod_i i^{a_i}a_i!}\cdot\prod_i y_i^{a_i}\Big)\cdot\frac{x^n}{n!},$$

and

$$\widetilde{Per_w}(x) = \sum_{n\geq 0}\Big(\sum_{a\dashv n}\prod_i y_i^{a_i}\Big)\cdot x^n = \sum_{n\geq 0}\prod_i\frac{1}{1-x^i y_i}.$$

◇

Exercises

Exercise 3.2.1 Check the equation 3.2.5.

Exercise 3.2.2 Verify the details of 11.6.1.

Exercise 3.2.3 Evaluate the cycle indicator polynomial of the action of the group $C_p \times C_p \times C_p$ on itself by left multiplication. Evaluate also the cycle indicator polynomial of the action of the following nonabelian group of order p^3, p being an odd prime, on itself by left multiplication:

$$G := \{(a, b) \mid a \in C_p \times C_p, b \in C_p\},$$

where (recall the definition of semidirect product, given above)

$$\hat{b}(a) := (x, xy), \text{ if } a = (x, y).$$

Exercise 3.2.4 Express the cycle indicator $C(S_n, n)$ in terms of the polynomials $C(S_k, k)$, $1 \leq k \leq n$.

Exercise 3.2.5 Prove the equation on the exterior cycle index, given in 3.2.12.

Exercise 3.2.6 Evaluate the characters of the natural actions of S_X on X^n and on X^n_{inj}.

Exercise 3.2.7 Prove that

$$\sum_{a\dashv n} y^{\Sigma_i a_i} = [y]^n.$$

3.3 Sums of Cycle Indicators, Recursive Methods

If A_X denotes the alternating group on the finite set X then, by 3.1.6, the coefficient of the monomial $\prod_y y^{c(f,y)}$ in $C(A_X, X \mid \sum y)$ is equal to the number of A_X–classes of mappings with the same content as f. If we exclude the trivial case $|X| = 1$, then the A_X–class of f differs from its S_X–class if and only if its stabilizers $(S_X)_f$ and $(A_X)_f$ are equal, and if these classes differ, then the S_X–class of f consists of *two* A_X–classes.

The stabilizer of $f \in Y^X$ in S_X is the following direct sum of the symmetric groups on the inverse images of the $y \in Y$:

3.3.1
$$(S_X)_f = \bigoplus_y S_{X_y}, \text{ where } X_y := f^{-1}(y).$$

The stabilizer of f in A_X is the intersection of $(S_X)_f$ with A_X. As each noninjective f is left fixed by a transposition while each injective f has the identity subgroup as its stabilizer, we obtain

3.3.2 Corollary *The stabilizers of $f \in Y^X$ in S_X and in A_X are the same, or, equivalently, they both are equal to the identity subgroup, if and only if f is injective.*

Thus the difference of cycle indicators of alternating and symmetric groups has the following useful interpretation (recall figure 2.1):

3.3.3 Corollary *For $|X| > 1$ the polynomial*

$$C(A_X, X \mid \sum y) - C(S_X, X \mid \sum y)$$

is the generating function for the enumeration of the injective S_X–classes on Y^X by weight.

Corresponding results hold for cyclic and dihedral groups. Also these arguments can be considered as a certain kind of involution principle. In fact we obtain in the same way the following generalization of 3.3.3 (recall 2.1.15):

3.3.4 Corollary *If $_GX$ denotes a chiral action, then*

$$C(G^+, X \mid \sum y) - C(G, X \mid \sum y)$$

is the generating function for the enumeration by weight of the G–classes on Y^X which split over G^+.

Besides these differences of cycle indicator polynomials we can form sums of cycle indicator polynomials of series of groups, e. g. we obtain (by simply comparing coefficients) the equation

3.3.5
$$\sum_{n=0}^{\infty} C(S_n, n) = \exp \sum_{k=1}^{\infty} \frac{z_k}{k} \in \mathbb{Q}[\![z_1, z_2, \ldots]\!],$$

if $C(S_0, 0 \mid p(u)) := 1$ and where $\mathbb{Q}[[z_1, z_2, \ldots]]$ denotes the ring of formal power series over $\mathbb{Q}$ in the indeterminates $z_1, z_2, \ldots$. An immediate consequence is

3.3.6 $$\sum_{n=0}^{\infty} C(S_n, n \mid p(u)) = \exp \sum_{k=1}^{\infty} \frac{1}{k} p(u^k) \in \mathbb{Q}[[u]],$$

This sum is of great interest since it can be used for a recursion method which we are going to describe next.

3.3.7 Application to graph theory (trees) A graph has been called a *tree* if and only if it is connected and does not contain any cycle. A tree with a single distinguished vertex is called a *rooted* tree. Figure 3.2 shows the smallest rooted trees. The root is distinguished by indicating it as a circle while the other vertices are indicated by a bullet.

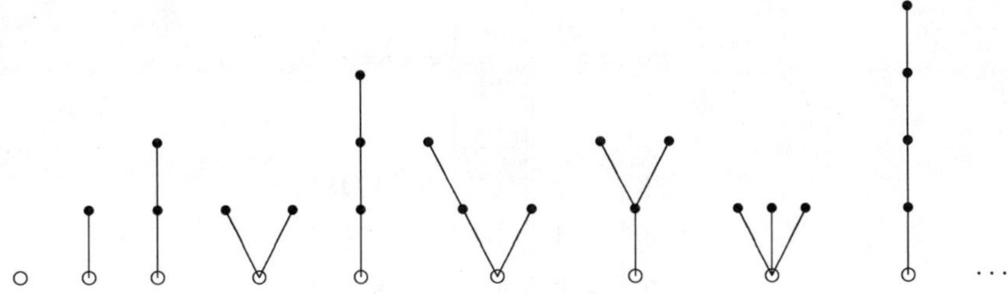

Fig. 3.2. The smallest rooted trees

According to figure 3.2, the generating function for the enumeration of rooted trees by their number of vertices is

$$\widetilde{\mathcal{T}^r}(x) = x + x^2 + 2x^3 + 4x^4 + \ldots.$$

We claim that this formal power series satisfies a recursion formula in terms of cycle indicator polynomials of symmetric groups.

3.3.8 $$\widetilde{\mathcal{T}^r}(x) = x \sum_{n=0}^{\infty} C\big(S_n, n \mid \widetilde{\mathcal{T}^r}(x)\big).$$

Proof: It is clear that the generating function for the enumeration of rooted trees is equal to the sum over all $n \in \mathbb{N}$ of the generating functions of rooted trees with root degree n, i. e. where the root is incident with exactly n edges. It therefore remains to evaluate the generating function for the enumeration of trees with root degree n.

Denote by $\mathcal{R}$ the set of all the rooted trees. A rooted tree with root degree n can be considered as an orbit of S_n on the set $\mathcal{R}^n$. If we now give an element of $\mathcal{R}$ the weight x^v, where v denotes the number of vertices of the rooted tree in question, then we obtain the power series $xC(S_n, n \mid \widetilde{\mathcal{T}^r}(x))$ which therefore is the desired generating function for the enumeration of rooted trees with root degree n by their number of vertices. This completes the proof. □

An application of the above exponential expressions 3.3.5 and 3.3.6 for the sum of the cycle indicator polynomials of finite symmetric groups as an exponential formal power series allows to rewrite the recursion formula for the generating function of the rooted trees in the following way:

3.3.9 Corollary *The formal power series $r(x)$, which generates the numbers of rooted trees by number of vertices, satisfies the recursion formula*

$$r(x) = x \cdot \exp \sum_{k=1}^{\infty} \frac{r(x^k)}{k}.$$

This recursion together with a suitable program system, like MAPLE or MACSYMA, allows to evaluate the smallest numbers of rooted trees, some of which are shown in the following table:

n	$\widetilde{t}^r_n$	n	$\widetilde{t}^r_n$
1	1	11	1842
2	1	12	4766
3	2	13	12486
4	4	14	32973
5	9	15	87811
6	20	16	235381
7	48	17	634847
8	115	18	1721159
9	286	19	4688676
10	719	20	12826228

◇

Similar arguments apply in each case when the structure in question consists of a certain and well defined number of structures of the same kind. There are various approaches, the books are full of them, for example the book of Goulden/Jackson ([61]) gives an approach that covers very many cases in a different and more elementary way compared to the theory of species which was briefly indicated above.

3.3.10 Application to graph theory (connected graphs) Each graph is a disjoint union of *connected* graphs. Recall from above that a graph is called *connected* if and only if from each of its vertices we can reach any other vertex by walking along suitably chosen edges. These connected subgraphs are well defined, and are called the *connected components*. Thus the generating function $\widetilde{\mathcal{G}}(x) = \sum_n \widetilde{g}_n x^n$ of the graphs and the generating function $\widetilde{\mathcal{G}}^c(x) = \sum_n \widetilde{g}^c_n x^n$ for the connected graphs are related as follows:

3.3.11
$$\widetilde{\mathcal{G}}(x) = \sum_{n=0}^{\infty} C(S_n, n \mid \widetilde{\mathcal{G}}^c(x)) = \exp \sum_{k=1}^{\infty} \frac{\widetilde{\mathcal{G}}^c(x^k)}{k}.$$

As we already know (see 2.3.4) that

$$\widetilde{g}_n = \frac{1}{n!} \sum_{\pi \in S_n} 2^{c(\bar{\pi})},$$

we can use 3.3.11 in order to evaluate the entries of the following table:

n	$\widetilde{g}_n$	$\widetilde{g}_n^c$
1	1	1
2	2	1
3	4	2
4	11	6
5	34	21
6	156	112
7	1044	853
8	12346	11117
9	274668	261080
10	12005168	11716571

◇

3.4 Applications to Chemistry

Further interesting examples of actions and of symmetry classes show up in the sciences. It was already briefly mentioned that even the origins of this theory of enumeration lie in chemistry and the problem of isomerism. A few remarks concerning the history are therefore in order. It was already in the 18th century (cf. [100]) when Alexander von Humboldt (a German who later on became very famous for his scientific expeditions to South-America) conjectured that there might be chemical substances which are composed by the same set of atoms but have different properties. In his book with the title "Versuche über die gereizte Muskel- und Nervenfaser, nebst Vermutungen über den chemischen Prozeß in der Tier- und Pflanzenwelt", published in 1797, he writes on page 128 of volume I of [71]:

> Drei Körper a, b und c können aus *gleichen* Quantitäten Sauerstoff, Wasserstoff, Kohlenstoff, Stickstoff und Metall zusammengesetzt und in ihrer Natur doch unendlich *verschieden* seyn.

But it needed a quarter of a century to develop the analytical methods that allowed to find out what the atomic constituents of a chemical molecule are and to realize that in fact there are different molecules with the same atomic constituents. These methods were developed in particular by J. L. Gay–Lussac (a close friend of Humboldt) and

J. von Liebig (whom Humboldt recommended for a chair for chemistry in Gießen), who were the first to prove von Humboldt's conjecture to be true. Here is a sentence taken from a paper of Gay–Lussac that describes their discovery:

> comme ces deux acides son très différents, il faudrait pour expliquer leur différence admettre entre leurs éléments un mode de combinaison différent. C'est un objet qui appelle un nouveau examen.

After a while it was accepted that a new phenomenon had been discovered, Berzelius gave it the name *isomerism*. Chemists tried to find out what the reason is by sketching molecules. Here are three of these attempts to draw the molecule of C_2H_5OH: This one is due to Kekulé:

The following vizualization is due to Couper:

$$\begin{array}{ll} C & \left\{ \begin{array}{l} O \cdots OH \\ H_2 \end{array} \right. \\ \vdots & \\ C & \quad \cdots H_3 \end{array}$$

The next one is the way how Loschmidt drew this molecule:

The breakthrough towards a solution of this problem is due to Couper, Loschmidt and to Alexander Crum Brown who made the essentials (which are the touching points) even clearer. Alexander Crum Brown used a variation of Loschmidt's method, putting the touching circles of atoms apart, replacing the touching points by edges (which were already indicated by Couper) in order to emphasize the connection, the *covalent bonds* as they are called today. Here are his drawings for the alcohol C_3H_7OH:

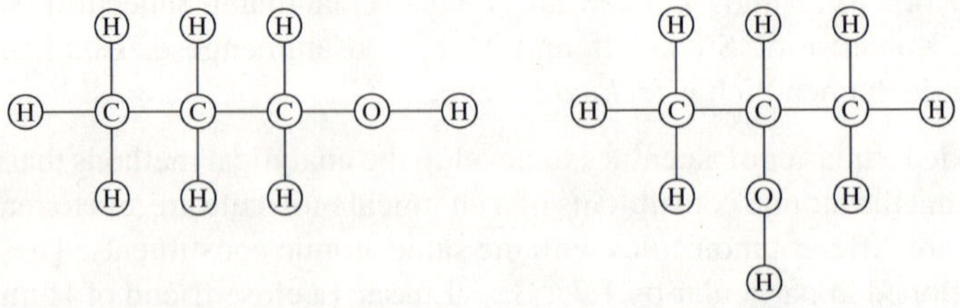

This introduction of *molecular graphs* solved the problem of isomerism by showing that there may exist different graphs that correspond to a given chemical formula (which prescribes in fact the degree sequence, which is the sequence of valencies of the atoms in the given formula). Moreover it gave rise to the development of graph theory (there are of course also other birthdays of graph theory known, for example Euler's solution of the Königsberg bridge problem, and Kirchhoff's description of electric circuits as well as Hamiltons game called "trip around the world"). It stimulated the *combinatorial theory of enumeration* since the question of chemical isomerism is at first glance equivalent to the problem of *constructing all the connected multigraphs without loops and which have a given edge degree sequence.* At second glance it is more complicated since the vertices of the graphs carry atom names, and so things become considerably more complicated if several atoms with the same valence occur.

Even the *name* graph comes from chemistry. It was introduced by J. J. Sylvester in a publication entitled *Chemistry and Algebra,* which appeared in *Nature* **17** (1877-1878), where he connects algebraic invariants and chemical molecules, and where he writes (p. 284):

> Every invariant and covariant thus becomes expressible by a *graph* precisely identical with a Kekuléan diagram or chemicograph.

Let us consider, for example, the chemical formula C_3H_7OH again, or, more generally the formula $C_nH_{2n+1}OH$. The problem is to construct all the graphs which correspond to each of these formulae, i. e. all the graphs that consist of n vertices of degree 4 (the carbon atoms are of that valency) together with $2n+2$ vertices of degree 1 (the hydrogen) and one of degree 2 (the oxygen) which must have a neighbour of degree 1 (so that a hydroxyl group $-OH$ occurs). This problem amounts to the construction of rooted trees, as, by a well known result of graph theory, such degree sequences can be satisfied by trees only (the cyclomatic number is zero in this case). The root represents the substructure $\equiv C-O-H$, so that the root degree is ≤ 3. Consider the two examples shown above which correspond to the formula C_3H_7OH. As the carbon atoms are of valency 4, we can make life easier by neglecting the hydrogen atoms, so that the *skeletons* remain, which correspond to the rooted trees with 3 vertices. From each one of these structures we can reconstruct the original molecular graph and hence the desired generating function $a(x)$ for the numbers of alcohols satisfies the recursion

3.4.1 $$a(x) = x\sum_{n=0}^{3} C(S_n, n \mid a(x)).$$

The construction of all the molecular graphs corresponding to a given chemical formula is an enourmously complex problem. As we already mentioned, a simple formula like that of benzene corresponds to 217 graphs already. Another example is $C_8H_{16}O_2$, say, it has 13,190 isomers, and it is clearly easy to give chemical formulae that have billions of corresponding molecular graphs. Here is a table of numbers of isomers with the chemical formula C_jH_i,

i/j	1	2	3	4	5	6	7	8	9	10	11
0			1	3	6	19	50	204	832	4330	25227
2		1	2	7	21	85	356	1804	10064	64352	455822
4	1	1	3	11	40	185	920	5308	33860	241297	1885531
6		1	2	9	40	217	1230	7982	56437	439373	3717018
8			1	5	26	159	1031	7437	57771	488125	4442438
10				2	10	77	575	4679	40139	369067	3614427
12					3	25	222	2082	19983	201578	2135717
14						5	56	654	7244	81909	950064
16							9	139	1902	24938	323512
18								18	338	5568	84051
20									35	852	16216
22										75	2145
24											159

(It should be mentioned here that, of course, there are restrictions known that come from stability considerations. In the case of the benzene, for example, you can use a *permanent badlist,* that cuts the number of isomers down to approximately 110 from 217 isomers without any forbidden substructure.)

It is therefore crucial to reduce the chemical formula with the aid of prescribed substructures. As a simple and famous example, we will take the isomers of dioxin. The chemical formula is $C_{12}O_2Cl_4H_4$. And dioxin means that there is a substructure of the following form:

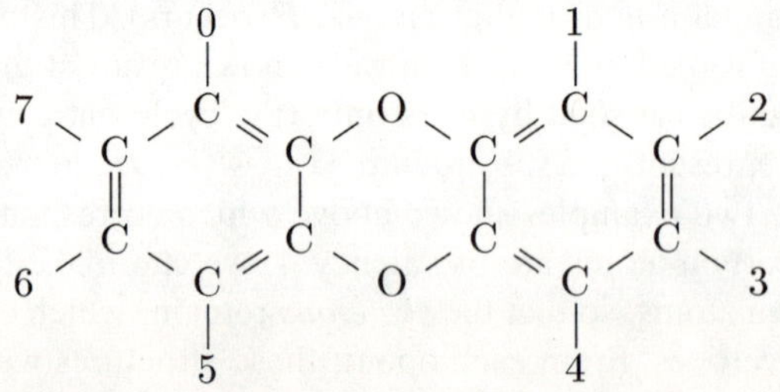

```
          0                         1
          |                         |
 7        C         O         C         2
   \    /   \\    /   \    //   \    /
     C        C         C         C
     ||       |         |         ||
     C        C         C         C
   /    \   //   \    /   \\    /   \
 6        C         O         C         3
          |                   |
          5                   4
```

where we have numbered the free valences from 0 to 7. The problem is to attach 4 hydrogen and 4 chlorine atoms in all the essentially different ways over these eight free places. In other words, we have to evaluate a transversal of the orbits of the symmetry group G of the skeleton on the set of mappings

$$Y^X := \{H, Cl\}^8.$$

The skeleton is supposed to be planar. Hence the symmetry group is the Kleinian four group V_4, the direct product of two reflections:

$$V_4 = \langle (05)(14)(23)(67), (01)(27)(36)(45) \rangle.$$

Hence the set of molecules which can be obtained by attaching hydrogen or chlorine molecules to the skeleton of the dioxin is bijective to the set of orbits

$$V_4 \backslash\backslash \{H, Cl\}^8.$$

The weighted enumeration of these molecules has the generating function

$$C(V_4, 8 \mid 1 + y) = \frac{1}{4}(z_1^8 + 3z_2^4)\Big|_{z_i := 1+y^i} = \frac{1}{4}((1 + y)^8 + 3(1 + y^2)^4)$$

$$= 1 + 2y + 10y^2 + 14y^3 + 22y^4 + 14y^5 + 10y^6 + 2y^7 + y^8.$$

The middle term shows that there are exactly 22 isomers of dioxin.

In case we want the isomers of dioxin themselves, then, according to 1.2.12, we restrict attention to the set of molecules that contain exactly 4 hydrogen and exactly 4 chlorine atoms, i. e. we restrict attention to the orbits of weight (4, 4). We already know that they are bijective to the set of double cosets

$$V_4 \backslash S_8 / S_4 \oplus S_4.$$

Down below we shall describe methods that allow systematically to evaluate transversals of such sets of double cosets. In the present case we can do this by hand. There are exactly 120 left cosets of $S_4 \oplus S_4$ in S_8 which we can group into double cosets with respect to the Kleinian four group, and then retransform the representatives in to mappings, i. e. into molecules. Two of them are here:

```
              Cl                    Cl
              |                     |
    H         C           O         C           H
      \     /   \\      /   \     //  \       /
        C           C           C         C
        ||          |           |         ||
        C           C           C         C
      /   \      //   \      /    \\    /   \
    H         C           O         C          H
              |                     |
              Cl                    Cl

              H                     Cl
              |                     |
    Cl        C           O         C           H
      \     /   \\      /   \     //  \       /
        C           C           C         C
        ||          |           |         ||
        C           C           C         C
      /   \      //   \      /    \\    /   \
    H         C           O         C          H
              |                     |
              Cl                    Cl
```

The very same methods can be applied in *combinatorial chemistry*, a new and promising technology for mass synthesis. We should like to discuss a prominent example introduced in [31],[32]. The method used for the construction of a *library* of molecules is to start off from a *central molecule,* for example from a benzene triacid chloride, which looks as follows:

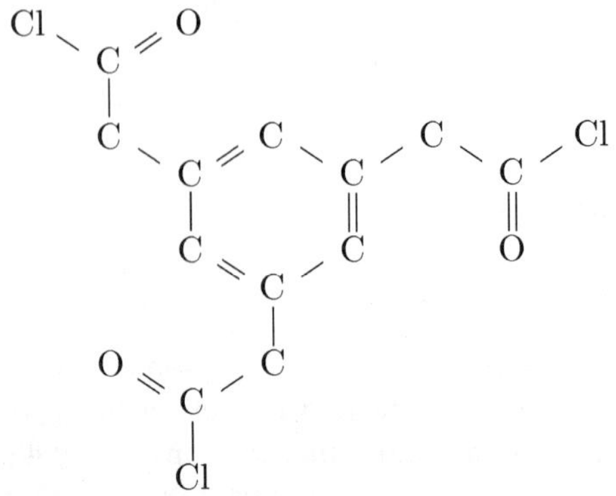

```
        Cl      O
          \   //
            C
            |
            C           C           C         Cl
              \       //  \       /   \     /
                C           C           C
                |           ||          ||
                C           C           O
                  \\      /
                      C
                      |
        O             C
          \\        /
              C
              |
              Cl
```

This central molecule is supposed to react with *amino acids,* i. e. with certain molecules containing a substructure of the form

```
          N     O
          |     ||
    * -   C  -  C  -  O
```

(The star $*$ stands for the rest of the amino acid.) This substructure reacts with the *active sites* of the central molecule, i. e. with its substructures

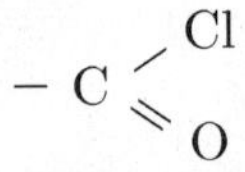

```
            Cl
          /
    - C
          \\
            O
```

The reaction is as follows (in graph theoretical terms): the carbon atom of the active site couples with the nitrogene atom of the given substructure of the amino acid while the chlorine atom of the active site decouples. After that reaction, the active site of the cubane, xanthene or triacid now becomes

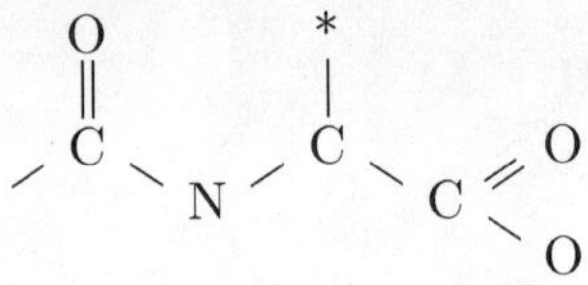

```
        O           *
        ||          |
        C           C           O
      /   \       /   \       //
              N           C
                            \
                              O
```

The desired combinatorial library can be identified with the set of orbits of the subgroup of proper rotations in the geometric symmetry group G of the central molecule on the set of mappings Y^X, where X means the set of active sites of the central molecule, while Y means the set of admissible amino acids. For example, in the case of the triacid, the group of proper rotations is the full symmetric group S_3. This is true, since the molecule is a three-dimensional entity, and so the symmetry operations which induce reflections of the plane in which we draw the molecule, are in fact due to rotations in space. Thus the combinatorial library can be identified with the set of orbits

$$S_3 \backslash\backslash Y^3,$$

the order of which is

$$\frac{1}{6}\left(|Y|^3 + 3|Y|^2 + 2|Y|\right),$$

while the generating function is equal to

$$C(S_3, 3 \mid \sum_y y) = \frac{1}{6}\Big(\big(\sum_y y\big)^3 + 3\big(\sum_y y\big)\big(\sum_y y^2\big) + 2\big(\sum_y y^3\big)\Big).$$

For example, if we allow up to three different amino acids, we obtain the following generating function for the enumeration of the elements in the library by weight:

$$y_0^3 + y_0y_1^2 + y_0y_1y_2 + y_0y_2^2 + y_0^2y_1 + y_0^2y_2 + y_1^3 + y_1^2y_2 + y_1y_2^2 + y_2^3.$$

The fact that each weight occurs exactly once is, of course, directly clear from the symmetry. A slightly less trivial example can be found in the exercise.

Exercises

Exercise 3.4.1 Evaluate the order of combinatorial libraries arising from the central molecule xanthene by reactions with amino acids (see the example with the central triacid, and give the formula in terms of the number $|Y|$ of admissible amino acids)

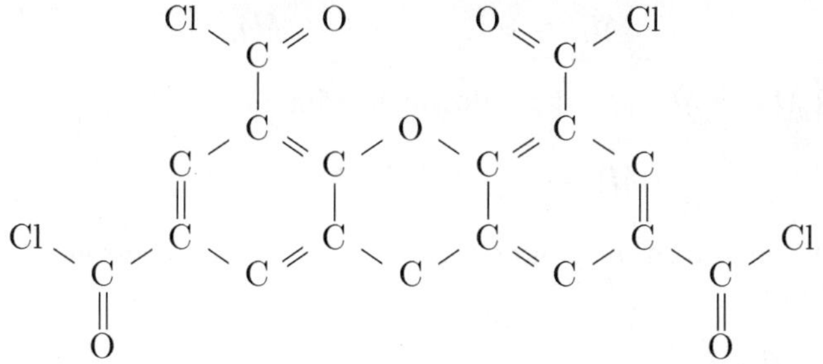

```
                Cl     O           O     Cl
                  \   //            \\   /
                    C                  C
                    |                  |
                    C        O         C
                 /    \\   /   \    //   \
               C        C         C        C
               ||       |         |        ||
     Cl        C        C         C        C        Cl
       \     /   \    //  \     /   \\   /   \     /
         C          C        C         C         C
         ||                                      ||
         O                                       O
```

Derive the generating function for the case when up to two different amino acids are allowed.

3.5 A Generalization

Ten years before G. Pólya published his pioneering work, a paper by J. H. Redfield appeared concerning superpositions of graphs. But it was overlooked for many years (a second paper which went even further was rejected and printed only recently after it had been discovered among his mathematical papers). It is difficult to read, but it already contains the main results on enumeration by weight, moreover it presents a more general approach which we are going to describe next. It will not be necessary to give a formal definition of *superposition*, since the reader will clearly see what is meant from an example, and we shall examine a more general situation later on anyway. Two graphs on 5 vertices together with one (out of 4) superpositions of these graphs are shown below. The edges of one of the two superimposed graphs are dotted.

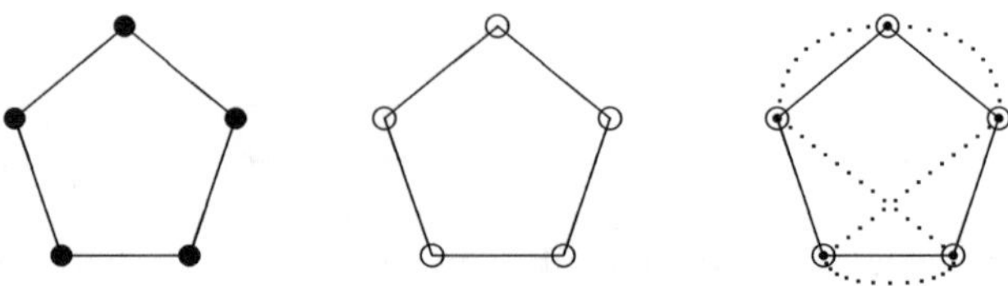

Now we introduce an enumeration problem, a special case of which is the enumeration of superpositions of graphs. Consider, for given $m, n \in \mathbb{N}^*$, the set M of all the $n \times m$–matrices $A = (a_{ik})$ which contain in each row all the elements of m, i. e. each row of $A \in M$ is an injective m–tuple over m. Now we establish on M an equivalence relation $\sim_c$ by saying that $A, A' \in M$ are *column equivalent* if A' arises from A by applying a suitable column permutation σ:

$$(a_{ik}) \sim_c (a'_{ik}) :\Longleftrightarrow \exists\, \sigma \in S_m \; \forall\, i, k \colon\; a'_{ik} = a_{i,\sigma^{-1}k}.$$

The equivalence class of (a_{ik}) will be denoted by $(a_{ik})_c$, and we put

$$M_c := \{(a_{ik})_c \mid (a_{ik}) \in M\}.$$

It is clear that M and M_c have the following cardinalities:

3.5.1 $$|M| = m!^n, \text{ and } |M_c| = m!^{n-1}.$$

Assume that we are given subgroups $H_0, \ldots, H_{n-1} \leq S_m$. Their direct product acts on M_c as follows:

3.5.2 $$(\times_i H_i) \times M_c \to M_c \colon ((\pi_0, \ldots, \pi_{n-1}), (a_{ik})_c) \mapsto (\pi_i a_{ik})_c,$$

and we ask for the number of orbits. One of the most important results of Redfield's paper is the proof of the fact that the number of superpositions of n graphs with m vertices and automorphism groups $H_0, \ldots, H_{n-1}$ is

3.5.3 $$|(\times_i H_i) \backslash\!\backslash M_c|.$$

This result is mostly called the *Redfield-Read Superposition Theorem.* Since we are not particularly interested in this number of graphs, we go further and formulate and solve a counting problem which generalizes both Redfield's superposition problem and the enumeration of symmetry classes of mappings. To begin with we note that $f \in m^n$ can be identified with the one column matrix

$$\begin{pmatrix} f(0) \\ \vdots \\ f(n-1) \end{pmatrix}.$$

In addition we recall the action 1.3.2 of the exponentiation group. For subgroups $H \leq S_m$, $G \leq S_n$, the image $\tilde{f} := (\psi, \pi) f$, for (ψ, π) in $H \wr G$, satisfies

$$\begin{pmatrix} \tilde{f}(0) \\ \vdots \\ \tilde{f}(n-1) \end{pmatrix} = \begin{pmatrix} \psi(0) f(\pi^{-1}0) \\ \vdots \\ \psi(n-1) f(\pi^{-1}(n-1)) \end{pmatrix}.$$

We compare this with Redfield's approach and seek for a generalization which covers this and the former approach. We are led to the following *Ansatz*. Consider, instead of $n \times m$–matrices, the $n \times s$–matrices A, for a fixed $s \in \underline{m}$, where each row of A is an injective s–tuple over m:

$$M^{<s>} := \{A \in m^{n \times s} \mid \text{each row lies in } m^s_{inj}\}.$$

The order of this set is obviously

$$|M^{<s>}| = \left(\binom{m}{s} s!\right)^n = ([m]_s)^n \,, \, [m]_s := m(m-1)\cdots(m-s+1).$$

In order to introduce a suitable group which acts on $M^{<s>}$, we define the following *generalized wreath product.* Assume that $G \leq S_n$ has the orbits $\omega_0, \ldots, \omega_{r-1} \subseteq n$. We put, for $H_i \leq S_m, i \in r$,

$$(\times_{i \in r} H_i) \wr G := \{(\psi, \pi) \mid \psi \in (\cup_i H_i)^n, \pi \in G, i \in \omega_j \Rightarrow \psi(i) \in H_j\}.$$

The multiplication is still defined by $(\psi, \pi)(\varphi, \rho) := (\psi\varphi_\pi, \pi\rho)$, and this group acts on $M^{<s>}$ in the following way:

3.5.4 $$(\times_i H_i) \wr G \times M^{<s>} \to M^{<s>} : ((\psi, \pi), (a_{ik})) \mapsto (\psi(i) a_{\pi^{-1}i,k}).$$

As we have just seen, the particular case $s := 1$ yields the exponentiation group action on m^n. In order to cover also Redfield's column equivalence relation we furthermore introduce the corresponding action of S_s:

3.5.5 $$S_s \times M^{<s>} \to M^{<s>} : (\sigma, (a_{ik})) \mapsto (a_{i,\sigma^{-1}k}),$$

and note that these two actions commute:

3.5.6 $$(\psi, \pi)\sigma A = \sigma(\psi, \pi) A.$$

Hence we have in fact obtained an action of the direct product:

$$((\times_i H_i) \wr G \times S_s) \times M^{<s>} \to M^{<s>},$$

defined by

$$(((\psi, \pi), \sigma), (a_{ik})) \mapsto (\psi(i) a_{\pi^{-1}i, \sigma^{-1}k}).$$

The character of this action is given in

3.5.7 Lemma

$$a_1((\psi,\pi),\sigma) = \prod_{\nu\in c(\pi)} \prod_{k\in \underline{s}} \binom{a_k(h_\nu(\psi,\pi))}{a_k(\sigma^{l_\nu})} a_k(\sigma^{l_\nu})! \cdot k^{a_k(\sigma^{l_\nu})}.$$

Proof: We start by considering the particular case $n = 1$, in which the set $M^{<s>}$ is just the set of injective s–tuples over m, while the product $((\times_i H_i) \wr G) \times S_s$ is isomorphic to the direct product $H_0 \times S_s$. Hence, for this case, we obtain the assertion from 2.4.11. For the general case $n \geq 1$ we remark that A is a fixed point of $((\psi,\pi),\sigma)$ if and only if its elements satisfy the equations

$$a_{ik} = \psi(i) a_{\pi^{-1}i,\sigma^{-1}k}.$$

Hence, for each $t \in \mathbb{N}^*$, the following is true:

3.5.8 $$a_{ik} = \psi\psi_\pi \ldots \psi_{\pi^{t-1}}(i) a_{\pi^{-t}i,\sigma^{-t}k}.$$

Now we consider a fixed $j \in n$, and l, the length of its cyclic factor in π, which satisfy:

3.5.9 $$\forall\, k \in s\colon\ a_{jk} = \psi\psi_\pi \ldots \psi_{\pi^{l-1}}(j) a_{j,\sigma^{-l}k}.$$

Equation 3.5.8 means that the rows of index i lying in the cycle of j, are completely determined by the entries of the j-th row. Furthermore the j-th row of A, considered as an element of $m^{\underline{s}}_{inj}$, must be a fixed point of

$$(\psi\psi_\pi \ldots \psi_{\pi^{l-1}}(j), \sigma^l),$$

by 3.5.9. In order to evaluate the number of fixed points of $((\psi,\pi),\sigma)$, we consider the standard cycle notation for π, say

$$\pi = \prod_{\nu\in c(\pi)} (j_\nu \ldots \pi^{l_\nu - 1}(j_\nu)).$$

For each one of its cyclic factors $(j_\nu \ldots \pi^{l_\nu-1}(j_\nu))$ we can freely choose an injective s–tuple over m among the fixed points of $(h_\nu(\psi,\pi),\sigma^{l_\nu})$. If we now form the product over all the cycles of π, taking the number of fixed points of $(h_\nu(\psi,\pi),\sigma^{l_\nu})$ as the factor corresponding to the ν–th cyclic factor, then we obtain the desired number, and this proves the statement, as can be seen from the foregoing argument in the case $n = 1$. □

In order to derive Redfield's results from this more general one, we need another tool:

3.5.10 de Bruijn's Lemma *Assume that X is both a G– and an H–set, and that we have*

$$\forall\, g \in G, h \in H\ \exists\, h' \in H\ \forall\, x\colon\ g(hx) = h'(gx).$$

Then G acts on $H \backslash\backslash X$ as follows:

$$G \times (H \backslash\!\backslash X) \to H \backslash\!\backslash X \colon (g, H(x)) \mapsto H(gx),$$

and the character of this action is

$$a_1(g) := |(H \backslash\!\backslash X)_g| = \frac{1}{|H|} \sum_{h \in H} |X_{gh}| =: \frac{1}{|H|} \sum_{h \in H} a_1(gh).$$

Proof: It is easy to check that $(g, H(x)) \mapsto H(gx)$ is an action. Furthermore we have that

$$\begin{aligned}
\sum_{h \in H} |X_{gh}| &= \sum_{h \in H} \sum_{x \in X_{gh}} 1 \\
&= \sum_{x\colon gH(x)=H(x)} \sum_{h\colon hx=g^{-1}x} 1 \\
&= \sum_{x\colon gH(x)=H(x)} |H_x| \\
&= |H| \sum_{x\colon gH(x)=H(x)} |H(x)|^{-1} \\
&= |H| \sum_{\omega \in H \backslash\!\backslash X\colon g\omega=\omega} 1 \\
&= |H||(H \backslash\!\backslash X)_g|.
\end{aligned}$$

□

In order to derive from this the character of the action 3.5.2 we use 3.5.7 and 3.5.10 which give

3.5.11 $$a_1\big((\pi_0, \ldots, \pi_{n-1})\big) = \frac{1}{m!} \sum_{\sigma \in S_m} \prod_{\nu \in n} \prod_{k \in \underline{m}} \binom{a_k(\pi_\nu)}{a_k(\sigma)} a_k(\sigma)!\, k^{a_k(\sigma)}.$$

Now we observe that the summand on the right hand side is nonzero only if all the π_ν are of the same cycle type as σ, in which case the double product takes the value

$$\Big(\prod_{k \in \underline{m}} a_k(\sigma)! k^{a_k(\sigma)}\Big)^n.$$

Applying this to 3.5.11 we get

$$a_1((\pi_0, \ldots, \pi_{n-1})) = \begin{cases} \left(\prod_k a_k! k^{a_k}\right)^{n-1}, & \text{if each } a(\pi_\nu) = (a_1, \ldots, a_n), \\ 0, & \text{otherwise.} \end{cases}$$

This result allows to express the number of orbits of $\times H_i$ on M_c in terms of the cycle indicator polynomials of the H_i. We introduce Redfield's *cap operation* $\cap$ on the set $\mathbb{Q}[Z]$ of polynomials over $\mathbb{Q}$ in the set $Z := \{z_0, \ldots, z_{m-1}\}$ of indeterminates. It is defined to be the linear extension of the following operation on monomials. Using the abbreviation $Z^b := z_0^{b_0} \cdots z_{m-1}^{b_{m-1}}$, for $b := (b_0, \ldots, b_{m-1})$, where $b_i \in \mathbb{N}$, we put:

$$\underbrace{Z^c \cap Z^d \cap \ldots \cap Z^e}_{n \text{ terms}} = \begin{cases} \left(\prod_k a_k! k^{a_k}\right)^{n-1}, & c_i = d_i = \ldots = e_i = a_i, \\ 0, & \text{otherwise.} \end{cases}$$

This together with 3.5.11 yields

3.5.12 Redfield's Counting Theorem *The number of orbits of the direct product $\times_i H_i$ on the set M_c of equivalence classes of matrices is equal to the cap product of the corresponding cycle indicator polynomials:*

$$|(\times_{i \in n} H_i) \backslash\!\backslash M_c| = \bigcap_{i \in n} C(H_i, m).$$

In order to provide an example we consider the dihedral group on 5 points and put $H_0 := H_1 := D_5$. This group D_5 is the automorphism group of a regular 5-gon. According to 3.2.3, the cycle indicator polynomial of this group is $C(D_5, 5) = \frac{1}{10}(z_0^5 + 4z_4 + 5z_0z_1^2)$. Hence, applying Redfield's Counting Theorem, we obtain for the number of superpositions of two such graphs:

$$\begin{aligned} C(D_5, 5) \cap C(D_5, 5) &= \frac{1}{10}(z_0^5 + 4z_4 + 5z_0z_1^2) \cap \frac{1}{10}(z_0^5 + 4z_4 + 5z_0z_1^2) \\ &= \frac{1}{100}(z_0^5 \cap z_0^5 + 16(z_4 \cap z_4) + 25(z_0z_1^2 \cap z_0z_1^2)) \\ &= \frac{1}{100}(5! + 16 \cdot 5 + 25 \cdot 8) = 4, \end{aligned}$$

which means that there are exactly 4 superpositions of two such graphs (see exercise 3.5.1). We can easily generalize this example since we know the cycle structure of the elements in dihedral groups (see 3.2.3). For example, if p denotes an odd prime, then the cap product of two cycle indicator polynomials of D_p is equal to

$$\frac{1}{4p^2}\left(p! + (p-1)^2 \cdot p + p^2((p-1)/2)! 2^{(p-1)/2}\right).$$

An immediate consequence is

3.5.13 Corollary *Each prime $p > 3$ satifies the congruence*

$$(p-1)! + p(p-2) + 1 \equiv 0 \ (4p).$$

Exercises

Exercise 3.5.1 Draw the four different superpositions of the two pentagonal graphs shown above.

Exercise 3.5.2 Derive from 3.5.7 that $|H \wr G \times S_s \backslash\backslash M^{<s>}|$ is equal to

$$\frac{1}{s!}\sum_{\pi\in S_s} C\big(G, n; C(H,m)\cap Z^{a(\pi)},\ldots,C(H,m)\cap Z^{a(\pi^m)}\big).$$

3.6 The Decomposition Theorem

Redfield's theorem 3.5.12 allows to express the solutions of many enumeration problems as cap products of cycle indicator polynomials. We consider a few examples.

3.6.1 Lemma *The number of orbits of H^G on n^n_{inj}, where $H, G \leq S_n$, is equal to the cap–product*

$$C(H,n)\cap C(G,n).$$

Proof: The set of bijections n^n_{inj} can be identified with the set M_c of classes of column equivalent $2\times n$–matrices containing injective n–tuples over n as rows, so that Redfield's theorem yields the statement. □

Noticing that bijections are permutations, we can reformulate 3.6.1 as follows:

3.6.2 Corollary *The cap product $C(H,n)\cap C(G,n)$ is equal to the number of classes of elements $\sigma\in S_n$ with respect to the following equivalence relation:*

$$\sigma_1\sim\sigma_2 \iff \exists\,\pi\in H, \rho\in G\colon \sigma_1=\pi\sigma_2\rho^{-1}.$$

In other words, this cap product is equal to the number of (H,G)–double cosets in S_n. More explicitly, in terms of intersections of conjugacy classes C^a of the symmetric group S_n with H and G, this cap product is equal to

$$\frac{1}{|H||G|}\sum_{a\dashv n}|C^a\cap H||C^a\cap G|\prod_k a_k!k^{a_k}=\frac{n!}{|H||G|}\sum_{a\dashv n}\frac{|C^a\cap H||C^a\cap G|}{|C^a|}.$$

An expression for the number of graphs can also be obtained from Redfield's Counting Theorem as follows. A labeled graph with v vertices and e edges can be considered as an injective mapping f from $\binom{v}{2}$ to $\binom{v}{2}$, where the i-th pair of vertices is connected by an edge if and only if $f(i)\leq e$. This together with 3.6.1 gives:

3.6.3 Corollary *The number of graphs with v vertices and e edges is equal to the cap product*

$$C\left(S_v,\binom{v}{2}\right)\cap C\left(S_e\oplus S_{\binom{v}{2}\backslash e},\binom{v}{2}\right).$$

More generally we obtain for the number of G–classes on n^n, which consist of mappings with prescribed content the following result:

3.6.4 Corollary *The number of G–classes of content* $(b_0, \ldots, b_{m-1})$ *on* m^n *is equal to the cap product*

$$C(G, n) \cap C(S_{b_0} \oplus \ldots \oplus S_{b_{m-1}}, n).$$

We shall return to this later. Tools from linear representation theory will allow a further elucidation of Redfield's methods and results. Moreover, we can derive an algorithm for a redundancy free construction of representatives of the G–classes on Y^X from 3.6.3.

Besides this useful cap operation Redfield introduced a *cup operation* $\cup$, defined to be the linear extension of the following operation on monomials:

$$Y^c \cup \ldots \cup Y^d := (Y^c \cap \ldots \cap Y^d) Y^d.$$

It clearly has the following property:

3.6.5 Lemma *The cap product* $C(H, n) \cap C(G, n)$ *is equal to the sum of the coefficients in the cup product* $C(H, n) \cup C(G, n)$.

This shows that it should be useful to examine cup products of cycle indicator polynomials. In fact it is possible to write each such cup product as a sum of cycle indicators of stabilizers. We shall prove this result of Redfield with the aid of the following lemma:

3.6.6 Redfield's Lemma *Let* ${}_GX$ *be a finite action,* $\omega_0, \ldots, \omega_{r-1}$ *its orbits,* $x_i \in \omega_i$ *and* $\chi\colon G \to \mathbb{Q}[Y]$ *a class function. Then we have for the stabilizers* $G_i := G_{x_i}$ *of these representatives that*

$$\frac{1}{|G|} \sum_{g \in G} \chi(g) |X_g| = \sum_{i=0}^{r-1} \frac{1}{|G_i|} \sum_{g \in G_i} \chi(g).$$

Proof:

$$|G| \sum_i \frac{1}{|G_i|} \sum_{g \in G_i} \chi(g) = \sum_i |\omega_i| \sum_{g \in G_i} \chi(g) = \sum_{x \in X} \sum_{g \in G_x} \chi(g) = \sum_g \chi(g) |X_g|.$$

□

We are now in a position to prove the main theorem of Redfield's first paper:

3.6.7 Redfield's Decomposition Theorem *Assume that we are given subgroups* $H_0, \ldots, H_{m-1}$ *of* S_n *and an action of* $\times_i H_i$ *on* M_c. *Let* $\omega_0, \ldots, \omega_{r-1}$ *denote the orbits, choose representatives* $A_i \in \omega_i$, *and denote by* G_i *their stabilizers. Then each* G_i *is conjugate in* $\times^m S_n$ *to a subgroup of the diagonal* $\Delta(\times^m S_n)$, *and the cup product of the cycle indicator polynomials of the* H_i *can be decomposed with the aid of the* G_i *as follows:*

$$C(H_0, n) \cup \ldots \cup C(H_{m-1}, n) = \sum_{i=0}^{r-1} \frac{1}{|G_i|} \sum_{(\pi_0, \ldots, \pi_{m-1}) \in G_i} \prod_{k=1}^{n} y_k^{a_k(\pi_0)}.$$

Proof: Recall the action in question:

$$(\times_i H_i) \times M_c \to M_c \colon ((\pi_0, \ldots, \pi_{m-1}), (a_{ik})_c) \mapsto (\pi_i a_{ik})_c.$$

Then A is a matrix, the rows of which are injective n–tuples over the set n, i. e. they can be identified with elements of S_n, and hence M can be identified with $\times^m S_n$. The equivalence relation $\sim_c$ can therefore be expressed in group theoretical terms as follows. If $A \in M$ is identified with $(\rho_0, \ldots, \rho_{m-1})$, $B \in M$ with $(\rho_0', \ldots, \rho_{m-1}')$, then

$$A \sim_c B \iff \exists\, \sigma \in S_n \colon (\rho_0', \ldots, \rho_{m-1}') = (\rho_0\sigma, \ldots, \rho_{m-1}\sigma).$$

Thus the equivalence class A_c can be identified with the following left coset of the diagonal subgroup:

$$\{(\rho_0\sigma, \ldots, \rho_{m-1}\sigma) \mid \sigma \in S_n\} = (\rho_0, \ldots, \rho_{m-1})\Delta(\times^m S_n).$$

Analogously we can rewrite Redfield's action of $\times_i H_i$ on M_c as follows:

$$((\pi_0, \ldots, \pi_{m-1}), (\rho_0, \ldots, \rho_{m-1})_c) \mapsto (\pi_0\rho_0, \ldots, \pi_{m-1}\rho_{m-1})_c.$$

The stabilizer of $A_c = (\rho_0, \ldots, \rho_{m-1})_c$ is the subgroup

$$H_0 \times \cdots \times H_{m-1} \cap (\rho_0, \ldots, \rho_{m-1})\Delta(\times^m S_n)(\rho_0, \ldots, \rho_{m-1})^{-1},$$

and hence is conjugate to a subgroup of the diagonal, as is claimed.

In order to prove the second part of the assertion, we first of all note that by the definition of the cup operation we have that

3.6.8
$$\bigcup_{i \in m} C(H_i, n) = \frac{1}{\prod_i |H_i|} \sum_a \Big(\prod_{k \in \underline{n}} a_k! k^{a_k}\Big)^{m-1} \prod_{k \in \underline{n}} y_k^{a_k},$$

where the sum is taken over all the $(\pi_0, \ldots, \pi_{m-1}) \in \times_i H_i$ such that $a(\pi_0) = \ldots = a(\pi_{m-1}) = (a_1, \ldots, a_n)$. In order to show that this cup product can be decomposed as claimed, we apply Redfield's Lemma 3.6.6. We therefore introduce the class function $\chi \colon \times_i H_i \to \mathbb{Q}[Y]$, defined by

$$\chi((\pi_0, \ldots, \pi_{m-1})) := \begin{cases} \prod_{k \in \underline{n}} y_k^{a_k}, & \text{for each } i \in m \colon a(\pi_i) = (a_1, \ldots, a_n), \\ 0, & \text{otherwise.} \end{cases}$$

Then 3.6.8 together with 3.6.6 yields the statement. □

Putting $m := 1$, we obtain the following decomposition of the cycle indicator polynomial according to the stabilizers of the orbits:

3.6.9 Corollary *Consider $H \le S_n$ with orbits $\omega_0, \ldots, \omega_{r-1}$ on Y^X, representatives $f_i \in \omega_i$ and stabilizers G_i of these representatives. Then the cycle indicator polynomial $C(H, n)$ is equal to the following sum:*

$$C(H, n) = \sum_{i \in r} \frac{1}{|G_i|} \sum_{\pi \in G_i} \prod_{k \in \underline{n}} y_k^{a_k(\pi)}.$$

3.6.10 Examples Let us also consider a few examples of the particular case $m := 2$. Redfield's Decomposition Theorem yields

$$C(H_0, n) \cup C(H_1, n) = \sum_{i \in r} \frac{1}{|G_i|} \sum_{(\pi_0,\pi_1)\in G_i} \prod_{k \in \underline{n}} y_k^{a_k(\pi_0)}.$$

The mapping

3.6.11 $$\varphi\colon M_c \to S_n\colon (\rho_0, \rho_1)_c \mapsto \rho_0\rho_1^{-1}$$

is clearly a bijection, and it has the property

$$\varphi\big((\pi_0, \pi_1)(\rho_0, \rho_1)_c\big) = \pi_0\rho_0\rho_1^{-1}\pi_1^{-1},$$

which means that the orbit of $(\rho_0, \rho_1)_c$ is mapped onto the (H_0, H_1)–double coset $H_0\rho_0\rho_1^{-1}H_1$. The stabilizer of $(\rho, 1)_c, \rho \in S_n$, is conjugate to the intersection $(\rho^{-1}H_0\rho \times H_1) \cap \Delta(S_n \times S_n)$, and we have obtained:

3.6.12 Corollary *If $\{\pi_0, \dots, \pi_{r-1}\}$ is a transversal of the (H_0, H_1)–double cosets of S_n, then, by Redfield's Decomposition Theorem, we have*

$$C(H_0, n) \cup C(H_1, n) = \sum_{i=0}^{r-1} C(\pi_i^{-1} H_0\pi_i \cap H_1, n).$$

Special cases are

3.6.13 $$C(A_n, n) \cup C(A_n, n) = |A_n\backslash S_n/A_n| \cdot C(A_n, n),$$

and (exercise 3.6.1)

3.6.14 $$C(S_{n-1}, n) \cup C(S_{n-1}, n) = C(S_{n-1}, n) + C(S_{n-2}, n).$$

◇

Redfield's Decomposition Theorem shows that a good knowledge of the conjugacy classes of subgroups of $\Delta(\times^m S_n)$ is important. We can obtain the desired information from a transversal of the left cosets of the normalizer of $\Delta(\times^m S_n)$ in $\times^m S_n$. In order to get such a transversal we derive the following theorem (due to Thürlings [154]) which describes, for an arbitrary finite G, the lattice of subgroups in between $\Delta(\times^m G)$ and $\times^m G$. Then we shall evaluate, for particular cases, the normalizer of $\Delta(\times^m G)$ and consider applications.

3.6.15 Theorem *Let G be a finite group and $l \in \mathbb{N}^*$. The lattice of subgroups of $\times^l G$, the normalizer of which contains $\Delta(\times^l G)$, is isomorphic to the lattice of subgroups in between $\Delta(\times^{l+1} G)$ and $\times^{l+1} G$. A lattice isomorphism is provided by the mapping*

$$U \mapsto (U|G) := \{(u_0 g, \dots, u_{l-1} g, g) \mid (u_0, \dots, u_{l-1}) \in U, g \in G\}.$$

Proof: It is easy to check that $U \mapsto (U|G)$ is of the desired form and injective. We have to show that it is surjective and that it respects the lattice operations. In order to do this we consider the action of $(\times^{l+1}G)$ on $(\times^l G)$, defined by

$$((g_0, \ldots, g_l), (g'_0, \ldots, g'_{l-1})) \mapsto (\ldots, g_i g'_i g_l^{-1}, \ldots).$$

This action is clearly transitive, and the stabilizer of the identity $(1, \ldots, 1)$ is $\Delta(\times^{l+1}G)$. Consider a subgroup H, where

$$\Delta(\times^{l+1}G) \leq H \leq \times^{l+1}G,$$

and the orbit ω of 1 under H, i. e. consider $\omega := H((1, \ldots, 1))$. For any element $(g'_0, \ldots, g'_{l-1}) \in \omega$ there exists an element $(h_0, \ldots, h_l) \in H$ such that $(g'_0, \ldots, g'_{l-1}) = (h_0 h_l^{-1}, \ldots, h_{l-1} h_l^{-1})$. And as H contains the diagonal subgroup $\Delta(\times^{l+1}G)$, we have that

$$(h_0, \ldots, h_{l-1}, 1) \in H \iff (h_0, \ldots, h_{l-1}) \in \omega.$$

Using this remark it is easy to check that ω is a subgroup of $\times^l G$, the normalizer of which in $\times^l G$ contains the diagonal of $\times^l G$, and that furthermore $H = (\omega|G)$ holds. Hence the mapping $U \mapsto (U|G)$ is not only injective but also surjective and compatible with the lattice structure. □

This theorem has the following consequence:

3.6.16 Corollary *The lattice of subgroups in between $\Delta(G \times G)$ and $G \times G$, where G is a finite group, is isomorphic to the lattice of normal subgroups of G.*

Recall now the definition of $Z(G)$, the *center* of G:

$$Z(G) := \{h \in G \mid hg = gh, \text{ for each } g \in G\}.$$

One easily checks that

3.6.17 $$Z(\times^m G) = \times^m Z(G).$$

3.6.18 Theorem *The normalizer of $\Delta(\times^{l+1}G)$ in $\times^{l+1}G$ is the subgroup*

$$(Z(\times^l G)|G) = \{(g_0 g, \ldots, g_{l-1} g, g) \mid g_i \in Z(G), g \in G\}.$$

Proof: It is obvious, that $(Z(\times^l G)|G)$ is a subgroup of the normalizer. Consider $(h_0, \ldots, h_{l-1}) \notin Z(\times^l G)$. There exist $i \in l$ and $g \in G$ such that $h_i g h_i^{-1} \neq g$. The last component of

$$h^* := (h_0, \ldots, h_{l-1}, 1)(g, \ldots, g)(h_0, \ldots, h_{l-1}, 1)^{-1}$$

shows that $h^* \notin \Delta(\times^{l+1}G)$. But each subgroup between $\Delta(\times^{l+1}G)$ and $\times^{l+1}G$ is of the form $(U|G)$ described in theorem 3.6.16. In particular the normalizer of $\Delta(\times^{l+1}G)$ has this form, and therefore it is the normalizer of a subgroup of $(Z(\times^l G)|G)$. □

As the normalizer $N_G(U)$ of $U \leq G$ is the stabilizer of U under the conjugation operation of G on the subgroup lattice $L(G)$, a knowledge of a transversal of the left cosets of $N_G(U)$ in G suffices to describe the conjugacy class of U. If $\{g_0, \ldots, g_{r-1}\}$ is such a transversal, then

$$\widetilde{U} = \{g_0 U g_0^{-1}, \ldots, g_{r-1} U g_{r-1}^{-1}\}.$$

Consider, for example, the symmetric group S_n. For $n > 2$, the center of S_n consists of 1 only, so that the diagonal subgroup $\Delta(\times^{l+1} S_n)$ is equal to its normalizer in $\times^{l+1} S_n$. A transversal of the left cosets of the diagonal subgroup of $\times^{l+1} S_n$ is

$$\{(\pi_0, \ldots, \pi_{l-1}, 1) \mid (\pi_0, \ldots, \pi_{l-1}) \in \times^l S_n\}.$$

Exercises

Exercise 3.6.1 Prove 3.6.14.

Exercise 3.6.2 Prove that

$$\underbrace{C(S_{n-1}, n) \cup \ldots \cup C(S_{n-1}, n)}_{m \text{ factors}} = \sum_{k=1}^{m} S(m, k) C(S_{n-k}, n),$$

where $S(m, k)$ denotes a Stirling number of the second kind. (Hint: Prove the following result of Foulkes: For each subgroup $H \leq S_n$

$$C(H, n) \cup \ldots \cup C(H, n) = \frac{1}{n!} \sum_{a \dashv n} \left(\frac{n! |C^a \cap H|}{|H||C^a|} \right)^m \prod_k x_k^{a_k}.$$

holds. Use the fact that $x^m = \sum_{k=0}^{m} S(m, k)[x]_k$, by exercise 2.4.7.)

4. Enumeration by Stabilizer Class

We now consider another refinement of the Cauchy-Frobenius Lemma. It is due to Burnside, and it allows to enumerate orbits which have a given conjugacy class of subgroups as stabilizers of their elements. For example, it allows us to count the orbits of maximal length $|G|$, the *asymmetric* orbits, since these are the orbits ω with trivial stabilizers $G_x = 1$, for each $x \in \omega$. A well known formula from Galois theory turns out to be such a number of asymmetric orbits.

The problem is that applications of Burnside's Lemma need certain matrices, the table of marks and its inverse, the calculation of which needs quite a good knowledge of the lattice $L(G)$ of subgroups of G. Fortunately there are program systems at hand (DISCRETA, GAP, MAGMA, SYMMETRICA) that allow to treat a lot of nontrivial cases successfully.

Finally we will combine Burnside's Lemma with the above results on weighted enumeration, so that we can enumerate orbits weight *and* stabilizer class. For example, we shall be able to count graphs by number of vertices, number of edges and symmetry group.

4.1 Counting by Stabilizer Class

The preceding chapter was devoted to the enumeration of orbits by weight, a problem that can be solved by an easy refinement of the Cauchy–Frobenius Lemma. In the present chapter we shall introduce another variation of this lemma in order to count orbits by stabilizer class.

Recall that the elements of an orbit have as their stabilizers a full conjugacy class of subgroups of G,

$$\forall\, x \in X\colon \{G_{x'} \mid x' \in G(x)\} = \widetilde{G_x} = \{gG_xg^{-1} \mid g \in G\}.$$

We say that $\widetilde{G_x}$ is the *type* of this orbit. The *set* of orbits of type $\widetilde{U}$, for a given subgroup U of G, is called the $\widetilde{U}$*–stratum* and indicated by

$$G \backslash\!\backslash_{\widetilde{U}} X.$$

Consider the lattice $L(G)$ of subgroups of G. The group G is supposed to be finite and hence $L(G)$ is a locally finite poset with respect to the inclusion order $\subseteq$. Hence we shall be able to apply in particular the Möbius Inversion, using the

Möbius function $\mu(-,-)$ on $(L(G),\leq)$. We assume that the subgroups U^i of G are numbered in such a way that

4.1.1 $$U^i \subseteq U^k \implies i \leq k.$$

In this case the *zeta–matrix* as well as the *Möbius–matrix* of G, defined by

$$\zeta(G) := \Big(\zeta(U^i, U^k)\Big), \text{ and } \mu(G) := \Big(\mu(U^i, U^k)\Big) = \zeta(G)^{-1},$$

are upper triangular. Let us consider an easy example, $G := S_3$. Here is the lattice of subgroups:

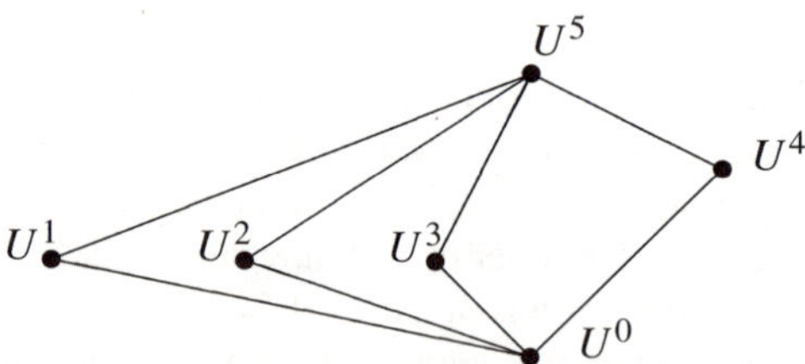

where $U^0 = \langle 1\rangle$, $U^1 = \langle(01)\rangle$, $U^2 = \langle(02)\rangle$, $U^3 = \langle(12)\rangle$, $U^4 = \langle(012)\rangle$, $U^5 = S_3$. This gives the zeta-matrix

$$\zeta(S_3) = \begin{pmatrix} 1 & 1 & 1 & 1 & 1 & 1 \\ & 1 & 0 & 0 & 0 & 1 \\ & & 1 & 0 & 0 & 1 \\ & & & 1 & 0 & 1 \\ & & & & 1 & 1 \\ & & & & & 1 \end{pmatrix},$$

and so we obtain, by invertion, the Möbius-matrix

$$\mu(S_3) = \begin{pmatrix} 1 & -1 & -1 & -1 & -1 & 3 \\ & 1 & 0 & 0 & 0 & -1 \\ & & 1 & 0 & 0 & -1 \\ & & & 1 & 0 & -1 \\ & & & & 1 & -1 \\ & & & & & 1 \end{pmatrix}.$$

In order to evaluate $|G \backslash\!\backslash_{\widetilde{U}} X|$, we consider, for a finite action ${}_GX$ and for the subgroups U of G, the sets

$$X_U := \{x \in X \mid \forall\, g \in U\colon\; gx = x\}.$$

Burnside called the order $|X_U|$ of this set the *mark* of U on X. We express it in terms of lengths of strata:

$$\begin{aligned}
|X_U| &= \sum_{x:\, U\le G_x} 1 \\
&= \sum_{V:\, U\le V\le G}\ \sum_{x:\, V=G_x} 1 \\
&= \sum_{V:\, U\le V\le G} \frac{1}{|\widetilde{V}|} \sum_{x:\, G_x\in\widetilde{V}} 1 \\
&= \sum_{V:\, U\le V\le G} \frac{|G/V|}{|\widetilde{V}|} \sum_{t\in T:G_t\in\widetilde{V}} 1 \\
&= \sum_{V:\, U\le V\le G} \frac{|G/V|}{|\widetilde{V}|} \left| G\backslash\!\backslash_{\widetilde{V}} X \right|, \\
&= \sum_{V} \zeta(U,V)\frac{|G/V|}{|\widetilde{V}|} \left| G\backslash\!\backslash_{\widetilde{V}} X \right|,
\end{aligned}$$

where T denotes a transversal of the orbits of G on X. By Möbius Inversion this equation is equivalent to:

4.1.2 $$\left| G\backslash\!\backslash_{\widetilde{U}} X \right| = \frac{|\widetilde{U}|}{|G/U|}\sum_V \mu(U,V)|X_V|.$$

In order to simplify this expression we consider the set $\widetilde{L}(G)$ consisting of the conjugacy classes of subgroups of G:

$$\widetilde{L}(G) := \left\{\widetilde{U}_0,\ldots,\widetilde{U}_{d-1}\right\},\ \text{with representatives } U_i\in\widetilde{U}_i.$$

U^i must not be mixed up with U_i, and d will denote for the rest of this chapter the number of conjugacy classes of subgroups of G. Putting

4.1.3 $$b_{ik} := \frac{|\widetilde{U}_i|}{|G/U_i|}\sum_{V\in\widetilde{U}_k}\mu(U_i,V) = \frac{|\widetilde{U}_i|}{|G/U_i|}\mu(U_i,\widetilde{U}_k),$$

where $\mu(U_i,\widetilde{U}_k) := \sum_{V\in\widetilde{U}_k}\mu(U_i,V)$, we can introduce the matrix

4.1.4 $$B(G) := (b_{ik}) = \begin{pmatrix} \ddots & & 0 \\ & |N_G(U_i)/U_i|^{-1} & \\ 0 & & \ddots \end{pmatrix} \left(\mu(U_i,\widetilde{U}_k)\right).$$

I suggest to call this matrix the *Burnside matrix* of G, although Burnside considered in fact the inverse of $B(G)$,

$$M(G) := B(G)^{-1} = \left(\zeta(U_i,\widetilde{U}_k)\right)\begin{pmatrix} \ddots & & 0 \\ & |N_G(U_i)/U_i| & \\ 0 & & \ddots \end{pmatrix},$$

(exercise 4.1.1), as Möbius inversion was not well known at that time. He called the inverse of $B(G)$ the table of marks; we shall return to this matrix later.

Using the Burnside matrix of G we can now reformulate 4.1.2, since the number $|X_V|$ of invariants of the subgroup V is clearly constant on the conjugacy class $\widetilde{V}$,

4.1.5 $$X_U \to X_{gUg^{-1}} : x \mapsto gx \text{ is a bijection.}$$

It is in fact the following lemma (and *not* the lemma of Cauchy–Frobenius that is quite often named after Burnside, see the remarks on history in the chapter containing the comments and references) which is

4.1.6 Burnside's Lemma *If ${}_GX$ is a finite action, then the vector of the lengths $|G \backslash\!\backslash_{\widetilde{U}_i} X|$ of the strata of G on X satisfies the equation*

$$\begin{pmatrix} \vdots \\ |G \backslash\!\backslash_{\widetilde{U}_i} X| \\ \vdots \end{pmatrix} = B(G) \cdot \begin{pmatrix} \vdots \\ |X_{U_i}| \\ \vdots \end{pmatrix}.$$

We can apply this now in order to enumerate the G–classes on Y^X by type. Since $f \in Y^X$ is fixed under each $g \in U_i$ if and only if f is constant on each orbit of U_i on X, we obtain:

4.1.7 Corollary *The number of symmetry classes of G on Y^X of type $\widetilde{U}_i$ is the i-th entry of the one column matrix*

$$B(G) \cdot \begin{pmatrix} \vdots \\ |Y|^{|U_i \backslash\!\backslash X|} \\ \vdots \end{pmatrix}.$$

Thus, for example, the number of k–graphs on v vertices which are of type $\widetilde{U}_j$, for a subgroup U_j of S_v, is the element in the j-th row of the matrix

$$B(S_v) \cdot \begin{pmatrix} \vdots \\ (k+1)^{|U_i \backslash\!\backslash \binom{v}{2}|} \\ \vdots \end{pmatrix}.$$

For $v := 4$, an application of the Burnside matrix shown in the appendix yields the table 4.1. The last column of this table shows among other things that there exist altogether 900 4–graphs on 4 vertices, and that exactly 465 of them are of type $\widetilde{U}_0$. Furthermore we note that the fourth, the sixth and the tenth row of this table consist of zeros only, which stimulates the question whether this also holds for bigger k as well. These rows belong to $U_3 \simeq A_3$, $U_5 \simeq C_4$ and $U_9 \simeq A_4$. A shorter formulation of this question reads as follows: Do A_3, C_4 or A_4 occur as automorphism groups of multigraphs? More generally: *Which subgroups of G occur as stabilizers of mappings $f \in Y^X$?* In order to answer this question we recall 3.3.1 which says that the stabilizer of f in the symmetric group S_X is the Young subgroup

type\k	0	1	2	3	4
$\widetilde{U}_0$	0	0	11	100	465
$\widetilde{U}_1$	0	2	21	84	230
$\widetilde{U}_2$	0	1	9	36	100
$\widetilde{U}_3$	0	0	0	0	0
$\widetilde{U}_4$	0	2	9	24	50
$\widetilde{U}_5$	0	0	0	0	0
$\widetilde{U}_6$	0	0	1	4	10
$\widetilde{U}_7$	0	2	6	12	20
$\widetilde{U}_8$	0	2	6	12	20
$\widetilde{U}_9$	0	0	0	0	0
$\widetilde{U}_{10}$	1	2	3	4	5
$\sum$	1	11	66	276	900

Table 4.1. Numbers of multigraphs on 4 vertices, counted by type

$$(S_X)_f = \bigoplus_{y \in Y} S_{X_y}, \text{ where } X_y := f^{-1}(y).$$

This implies

4.1.8 Corollary *The stabilizers of functions $f \in Y^X$ are the subgroups $U \leq G$, the corresponding permutation group $\bar{U}$ of which satisfies*

$$\bar{U} = S_\alpha \cap \bar{G},$$

where $\bar{G}$ corresponds to G, and S_α is the direct sum of the symmetric groups on the orbits of U:

$$S_\alpha := \bigoplus_{\omega \in U \backslash\backslash X} S_\omega.$$

We can now answer one of the questions posed above concerning automorphism groups of graphs.

4.1.9 Corollary *The alternating group A_v does not occur as an automorphism group of a multigraph on v vertices, if $v \geq 3$.*

Proof: The assumption $\bar{A}_v = S_\alpha \cap \bar{S}_v$ yields, as A_v acts transitively on $\binom{v}{2}$, that S_α is the full symmetric group on $\binom{v}{2}$, and hence $\bar{A}_v = \bar{S}_v$, which is a contradiction if $v > 2$. □

A similar result holds for the cyclic group C_v (see exercise 4.1.3) and the alternating subgroup A_{v-1} (exercise 4.1.4).

Exercises

Exercise 4.1.1 Verify the equality

$$\left(\zeta(U_i, \widetilde{U}_k)\right)^{-1} = \left(\mu(U_i, \widetilde{U}_k)\right).$$

Exercise 4.1.2 Prove the following identity for the elements of the Burnside matrix:

$$b_{ik} = \mu(\widetilde{U}_i, U_k)\frac{|\widetilde{U}_k|}{|G/U_i|}, \text{ where } \mu(\widetilde{U}_i, U_k) := \sum_{V \in \widetilde{U}_i} \mu(V, U_k).$$

Exercise 4.1.3 Prove that the cyclic group C_v does not occur as an automorphism group of a multigraph on v vertices, if $v \geq 2$.

Exercise 4.1.4 Prove that the subgroup A_{v-1} of A_v does not occur as an automorphism group of a multigraph on v vertices, if $v \geq 3$.

4.2 Asymmetric Orbits, Lyndon Words, the Cyclotomic Identity

Another immediate consequence of 4.1.2 is an expression for the number of orbits of prescribed length. We derive from 4.1.2

4.2.1 Corollary *If $_GX$ is a finite action, then the number of orbits of length k of G on X is equal to*

$$\frac{1}{k} \sum_{i:\ |G/U_i|=k} |\widetilde{U}_i| \sum_j \mu(U_i, \widetilde{U}_j)|X_{U_j}|,$$

and in particular the number of asymmetric orbits, *i. e. of orbits of type $\widetilde{1}$ or of maximal length $|G|$, is equal to*

$$\frac{1}{|G|} \sum_j \mu(1, \widetilde{U}_j)|X_{U_j}|.$$

We consider a well known example from Galois theory.

4.2.2 Example Consider the finite field $GF(q), q := p^m$, p a prime number, and its extension field $GF(q^n)$ of degree n. By Galois theory its Galois group G is generated by the *Frobenius isomorphism*

$$\sigma: GF(q^n) \to GF(q^n): x \mapsto x^q,$$

which is of order n:

$$G = \langle\sigma\rangle = \{1, \sigma, \ldots, \sigma^{n-1}\} \simeq C_n.$$

Moreover $GF(q^n)$ possesses a *normal basis*, which means that there exist x in $GF(q^n)$ such that $\{x, \sigma x, \ldots, \sigma^{n-1}x\}$ is a $GF(q)$–basis of $GF(q^n)$. Expressing the elements of $GF(q^n)$ in terms of this basis as $GF(q)$–linear combinations

$$\sum_{i=0}^{n-1} f(i)\sigma^i x, \ f(i) \in GF(q),$$

we can identify them with vectors $f = (f(0), \ldots, f(n-1)) \in q^n$ on which the Galois group generator σ acts as a cyclic shift. Hence the existence of a normal basis shows that the action of the Galois group G on $GF(q^n)$ is isomorphic to the canonical action of the cyclic group C_n on q^n. Denoting by

$$G \setminus\!\!\setminus_{|G|} X = G \setminus\!\!\setminus_{\tilde{1}} X$$

the set of orbits of (maximal) length $|G|$, we now evaluate the number

$$\left| G \setminus\!\!\setminus_{|G|} GF(q^n) \right| = \left| C_n \setminus\!\!\setminus_n q^n \right|$$

of orbits of maximal length n of the Galois group. As the cyclic group C_n has a subgroup lattice isomorphic to the lattice of divisors d of n, the Möbius function μ of $L(C_n)$ is the number theoretic Möbius function. Moreover the unique subgroup of order d, where d is a divisor of n, in the cyclic group C_n is generated by the power $(0, \ldots, n-1)^{n/d}$ of the full cycle. By 11.2.12, this power consists of n/d disjoint cycles, each of which has the length d. Furthermore $f \in q^n$ is fixed under this subgroup $\langle (0, \ldots, n-1)^{n/d} \rangle$ if and only if f is constant on the disjoint cyclic factors of $(0, \ldots, n-1)^{n/d}$. Hence, by 4.2.1, we obtain

4.2.3 $$\left| G \setminus\!\!\setminus_{|G|} GF(q^n) \right| = \left| C_n \setminus\!\!\setminus_n q^n \right| = \frac{1}{n} \sum_{d|n} \mu(d) q^{n/d},$$

where μ denotes the number theoretic Möbius function. Note that this number of asymmetric orbits is also equal to the number of irreducible monic polynomials with zeros in $GF(q^n)$ and coefficients in $GF(q)$, since by Galois theory a root of a polynomial is mapped onto a root by each element of the Galois group. ◇

More generally, we can consider the orbits of maximal length of C_n on m^n. The numbers

$$l_{mn} := \left| C_n \setminus\!\!\setminus_n m^n \right| = \frac{1}{n} \sum_{d|n} \mu(d) m^{n/d}$$

are called *Dedekind numbers*. Some of the smaller ones are shown in the following table:

$m \backslash n$	1	2	3	4	5	6	7
1	1	0	0	0	0	0	0
2	2	1	2	3	6	9	18
3	3	3	8	18	48	116	312
4	4	6	20	60	204	670	2340
5	5	10	40	150	624	2580	11160
6	6	15	70	315	1554	7735	39990
7	7	21	112	588	3360	19544	117648

.

The $f \in m^n$ can be considered as *words*

$$f = f(0) \ldots f(n-1)$$

of *length* n over the *alphabet* m. The *lexicographically smallest* elements in the orbits (for short: $f \leq C_n(f)$) of length n of C_n on m^n form the *canonic transversal*

$$L_n(m) := \{ f \in m^n \mid |C_n(f)| = n,\ f \leq C_n(f) \}.$$

Its elements are called the *Lyndon words* of length n over the alphabet m. The union of all these sets

$$L(m) := \bigcup_n L_n(m)$$

is called the *Lyndon set* over the alphabet m.

Now we consider an arbitrary word $f \in m^n$. As each letter $i \in m$ forms a Lyndon word of length 1, there exists a uniquely determined longest left factor $l_0(f) = f(0) \ldots$ of f which is a Lyndon word. The same holds for the remaining part w of $f = l_0(f)w$, and hence there is a unique decomposition

4.2.4
$$f = l_0(f) l_1(f) \ldots l_{\lambda(f)-1}(f)$$

of f into Lyndon words such that in the right part $l_i(f) \ldots l_{\lambda(f)-1}(f)$ the Lyndon word $l_i(f)$ is the maximal left Lyndon factor. This decomposition is called the *Lyndon decomposition* of f, and the number $\lambda(f)$ of factors is called the *Lyndon length* of f. Even more, the Lyndon decomposition can also be described as follows (exercise 4.2.1)

4.2.5 Lyndon's Theorem *The factors $l_i(f)$ of the decomposition 4.2.4 are lexicographically decreasing:*

$$l_0(f) \geq l_1(f) \geq \ldots \geq l_{\lambda(f)-1}(f).$$

Conversely, each decomposition into (weakly) decreasing Lyndon words is the decomposition 4.2.4.

Using the Lyndon length we can define the following generating function:

$$l_n(x, m) := \sum_{f \in m^n} x^{\lambda(f)},$$

which generates the numbers of words of length n over the Alphabet m with prescribed Lyndon length. V. Strehl has recently noticed and proved the remarkable fact that this function is symmetric in x and m. For example

$$l_3(x, m) = \frac{1}{3!}\big(x^3m^3 + 3(x^3m^2 + x^2m^3 - x^2m^2) + 2(x^3m + m^3x) - 2xm\big).$$

In order to prove this it is enough to show that, for all k, m, n in $\mathbb{N}^*$, the identity $l_n(k, m) = l_n(m, k)$ is true. We consider a second alphabet k, say, and the corresponding set k^* of all the words over k. Let $\Phi(m, k)$ denote the set of all the mappings $\varphi: L(m) \to k^*$ such that all but a finite number of values $\varphi(f)$ are equal to the empty word $\varepsilon \in k^*$ and assume that $f_0, \ldots, f_{n-1}$ are the Lyndon words with $\varphi(f_i) \neq \epsilon$. We map φ onto $\hat{\varphi}: L(k) \to m^*$, defined as follows: Let $\hat{f} \in L(k)$ occur in the Lyndon decompositions of $\varphi(f_i)$, say with the multiplicity $m\big(\varphi(f_i), \hat{f}\big)$. We can assume without restriction that $f_i \geq f_{i+1}$, for each i. Then put

$$\hat{\varphi}(\hat{f}) := f_0 \cdots \underbrace{f_i \cdots f_i}_{m(\varphi(f_i),\hat{f})} \cdots f_{n-1}.$$

Let us consider an example. Assume that $\varphi: L(3) \to 2^*$ has the following nonempty images (we separate Lyndon factors by dots):

$$\varphi(1) = 1 \cdot 01 \cdot 0,\ \varphi(01) = 1 \cdot 01,\ \varphi(002) = 01 \cdot 0 \cdot 0,$$

then $\hat{\varphi}$ has just the following nonempty images:

$$\hat{\varphi}(0) = 1 \cdot 002 \cdot 002,\ \hat{\varphi}(01) = 1 \cdot 01 \cdot 002,\ \hat{\varphi}(1) = 1 \cdot 01.$$

It is clear that φ can be reconstructed from $\hat{\varphi}$, and Lyndon's Theorem assures that this is true also in general. Hence $\varphi \mapsto \hat{\varphi}$ is a bijection between $\Phi(m,k)$ and $\Phi(k,m)$. Now we introduce a *norm* by putting

$$\|\varphi\| := \sum_{f \in L(m)} |f||\varphi(f)|,$$

where $|w|$ denotes the length (number of letters) of the word w. An important property of the norm is the identity

$$\|\varphi\| = \sum_{f \in L(m)} |f| \sum_{\hat{f} \in L(k)} |\hat{f}| \cdot m\big(\varphi(f), \hat{f}\big) = \|\hat{\varphi}\|,$$

where the last equation holds since $m\big(\varphi(f), \hat{f}\big) = m\big(\hat{\varphi}(\hat{f}), f\big)$, by definition of $\hat{f}$.

4.2.6 Strehl's Lemma *For all $k, m, n \in \mathbb{N}^*$ we have the identity*

$$l_n(k,m) = l_n(m,k).$$

Proof: Consider $f \in m^n$ and the corresponding subset $\Lambda_f \subseteq \Phi(m,k)$, consisting of the following φ: $\varphi(f') = \varepsilon$ if and only if the Lyndon word f' is not a factor of f, while otherwise $\varphi(f') \in k^{m(f,f')}$. Easy checks show that $|\Lambda_f| = k^{\lambda(f)}$ and

$$\varphi \in \bigcup_{f \in m^n} \Lambda_f \iff \|\varphi\| = n,$$

which allows to complete the proof:

$$l_n(k,m) = \sum_{f \in m^n} k^{\lambda(f)} = \big|\{\varphi \in \Phi(m,k) \,\big|\, \|\varphi\| = n\}\big|$$

$$= \big|\{\hat{\varphi} \in \Phi(k,m) \,\big|\, \|\hat{\varphi}\| = n\}\big| = \sum_{\hat{f} \in k^n} m^{\lambda(\hat{f})} = l_n(m,k).$$

□

Now we use a further indeterminate z in order to define the formal power series

$$L(x,m) := \sum_n l_n(x,m)z^n.$$

Lyndon's Theorem shows that the following is true:

$$L(x,m) = \prod_n \left(\frac{1}{1-x\cdot z^n}\right)^{l_{mn}},$$

so that we obtain the following surprising result:

$$\prod_n \left(\frac{1}{1-k\cdot z^n}\right)^{l_{mn}} = \prod_n \left(\frac{1}{1-m\cdot z^n}\right)^{l_{kn}},$$

which implies in particular (put $k := 1$ and note that $l_{1n} = 1$ if $n = 1$, and $l_{1n} = 0$, otherwise):

4.2.7 The Cyclotomic Identity *For all* $m, n \in \mathbb{N}^*$ *and the Dedekind numbers* $l_{mn} = \frac{1}{n}\sum_{d|n}\mu(d)m^{n/d}$ *we have the identity*

$$\prod_n \left(\frac{1}{1-z^n}\right)^{l_{mn}} = \frac{1}{1-m\cdot z}.$$

Besides the set of asymmetric orbits, we shall also consider the set of *asymmetric elements:*

$$\overline{_GX} := \{x \in X \mid G_x = \{1\}\}.$$

It is the set theoretic union of the asymmetric orbits, and so it has the order

$$|\overline{_GX}| = |G|\cdot|G\backslash\!\!\backslash_{\tilde{1}}X| = \sum_i \mu(1,\tilde{U}_i)|X_{U_i}|. \tag{4.2.8}$$

4.2.9 Application (asymmetric structures) The species *asymmetric S-structures* will be indicated by $\overline{\mathcal{S}}$, it is defined by

$$\overline{\mathcal{S}}[n] := \{\sigma \in \mathcal{S}[n] \mid (S_n)_\sigma = \{1\}\}.$$

For its cardinality series we obtain from 4.2.8:

$$\overline{\mathcal{S}}(x) = \sum_{n\geq 0}\sum_{i\in d}\mu(1,\tilde{U}_i)|\mathcal{S}[n]_{U_i}|\cdot\frac{x^n}{n!},$$

where the inner sum is taken over the numbers i of conjugacy classes $\tilde{U}_i$ of subgroups of S_n. The type series is

$$\widetilde{\overline{\mathcal{S}}}(x) = \sum_{n\geq 0}\frac{1}{n!}\sum_{i\in d}\mu(1,\tilde{U}_i)|\mathcal{S}[n]_{U_i}|\cdot x^n = \overline{\mathcal{S}}(x).$$

We note that in this case the cardinality and the type series are equal, which may astonish at first sight, but not at second sight, since the first is written in exponential form contrary to the second, and we are dealing with actions of symmetric groups and we are considering the asymmetric orbits only. ◇

Exercises

Exercise 4.2.1 Prove 4.2.5.

4.3 Tables of Marks and Burnside Matrices

The inverse $M(G) = (m_{ik})$ of the Burnside matrix $B(G) = (b_{ik})$ was introduced by Burnside (1911,[20]) and called the *table of marks,* as it was mentioned already. Sometimes it is also called the *supercharacter table* for a reason which will become clear later. The definition of b_{ik} together with exercise 4.1.1 and the identity

4.3.1 $$|\widetilde{U}_i|\zeta(U_i, \widetilde{U}_k) = |\widetilde{U}_k|\zeta(\widetilde{U}_i, U_k),$$

(which is immediate from the definition of the zeta function) shows that the elements of the table of marks are of the following form:

4.3.2 $$m_{ik} = \frac{|G/U_k|}{|\widetilde{U}_k|}\zeta(U_i, \widetilde{U}_k) = \frac{|G/U_k|}{|\widetilde{U}_i|}\zeta(\widetilde{U}_i, U_k) \in \mathbb{N},$$

where $\zeta(\widetilde{U}, V) := \sum_{W \in \widetilde{U}} \zeta(W, V)$. The claim that $m_{ik} \in \mathbb{N}$ follows from the fact that the order of a conjugacy class of a subgroup is equal to the index of the normalizer, and so

4.3.3 $$\frac{|G/U_k|}{|\widetilde{U}_k|} = \frac{|N_G(U_k)|}{|U_k|} \in \mathbb{N}.$$

Equation 4.3.2 shows that the entries of $M(G)$ describe the poset

$$(\widetilde{L}(G), \preceq)$$

which consists of the conjugacy classes $\widetilde{U}_i$ of subgroups and the partial order

4.3.4 $$\widetilde{U}_i \preceq \widetilde{U}_k :\Longleftrightarrow \exists\, U \in \widetilde{U}_i, V \in \widetilde{U}_k : U \leq V.$$

4.3.2 implies the following equivalence:

4.3.5 $$m_{ik} \neq 0 \iff \widetilde{U}_i \preceq \widetilde{U}_k.$$

Burnside called $M(G)$ the table of marks for the following reason:

4.3.6 Lemma *The entry m_{ik} is the number of left cosets of U_k in G which remain fixed under left multiplication by the elements of U_i.*

Proof:

$$\big|\{gU_k \mid \forall\, x \in U_i : xgU_k = gU_k\}\big| = \frac{1}{|U_k|}\big|\{g \in G \mid g^{-1}U_ig \leq U_k\}\big|$$

$$= \frac{|G|}{|U_k||\widetilde{U}_i|}\big|\{V \in \widetilde{U}_i \mid V \leq U_k\}\big| = \frac{|G/U_k|}{|\widetilde{U}_i|}\zeta(\widetilde{U}_i, U_k) = m_{ik}.$$

□

Hence m_{ik} is, so to speak, the *mark* which U_i leaves when it is acting on the left cosets of U_k. We now derive further properties of these elements. As G is assumed to be finite, we can choose a numbering of the conjugacy classes $\widetilde{U}_i$ such that the following implication holds:

4.3.7
$$|U_i| < |U_k| \implies i < k.$$

This guarantees in particular that the partial order is respected:

4.3.8
$$\widetilde{U}_i \preceq \widetilde{U}_k \implies i \leq k.$$

Under the assumptions of 4.3.7, $M(G)$ is upper triangular, $U_0 = \{1\}$, $U_{d-1} = G$, and so the table of marks takes the following form:

4.3.9
$$M(G) = \begin{pmatrix} |G| & \dots & |G/U_k| & \dots & 1 \\ & \ddots & & * & \vdots \\ & & |N_G(U_i)/U_i| & & \vdots \\ & 0 & & \ddots & \vdots \\ & & & & 1 \end{pmatrix}.$$

Other consequences of 4.3.2 are divisibility properties:

4.3.10 Lemma *If $\widetilde{U}_i \preceq \widetilde{U}_k$, then $m_{ik} = m_{kk}\zeta(U_i, \widetilde{U}_k)$, and so m_{kk} divides all the m_{ik} in the same column.*

Moreover, certain nonzero elements in a column form a monotone sequence, for $m_{ik} \neq 0 \neq m_{jk}$ means that $\widetilde{U}_i \preceq \widetilde{U}_k \succeq \widetilde{U}_j$. If in addition $\widetilde{U}_i \preceq \widetilde{U}_j$ holds, which implies that $i \leq j$, then we have

$$\widetilde{U}_i \preceq \widetilde{U}_j \preceq \widetilde{U}_k.$$

As we may assume $U_i \leq U_j$ without restriction, we obtain:

$$gU_jg^{-1} \leq U_k \implies gU_ig^{-1} \leq U_k,$$

so that an application of

4.3.11
$$m_{ik} = \frac{1}{|U_k|} \big| \{g \in G \mid gU_ig^{-1} \leq U_k\} \big|$$

finally yields

4.3.12 Corollary *If $\widetilde{U}_i \preceq \widetilde{U}_j \preceq \widetilde{U}_k$, then the corresponding elements in the k-th column of the table of marks of G satisfy*

$$m_{ik} \geq m_{jk} \geq m_{kk}.$$

The evaluation of the table of marks is usually quite difficult since one has to know $L(G)$, its Hasse diagram, and the orders of the U_i and their normalizers. Fortunately there exists several subgroup lattice programs which can be used in order to evaluate the tables in the appendix.

Burnside's original motivation for introducing the table of marks was the problem of decomposing a given action into its orbits or, in other words, to decompose a permutation representation into its *transitive constituents*. The question was whether it suffices to consider only the *character* of the action ${}_GX$, i. e. the function $\chi : g \mapsto |X_g|$. In order to explain character theoretically what is meant by the decomposition of the action ${}_GX$ into its transitive constituents we recall that there exists a natural equivalence relation on the set of actions of G on finite sets. Two actions, ${}_GX$ and ${}_GY$, say, were called *similar* if and only if there exists a bijection $\varphi : X \to Y$ which is *G–invariant,* i. e. for which the following holds:

$$\forall\, x \in X, g \in G : \ \varphi(gx) = g\varphi(x).$$

We know from 1.2.6 that there are exactly as many similarity classes of transitive actions as there are conjugacy classes of subgroups. An immediate implication is

4.3.13 Lemma *The set $\{\, {}_G(G/U_i) \mid i \in d\}$ is a transversal of the similarity classes of transitive actions of G.*

Proof: The actions ${}_G(G/U_i)$ and ${}_G(G/U_k)$, $i \neq k$, are not similar, since the respective stabilizers are U_i and U_k, which are not conjugate. □

4.3.14 Corollary *The characters*

$$\chi_i : G \to \mathbb{C} : g \mapsto |(G/U_i)_g|,$$

of the actions ${}_G(G/U_i)$, $U_i \in \widetilde{U_i}$, $i \in d$, are the transitive characters of G. These characters have the following values (recall that $C^G(g)$ denotes the conjugacy class and $C_G(g)$ the centralizer of g):

$$\chi_i(g) \ = \ \frac{|G|}{|U_i|}\frac{|C^G(g) \cap U_i|}{|C^G(g)|}.$$

Proof: Consider a transversal T of the left cosets of U_i. By definition of χ_i, we have

$$\chi_i(g) = \Big|\{t \in T \mid gtU_i = tU_i\}\Big| = \Big|\{t \in T \mid t^{-1}gt \in U_i\}\Big|$$

$$= \Big|\{g' \in G \mid g'gg'^{-1} \in U_i\}\Big|\frac{1}{|U_i|} = \frac{|C^G(g) \cap U_i|}{|U_i|}|C_G(g)|,$$

and the result follows since $|G/C_G(g)| = |C^G(g)|$. □

Burnside saw that a knowledge of the character χ of ${}_GX$ together with a table of the χ_i does *not* suffice to decompose χ into its transitive constituents. Such a decomposition is equivalent to the evaluation of the coefficients $n_i \in \mathbb{N}$ in the equation

$$\chi = \sum_{i \in d} n_i \chi_i,$$

where n_i is the number of orbits of G on X similar to ${}_G(G/U_i)$, as G–sets.

Later on we shall provide an example which shows that *this linear combination is not uniquely determined. But it turns out that replacing the χ_i by the rows of the table of marks we get a unique linear combination.* We show that the χ_i form part of the table of marks.

4.3.15 Lemma *If $U_i = \langle g \rangle$ is a cyclic subgroup of G and χ_k the character of ${}_G(G/U_k)$, then $m_{ik} = \chi_k(g)$.*

Proof: By an application of 4.3.11 we obtain

$$m_{ik} = \frac{1}{|U_k|} \left| \{ g' \in G \mid g' g g'^{-1} \in U_k \} \right| = \chi_k(g).$$

□

This is the reason why $M(G)$ is sometimes called the table of *supercharacters* of G, and it also shows that it is helpful to indicate the columns of $M(G)$ which belong to cyclic subgroups so that we can easily identify the transitive characters from $M(G)$. For example the table of marks of S_4 is

4.3.16 $$M(S_4) = \begin{pmatrix} 24 & 12 & 12 & 8 & 6 & 6 & 6 & 4 & 3 & 2 & 1 \\ & 2 & . & . & 2 & . & . & 2 & 1 & . & 1 \\ & & 4 & . & 2 & 2 & 6 & . & 3 & 2 & 1 \\ & & & 2 & . & . & . & 1 & . & 2 & 1 \\ & & & & 2 & . & . & . & 1 & . & 1 \\ & & & & & 2 & . & . & 1 & . & 1 \\ & & & & & & 6 & . & 3 & 2 & 1 \\ & & & & & & & 1 & . & . & 1 \\ & & & & & & & & 1 & . & 1 \\ & & & & & & & & & 2 & 1 \\ & & & & & & & & & & 1 \end{pmatrix} \begin{matrix} \leftarrow \\ \leftarrow \\ \leftarrow \\ \leftarrow \\ \\ \leftarrow \\ \\ \\ \\ \\ \\ \end{matrix}$$

a table which corresponds to the following numbering of subgroups U_i:

$$U_0 = \langle 1 \rangle,\ U_1 = \langle (13) \rangle,\ U_2 = \langle (02)(13) \rangle,\ U_3 = \langle (021) \rangle,$$

$$U_4 = \langle (02), (13) \rangle,\ U_5 = \langle (0123) \rangle,\ U_6 = \langle (01)(23), (03)(12) \rangle,$$

$$U_7 = \langle (021), (02) \rangle,\ U_8 = \langle (0123), (13) \rangle,\ U_9 = \langle (021), (031) \rangle,$$

$$U_{10} = \langle (0213), (0231) \rangle.$$

The rows which correspond to cyclic groups are marked by an arrow $\leftarrow$, and hence the table of transitive permutation characters of S_4 is as follows:

	(1^4)	$(1^2 2)$	(2^2)	(13)	(4)
χ_0	24	.	.	.	.
χ_1	12	2	.	.	.
χ_2	12	.	4	.	.
χ_3	8	.	.	2	.
χ_4	6	2	2	.	.
χ_5	6	.	2	.	2
χ_6	6	.	6	.	.
χ_7	4	2	.	1	.
χ_8	3	1	3	.	1
χ_9	2	.	2	2	.
χ_{10}	1	1	1	1	1

A few words on the entries b_{ik} of the Burnside matrix $B(G) = M(G)^{-1}$ are in order since in fact it is this matrix which we usually apply for enumerative purposes (see section 3.1). By the definition of b_{ik} we have

4.3.17
$$b_{ik} \in \mathbb{Q}.$$

If again ϕ denotes the Euler function, then we have for the row and column sums of $B(G)$:

4.3.18 Lemma *The column sums of the Burnside matrix $B(G)$ of a finite group G satisfy*

$$\sum_i b_{ik} = \begin{cases} \phi(|U_k|)/|N_G(U_k)|, & \textit{if } U_k \textit{ is cyclic,} \\ 0, & \textit{otherwise,} \end{cases}$$

while the row sums are equal to

$$\sum_k b_{ik} = \begin{cases} 1, & \textit{if } U_i = G, \\ 0, & \textit{otherwise.} \end{cases}$$

Hence the sum of all the elements of the Burnside matrix is

$$\sum_{i,k} b_{ik} = 1.$$

Proof: We put

$$x_k := \begin{cases} \phi(|U_k|)/|N_G(U_k)|, & \text{if } U_k \text{ is cyclic,} \\ 0, & \text{otherwise.} \end{cases}$$

Then we note that it is sufficient to prove:

4.3.19
$$1 = \sum_j x_j m_{ji}.$$

This is true since multiplication of both sides by b_{ik} and summation over i yields what we want to prove, namely

$$\sum_i b_{ik} = \sum_j x_j \underbrace{\sum_i m_{ji} b_{ik}}_{=\delta_{jk}} = x_k.$$

For the remaining verification of 4.3.19 we use

$$\sum_j x_j m_{ji} = \sum_{U_j cyclic} \frac{\phi(|U_j|)}{|N_G(U_j)|} \frac{|N_G(U_j)|}{|U_i|} \left| \{U \in \widetilde{U}_j \mid U \leq U_i\} \right|$$

$$= \frac{1}{|U_i|} \sum_{U_j cyclic} \phi(|U_j|) \left| \{U \in \widetilde{U}_j \mid U \leq U_i\} \right|.$$

The right hand side is equal to 1, for each $g \in U_i$ generates a cyclic subgroup $\langle g \rangle$, this subgroup is contained in a class $\widetilde{U}_j$, where U_j is cyclic, and, being cyclic, it possesses exactly $\phi(|\langle g \rangle|) = \phi(|U_j|)$ different generators by exercise 4.3.1. Therefore the sum is just $|U_i|$.

The statement that the row sums are equal to 0 except for the last row, where it is 1 can be obtained simply by considering a trivial action of G. Take a one element set X and let G act *trivially* on it:

$$G \times \{x\} \to \{x\} : (g, x) \mapsto gx := x.$$

Then the only orbit $\{x\}$ is of type $\widetilde{G}$, so that Burnside's Lemma yields the equation

$$B(G) \cdot \begin{pmatrix} 1 \\ \vdots \\ 1 \\ 1 \end{pmatrix} = \begin{pmatrix} 0 \\ \vdots \\ 0 \\ 1 \end{pmatrix},$$

from which the statement follows.

The total sum of entries is therefore equal to 1, and this completes the proof. □

As $M(G)$ is upper triangular if the numbering of the conjugacy classes of subgroups satisfies 4.1.1, also $B(G)$, its inverse, is upper triangular. Furthermore several of its entries can be made explicit in terms of the Möbius function of the lattice of subgroups $L(G)$:

4.3.20 Corollary *If the numbering of the conjugacy classes of subgroups of G satisfies 4.1.1, then we have for the Burnside matrix of G:*

$$B(G) = \begin{pmatrix} \frac{1}{|G|} & \cdots & \frac{\mu(1,U_k)}{|N_G(U_k)|} & \cdots & \frac{\mu(1,G)}{|G|} \\ & \ddots & & * & \vdots \\ & & \frac{|U_i|}{|N_G(U_i)|} & & \frac{\mu(U_i,G)}{|N_G(U_i)/U_i|} \\ & 0 & & \ddots & \vdots \\ & & & & 1 \end{pmatrix}.$$

From the definitions of m_{ik} and b_{ik} we obtain interesting relations between various elements or products of elements of these two matrices, e. g. that the following products are rational integral:

4.3.21
$$m_{ik}b_{kj} = \zeta(U_i, \widetilde{U}_k)\mu(U_k, \widetilde{U}_j) \in \mathbb{Z}.$$

Exercises

Exercise 4.3.1 Show that the number of generators of a cyclic group of order n is equal to $\phi(n)$.

4.4 Weighted Enumeration by Stabilizer Class

The proof of the weighted form 3.1.1 of the Cauchy-Frobenius Lemma was as easy as the proof of its constant form 2.1.1. The same holds for the corresponding weighted form of Burnside's Lemma, which we are going to introduce next. Recall that, by Burnside's result 4.1.2, we obtain for a transversal T of $G \backslash\backslash X$:

$$|G \backslash\backslash_{\widetilde{U}} X| = \sum_{t \in T:\, G_t \in \widetilde{U}} 1 = \frac{|\widetilde{U}|}{|G/U|} \sum_{V \leq G} \mu(U, V) \sum_{x \in X_V} 1.$$

If we now replace the 1's on both sides by the weight of the element to which they correspond, then we obtain the identity

$$\sum_{t \in T:\, G_t \in \widetilde{U}} w(t) = \frac{|\widetilde{U}|}{|G/U|} \sum_{V \leq G} \mu(U, V) \sum_{x \in X_V} w(x).$$

Since the bijection described in 4.1.5 is weight preserving, we get

4.4.1
$$\sum_{t \in T:\, G_t \in \widetilde{U}_i} w(t) = \frac{|\widetilde{U}_i|}{|G/U_i|} \sum_k \mu(U_i, \widetilde{U}_k) \sum_{x \in X_{U_k}} w(x).$$

A direct consequence is the desired result on the enumeration by weight and stabilizer class:

4.4.2 Burnside's Lemma, weighted form *Let $_GX$ denote a finite action and $w: X \to R$ a weight function from X into a commutative ring R which contains $\mathbb{Q}$ as a subring. If w is constant on the orbits of X, then we have, for the elementes t of a transversal T of the orbits and the vector of the sums of weights of transversals of strata $G \backslash\backslash_{\widetilde{U}_i} X$ of G on X, the equation*

$$\begin{pmatrix} \vdots \\ \sum_{t:G_t \in \widetilde{U}_i} w(t) \\ \vdots \end{pmatrix} = B(G) \cdot \begin{pmatrix} \vdots \\ \sum_{x:U_i \leq G_x} w(x) \\ \vdots \end{pmatrix}.$$

This weighted form of Burnside's Lemma was, as far as I know, first stated and proved in P. Stockmeyer's thesis ([148]). He provided applications of the following immediate consequence to the enumeration of graphs (resp. symmetry classes of mappings) by weight and type:

4.4.3 Corollary *The generating function for the enumeration of G–classes on Y^X of type $\widetilde{U}_j$ by weight w: $f \mapsto \prod f(x) \in \mathbb{Q}[Y]$ is the j-th row of the following one column matrix:*

$$B(G) \cdot \begin{pmatrix} \vdots \\ \prod_{\nu \in |U_i \backslash\backslash X|} \sum_y y^{l_\nu(U_i)} \\ \vdots \end{pmatrix},$$

where $l_\nu(U_i)$ denotes the length of the ν-th orbit of U_i on X.

Thus, for example, using the same order as above, we obtain for the graphs on $v = 4$ vertices the column;

$$B(S_4) \cdot \begin{pmatrix} (y_0+y_1)^6 \\ (y_0+y_1)^2(y_0^2+y_1^2)^2 \\ (y_0+y_1)^2(y_0^2+y_1^2)^2 \\ (y_0^3+y_1^3)^2 \\ (y_0+y_1)^2(y_0^4+y_1^4) \\ (y_0^4+y_1^4)(y_0^2+y_1^2) \\ (y_0^2+y_1^2)^3 \\ (y_0^3+y_1^3)^2 \\ (y_0^4+y_1^4)(y_0^2+y_1^2) \\ y_0^6+y_1^6 \\ y_0^6+y_1^6 \end{pmatrix} = \begin{pmatrix} 0 \\ y_0^4y_1^2+y_0^2y_1^4 \\ y_0^3y_1^3 \\ 0 \\ y_0^5y_1+y_0y_1^5 \\ 0 \\ 0 \\ 2y_0^3y_1^3 \\ y_0^4y_1^2+y_0^2y_1^4 \\ 0 \\ y_0^6+y_1^6 \end{pmatrix}.$$

This refines the second column of table 4.1 by showing, for example, that the two graphs of type $\widetilde{U}_7$ both contain exactly 3 edges.

A further example is the action of a finite group G on the set $Y^X := m^G$ induced by the regular action ${}_GG$ of G on itself. According to 4.4.3 we obtain the generating function for the orbits of type $\widetilde{U}_i$ by weight:

$$\sum_k b_{ik} \prod_{\nu \in |U_k \backslash\backslash G|} \sum_{j \in m} y_j^{l_\nu(U_k)}.$$

The orbits of U_k on G are the right cosets of U_k, hence $|U_k \backslash\backslash G| = |G/U_k|$ and $l_\nu(U_k) = |U_k|$, so that we have proved

4.4.4 Corollary *The coefficient of the monomial $y_0^{r_0} \dots y_{m-1}^{r_{m-1}}$ in the polynomial*

$$\sum_k b_{ik} \Big(\sum_{j \in m} y_j^{|U_k|} \Big)^{|G/U_i|}$$

is equal to the number of orbits of type $\widetilde{U}_i$ of G on the set m^G which are of content $(r_0, \dots, r_{m-1})$.

Now we note that, for $f \in m^G$, U is a subgroup of its stabilizer G_f, if and only if f is constant on the right cosets of U in G. Hence G_f is just the *maximal subgroup* of G such that f is constant on its right cosets. We consider the particular case $m := 2$, where m^G can be identified with the *subsets* of G. We derive from 4.4.4 the result:

4.4.5 Corollary *The coefficient of $y_0^r y_1^{|G|-r}$ in the polynomial*

$$|G/U_i| \sum_k b_{ik} \left(y_0^{|U_k|} + y_1^{|U_k|}\right)^{|G/U_i|}$$

is equal to the number of subsets $S \subseteq G$ of order r with the property that the maximal subgroup $U \leq G$ such that S is a union of right cosets of U in G, belongs to $\widetilde{U}_i$.

Let us return to the weighted form of Burnside's Lemma in order to discuss an example in detail.

4.4.6 Application (asymmetric symmetry classes by weight) A case of particular interest is the enumeration by weight of *asymmetric* symmetry classes, i. e. of orbits with trivial stabilizer class $\widetilde{1}$. The corresponding generating function is direct from 4.4.1:

4.4.7
$$\sum_{t \in T:\, G_t = 1} w(t) = \sum_k \frac{\mu(1, U_k)}{|N_G(U_k)|} \sum_{x:\, U_k \leq G_x} w(x).$$

This series is called the *asymmetry series.* For example, the asymmetry series of the action ${}_G(Y^X)$ can be obtained by Pólya–substitution of $\sum y$ in the *asymmetry indicator*

4.4.8
$$A(G, X) := \sum_k \frac{\mu(1, U_k)}{|N_G(U_k)|} \prod_i z_i^{|U_k \backslash\backslash_i X|},$$

where $U_k \backslash\backslash_i X$ denotes the set of orbits of length i of U_k on X.

4.4.9 Corollary *The generating function for the enumeration of asymmetric G-classes on Y^X by multiplicative weight is*

$$A\big(G, X \mid \textstyle\sum y\big).$$

In the case of the cyclic group the asymmetry indicator is

4.4.10
$$A(C_n, n) = \frac{1}{n} \sum_{d|n} \mu(d) z_d^{n/d}.$$

In the case of the symmetric group we obtain the following asymmetry indicator:

$$A(S_n, n) = \frac{1}{n!} \sum_{V \leq S_n} \mu(1, V) \prod_i z_i^{|V \backslash\backslash_i n|}.$$

In this particular example we have a second expression for this series, which we can obtain from 3.3.3: $A(S_n, n) = C(S_n, n) - C(A_n, n)$. Comparing the coefficients of the monomial z_n on both sides of this equation we get the following result on the Möbius function on the subgroup lattice of the symmetric group:

4.4.11
$$\sum_{V \; transitive} \mu(1, V) = (-1)^{n-1}(n-1)!.$$

◇

Another consequence of 4.4.1 is

4.4.12
$$\sum_k \sum_{V \in \widetilde{U}_k} \mu(U, V) \sum_{x:\, U_k \leq G_x} w(x) \equiv 0 \left(\frac{|G/U|}{|\widetilde{U}|}\right).$$

A particular case of this is (put $U := \{1\}$):

4.4.13
$$\sum_k \sum_{V \in \widetilde{U}_k} \mu(1, V) \sum_{x:\, U_k \leq G_x} w(x) \equiv 0\ (|G|).$$

4.4.14 Application (congruences between multinomials) Applying 4.4.13 to actions of the form ${}_G(Y^X)$ and the corresponding multiplicative weight we obtain (recall 4.4.3):

4.4.15
$$\sum_{U \leq G} \mu(1, U) \prod_{\nu \in |U \,\backslash\backslash X|} \sum_{y \in Y} y^{l_\nu(U)} \equiv 0\ (|G|).$$

In the case when G is the cyclic group of order p^m, acting on p^m, this congruence reads as follows:

4.4.16
$$\left(\sum y\right)^{(p^m)} \equiv \left(\sum y^p\right)^{(p^{m-1})} (p^m).$$

Being a congruence between coefficients of corresponding monomial summands, this shows that for multinomial coefficients the following is true:

4.4.17
$$\binom{p^m}{n_0, \ldots, n_{|Y|-1}} \equiv \binom{p^{m-1}}{n_0/p, \ldots, n_{|Y|-1}/p} (p^m),$$

where, by convention, a multinomial coefficient containing a non integral n_i/p is zero. This can be iterated, in order to get

$$\begin{aligned}\binom{p^m}{n_0, \ldots, n_{|Y|-1}} &\equiv \binom{p^{m-k}}{n_0/p^k, \ldots, n_{|Y|-1}/p^k} (p^{m-k+1}) \\ &\equiv 0\ (p^{m-k}),\end{aligned}$$

if p^k is the greatest common divisor of $\{n_0, \ldots, n_{|Y|-1}, p^m\}$. ◇

Exercises

Exercise 4.4.1 Prove that for natural numbers n and the number theoretic Möbius function μ the following is true:

$$\sum_{t|n} \mu(t) \binom{n/t}{n_0/t, \ldots, n_{|Y|-1}/t} \equiv 0\ (n).$$

5. Poset and Semigroup Actions

We are now going to discuss finite actions ${}_GX$, where X is a poset and where G respects the order: $x < x' \Rightarrow gx < gx'$. Thus G acts as a group of automorphisms on $(X, \leq)$. This yields generalizations of several notions introduced in the preceding chapter and it gives further insight. We shall also consider actions that respect the multiplication of a semigroup. Hence in particular lattice actions are considered as well as groups of automorphisms of partial orders, semigroups or other algebraic structures.

Nice examples are the actions of the symmetry groups of Platonic solids, actions on power sets, lattices of subspaces of finite vector spaces, and so on. The construction of $t-(v, k, \lambda)$-designs is another application.

The introduction of the Burnside ring, which is the ring generated by the isomorphism classes of finite transitive actions of a given group helps to explain the table of marks by the conjugation action of the group on its subgroup lattice.

5.1 Actions on Posets, Semigroups, Lattices

We have discussed the Burnside matrix and the table of marks. They belong (in a certain sense which will become clear later) to the subgroup lattice $L(G)$ on which G acts by conjugation of subgroups:

5.1.1 $$G \times L(G) \to L(G) \colon (g, U) \mapsto gUg^{-1}.$$

This operation respects the partial order $\leq$ of $L(G)$:

5.1.2 $$U \leq U' \implies gUg^{-1} \leq gU'g^{-1},$$

and it respects the operations $\wedge$ and $\vee$ on $L(G)$:

$$g(U \wedge U')g^{-1} = g(U \cap U')g^{-1} = gUg^{-1} \wedge gU'g^{-1},$$

$$g(U \vee U')g^{-1} = g\langle U \cup U'\rangle g^{-1} = gUg^{-1} \vee gU'g^{-1}.$$

We therefore consider actions of finite groups which respect a partial order or a multiplication or both the operations $\wedge$ and $\vee$ of a finite lattice. In other words, we are going to consider *actions of finite groups as groups of automorphisms of posets, semigroups and lattices*, in the following sense. Let $(X, \leq)$ denote a poset. G acts on X as a group of *automorphisms* if and only if

5.1.3 $$\forall\, g \in G,\ x, x' \in X\ (x \leq x' \iff gx \leq gx').$$

We shall abbreviate this by writing

$$_G(X, \leq),$$

and we shall call this a *poset action*.

5.1.4 Examples We met interesting posets defined by $_GX$ when we discussed the Galois theory of a finite action in 1.1.13, based on the consideration of the functions

$$\mathrm{orb}_G\colon L(G) \to Par[X]\colon U \mapsto U \backslash\!\backslash X,\ \ \mathrm{stab}_G\colon Par[X] \to L(G)\colon p \mapsto G_p.$$

orb_G turned out to be a Galois function, $\mathrm{stab}_G \circ \mathrm{orb}_G$ is a closure operator on $L(G)$, while $\mathrm{orb}_G \circ \mathrm{stab}_G$ is a co-closure on $Par[X]$. The sets of closed, respectively co-closed elements are the posets

$$(\{G_p \mid p \in Par[X]\}, \leq), (\{U \backslash\!\backslash X \mid U \leq G\}, \leq),$$

where $\leq$ means inclusion of subgroups or refinement of partitions, respectively. On these sets we have natural actions of G by conjugation and by application,

$$G \times \{G_p \mid p \in Par[X]\} \to \{G_p \mid p \in Par[X]\}\colon (g, G_p) \mapsto gG_pg^{-1},$$

and

$$G \times \{U \backslash\!\backslash X \mid U \leq G\} \to \{U \backslash\!\backslash X \mid U \leq G\}\colon (g, p) \mapsto gp := \{gp_0, \ldots\}.$$

Both these actions are obviously poset actions, and they are similar:

$$_G\big(\{G_p \mid p \in Par[X]\}, \leq\big) \approx {}_G\big(\{U \backslash\!\backslash X \mid U \leq G\}, \leq\big).$$

An interesting particular case will be considered in the application 5.1.8. ◇

5.1.5 Lemma *If $_G(X, \leq)$ denotes a finite poset action, then the following holds:*

- *Any two elements in the same orbit are incomparable, i. e. the orbits are antichains.*
- *The orbits ω_i of G on X can be numbered in such a way that*

$$\omega_i \ni x \leq x' \in \omega_k \Longrightarrow i \leq k.$$

- *For any orbit ω and a fixed $x \in X$, the numbers*

$$\big|\{x' \in \omega \mid x \leq x'\}\big| \ and \ \big|\{x' \in \omega \mid x \geq x'\}\big|$$

depend only on the orbit containing x and not on the chosen representative.

– *For any $x, x' \in X$, we have*

$$|G(x)| \cdot |\{x'' \in X \mid x \leq x'' \in G(x')\}| = |G(x')| \cdot |\{x''' \in X \mid x' \geq x''' \in G(x)\}|.$$

Proof:

i) If $x \in X$ were comparable with $gx \neq x$, say (without restriction) $x < gx$, then we had

$$x < gx < g^2x < \ldots < g^{-1}x < x,$$

which is a contradiction.

ii) Suppose $x_0, x_1 \in \omega_i, x_0', x_1' \in \omega_k, i \neq k$. Assume that $x_0 < x_0'$ and that x_1 and x_1' are comparable. Then $x_1 > x_1'$ would yield, for suitable $g, g' \in G$: $gx_0 = x_1 > x_1' = g'x_0'$, and hence also $x_0 > g^{-1}g'x_0'$, which implies the contradiction $x_0' > g^{-1}g'x_0'$. Thus the partial order can be embedded into a total order that respects the partial order:

$$\underbrace{x_0, x_1, \ldots, x_{|\omega_0|-1}}_{\in\, \omega_0}, \underbrace{x_{|\omega_0|}, \ldots, x_{|\omega_0|+|\omega_1|-1}}_{\in\, \omega_1}, \ldots, \text{ where } x_i < x_k \text{ implies that } i < k.$$

iii) is clear from 5.1.3.

iv) follows from a trivial "double count". We consider the bipartite graph consisting of the two orbits $G(x)$ and $G(x')$, where comparable elements are joint by an edge:

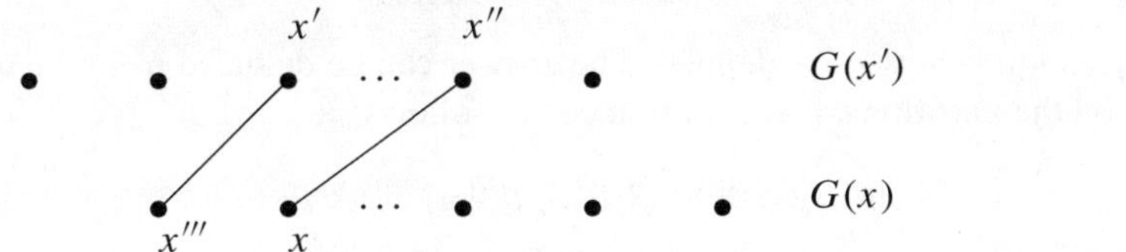

□

5.1.6 Corollary *A poset action $_G(X, \leq)$ induces a poset structure on the set $G \backslash\backslash X$ of orbits:*

$$(G \backslash\backslash X, \leq), \text{ where } \omega \leq \omega' \iff \exists\, x \in \omega, x' \in \omega' \colon x \leq x'.$$

5.1.7 Example We have just seen that the Galois theory of a group action $_GX$ leads to the similar poset actions

$$_G\big(\{G_p \mid p \in Par[X]\}, \leq\big) \approx {_G}\big(\{U \backslash\backslash X \mid U \leq G\}, \leq\big).$$

The corollary 5.1.6 shows that these poset actions lead to two posets of orbits which, of course, are essentially the same, they are order isomorphic. One of them is the poset of conjugacy classes of stabilizers:

$$\big(\{\widetilde{G}_p \mid p \in Par[X]\}, \leq\big),$$

and the other one is the poset

$$\big(G \backslash\backslash\{U \backslash\backslash X \mid U \leq G\}, \leq\big).$$

◇

Here is an interesting example:

5.1.8 Application (a suborder of the dominance order of partitions) We consider the natural action of the symmetric group S_n on the set n and the corresponding posets of closed and co-closed elements according to 5.1.4.

The stabilizer of a partition $p = \{p_0, \ldots\}$ of the set n in the symmetric group S_n is obviously the direct sum $\oplus_i S_{p_i}$ of the symmetric groups S_{p_i} on the blocks p_i of p. Thus $\{(S_n)_p \mid p \in Par[n]\}$, the set of closed elements, consists of these direct sums, which are the Young subgroups S_λ of S_n. On the other hand, each element of the poset $(Par[n], \leq)$ is co-closed. Hence, according to 5.1.4, we obtain the similar poset actions

$$_{S_n}\big(\{S_\lambda \mid \lambda \models n\}, \leq\big) \approx {}_{S_n}\big(Par[n], \leq\big).$$

This leads to the posets of orbits

$$\big(\{\widetilde{S_\lambda} \mid \lambda \models n\}, \leq\big), \text{ and } \big(S_n \backslash\!\backslash Par[n], \leq\big),$$

which are *order isomorphic.* In both cases the orbits are characterized by the proper partitions $\alpha \vdash n$. The orbit $\widetilde{S_\lambda}$ of S_λ corresponds to the proper partition λ^*, obtained from λ by reordering its entries λ_i in decreasing order, while the orbit of $p \in Par[n]$ corresponds to the proper partition β, obtained by ordering the block lengths $|p_i|$ in decreasing order. Let us denote the corresponding poset of proper partitions by

$$(\{\alpha \mid \alpha \vdash n\}, \trianglelefteq').$$

The question is how $\trianglelefteq'$ is defined. The answer can be deduced from the refinement order on the partitions. It was introduced as follows:

$$p \leq q :\iff \forall\, p_i \in p\; \exists\, q_j \in q\colon\; p_i \subseteq q_j.$$

Assume that α is the partition of the block lengths of p, while β is the partition of the block lengths of q. If $\alpha_0 = |p_i|$ and $p_i \subseteq q_j$, then, as $\beta_0 \geq |q_j|$, we have that $\alpha_0 \leq \beta_0$. Moreover, if $\alpha_1 = |p_k|$ and $p_k \subseteq q_l$, then, by $\beta_0 + \beta_1 \geq |q_j| + |q_l|$, we can deduce that $\alpha_0 + \alpha_1 \leq \beta_0 + \beta_1$, and so on. We find that

$$\alpha \trianglelefteq' \beta \Longrightarrow \forall\, i\colon \sum_{j=0}^{i} \alpha_j \leq \sum_{j=0}^{i} \beta_j.$$

This shows that the present order is a suborder of the well known *dominance order* of partitions which is defined by

5.1.9 $$\alpha \trianglelefteq \beta :\iff \forall\, i\colon \sum_{j=0}^{i} \alpha_j \leq \sum_{j=0}^{i} \beta_j.$$

We shall meet the dominance order later on in the context of representation theory of symmetric groups. It turned out to be crucial for a modern approach to this theory, it is in fact *the* partial order on partitions of natural numbers. The dominance order shows up already 1903 in the book by Grace and Young ([63]) on *The Algebra of Invariants* (section 280). ◇

Lemma 5.1.5 shows that, for each finite poset action ${}_G(X, \leq)$ and every numbering of its orbits ω_i (and representatives $x_i \in \omega_i$) that satisfies the conditions of the last item in the lemma, we obtain two matrices

$$A^{\leq} := \left(a_{ik}^{\leq}\right) \text{ and } A^{\geq} := \left(a_{ik}^{\geq}\right),$$

defined by

$$a_{ik}^{\leq} := \left| \{x'' \in \omega_k \mid x_i \leq x''\} \right| \text{ and } a_{ik}^{\geq} := \left| \{x'' \in \omega_k \mid x_i \geq x''\} \right|.$$

Their entries are related by

5.1.10 $$|\omega_i| \cdot a_{ik}^{\leq} = |\omega_k| \cdot a_{ki}^{\geq}.$$

We repeat that these matrices do *not* depend on the choice of the representatives, but, of course, on the numbering of the orbits.

Now we are going to apply this to particular posets. A lattice $(L, \wedge, \vee)$ defines a poset $(L, \leq)$ and besides this it yields two *semigroups* $(L, \wedge)$ and $(L, \vee)$, as both these compositions are associative by the definition of a lattice.

5.1.11 Lemma *Let ${}_GL$ denote a finite group action and assume that $(L, \wedge, \vee)$ is a lattice. Then the following three conditions are equivalent:*

i) $\forall\, x, x', g : x < x' \Longrightarrow gx < gx'$,

ii) $\forall\, x, x', g : g(x \wedge x') = gx \wedge gx'$,

iii) $\forall\, x, x', g : g(x \vee x') = gx \vee gx'$.

Proof:

The first item implies the second and the third item: As $x \wedge x'$ is less than or equal to both x and x', we obtain from i) that $g(x \wedge x')$ is less than or equal to both gx and gx'. This yields

$$g(x \wedge x') \leq gx \wedge gx'. \qquad (\star)$$

If now $g(x \wedge x')$ were strictly less than $gx \wedge gx'$, we also had, by i):

$$g^2(x \wedge x') < g(gx \wedge gx') \leq g^2x \wedge g^2x',$$

where the last inequality comes from $(\star)$. Hence, for each $n \in \mathbb{N}$, we obtain

$$g^n(x \wedge x') < g^nx \wedge g^nx',$$

which yields a contradiction by putting $n := |\langle g \rangle|$. Thus $g(x \wedge x') = gx \wedge gx'$ must hold. The third item follows quite analogously.

The second item yields the first: The assumption $x < x'$ yields $x \wedge x' = x$, so that $g(x \wedge x') = gx$, and hence, by ii), $gx \leq gx'$. This implies $gx < gx'$, as $x \neq x'$.

The fact that the third item gives the first follows similarly, and this completes the proof. □

Lemma 5.1.11 means that we may either check if the action respects the partial order or one of the two compositions $\wedge$ or $\vee$. In each of these cases we shall use either the notation ${}_G(L, \leq)$ or we shall indicate this situation by

$$ {}_G(L, \wedge, \vee) $$

and call such actions *lattice actions*.

5.1.12 Example A very interesting lattice is the *lattice of concepts* which we meet in *concept analysis,* a new and application oriented branch of lattice theory devoted to the analysis of knowledge which is given in form of a context. It was introduced by Wille, elaborated and applied by him and his co-workers, and the first book on that theory recently appeared ([58], by Ganter and Wille).

A *context* is a 0-1-matrix the i-th row of which corresponds to an *object* $o_i \in \mathcal{O}, i \in m$, and the k-th column of which corresponds to the *property* $p_k \in \mathcal{P}, k \in n$. The entry in the i-th row and k-th column is 1 if we know that o_i has the property p_k, otherwise we put a 0 there. Hence such a context C can be considered as an incidence relation $I \subseteq \mathcal{O} \times \mathcal{P}$ between $\mathcal{O}$ and $\mathcal{P}$, or as a mapping

$$ C \in 2^{m \times n}. $$

According to French philosophy of 17-th century (see e. g. [5],[4]) a concept consists of a set of objects, the *extent,* and a set of properties, the *intent,* such that neither an element of the extent nor an element of the intent can be left out without disturbing the notion of the concept in question. Wille formalized that as follows. He introduced the *derivatives* of sets $O \subseteq \mathcal{O}$ of objects and of sets $P \subseteq \mathcal{P}$ of properties by putting

$$ O' := \{p \in \mathcal{P} \mid \forall\, o \in O\colon\ p(o)\},\ \ P' := \{o \in \mathcal{O} \mid \forall\, p \in P\colon\ p(o)\}, $$

where $p(o)$ means that the object o has the property p. A *concept* is defined to be a pair $\mathcal{C} := (O, P)$ such that $O' = P$, and $P' = O$.

The set

$$ \mathcal{CL}(C) := \{\mathcal{C} = (O, P) \mid O' = P, P' = O\} $$

of all the concepts of the given context C carries a natural partial order:

$$ (O, P) \leq (\widetilde{O}, \widetilde{P}) :\iff O \subseteq \widetilde{O} \iff \widetilde{P} \subseteq P. $$

If $(O, P) \leq (\widetilde{O}, \widetilde{P})$, then we say that (O, P) is a *subconcept* or a *subsumption* of $(\widetilde{O}, \widetilde{P})$, or, equivalently, that $(\widetilde{O}, \widetilde{P})$ is a *superconcept* or a *generalization* of (O, P).

Two concepts have an infimum (exercise)

$$ (O, P) \wedge (\widetilde{O}, \widetilde{P}) := (O \cap \widetilde{O}, (O \cap \widetilde{O})') = (O \cap \widetilde{O}, (P \cup \widetilde{P})'') $$

and a supremum

$$ (O, P) \vee (\widetilde{O}, \widetilde{P}) := ((P \cup \widetilde{P})', P \cup \widetilde{P}) = ((O \cup \widetilde{O})'', P \cap \widetilde{P}). $$

Thus $\mathcal{CL}(C)$ is a complete lattice (exercise 5.1.1) which allows to reconstruct the context (up to redundancy, for example equal rows or columns etc.). It is called the *lattice of concepts.*

Here is a very easy example, a context that reflects a few basic notions of algebra: the property r means *rational, i* means *irrational, t* means *transcendental* and a means *algebraic.* The objects are the real numbers 2, $\sqrt{2}$ and π.

$C :=$

$o_i \backslash p_k$	r	i	a	t
2	1	0	1	0
$\sqrt{2}$	0	1	1	0
π	0	1	0	1

.

The corresponding lattice of concepts has the following Hasse diagram (where we indicated the concepts generated by a single objects (i. e. the concepts of the form $(\{o\}'', \{o\}')$):

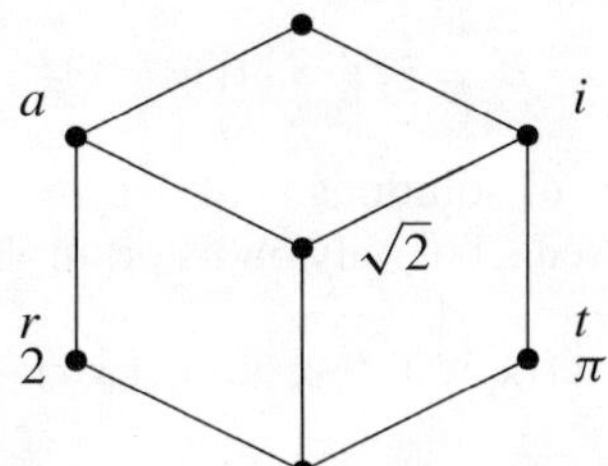

It nicely illustrates the properties of the given objects $2, \sqrt{2}, \pi$.

The automorphism group $Aut(C)$ of the context C is the stabilizer of the incidence relation, i. e. s

$$Aut(C) = \{(\pi, \rho) \in S_{\mathcal{O} \times \mathcal{P}} \mid (o, p) \in I \iff (\pi o, \rho p) \in I\}.$$

It is obvious that its action on the concept lattice is a lattic action:

$$_{Aut(C)}\big(\mathcal{CL}(C), \wedge, \vee\big).$$

◇

An immediate consequence of 5.1.5 is that the following numbers (we use $\wedge$ instead of $\leq$ and $\vee$ instead of $\geq$ in order to emphasize that we are now dealing with a lattice) do not depend on the choice of representatives of the orbits but only on the chosen numbering which again is assumed to fulfil item iv) of the lemma:

5.1.13 $\quad a_{ik}^{\wedge} := \big| \{x' \in \omega_k \mid x_i \leq x'\} \big|$ and $a_{ik}^{\vee} := \big| \{x' \in \omega_k \mid x_i \geq x'\} \big|$,

where x_i is an element of ω_i. The $a_{ik}^{\wedge}$ form an upper triangular matrix $A^{\wedge} := (a_{ik}^{\wedge})$, while the $a_{ik}^{\vee}$ form a lower triangular matrix $A^{\vee} := (a_{ik}^{\vee})$. The main diagonals consist of ones, and hence both these matrices are invertible over $\mathbb{Z}$. Furthermore, by 5.1.10, the elements of these matrices are related by the equations

$$|\omega_i| \cdot a_{ik}^{\wedge} = |\omega_k| \cdot a_{ki}^{\vee}.$$

Now we consider the more general case, where a group G acts as a group of automorphisms on a *semigroup* $(X, \cdot)$, i. e. we assume that the action also satisfies

$$\forall\, x, x', g : \ g(x \cdot x') = gx \cdot gx'.$$

Such actions are called *semigroup actions* and they are indicated by

$$_G(X, \cdot).$$

We denote the orbits of G on X by

$$\omega_0, \ldots, \omega_{d-1}.$$

A trivial but important remark is (exercise 5.1.2):

5.1.14 Lemma *For each $i, j, k \in d$ and any $z, z' \in \omega_k$ we have*

$$\left|\{(x, x') \in \omega_i \times \omega_j \mid x \cdot x' = z\}\right| = \left|\{(x, x') \in \omega_i \times \omega_j \mid x \cdot x' = z'\}\right|.$$

In other words: The number of solutions $(x, x') \in \omega_i \times \omega_j$ of $x \cdot x' = z, z \in \omega_k$, does *not* depend on the chosen z but only on its orbit. We can therefore put

5.1.15
$$a_{ijk}^{\cdot} := \left|\{(x, x') \in \omega_i \times \omega_j \mid x \cdot x' = z\}\right|,$$

for a fixed $z \in \omega_k$. (The reader should note the upper index "$\cdot$" which indicates the multiplication in question, and which therefore is *not* a fly blow.)

Now we introduce a ring which has these numbers as its structure constants. To do this we start from the *semigroup ring* of X over $\mathbb{Z}$, which is the set

$$\mathbb{Z}^X = \{f \mid f : X \to \mathbb{Z}\},$$

together with addition and multiplication defined by:

$$(f + f')(x) := f(x) + f'(x), \ (f \star f')(x) := \sum_{x' \cdot x'' = x} f(x') f'(x'').$$

We denote the resulting *ring* by $\mathbb{Z}^{X,\cdot}$. Its elements will be written as "formal sums"

$$f = \sum_{x \in X} f_x x, \ \text{where } f_x := f(x).$$

If, in addition, we are given an action $_GX$, then we call the f that are fixed under each $g \in G$, for short: the $f \in \mathbb{Z}_G^X$, the *G–invariants* or the *G–invariant mappings* (recall the notation X_G introduced in the first section). They form an important subring (exercise 5.1.3), the main properties of which are gathered in the following lemma:

5.1.16 Lemma

- *The G–invariants* $f \in \mathbb{Z}^X$ *form the subring:*
$$\mathbb{Z}_G^{X,\cdot} := \{f : X \to \mathbb{Z} \mid \forall g \in G\colon f = f \circ \bar{g}^{-1}\}.$$
- *This subring has as a* $\mathbb{Z}$*–basis the* orbit sums
$$\underline{\omega_i} := \sum_{x \in \omega_i} x \in \mathbb{Z}_G^{X,\cdot}.$$
- *The structure constants of* $\mathbb{Z}_G^{X,\cdot}$ *are the* $a^{\cdot}_{ijk}$ *defined in 5.1.15, i. e. we have for the product of basis elements*
$$\underline{\omega_i} \cdot \underline{\omega_j} = \sum_k a^{\cdot}_{ijk}\, \underline{\omega_k}.$$

In this way each action of a finite group G on a finite lattice $(L, \wedge, \vee)$ *as a group of lattice automorphisms yields the two rings* $\mathbb{Z}^{L,\wedge}$ *and* $\mathbb{Z}^{L,\vee}$ *together with their subrings*
$$\mathbb{Z}_G^{L,\wedge} \text{ and } \mathbb{Z}_G^{L,\vee},$$
the multiplicative structure of which is described by the constants
$$a^{\wedge}_{ijk} := \left|\{(x, x') \in \omega_i \times \omega_j \mid x \wedge x' = z\}\right|,$$
$$a^{\vee}_{ijk} := \left|\{(x, x') \in \omega_i \times \omega_j \mid x \vee x' = z\}\right|,$$
for a fixed $z \in \omega_k$.

A paradigmatic example is formed by the subgroup lattice $L := L(G)$ and the action of G on it by conjugation. We are now in a position to state and prove the main theorem of this section ([116]):

5.1.17 Plesken's Theorem *Let* ${}_G(L, \wedge, \vee)$ *denote a finite lattice action of G. Assume that* $\omega_0, \ldots, \omega_{d-1}$ *are the orbits,* $\underline{\omega_0}, \ldots, \underline{\omega_{d-1}}$ *their sums, numbered according to 5.1.5.*

- *The mapping*
$$\underline{\omega_k} \mapsto \begin{pmatrix} a^{\wedge}_{0,k} \\ \vdots \\ a^{\wedge}_{d-1,k} \end{pmatrix}$$
defines a ring isomorphism between $\mathbb{Z}_G^{L,\wedge}$ *and* $\mathbb{Z}^d$*, while*
- *the mapping*
$$\underline{\omega_k} \mapsto \begin{pmatrix} a^{\vee}_{0,k} \\ \vdots \\ a^{\vee}_{d-1,k} \end{pmatrix}$$
defines a ring isomorphism between $\mathbb{Z}_G^{L,\vee}$ *and* $\mathbb{Z}^d$*, where* $\mathbb{Z}^d$ *is equipped with pointwise addition and multiplication.*

Proof: In order to check the homomorphy first, we consider the product $a_i^\wedge \cdot a_j^\wedge$ of the i–th and j–th column of $A^\wedge$. We want to verify that it satisfies $a_i^\wedge \cdot a_j^\wedge = \sum_k a_{ijk}^\wedge a_k^\wedge$. From the definition of $a_{li}^\wedge$ we obtain (for a fixed $x \in \omega_l$):

$$a_{li}^\wedge \cdot a_{lj}^\wedge = |\{y \in \omega_i \mid y \geq x\}| \cdot |\{z \in \omega_j \mid z \geq x\}|$$

$$= |\{(y,z) \in \omega_i \times \omega_j \mid x \leq (y \wedge z)\}| = \sum_k a_{ijk}^\wedge a_{lk}^\wedge,$$

which proves homomorphy. In order to check the isomorphy we use that both $A^\wedge := (a_{ik}^\wedge)$ and $A^\vee := (a_{ik}^\vee)$ are triangular by 5.1.13 and contain only 1's along their main diagonal (and hence are invertible over $\mathbb{Z}$). This shows that the above mappings are even $\mathbb{Z}$–isomorphisms, which completes the proof of the first statement, the second follows analogously. □

These results show the fundamental importance of the *columns* of $A^\wedge$. But the *rows* also have interesting properties (see exercise 5.1.4).

Exercises

Exercise 5.1.1 Prove that $\mathcal{CL}(C)$ is in fact a complete lattice, i. e. each of its subsets of $\mathcal{O}$ and of $\mathcal{P}$ has both an infimum and a supremum.

Exercise 5.1.2 Prove 5.1.14.

Exercise 5.1.3 Check the first statement of 5.1.16.

Exercise 5.1.4 Prove that the matrices $A_i^\wedge := (a_{ijk}^\wedge)$ have the rows of $A^\wedge$ as eigenvectors. What are the eigenvalues?

5.2 Examples

Decorative lattices on which finite groups act as groups of automorphisms are formed by the faces, edges and vertices of the regular polyhedra: the tetrahedron, the cube, the octahedron, the dodecahedron and the icosahedron. Let us consider the easiest case, the *tetrahedron* (Figure 5.1). Its four faces, six edges and four vertices together with T itself and $\emptyset$ form a lattice L of order 16, which is ordered by the incidence relation. It is obvious that the alternating group A_4 acts on L and respects incidence, i. e. it acts on L as a group of automorphisms. The corresponding matrices $A^\wedge$ and $A^\vee$ are the following ones:

$$A^\wedge = \begin{pmatrix} 1 & 4 & 6 & 4 & 1 \\ & 1 & 3 & 3 & 1 \\ & & 1 & 2 & 1 \\ & & & 1 & 1 \\ & & & & 1 \end{pmatrix}, \quad A^\vee = \begin{pmatrix} 1 & & & & \\ 1 & 1 & & & \\ 1 & 2 & 1 & & \\ 1 & 3 & 3 & 1 & \\ 1 & 4 & 6 & 4 & 1 \end{pmatrix}.$$

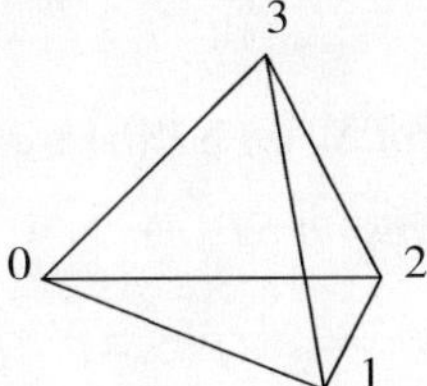

Fig. 5.1. The tetrahedron

A product of columns $a_i^\wedge, 0 \leq i \leq 4$, is e. g.

$$a_2^\wedge \cdot a_3^\wedge = \begin{pmatrix} 24 \\ 9 \\ 2 \\ 0 \\ 0 \end{pmatrix} = 2a_2^\wedge + 3a_1^\wedge.$$

This identity shows that $a_{232}^\wedge = 2$ and $a_{231}^\wedge = 3$, which means that each edge can be represented in exactly two ways as the infimum of a facet and an edge, while a vertex can be represented in exactly three ways as such an infimum.

Our next example is the lattice $SP(n)$ consisting of the *set partitions* of $n = \{0, \ldots, n-1\}$, ordered by *refinement*. This lattice is called the *partition lattice* of n, the particular case $SP(4)$ is shown in Figure 5.2, where the distribution of the four elements into blocks is indicated by "/".

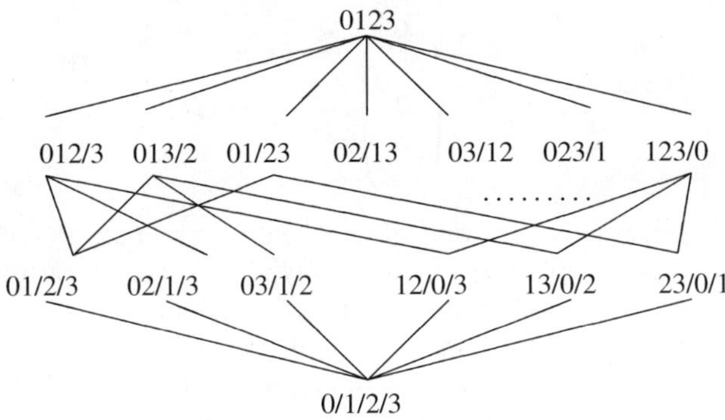

Fig. 5.2. The lattice $SP(4)$

If we order the lengths of the blocks of $p \in SP(n)$ into a nonincreasing sequence α, then we obtain a proper partition

$$\alpha(p) = (\alpha_0, \alpha_1, \ldots) \vdash n,$$

which we call the *(partition) type* of p. The following facts are easily checked (exercise 5.2.1):

5.2.1 Lemma *For the natural action of S_n on $SP(n)$ we have:*

- *S_n acts as a group of automorphisms on $SP(n)$:*

$$p \leq p' \Longleftrightarrow \pi p \leq \pi p'.$$

- *The lattice $(SP(n), \leq)$ is isomorphic to the sublattice of $L(S_n)$ consisting of the stabilizers of the set partitions.*
- *The orbits of S_n on $SP(n)$ are the subsets of partitions of the same type:*

$$S_n(p) = S_n(p') \Longleftrightarrow \alpha(p) = \alpha(p').$$

If we choose the following numbering for the orbits of S_4 on the set partitions of 4,

$$\omega_0 := S_4(0/1/2/3),\ \omega_1 := S_4(01/2/3),\ \omega_2 := S_4(01/23),$$

$$\omega_3 := S_4(012/3),\ \omega_4 := S_4(0123),$$

then the matrices $A^\wedge$ and $A^\vee$ of $SP(4)$ take the following form:

$$A^\wedge = \begin{pmatrix} 1 & 6 & 3 & 4 & 1 \\ & 1 & 1 & 2 & 1 \\ & & 1 & 0 & 1 \\ & & & 1 & 1 \\ & & & & 1 \end{pmatrix},\quad A^\vee = \begin{pmatrix} 1 & & & & \\ 1 & 1 & & & \\ 1 & 2 & 1 & & \\ 1 & 3 & 0 & 1 & \\ 1 & 6 & 3 & 4 & 1 \end{pmatrix}.$$

Again the products of columns $a_i^\wedge, 0 \leq i \leq 4$, yield interesting enumerative results, for example

$$a_1^\vee \cdot a_2^\vee = \begin{pmatrix} 0 \\ 0 \\ 2 \\ 0 \\ 18 \end{pmatrix} = 2a_2^\vee + 12a_4^\vee$$

shows that there are exactly two ways of representing a set partition of type (2^2) of the set 4 as supremum $p \vee p'$ of a partition p of type $(2, 1^2)$ and a partition p' of type (2^2). The equation shows furthermore that the set partition 0123 of type (4) can be represented in exactly 12 ways as the supremum $p \vee p'$, where $\alpha(p) = (2, 1^2)$ and $\alpha(p') = (2^2)$. This can be checked easily in Figure 5.2. The lattice $SP(n)$ possesses a *rank function* (for the basic notions of lattice theory see e. g. the book [1] by Aigner), the rank of $p \in SP(n)$ is

5.2.2 $$r(p) = n - l(p),$$

where $l(p)$ means the *length* of p i. e. the number of blocks. The order of the $(n-k)$-th *level* of $SP(n)$, from bottom to top, is the number of set partitions of n consisting of k blocks, i. e. the Stirling number of the second kind $S(n, k)$ (cf. 2.4.16):

5.2.3 $$|\{p \in SP(n) \mid r(p) = n - k\}| = S(n,k).$$

Correspondingly the total order $|SP(n)|$ of this lattice is the n-th Bell number (recall 2.4.17):

5.2.4 $$|SP(n)| = B_n.$$

Further important finite lattices are the subspace lattices of finite vector spaces. We denote by

$$\mathcal{L}(d,q)$$

the lattice of subspaces of the d–dimensional vector space $V := GF(q)^d$ over the Galois field $GF(q)$. The general linear group $GL(V)$, which is $GL_d(q)$, acts on it as a group of automorphisms, and the orbits are clearly formed by the subspaces of the same dimension. The rank function is

5.2.5 $$r(U) = \dim(U),$$

and hence the order of the k-th level of $\mathcal{L}(d,q)$ is equal to the number of k–dimensional subspaces of $V = GF(q)^d$. For technical reasons we put $d = m + n$ and state:

5.2.6 Lemma *For natural numbers m and n, where $0 < n$, the number of subspaces $U \leq GF(q)^{m+n}$ of dimension n is equal to*

$$\frac{(q^{m+n}-1)\cdot\ldots\cdot(q^{m+1}-1)}{(q^n-1)\cdot\ldots\cdot(q-1)} = \begin{bmatrix} m+n \\ n \end{bmatrix}_q.$$

Proof: It is easy to see that the number of n–tuples consisting of linearly independent elements of $GF(q)^{m+n}$ is equal to

$$(q^{m+n}-1)(q^{m+n}-q)\cdot\ldots\cdot(q^{m+n}-q^{n-1}).$$

In the same way we see that the number of different bases of $GF(q)^n$ is equal to

$$(q^n-1)(q^n-q)\cdot\ldots\cdot(q^n-q^{n-1}).$$

Hence the number of subspaces of dimension n is equal to the quotient of these two cardinalities and the statement is obtained by cancelling powers of q and recalling the definition of q–binomial numbers from 3.1.11. □

Now we introduce the *Gaussian polynomials*:

5.2.7 $$G_{m,n} := \sum_{k=0}^{mn} p(k;m,n)x^k,$$

where $p(k;m,n)$ is defined to be the number of proper partitions $\alpha \vdash k$ into at most m parts $\alpha_i > 0$ and subject to the condition that each $\alpha_i \leq n$.

5.2.8 Lemma

$$G_{m,n} = \begin{bmatrix} m+n \\ n \end{bmatrix}.$$

Proof: First, note that

$$\begin{bmatrix} m+n \\ n \end{bmatrix} = x^n \begin{bmatrix} (m-1)+n \\ n \end{bmatrix} + \begin{bmatrix} m+(n-1) \\ n-1 \end{bmatrix};$$

this is immediate from the definition of the binomial function. Moreover we have that

$$\begin{bmatrix} m+0 \\ 0 \end{bmatrix} = \begin{bmatrix} 0+n \\ n \end{bmatrix} = 1 = G_{m0} = G_{0n}.$$

It therefore suffices to show that the $G_{m,n}$ satisfy the recursion

5.2.9 $$G_{m,n} = x^n G_{m-1,n} + G_{m,n-1}, \text{ if } 0 < m, n.$$

In order to prove this we apply the following identity for $k \geq n$: Putting $p(r; s, t) := 0$, if $r < 0$, we have

$$p(k; m, n) = p(k-n; m-1, n) + p(k; m, n-1).$$

(The first summand on the right hand side describes the number of partitions α of k with not more than m parts and $\alpha_0 = n$. The second summand is the number of $\alpha \vdash k$ such that $\alpha_0 < n$ while the number of parts is less than or equal to m.) This identity yields

$$G_{m,n} = \sum_{k=0}^{mn} p(k-n; m-1, n)x^k + \sum_{k=0}^{mn} p(k; m, n-1)x^k.$$

Since $p(k-n; m-1, n) = 0$ if $k < n$ and $p(k; m, n-1) = 0$ if $k > m(n-1)$, we may proceed as follows:

$$\begin{aligned} G_{m,n} &= \sum_{k=n}^{mn} p(k-n; m-1, n)x^k + \sum_{k=0}^{m(n-1)} p(k; m, n-1)x^k \\ &= x^n \sum_{k=n}^{mn} p(k-n; m-1, n)x^{k-n} + G_{m,n-1} = x^n G_{m-1,n} + G_{m,n-1}. \end{aligned}$$

□

An immediate consequence is

5.2.10 Corollary *For natural numbers m, n and prime powers q*

$$G_{m,n}(q) = \begin{bmatrix} m+n \\ n \end{bmatrix}_q = \sum_{k=0}^{mn} p(k; m, n)q^k$$

is equal to the order of the n-th level of the subspace lattice $\mathcal{L}(m+n, q)$.

The last equation

$$\begin{bmatrix} m+n \\ n \end{bmatrix}_q = \sum_{k=0}^{mn} p(k;m,n)q^k$$

can be made explicit by associating with each subspace U of dimension n in $GF(q)^{m+n}$ a *canonic basis* in the following way. The basis in question consists of the n columns of a matrix with $m+n$ rows and n columns which we are going to construct now. The columns are vectors in $GF(q)^{m+n}$ with respect to the standard basis of $GF(q)^{m+n}$. They are obtained from U by successively picking elements and manipulating the elements which were already picked. In the first step we take an element from U that has a nonzero entry as far down in the column as possible, and without restriction we can assume that this entry is 1. Now we subtract all the multiples of that vector from U and pick from the rest again a vector that has a nonzero entry as far down as possible. Moreover, we manipulate the vector picked first by subtracting a suitable multiple of the second vector, so that in the position of the lowest nonzero entry of the second vector there is a zero in the first one, and so on. Here is a picture of the resulting matrix the columns of which contain that basis of U. The stars mean entries belonging to the ground field:

5.2.11
$$\begin{pmatrix}
* & * & * & \dots & * \\
\dots & \dots & \dots & \dots & \dots \\
* & * & * & \dots & * \\
0 & 0 & 0 & \dots & 1 \\
* & * & * & \dots & 0 \\
\dots & \dots & \dots & \dots & \dots \\
* & * & * & \dots & 0 \\
0 & 0 & 1 & \dots & 0 \\
* & * & 0 & \dots & 0 \\
\dots & \dots & \dots & \dots & \dots \\
* & * & 0 & \dots & 0 \\
\dots & \dots & \dots & \dots & \dots \\
0 & 1 & 0 & \dots & 0 \\
* & 0 & 0 & \dots & 0 \\
\dots & \dots & \dots & \dots & \dots \\
* & 0 & 0 & \dots & 0 \\
1 & 0 & 0 & \dots & 0 \\
0 & 0 & 0 & \dots & 0 \\
\dots & \dots & \dots & \dots & \dots \\
0 & 0 & 0 & \dots & 0
\end{pmatrix}$$

It is easy to see that *each* way of replacing the entries $*$ by elements of $GF(q)$ yields a different subspace of dimension n. Moreover, the numbers α_i of the entries $*$ in the columns of that matrix form a proper partition α of the total number k of stars. Therefore, we can *construct* all the subspaces U by forming all the partitions of numbers k that can be put into a matrix with m rows and n columns, and by replacing the stars in all the possible ways by field elements, and finally entering the rows with the 1's and zeros in between.

The entries of $A^\vee$ and the entries of $A^\wedge$ are obtained by an application of 5.1.13:

5.2.12 Corollary *For the lattice $\mathcal{L}(d,q)$ we have*

$$A^\wedge = \left(\begin{bmatrix} d-i \\ k-i \end{bmatrix}_q\right)_{i,k\in\underline{d}}, \quad A^\vee = \left(\begin{bmatrix} i \\ k \end{bmatrix}_q\right)_{i,k\in\underline{d}}.$$

We can obtain the matrix $A^\vee$ for each prime power q by simply evaluating at q the matrix

5.2.13
$$A := \left(\begin{bmatrix} i \\ k \end{bmatrix}\right)_{i,k\in\mathbb{N}}.$$

Its upper left hand corner is

$$A = \begin{pmatrix} 1 & & & & & \\ G_{10} & G_{01} & & & & \\ G_{20} & G_{11} & G_{02} & & & \\ G_{30} & G_{21} & G_{12} & G_{03} & & \\ G_{40} & G_{31} & G_{22} & G_{13} & G_{04} & \\ \dots & \dots & \dots & \dots & \dots & \dots \end{pmatrix}$$

$$= \begin{pmatrix} 1 & & & \\ 1 & 1 & & \\ 1 & 1+x & 1 & \\ 1 & 1+x+x^2 & 1+x+x^2 & 1 \\ 1 & 1+x+x^2+x^3 & 1+x+2x^2+x^3+x^4 & \dots \\ \dots & \dots & \dots & \dots \end{pmatrix}.$$

Exercises

Exercise 5.2.1 Prove 5.2.1.

Exercise 5.2.2 Show that the matrix $A^\vee$ of the lattice $SP(n)$ can be evaluated recursively.

Exercise 5.2.3 Evaluate the matrices $A^\wedge$ and $A^\vee$ for the cube and the icosahedron, on which the groups S_4 and A_5 act, respectively.

Exercise 5.2.4 Evaluate these matrices for the action of S_n on the lattice formed by the subsets of n.

5.3 Application to Combinatorial Designs

Another interesting application of the foregoing results is the construction of $t-(v,k,\lambda)$-designs. In particular the recent discovery of the first 7- and the first 8-designs with small parameters (there is in fact an existence theorem for designs, [152], but it assumes astronomic sizes for the parameters) showed that the application of finite group actions to this topic led to a breakthrough. One of the two crucial points that made this construction feasible was the evaluation of Kramer–Mesner matrices via double coset methods. These matrices are matrices of the form $A^\wedge$. The other crucial point was a modern implementation of the LLL-algorithm which allowed to find 0-1-solutions of systems of linear equations with these particular matrices $A^\wedge$ as their matrices of coefficients.

In order to describe this in some detail, we consider the action of a subgroup G of the symmetric group S_V on a set V of vertices (of the design that will be introduced later on). This group later on will be supposed to be contained in the automorphism group of the designs that we want to construct, *and it is this assumption — of a prescribed group of automorphisms — that brings the construction within reach of computers of moderate power, since it allows an enormous data reduction.* The group G acts on V and therefore also on the power set 2^V of V. The action clearly respects the inclusion of sets, and so we obtain a lattice action

$$_G(2^V,\cap,\cup),$$

together with a matrix $A^\cap = (a_{ik})$, where a_{ik} denotes the number of elements in the k-th orbit of G on 2^V that contain a fixed element of the i-th orbit.

5.3.1 Example The symmetric group S_V is clearly transitive on the set $\binom{V}{i}$ of i-subsets of V so that we can number the orbits of S_V on 2^V by the number of their elements. This shows that the corresponding matrix $A^\wedge$ is of the form

$$A^\cap = \left(\binom{k}{i}\right)_{0\le i,k\le |V|}.$$

◇

A (simple) $t-(v,k,\lambda)$-*design* is a pair

$$(V,\mathcal{B}),$$

consisting of a set V of vertices and a set $\mathcal{B}$ of blocks, such that the parameters t, v, k and λ fulfil the following conditions: There are altogether v vertices, each block consists of k vertices, and every set of t vertices is contained in exactly λ blocks. More formally:

$$|V| = v,\ \mathcal{B} \subseteq \binom{V}{k}, \forall\, T \in \binom{V}{t}:\ \big|\{B \in \mathcal{B} \mid T \subseteq B\}\big| = \lambda.$$

Here is a well known example taken from geometry, the famous *Fano plane,*

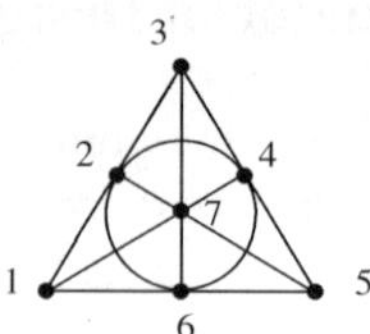

The Fano plane consists of the set of vertices $V = \{1, 2, 3, 4, 5, 6, 7\}$, together with the set of 7 blocks,

$$\mathcal{B} = \{\{1, 2, 3\}, \{3, 4, 5\}, \{5, 6, 1\}, \{1, 4, 7\}, \{2, 5, 7\}, \{3, 6, 7\}, \{2, 4, 6\}\}.$$

Hence $v = 7$ and $k = 3$. The Fano plane is a 1-design, with $\lambda = 3$, and it is a 2-design with $\lambda = 1$.

The designs on V can be described with the aid of the *incidence matrix* $M^v_{t,k}$ the rows of which correspond to the $T \in \binom{V}{t}$, and the columns of which correspond to the $B \in \binom{V}{k}$. The entries m^V_{TB} of $M^v_{t,k}$ are defined as follows:

$$m^V_{TB} := \begin{cases} 1, & \text{if } T \subseteq B, \\ 0, & \text{otherwise.} \end{cases}$$

Hence a $t - (v, k, \lambda)$-design $(V, \mathcal{B})$ is nothing but a selection of columns of that particular matrix, or, equivalently, a 0-1-vector x which solves the system of linear equations

5.3.2 Corollary *The set of $t - (v, k, \lambda)$-designs on V is the set of pairs $(V, \mathcal{B})$ that can be obtained from the 0-1-vectors x that solve the sytem of linear equations*

5.3.3
$$M^v_{t,k} \cdot x = \begin{pmatrix} \lambda \\ \vdots \\ \lambda \end{pmatrix}.$$

The set $\mathcal{B}$ of blocks B of the design corresponding to the solution x is

$$\mathcal{B} := \left\{ B \in \binom{V}{k} \,\middle|\, x_B = 1 \right\}.$$

There are, of course, several trivial cases, where solutions exist, but we are looking for *nontrivial* designs.

We note that the matrix $M^v_{t,k}$ is $\binom{v}{t} \times \binom{v}{k}$-matrix. After the construction of the first 6-design, people tried for many years to find 7-designs. A reasonable suspicion was that there should be 7-designs with $v = 33$ and $k = 8$. But in this case the matrix has about $6 \cdot 10^{13}$ entries, λ is unknown, and so it is impossible at present to find a 0-1-solution of such a big system of linear equations, even with a big computer.

The trick was (and is) to impose further conditions on the designs which we want to construct, namely to *prescribe a subgroup $G \leq S_V$ to be contained in the* *group of automorphisms.* This is of course risky, since there may very well exist *no such designs.* But in the case of 7-designs there was in fact also a suspicion for a certain group to occur as a group of automorphisms of designs, and in fact, it was. It will be described in a minute.

An element π of the symmetric group S_V is called an *automorphismus* of the design $(V, \mathcal{B})$, if it satisfies

$$\pi\mathcal{B} := \big\{\pi B := \{\pi b \mid b \in B\} \,\big|\, B \in \mathcal{B}\big\} = \mathcal{B}.$$

A subgroup G of the symmetric group consisting of such automorphisms is called *a group of automorphisms* of the design, and the maximal such group is called *the* or *the full* automorphism group:

$$Aut(V, \mathcal{B}) := \{\pi \in S_V \mid \pi\mathcal{B} = \mathcal{B}\}.$$

The task is to find at least one $t-(v, k, \lambda)$-desings containing G in its automorphism group and if possible to find all the $t-(v, k, \lambda)$-designs with G as their *full* group of automorphisms: $G = Aut(V, \mathcal{B})$. In order to bring this problem within reach of present PC's we consider, instead of $M^{v}_{t,k}$, a much smaller matrix which can be constructed with the aid of G. *This matrix is in fact a submatrix of the matrix* $A^{\wedge}$ *that corresponds to the lattice action* ${}_G(2^V, \cap, \cup)$ *mentioned above.* It is defined as follows:

$$M^{G}_{t,k} := (m^{G}_{T,K}), \text{ where } m^{G}_{T,K} := |\{K' \in G(K) \mid T \subseteq K'\}|,$$

T runs through a transversal of $G \backslash\backslash \binom{V}{t}$, K through a transversal of $G \backslash\backslash \binom{V}{k}$.

We recall that the evaluation of such transversals leads to double cosets: As S_V acts transitively both on $\binom{V}{k}$ and $\binom{V}{t}$, we obtain bijections

$$S_{(v-k,k)} \backslash S_V / G \to G \backslash\backslash \binom{V}{k}, \text{ and } S_{(v-t,t)} \backslash S_V / G \to G \backslash\backslash \binom{V}{t}.$$

In the chapter on construction and generation it will be described how these bijections can be used. The fundamental resultat is the following theorem. It is based on the fact that the prescription of G means that we can restrict attention to the orbit sets $G \backslash\backslash \binom{V}{k}$ and $G \backslash\backslash \binom{V}{t}$ instead of looking at single T or K.

5.3.4 The Theorem of Kramer and Mesner *The set of all the* $t-(v, k, \lambda)$*-designs with* $G \leq S_V$ *contained in their group of automorphisms can be obtained from the set of all the 0-1-solutions* x *of the linear system of equations*

$$M^{G}_{t,k} \cdot x = \begin{pmatrix} \lambda \\ \vdots \\ \lambda \end{pmatrix}.$$

The proof is clear from the foregoing characterization of designs as solutions of 5.3.3 and the definition of the matrix $M_{t,k}^G$ which says that a 0-1-solution of the system means just picking a subset of the set of orbits of G on $\binom{V}{k}$, i. e. to pick orbits of blocks.

5.3.5 Example Let us start with a small example. We take $t := \lambda := 1$, $v := 4$ and $k := 2$ as parameters, and as a group of automorphisms we prescribe

$$G := \langle (0123), (13) \rangle,$$

a subgoup of order 8 in S_4. This shows that we consider the set of vertices $V = 4$, together with the set of 2-subsets and the set of 1-subsets of V :

$$\binom{V}{k} = \big\{\{0,1\},\{0,2\},\{0,3\},\{1,2\},\{1,3\},\{2,3\}\big\}, \quad \binom{V}{t} = \big\{\{0\},\{1\},\{2\},\{3\}\big\}.$$

This yields the incidence matrix

$$M_{t,k}^v = M_{1,2}^4 = \begin{pmatrix} 1 & 1 & 1 & 0 & 0 & 0 \\ 1 & 0 & 0 & 1 & 1 & 0 \\ 0 & 1 & 0 & 1 & 0 & 1 \\ 0 & 0 & 1 & 0 & 1 & 1 \end{pmatrix}.$$

The sets of orbits are:

$$G \backslash\!\backslash \binom{V}{k} = \Big\{\big\{\{0,1\},\{0,3\},\{1,2\},\{2,3\}\big\},\big\{\{0,2\},\{1,3\}\big\}\Big\},$$

and

$$G \backslash\!\backslash \binom{V}{t} = \big\{\{0,1,2,3\}\big\}.$$

These orbits parametrize the rows and columns of the Kramer–Mesner matrix, which, according to its definition, has the following entries:

$$M_{t,k}^G = \begin{pmatrix} 2 & 1 \end{pmatrix}.$$

The number of entries of this matrix is 2, which is a reduction by the factor 12, compared with the number 24 of entries of the matrix $M_{t,k}^v$ shown above. Since λ is supposed to be 1, there exists exactly one 0-1-solution, namely $x = \binom{0}{1}$. The corresponding design is

$$(V, \mathcal{B}) = \big(\{0,1,2,3\}, \big\{\{0,2\},\{1,3\}\big\}\big).$$

Hence there exists exactly one $1-(4,2,1)$-design with G contained in its group of automorphisms. In order to check if this is the *full* group of automorphisms we use that it is a subgroup of order 8, i. e. of index 3, and so it is maximal. Hence it suffices to show that the symmetric group S_4 is *not* contained in the automorphism group. This is easy since clearly $\pi := (012)$ is not contained in G and does not fix any block of the design. ◇

The historically first 7-design ever constructed ([9]) was a $7-(33,8,10)$-design with prescribed group $A = P\Gamma L_2(32)$ of automorphisms. A permutation representation of this group on the set $\underline{33}$ of vertices is generated by the following two permutations:

$$\begin{aligned}\alpha &= (1\,2\,4\,8\,16)(3\,6\,12\,24\,17)(5\,10\,20\,9\,18)(7\,14\,28\,25\,19)\\ &\quad(11\,22\,13\,26\,21)(15\,30\,29\,27\,23)(31)(32)(33)\\ \beta &= (1\,18\,30)(2\,21\,12)(3\,10\,28)(4\,31\,32)(5\,24\,14)(6\,7\,17)(8\,25\,27)\\ &\quad(9\,19\,20)(11\,15\,13)(16\,23\,29)(22\,33\,26).\end{aligned}$$

The Kramer–Mesner matrix, which was already known to Magliveras and Leavitt, see [109], and which was recalculated by A. Betten, using double coset methods, is

```
2222222222220000000000000000000000000000000000000000000000000000000000000000000000000000000000000
2101110000000211112111111111111000000000000000000000000000000000000000000000000000000000000000000
1110000001000100200000000000100011111211211111110000000000000000000000000000000000000000000000000
0011000000000210010000000000001000110000001000121311111111110000000000000000000000000000000000000
0001200000000000000010000010011100000000100100100110001000011211111111000000000000000000000000000
0000110000000001000000100020000001001000001100111000000101001000000110021111110000000000000000000
0000013000110100100000101001001000100020001000100101010010000001000100010010001000000000000000000
0000004200000000020000200220000000020000002000000000000200000200000000000002000200000000000000000
0000000220000000002000002000002000002000020000000002000000000020000200002202000000000000000000000
0001000121000001100100000101000000001000100010100100001000000010011100000101001011100000000000000
0000002001200010011011010000110000001101000110010000010000000000100000000100001010011000000000000
0000000200020200000002200020000200000022000000020000200000000000000000000020000000020000000000000
0000000001012000000000101000022000000000000000011100000011001011000010010010001020010100000000000
0200000000002210020000101000001111000000000100000000000000000010001100101000000010000101110000000
0010100000001001001200100000010001000000200000000100101101100010000000000000101000000010211000000
0000110011000010100001020100000000000000010100000001010000000101001000000000100000000102110011100
0011020000000010010000000100100000001000100000000000010100001000001000110000100010102120000010010
0000000220000000000240002000000000000000000000000000000000000000002002200000000220200000000000200
0000010010011000010110010011001110000100000000100000101000100000000010000000101100010000000010001
0200000200000200000200000200020002000000000202000000000000000200000000000000000200020000000000000
0001000001010011000000000000100001100010001000010000100000100001100000010000000002001003101010010
0010100000011000102001001000000000100011000010110010011000000000000200000100000000000100000100000
0200000000000000100000100000000000102201000001100000000000010010010011000000100011010021000010000
0011000000000000010000100000000101000110100000010001000000010100000020100000010100000001010020001
0000000200000000000020020000000000000002000000002000000000200000000020020220000000000000200020000
0010000000000000000011000010100000000000001000000100101000020010000011011100110000000001110000002
0000000000100000100000110100000000000100010200000002001100002000000100100001110001001100010010100
0000000000000000000100101000101100001000001000010101011000000100010100020100110000000000000300110
0000000010000000100000000001010001101000100100001110010001000000100100011000100000000001110010110
0000000005000000000000000000000000500000000000050005000000000000000000000000000000000000000500000
0000000000000000000000000000000000000000000000005000000000500000000500050000000000050000000000000
0000000000000000000000000000000000000000000000005000000000000050000000000000000000000000500500000
```

The first 0-1-solutions of the system of linear equations with this matrix where found for $\lambda = 10$ and $\lambda = 16$, respectively. Here they are

```
0011100010100100110001101000010010000000101101100010111100000000100100101000011011101100101011100
1100011101011011001110010111101101111111010010011101000011111110110110101111001000100110101000111
```

They were obtained using the implementation of an improved version (by A. Wassermann) of the LLL-algorithm for the evaluation of short vectors.

The explicit version of the design follows from the 0-1–vector by taking the union of the orbits of blocks that correspond to the components of the vector which are equal to 1.

There are in fact 4 996 426 0-1-vectors that solve the above system for $\lambda = 10$. In the meantime 8-designs and 9-designs have been constructed, billions of them, see e. g. [10],[94].

5.4 The Burnside Ring

We saw how symmetry classes of mappings can be counted by weight and stabilizer class using the *Burnside matrix* which is the inverse of the table of marks. Moreover we met actions of groups on lattices $(L, \wedge, \vee)$, and the corresponding subrings $\mathbb{Z}_G^{L,\wedge}$ and $\mathbb{Z}_G^{L,\vee}$ of the semigroup rings $\mathbb{Z}^{L,\wedge}$ and $\mathbb{Z}^{L,\vee}$, where the matrices $A^\wedge = (a_{ik}^\wedge)$ and $A^\vee = (a_{ik}^\vee)$ played a central role. We now build a bridge between these two topics by introducing the Burnside ring $\Omega(G)$ of G which can be embedded into $\mathbb{Z}_G^{L(G),\wedge}$, the matrix $A^\wedge$ of which is closely connected with the table of marks. In order to do this we indicate first how a complete set of transitive G–sets can be obtained (exercise 5.4.1):

5.4.1 Lemma *Let G denote a finite group with its conjugacy classes of subgroups $\widetilde{U}_0, \dots, \widetilde{U}_{d-1}$ and representatives $U_i \in \widetilde{U}_i$. Then*

- *The sets $G/U_i := \{gU_i \mid g \in G\}, i \in d$, consisting of the left cosets gU_i of the subgroups U_i of G are G–sets with respect to the operation $g : xU_i \mapsto gxU_i$.*
- *The operations of G on these sets are transitive and pairwise dissimilar.*
- *$\{G/U_0, \dots, G/U_{d-1}\}$ is a complete system of transitive but pairwise dissimilar G–sets.*

Switching to the similarity classes

$$\overline{G/U_i} := \{{}_GX \mid {}_GX \approx {}_G(G/U_i)\},$$

we obtain the *complete system* $\Omega := \{\overline{G/U_0}, \dots, \overline{G/U_{d-1}}\}$ of G–similarity classes of transitive G–sets. Consider the free abelian group

$$\mathbb{Z}^\Omega := \{\psi \mid \psi : \Omega \to \mathbb{Z}\}, \text{ where } (\psi + \psi')(\overline{G/U_i}) := \psi(\overline{G/U_i}) + \psi'(\overline{G/U_i}).$$

As usual we will display the $\psi \in \mathbb{Z}^\Omega$ as "formal sums"

$$\psi = \sum_{i \in d} z_i \cdot \overline{G/U_i}, \text{ where } z_i := \psi(\overline{G/U_i}) \in \mathbb{Z}.$$

The addition of the two formal sums

$$\overline{G/U_i} := 1_{\mathbb{Z}} \cdot \overline{G/U_i} \text{ and } \overline{G/U_k} := 1_{\mathbb{Z}} \cdot \overline{G/U_k}$$

in $\mathbb{Z}^{\Omega}$ can be interpreted as first taking the disjoint union of G/U_i and G/U_k and afterwards switching to the G–similarity class $\overline{X}$ of the resulting G–set X:

$$\overline{G/U_i} + \overline{G/U_k} = \overline{G/U_i \dot{\cup} G/U_k}.$$

This is clear from the fact that each G–set possesses a unique decomposition into orbits (but note that also in the case when $i = k$ we have to form the *disjoint* union. For example, $|G/U_i \dot{\cup} G/U_i| = 2|G/U_i|$).

In this way, *each G–set X and its similarity class $\overline{X}$ can be identified with a uniquely determined element of the subset* $\mathbb{N}^{\Omega} \subseteq \mathbb{Z}^{\Omega}$:

5.4.2
$$\overline{X} = \sum_{i \in d} |G \backslash\!\backslash_{\widetilde{U}_i} X| \cdot \overline{G/U_i}.$$

Furthermore we can introduce on $\mathbb{Z}^{\Omega}$ a multiplication as linear extension of

$$\overline{G/U_i} \cdot \overline{G/U_j} := \sum_k b_{ijk} \overline{G/U_k}, \text{ if } \overline{G/U_i \times G/U_j} = \sum_k b_{ijk} \overline{G/U_k}.$$

Soon we shall see that this in fact yields a well defined multiplication. The corresponding structure constant b_{ijk} is equal to the number of orbits of G on $G/U_i \times G/U_j$ (with respect to the natural action $g(x, y) := (gx, gy)$) that belong to the class $\overline{G/U_k}$.

For example, the conjugacy classes of subgroups in $G := S_3$ are represented by the subgroups

$$U_0 := \{1\},\ U_1 := S_2,\ U_2 := A_3,\ U_3 := S_3,$$

and so we obtain the following transversal of the similarity classes of transitive S_3–sets:

$$G/U_0 = S_3/\{1\},\ G/U_1 = S_3/S_2,\ G/U_2 = S_3/A_3,\ G/U_3 = S_3/S_3,$$

which are of order 6,3,2,1, respectively. It is easy to check that S_3 acts, for example, transitively on $G/U_1 \times G/U_2$, so we obtain (already by checking cardinalities) that

$$\overline{G/U_1} \cdot \overline{G/U_2} = \overline{G/U_1 \times G/U_2} = \overline{G/U_0}.$$

5.4.3 Theorem *The following mapping defines an embedding of the ring* $(\mathbb{Z}^{\Omega}, +, \cdot)$ *into the ring* $\mathbb{Z}_G^{L(G),\wedge}$*:*

$$\mathbb{Z}^{\Omega} \hookrightarrow \mathbb{Z}_G^{L(G),\wedge} : \overline{G/U_i} \mapsto |N_G(U_i) : U_i| \cdot u_i,$$

where

$$u_i := \sum_{U \in \widetilde{U}_i} U \in \mathbb{Z}_G^{L(G),\wedge}.$$

Proof: The structure constants of $\mathbb{Z}^\Omega$ were denoted by b_{ijk}, while the structure constants of $\mathbb{Z}_G^{L(G),\wedge}$ are the $a_{ijk}^\wedge$ introduced in 5.1.16. It therefore remains to prove the following equation:

$$b_{ijk} = \frac{|N_G(U_i)/U_i||N_G(U_j)/U_j|}{|N_G(U_k)/U_k|} a_{ijk}^\wedge.$$

Since $|\widetilde{U}_k| = |G/N_G(U_k)|$ we have

$$b_{ijk} = \frac{1}{|G/U_k|} \left| \{(x, y) \in G/U_i \times G/U_j \mid G_{(x,y)} \in \widetilde{U}_k\} \right|$$

$$= \frac{|\widetilde{U}_k|}{|G/U_k|} \left| \{(x, y) \in G/U_i \times G/U_j \mid G_{(x,y)} = G_x \cap G_y = U_k\} \right|$$

$$= \frac{|U_k|}{|N_G(U_k)|} \left| \{(U, V) \in \widetilde{U}_i \times \widetilde{U}_j \mid U \cap V = U_k\} \right| \frac{|N_G(U_i)||N_G(U_j)|}{|U_i||U_j|}$$

$$= \frac{|N_G(U_i)/U_i||N_G(U_j)/U_j|}{|N_G(U_k)/U_k|} a_{ijk}^\wedge.$$

□

This shows in particular that

$$\Omega(G) := (\mathbb{Z}^\Omega, +, \cdot)$$

is a ring which we call the *Burnside ring* of G. The table of marks $M(G)$ of G is closely related to the table $A^\wedge = (a_{ik}^\wedge)$ of $\mathbb{Z}_G^{L(G),\wedge}$:

5.4.4 Theorem *The matrix $A^\wedge$ of $\mathbb{Z}_G^{L(G),\wedge}$ and the table $M(G)$ of marks of G satisfy the equation*

$$M(G) = A^\wedge \cdot \begin{pmatrix} \ddots & & 0 \\ & |N_G(U_k)/U_k| & \\ 0 & & \ddots \end{pmatrix}.$$

Proof: Using 4.3.2 and 5.1.13 we obtain:

$$m_{ik} = \frac{|G/U_k|}{|\widetilde{U}_k|} \left| \{V \in \widetilde{U}_k \mid U_i \le V\} \right| = |N_G(U_k)/U_k| \cdot a_{ik}^\wedge.$$

□

We have just seen how the Burnside ring $\Omega(G)$ can be embedded into $\mathbb{Z}_G^{L(G),\wedge}$, and we already know from 5.1.17 that $\mathbb{Z}_G^{L(G),\wedge}$ is isomorphic to $\mathbb{Z}^d$ via $u_i \mapsto a_i^\wedge$ so that we finally obtain

5.4.5 Corollary *The following mapping linearly extends to an embedding of rings:*

$$\varepsilon\colon \Omega(G) \hookrightarrow \mathbb{Z}^d : \overline{G/U_k} \mapsto \mu_k := \begin{pmatrix} m_{0k} \\ \vdots \\ m_{d-1,k} \end{pmatrix}.$$

We call these columns μ_k of $M(G)$ the *marks* of G.

In this way *we can identify the finite G–sets with the $\mathbb{N}$–linear combinations of the marks of G*. As 5.4.5 describes an embedding of rings, the product in $\Omega(G)$ corresponds to the pointwise product in $\mathbb{Z}^d$:

5.4.6 $$\varepsilon(\overline{G/U_i} \cdot \overline{G/U_j}) = \mu_i \cdot \mu_j.$$

Restricting attention to the i-th coordinate we obtain the mapping

$$\varepsilon_i\colon \Omega(G) \to \mathbb{Z}\colon \overline{X} \mapsto |X_{U_i}|,$$

and 5.4.5 says that ε_i is a homomorphism. Correspondingly we obtain, for *any* subgroup U of G, a *canonical homomorphism*

$$\varepsilon_U\colon \Omega(G) \to \mathbb{Z}\colon \overline{X} \mapsto |X_U|.$$

But note that the number of U–invariants $\varepsilon_U(\overline{X}) = |X_U|$ is *not* the desired coefficient of $\overline{G/U}$, when we express $\overline{X}$ as a $\mathbb{Z}$–linear combination of similarity classes $\overline{G/U_i}$ of transitive G–sets:

$$\overline{X} = \sum_{i\in d} |G \backslash\backslash_{\widetilde{U}_i} X| \cdot \overline{G/U_i}.$$

In certain cases $|G \backslash\backslash_{\widetilde{U}_i} X|$ can be obtained directly from $\varepsilon_U(\overline{X})$. Equation 4.1.2 shows that the following is true:

5.4.7 Lemma *The subgroup $U \in \widetilde{U}_i$ is maximal such that $\varepsilon_U(\overline{X}) \neq 0$ if and only if it is maximal such that $|G \backslash\backslash_{\widetilde{U}_i} X| \neq 0$, and in this case we have*

$$\varepsilon_U(\overline{X}) = |X_U| = |G \backslash\backslash_{\widetilde{U}} X||N_G(U)/U|.$$

Let us apply this to an example. Later on we shall use it in the proof of an astonishing theorem that leads to important applications.

5.4.8 Example Consider the set of mappings f from G to the set n which are of fixed weight t (we use the sum of the values as the weight of f):

$$n_t^G := \Big\{ f\colon G \to n \,\Big|\, \sum_g f(g) = t \Big\}.$$

It is easy to check that $f \in n_t^G$ remains fixed under each $g \in U$ if and only if f is constant on the right cosets of U. Hence such an invariant f takes each of its values with a multiplicity that is divisible by $|U|$, therefore $|U|$ divides t, and the invariants

$f \in n_t^G$ are in one-to-one correspondence with the elements of $n_{t/|U|}^{G/U}$. The order of this last set follows from exercise 2.4.6, and so we obtain

$$\varepsilon_U\left(\overline{n_t^G}\right) = \binom{|G/U| + t/|U| - 1}{t/|U|}.$$

In particular, if $|U| = t$, then $\varepsilon_U\left(\overline{n_t^G}\right) = |G/U|$, and we can apply 5.4.7, obtaining

$$U \in \widetilde{U}_i, |U| = t \Longrightarrow \left| G \backslash\backslash_{\widetilde{U}_i} \overline{n_t^G} \right| = \left| \widetilde{U}_i \right|.$$

In the case when the acting group is cyclic, these particular elements form a basis of the Burnside ring (see exercise 5.4.2 for another basis):

5.4.9 Lemma *The Burnside ring $\Omega(C_n)$ has the following $\mathbb{Z}$–basis:*

$$\left\{ \overline{n_t^{C_n}} \;\middle|\; t \text{ divides } n \right\}.$$

Proof: For each divisor s of n, we denote by $C(s)$ the unique subgroup of order s in C_n. The matrix (n_{st}) of the coefficients in

$$\overline{n_t^{C_n}} = \sum_{s|n} n_{st} \overline{C_n/C(s)}$$

is triangular with ones along their main diagonal, and hence invertible over $\mathbb{Z}$, which proves the statement. ◇

We recall that

$$\varepsilon_{U_i}\left(\overline{G/U_k}\right) = \varepsilon_i\left(\overline{G/U_k}\right) = m_{ik},$$

and so $\varepsilon_U = \varepsilon_V$ if and only if U and V are conjugate. Moreover,

$$\varepsilon_i\left(\overline{X}\right) = \varepsilon_i\Big(\sum_k |G \backslash\backslash_{\widetilde{U}_k} X| \cdot \overline{G/U_k}\Big) = \sum_k |G \backslash\backslash_{\widetilde{U}_k} X| \cdot m_{ik},$$

which shows that the desired numbers $|G \backslash\backslash_{\widetilde{U}_k} X|$ can be obtained from the numbers $\varepsilon_i\left(\overline{X}\right)$, since the table of marks is not singular. Hence we have proved

5.4.10 Corollary *We can identify the element $\overline{X}$ of the Burnside ring $\Omega(G)$ with the mapping $U \mapsto \varepsilon_U\left(\overline{X}\right)$, obtaining an embedding of $\Omega(G)$ into the following ring which is called the* ghost ring *of G:*

$$\widetilde{\Omega}(G) := \mathbb{Z}_{\sim}^{L(G)} := \left\{ f \in \mathbb{Z}^{L(G)} \;\middle|\; f \text{ constant on each conjugacy class} \right\}.$$

The ghost ring $\widetilde{\Omega}(C_n)$ of the cyclic group is equal to the Burnside ring of C_n, since C_n is abelian. Moreover, as C_n contains just one subgroup of order t, for each divisor t of n, both rings can be identified in a canonical way with $\mathbb{Z}^{T(n)}$, if $T(n)$ denotes the set of divisors of n. This leads to a canonic map into the ghost ring of G via the function $card$ that maps $U \in L(G)$ onto its cardinality:

$$\alpha: \mathbb{Z}^{T(|G|)} \to \mathbb{Z}^{L(G)}_{\sim}: f \mapsto f \circ card.$$

The crucial point is that this map is a *ring homomorphism*, which is clear from the following: First of all we know from 5.4.8 that, if $n := |G|$, then, if $C(|G|)$ denotes the cyclic group of order $|G|$:

$$\varepsilon_U\left(\overline{n_t^G}\right) = \varepsilon_{C(|U|)}\left(\overline{n_t^{C(|G|)}}\right).$$

This together with 5.4.7 implies

$$\alpha\left(\overline{n_t^{C(|G|)}}\right) = \overline{n_t^G}.$$

Finally we use that ε is a ring homomorphism, and therefore the following map must extend to a ring homomorphism:

$$\overline{n_t^{C(|G|)}} \mapsto \overline{n_t^G}.$$

Thus we have proved the following important result:

5.4.11 Theorem (Dress, Siebeneicher, Yoshida) *If G is a finite group of order n, then*

$$\alpha: \Omega(C(|G|)) \to \Omega(G): f \mapsto f \circ card,$$

is a ring homomorphism such that, for each finite $C(|G|)$–set X and any subgroup U of G, we have

$$\varepsilon_U\left(\alpha\left(\overline{X}\right)\right) = \varepsilon_{C(|U|)}\left(\overline{X}\right).$$

For example,

$$\varepsilon_{C(t')}\left(\overline{C_n/C(n/t)}\right) = \begin{cases} t, & \text{if } C(t') \subseteq C(n/t), \\ 0, & \text{otherwise.} \end{cases}$$

Hence, by the theorem,

$$\varepsilon_U\left(\alpha\left(\overline{C_n/C(n/t)}\right)\right) = \begin{cases} t, & \text{if } t \text{ divides } |G|/|U|, \\ 0, & \text{otherwise.} \end{cases}$$

An application of 5.4.7 now shows that

$$\left|G \backslash\!\backslash_{\widetilde{U}_i} \alpha\left(\overline{C_n/C(n/t)}\right)\right| = \begin{cases} |\widetilde{U}|, & \text{if } t = |G/U|, \\ 0, & \text{if } t \text{ does not divide } |G|/|U|. \end{cases}$$

5.4.12 Corollary *For each divisor t of $|G|$ there exists a G–set X_t, for which*

$$\varepsilon_U\left(\overline{X_t}\right) = \begin{cases} t, & \textit{if } t \textit{ divides } |G/U|, \\ 0, & \textit{otherwise,} \end{cases}$$

and therefore

$$|G \backslash\!\backslash_{\widetilde{U}_i} X_t| = \begin{cases} |\widetilde{U}_i|, & \textit{if } t = |G/U_i|, \\ 0, & \textit{if } t \textit{ does not divide } |G/U_i|. \end{cases}$$

This G–set X_t is of order t.

5.4.13 Application (Sylow's Theorem, revisited) A beautiful application is the following proof of Sylow's theorems due to Wagner ([157]) and Dress, Siebeneicher and Yoshida ([42]): From 5.4.12 we derive that

$$t = \sum_{i:\, t\ div.\ |G/U_i|} |\, G \backslash_{\widetilde{U}_i} X_t \,| \cdot |\, G/U_i \,| \in \sum_{i:\, t\ div.\ |G/U_i|} \mathbb{Z} \cdot |\, G/U_i \,|.$$

This shows that t is the greatest common divisor of the indices $|G/U_i|$ which are divisible by t. Hence, if $|G|$ is t times a prime power p^r, *there must exist a subgroup of order* p^r.

Dividing this equation by t and then reducing modulo p we obtain

$$1 \equiv |\{U \leq G \mid |U| = p^r\}| \ (p),$$

obtaining that *the number of subgroups of order* p^r *is congruent 1 modulo* p. The remaining Sylow theorem that each p–subgroup is subconjugate to each p–Sylow subgroup is left as exercise 5.4.4. ◇

Exercises

Exercise 5.4.1 Prove 5.4.1.

Exercise 5.4.2 Show that also the following set is a $\mathbb{Z}$–basis of $\Omega(C_n)$:

$$\left\{ \overline{\binom{C_n}{t}} \;\middle|\; t \text{ divides } n \right\}.$$

Exercise 5.4.3 Prove that for G–sets X of p–groups G the following congruence is true (if $U_d = G$):

$$\varepsilon_1\left(\overline{X}\right) \equiv n_d\left(\overline{X}\right) = \varepsilon_G\left(\overline{X}\right) \ (p).$$

Exercise 5.4.4 Use exercise 5.4.3 in order to derive that, for any p–Sylow subgroup P and each p–subgroup U, there exists a subgroup V of P which is conjugate to U. (Hint: Consider a p–subgroup V, a subgroup U for which p does not divide $|G/U|$, and examine $\varepsilon_V\left(\overline{G/U}\right)$.)

6. Representations

It will turn out to be useful to refine the preceding considerations of permutation representations. We consider linear representations on corresponding vector spaces and decompose them into their irreducible constituents. Having done this we can use all what is known on ordinary irreducible representations of finite groups, in particular we can use the results on ordinary representation theory of symmetric groups which will give us further insight into the problems.

For this purpose a self–contained introduction to the theory of ordinary representations of finite groups is given in the appendix. In this chapter, a problem–oriented introduction to the representation theory of finite symmetric groups is provided. It is based on symmetric and alternating group actions which yield two sequences of set theoretic bijections, so that the fundamental results can be formulated in terms of orbits, in terms of double cosets, in terms of matrices with natural entries and prescribed row and column sums, in terms of standard tableaux, and, last not least, in terms of standard bitableaux, which are polynomials.

This approach is carried out until we reach the point, where irreducible matrix representations can be evaluated. They allow to calculate symmetry adapted bases for vector spaces on which finite group representations act. The corresponding decompositions of vector spaces are analogous to the decomposition of sets on which finite groups act into their orbits.

6.1 Representations of Symmetric Groups

We shall characterize the ordinary irreducible representations of finite symmetric groups as constituents of certain permutation representations. In order to do this we introduce the notion of an *improper partition* λ of a natural number n, by which we mean a sequence $\lambda := (\lambda_0, \lambda_1, \ldots)$ of natural numbers λ_i such that $\sum_i \lambda_i = n$. We indicate this by writing

$$\lambda \models n.$$

Corresponding to $\lambda \models n$ let n^λ be a sequence of disjoint subsets $n_i^\lambda \subseteq n, i \in \mathbb{N}$, where

$$|n_i^\lambda| = \lambda_i,\ i \in \mathbb{N}, \text{ and } n = \bigcup_i n_i^\lambda.$$

Such an n^λ will be called a *decomposition* of n or, more explicitly, it will be called a λ*–decomposition* of n. Note that this concept is slightly more general than that of a set–partition since now *empty blocks are allowed.* We consider the direct sum of the symmetric groups of these blocks:

$$S(n^\lambda) := S_{n_0^\lambda} \oplus S_{n_1^\lambda} \oplus \ldots := \{\pi \in S_n \mid \forall\, i\colon \pi n_i^\lambda = n_i^\lambda\},$$

which is obviously isomorphic to the direct product $S_{\lambda_0} \times S_{\lambda_1} \times \ldots$. (Recall that $S_0 = \{1\}$.) These groups are Young subgroups. We can also directly associate with $\lambda \models n$ a *canonic decomposition* n^λ by putting

$$n_i^\lambda := \Big\{\sum_{j=0}^{i-1}\lambda_j, \sum_{j=0}^{i-1}\lambda_j + 1, \ldots, \sum_{j=0}^{i-1}\lambda_j + \lambda_i - 1\Big\}.$$

The corresponding Young subgroup will be called the *canonic Young subgroup* associated with λ and denoted by

$$S_\lambda.$$

It is important to realize the following consequence of 11.2.6:

6.1.1 Lemma *All the Young subgroups $S(n^\lambda)$ arising from an improper partition λ of the natural number n and corresponding λ–decompositions n^λ of the set n are conjugate subgroups of S_n.*

Thus all these subgroups corresponding to $\lambda \models n$ induce equivalent representations, and hence it suffices to consider the canonic Young subgroups S_α, α being a *proper* partition of n.

In the case when the numbering of the elements n_i^λ of n^λ matters, then we call the sequence $(n^\lambda) = (n_0^\lambda, n_1^\lambda, \ldots)$ a λ*–flag* and denote by

$$\lambda(n)$$

the set of *all* the λ–flags over n. The action

6.1.2 $$S_n \times \lambda(n) \to \lambda(n)\colon \big(\pi, (n^\lambda)\big) \mapsto (\pi n_0^\lambda, \pi n_1^\lambda, \ldots)$$

is transitive, and hence it is similar to the action of S_n on the set S_n/S_λ of left cosets of S_λ in S_n (cf. 1.2.6):

6.1.3 $${}_{S_n}\big(\lambda(n)\big) \approx {}_{S_n}\big(S_n/S_\lambda\big).$$

Thus we can apply Mackey's Theorem for group actions 1.2.16 to the corresponding action of S_n on $\lambda(n) \times \mu(n)$, where $\lambda, \mu \models n$, obtaining

6.1.4 Corollary *For any $\lambda, \mu \models n$, the corresponding canonic decompositions n^λ, n^μ and their stabilizers S_λ, S_μ we have the natural bijection*

$$S_n \backslash\backslash\big(\lambda(n) \times \mu(n)\big) \to S_\lambda\backslash S_n/S_\mu\colon S_n\big((n^\lambda), \pi(n^\mu)\big) \mapsto S_\lambda \pi S_\mu.$$

This together with 6.1.3 yields the natural bijections

$$S_n \backslash\backslash\big(S_n/S_\lambda \times S_n/S_\mu\big) \to S_n \backslash\backslash\big(\lambda(n) \times \mu(n)\big) \to S_\lambda\backslash S_n/S_\mu.$$

The permutation representation corresponding to 6.1.2 is equivalent to the representation $IS_\lambda \uparrow S_n$ induced by the identity representation of the Young subgroup (see the definition of induced representation in the appendix). We call such a representation a *Young representation* and denote its character by

$$\xi^\lambda.$$

These characters are called *Young characters* and 6.1.1 shows that the set

$$\{\xi^\alpha \mid \alpha \vdash n\}$$

is the *complete* set of Young characters of S_n. Later we shall in fact show that each ordinary irreducible character of S_n is a uniquely determined $\mathbb{Z}$–linear combination of Young characters and we shall evaluate the coefficients. The permutation representation corresponding to the action of S_n on $\lambda(n) \times \mu(n)$ is clearly equivalent to

$$IS_\lambda \uparrow S_n \otimes IS_\mu \uparrow S_n, \text{ with character } \xi^\lambda \xi^\mu.$$

Hence an application of exercise 11.5.8 yields, if we denote by ι the identity character:

6.1.5 Corollary *For any improper partitions $\lambda, \mu \models n$, the corresponding canonic Young subgroups $S_\lambda, S_\mu \leq S_n$, the Young characters ξ^λ, ξ^μ and the set $S_\lambda \backslash S_n / S_\mu$ of all the (S_λ, S_μ)–double cosets we have:*

$$[\xi^\lambda \mid \xi^\mu] = [\xi^\lambda \xi^\mu \mid \iota] = |S_n \backslash\!\backslash (\lambda(n) \times \mu(n))| = |S_\lambda \backslash S_n / S_\mu|.$$

Besides this we want to evaluate $[\xi^\lambda \xi^\mu \mid \epsilon]$, ϵ being the *alternating* or *sign* character, which maps each permutation onto its sign (cf. 11.2.16). This expression satisfies, by definition of the inner product of characters, the equation

$$[\xi^\lambda \xi^\mu \mid \epsilon] = [\xi^\lambda \xi^\mu \downarrow A_n \mid \iota \downarrow A_n] - [\xi^\lambda \xi^\mu \mid \iota],$$

and so it is in fact equal to the number of orbits of S_n on the set $\lambda(n) \times \mu(n)$ which *split over* A_n, i. e. which consist of several, and hence of exactly two, A_n–orbits (recall 2.1.13). But, for $n > 1$, an S_n–orbit splits over A_n if and only if the stabilizers of its elements are the same in both of these groups. Thus an application of 1.2.15 yields

6.1.6 $$[\xi^\lambda \xi^\mu \mid \epsilon] = \left|\left\{S_\lambda \pi S_\mu \in S_\lambda \backslash S_n / S_\mu \;\middle|\; S_\lambda \cap \pi S_\mu \pi^{-1} = \{1\}\right\}\right|.$$

In order to make these results more explicit we give a more detailed description of the double cosets $S_\lambda \pi S_\mu$:

6.1.7 Coleman's Lemma *For $\lambda, \mu \models n$ and corresponding flags (n^λ) and (n^μ) the following is true*

$$\pi \in S_\lambda \rho S_\mu \iff \forall\, i, j\colon |n_i^\lambda \cap \pi n_j^\mu| = |n_i^\lambda \cap \rho n_j^\mu|.$$

Proof:
i) If $\pi \in S_\lambda \rho S_\mu$, say $\pi = \sigma\rho\tau, \sigma \in S_\lambda, \tau \in S_\mu$, then, for each j,

$$\pi n_j^\mu = \sigma\rho\tau n_j^\mu = \sigma\rho n_j^\mu,$$

thus for each i and j we have:

$$n_i^\lambda \cap \pi n_j^\mu = n_i^\lambda \cap \sigma\rho n_j^\mu = \sigma\big[n_i^\lambda \cap \rho n_j^\mu\big].$$

This yields one half of the statement since σ is a bijection.
ii) The assumption

$$\forall\, i, j: \ |n_i^\lambda \cap \rho n_j^\mu| = |n_i^\lambda \cap \pi n_j^\mu|$$

implies that for fixed i the subsets $n_i^\lambda \cap \rho n_j^\mu$ and the subsets $n_i^\lambda \cap \pi n_j^\mu$ form two dissections of n_i^λ into subsets which can be collected into pairs

$$(n_i^\lambda \cap \rho n_j^\mu, n_i^\lambda \cap \pi n_j^\mu)$$

of subsets of equal order. Hence for each i there exists $\sigma_i \in S_{n_i^\lambda}$, which satisfies

$$\forall\, j: \ \sigma_i\big[n_i^\lambda \cap \rho n_j^\mu\big] = n_i^\lambda \cap \pi n_j^\mu.$$

The product $\sigma := \sigma_0\sigma_1 \ldots \in S_\lambda$ of such permutations σ_i then satisfies

$$\forall j: \sigma\rho n_j^\mu = \pi n_j^\mu.$$

Thus there exist $\tau \in S_\mu$ such that $\pi = \sigma\rho\tau$, as stated. □

The main point is the following direct consequence of Coleman's Lemma:

6.1.8 Corollary *If we denote by $M_{\lambda\mu}$ the set of matrices over $\mathbb{N}$ with row sums $\lambda_0, \lambda_1, \ldots$ and column sums $\mu_0, \mu_1, \ldots$, then we have the bijection*

$$S_\lambda \backslash S_n / S_\mu \longrightarrow M_{\lambda\mu}: S_\lambda \rho S_\mu \mapsto (z_{ij}),$$

where $z_{ij} := |n_i^\lambda \cap \rho n_j^\mu|$. The restriction of this mapping to the set

$$(S_\lambda \backslash S_n / S_\mu)' := \big\{S_\lambda \pi S_\mu \,\big|\, S_\lambda \cap \pi S_\mu \pi^{-1} = \{1\}\big\}$$

of double cosets with trivial intersection has as its image the set

$$M'_{\lambda\mu} := \big\{(z_{ij}) \in M_{\lambda\mu} \,\big|\, z_{ij} \in \{0, 1\}\big\}$$

of 0–1–matrices with row sums $\lambda_0, \lambda_1, \ldots$ and column sums $\mu_0, \mu_1, \ldots$.

Applying this to the above inner products of characters we obtain

6.1.9 Corollary *For $\lambda, \mu \models n$ we have*

$$[\xi^\lambda \mid \xi^\mu] = |M_{\lambda\mu}| \text{ and } [\xi^\lambda \mid \epsilon\xi^\mu] = |M'_{\lambda\mu}|.$$

We call the matrix (z_{ij}) the *double coset symbol* or, in short, the *dc–symbol* corresponding to $S_\lambda \rho S_\mu$.

Summarizing 6.1.4 and 6.1.8 and denoting by $(S_n \backslash\backslash \lambda(n) \times \mu(n))'$ the set of orbits which split over the alternating group, we get the following result:

6.1.10 Corollary *For $n > 1$ and $\lambda, \mu \models n$ we have natural bijections*

$$S_n \backslash\backslash S_n/S_\lambda \times S_n/S_\mu \to S_n \backslash\backslash \lambda(n) \times \mu(n) \to M_{\lambda\mu},$$

and

$$(S_n \backslash\backslash S_n/S_\lambda \times S_n/S_\mu)' \to (S_n \backslash\backslash \lambda(n) \times \mu(n))' \to M'_{\lambda\mu},$$

according to 6.1.8.

Now we restrict attention to specific pairs (α, β) of proper partitions of n, where β is closely related to α. In order to define this particular β we illustrate α by the corresponding *Young diagram* $[\alpha]$, which consists of n nodes $\times$ placed in rows. The i-th row of $[\alpha]$ consists of α_i nodes, and all the rows start in the same column. The partition $(3, 2, 1^2)$ for example is illustrated by $[3, 2, 1^2]$ (we write $[3, 2, 1^2]$ instead of $[(3, 2, 1^2)]$) which looks as follows:

$$\begin{array}{lll} \times & \times & \times \\ \times & \times & \\ \times & & \\ \times & & \end{array} .$$

Recalling that $\alpha_i \geq \alpha_{i+1}$, we see that the lengths α'_i of the columns of $[\alpha]$ form another partition α' of n:

$$\alpha' := (\alpha'_0, \alpha'_1, \ldots), \text{ where } \alpha'_i := \sum_{j, \alpha_j > i} 1.$$

This partition α' is called the *partition associated with* α. The Young diagram $[\alpha']$ arises from $[\alpha]$ by simply reflecting $[\alpha]$ in its main diagonal, e. g. $[(3, 2, 1^2)'] = [4, 2, 1]$. Partitions α and Young diagrams $[\alpha]$ where $\alpha = \alpha'$ are called *self–associated*.

It is an easy exercise to see that for any pair (α, α') of associated partitions there exists exactly one 0-1–matrix with row sums α_i and column sums α'_i :

6.1.11 $$\forall\, \alpha \vdash n\colon\ |M'_{\alpha\alpha'}| = 1.$$

This has the following important consequence:

6.1.12 $$\forall\, \alpha \vdash n\colon\ [\xi^\alpha \mid \epsilon\xi^{\alpha'}] = 1.$$

The representation of S_n which has the character $\epsilon\xi^{\alpha'}$ is the representation which is induced by the *alternating representation* A of $S_{\alpha'}$ (please check this carefully):

$$A\colon S_{\alpha'} \to GL(\mathbb{C})\colon \pi \mapsto \epsilon(\pi) id_{\mathbb{C}}.$$

Thus 6.1.12 yields (apply 11.5.2 and 11.5.3) the following result which is basic for the representation theory of the symmetric groups:

6.1.13 Theorem *For each proper partition $\alpha \vdash n$, the induced representations $IS_\alpha \uparrow S_n$ and $AS_{\alpha'} \uparrow S_n$ of S_n have exactly one irreducible representation as a common constituent, and they both contain it with multiplicity 1.*

We denote this constituent or its equivalence class by $[\alpha]$ and its character by ζ^α, so that, by slight abuse of the intersection symbol, we have

$$[\alpha] := IS_\alpha \uparrow S_n \cap AS_{\alpha'} \uparrow S_n, \text{ with character } \zeta^\alpha.$$

As $IS_{(n)} \uparrow S_n = IS_n$, the identity representation, $AS_{(n)} \uparrow S_n = AS_n$, the alternating representation, and $IS_{(1^n)} \uparrow S_n = AS_{(1^n)} \uparrow S_n = RS_n$, the regular representation, we obtain from 11.5.4 the following particular cases:

6.1.14 $$[n] = IS_n,\ [1^n] = AS_n,\ [\alpha'] = [1^n] \otimes [\alpha].$$

It is in fact our aim to show that $\{[\alpha] \mid \alpha \vdash n\}$ is a transversal of the equivalence classes of the ordinary irreducible representations of S_n. In order to prove this we introduce a total and a partial order on the sets

$$P(n) := \{\alpha \mid \alpha \vdash n\} \subseteq IP(n) := \{\lambda \mid \lambda \models n\}$$

of all the proper and all the improper partitions of n. A natural total order is the *lexicographic order*

$$\lambda < \mu :\Longleftrightarrow \exists\, i\colon \lambda_0 = \mu_0, \ldots, \lambda_{i-1} = \mu_{i-1}, \lambda_i < \mu_i.$$

The partial order $\trianglelefteq$ called the *dominance order*, that we met already is defined by

$$\lambda \trianglelefteq \mu :\Longleftrightarrow \forall\, i\colon \sum_0^i \lambda_\nu \leq \sum_0^i \mu_\nu.$$

It differs from the lexicographic order on $P(n)$ if and only if $n \geq 6$. The Hasse diagram of the poset $(P(6), \trianglelefteq)$ is shown here:

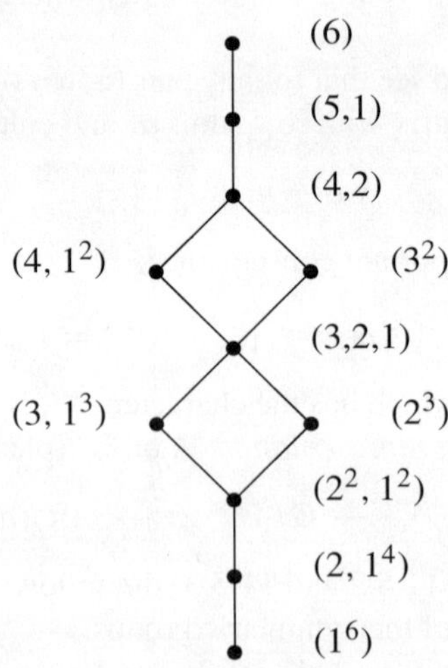

It will be useful to characterize when $\alpha \vdash n$ is *covered* by $\beta \vdash n$, i. e. when

$$\alpha \lhd \beta \text{ and } \nexists\, \gamma \vdash n\colon\ \alpha \lhd \gamma \lhd \beta.$$

We shall see that this can be expressed very nicely in terms of the corresponding Young diagrams $[\alpha]$ and $[\beta]$. In fact $[\beta]$ arises from $[\alpha]$ by moving one node from the end of a certain row of $[\alpha]$ upwards to the end of another row, say from the j-th row to the i-th row, as indicated here:

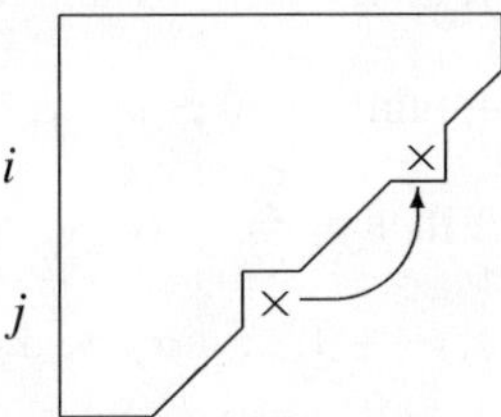

Moreover this step has to be as small a step as possible, which means that either $i = j - 1$ or $i < j - 1$ and $\alpha_i = \alpha_{i+1} = \ldots = \alpha_j$:

In formal terms this reads as follows:

6.1.15 Theorem *$\alpha \vdash n$ is covered by $\beta \vdash n$ if and only if there exist indices i and j such that $0 \leq i < j$ and*

i) $\beta_j = \alpha_j - 1$ and $\beta_i = \alpha_i + 1$, while, for $k \neq i, j$, we have $\alpha_k = \beta_k$,

ii) $i = j - 1$, or $\alpha_i = \alpha_j$.

Proof: Assuming first that α is covered by β, we put

$$i := \min\{\, k \mid \alpha_k \neq \beta_k \,\},\ \ j := \min\{\, t \mid \sum_0^t \alpha_k = \sum_0^t \beta_k, i < t \,\}.$$

Then obviously

$$0 \leq i < j \leq n,\ \text{ and } \alpha_j > \beta_j \geq 0, \beta_{j+1} \geq \alpha_{j+1}.$$

If $i > 0$, then $\alpha_i < \beta_i \leq \beta_{i-1} = \alpha_{i-1}$, and so in this case

$$\alpha_i + 1 \leq \beta_i \leq \beta_{i-1} = \alpha_{i-1}.$$

Also $\alpha_j > \beta_j \geq \beta_{j+1} \geq \alpha_{j+1}$, and therefore

$$\alpha_j - 1 \geq \alpha_{j+1}.$$

Hence in any case, whether $i > 0$ or not, we obtain

$$\alpha \lhd \gamma := (\alpha_0, \ldots, \alpha_{i-1}, \alpha_i + 1, \alpha_{i+1}, \ldots, \alpha_{j-1}, \alpha_j - 1, \alpha_{j+1}, \ldots) \unlhd \beta,$$

so that $\gamma = \beta$ by assumption, and $[\beta]$ arises from $[\alpha]$ by moving exactly one node upwards from the end of the j-th row to the end of the i-th row of $[\alpha]$.

If $i \neq j - 1$ and $\alpha_i \neq \alpha_j$, then $i < j - 1$ and $\alpha_i > \alpha_j$. We can therefore put

$$t := 1 + \min\{k \mid \alpha_k > \alpha_{k+1}, i \leq k < j\}.$$

Note that $i < t \leq j$. If $t = j$, then $\alpha_i = \ldots = \alpha_{j-1} > \alpha_j > 0$, and thus

$$\alpha \lhd (\alpha_0, \ldots, \alpha_{i-1}, \alpha_i + 1, \ldots, \alpha_{j-2}, \alpha_{j-1} - 1, \alpha_j, \ldots) \lhd \beta.$$

But this contradicts the assumption made that β covers α. If $t < j$, then $\alpha_i = \ldots = \alpha_{t-1} > \alpha_t \geq \ldots \geq \alpha_j > 0$, and so in this case we have

$$\alpha \lhd (\alpha_0, \ldots, \alpha_{t-1}, \alpha_t + 1, \ldots, \alpha_{j-1}, \alpha_j - 1, \alpha_{j+1}, \ldots) \lhd \beta,$$

which also contradicts the assumption.

Conversely let α, β, i and j satisfy the conditions of the theorem, and let γ be a partition of n with the property that γ covers α and $\gamma \unlhd \beta$. Then, for the parts of γ, we have:

$$\forall\, k < i : \alpha_k = \gamma_k = \beta_k, \text{ and } \forall\, k > j : \alpha_k = \gamma_k = \beta_k.$$

Hence, if $i = j - 1$, the fact that γ covers α implies that $\gamma_i = \alpha_i + 1$, and $\gamma_{i+1} = \alpha_i - 1$, and then $\gamma = \beta$. In the case when $i \neq j - 1$, we obtain from the assumption that $\alpha_i = \ldots = \alpha_j$. Furthermore, by the same arguments which were used in the first part of this proof, there exist k and l such that

$$\gamma_k = \alpha_k + 1, \gamma_l = \alpha_l - 1, \gamma_\nu = \alpha_\nu \text{ for } \nu \neq k, l.$$

But this can hold only if $k = i$ and $l = j$, for otherwise

$$\alpha_k = \gamma_k - 1 \leq \gamma_{k-1} - 1 = \alpha_{k-1} - 1 < \alpha_{k-1} = \alpha_i,$$

or

$$\alpha_l = \gamma_l + 1 \geq \gamma_{l+1} + 1 = \alpha_{l+1} + 1 > \alpha_{l+1} = \alpha_j$$

would contradict $\alpha_i = \ldots = \alpha_j$. Therefore $\gamma = \beta$ also in this case. □

We shall use this in the proof of the following very important fact:

6.1.16 Lemma *For any proper partitions $\alpha, \beta \vdash n$ we have:*

$$\alpha \unlhd \beta \iff \beta' \unlhd \alpha'.$$

Proof: If $\alpha \lhd \beta$, then there exist partitions $\alpha^\nu \vdash n, 0 \leq \nu \leq r$, which satisfy

$$\alpha = \alpha^0 \lhd \alpha^1 \lhd \ldots \lhd \alpha^r = \beta,$$

and where α^i *covers* α^{i-1}. Now 6.1.15 shows that

$$(\alpha^i)' \lhd (\alpha^{i-1})'.$$

Thus we obtain the chain of conjugate partitions

$$\beta' = (\alpha^r)' \lhd \ldots \lhd (\alpha^0)' = \alpha',$$

which shows that $\beta' \lhd \alpha'$. The converse is true by symmetry. □

Obviously $\unlhd$ can be embedded into the lexicographic order:

6.1.17 $$\alpha \unlhd \beta \Longrightarrow \alpha \leq \beta.$$

The dominance order on $P(n)$ is defined in terms of the partial sums

$$s_i^\alpha := \sum_0^i \alpha_j$$

of the parts of α. Using these expressions we can define, for any $\alpha, \beta \vdash n$, the *infimum* $\alpha \wedge \beta$ by

$$\alpha \wedge \beta := \gamma, \text{ where } s_i^\gamma := \min\{\, s_i^\alpha, s_i^\beta \,\},$$

and

$$\alpha \vee \beta := (\alpha' \wedge \beta')'.$$

It is easy to check that $\alpha \vee \beta$ is the *supremum* and $\alpha \wedge \beta$ the infimum of α and β with respect to the dominance order, so that we have obtained (cf. 6.1.16):

6.1.18 Corollary *The dominance order $\unlhd$ induces a lattice structure on $P(n)$. The mapping*

$$-': P(n) \to P(n): \alpha \mapsto \alpha'$$

is an order antiisomorphism on $P(n)$.

This lattice $(P(n), \wedge, \vee)$ is a very important combinatorial structure. We call it the *diagram lattice* of order n, since the name "partition lattice" would be misleading, as it is already the standard name of a different lattice structure.

We return to the matrices with prescribed row and column sums.

6.1.19 Lemma *For any proper partitions $\alpha, \beta \vdash n$ we have:*

$$\left| M'_{\alpha\beta'} \right| > 0 \Longrightarrow \alpha \unlhd \beta.$$

Proof: Assume that $A \in M'_{\alpha\beta'}$ and shift its nonzero entries in their rows as far to the left as possible. The resulting matrix A' contains in its 0-th column α'_0 elements 1. Hence $\alpha'_0 \geq \beta'_0$. Moreover the first two columns of A' contain $\alpha'_0 + \alpha'_1$ entries 1, so that also $\alpha'_0 + \alpha'_1 \geq \beta'_0 + \beta'_1$, and so on. Thus $\alpha' \trianglerighteq \beta'$, which is equivalent to the assertion, by 6.1.16. □

6.1.20 Lemma $[\xi^\alpha \mid \zeta^\beta] > 0$ *implies that* $\alpha \trianglelefteq \beta$.

Proof: The equation $[\epsilon\xi^{\beta'} \mid \zeta^\beta] = 1$ together with the above assumption yields that

$$0 < [\xi^\alpha \mid \epsilon\xi^{\beta'}] = \left| M'_{\alpha\beta'} \right|,$$

which implies the statement by 6.1.19. □

We are now in a position to prove the main theorem of the ordinary representation theory of finite symmetric groups:

6.1.21 Theorem *The set* $\{[\alpha] \mid \alpha \vdash n\}$ *is a transversal of the equivalence classes of ordinary irreducible representations of* S_n.

Proof: We need only show that $[\alpha] = [\beta]$ implies $\alpha = \beta$, since the cardinality of $\{[\alpha] \mid \alpha \vdash n\}$ is equal to the number of conjugacy classes of S_n, and so this system must be complete. But if $[\alpha] = [\beta]$, we can argue in the following way:

$$(IS_\alpha \uparrow S_n, [\beta]) = (IS_\alpha \uparrow S_n, [\alpha]) = 1 = (IS_\beta \uparrow S_n, [\beta]) = (IS_\beta \uparrow S_n, [\alpha]).$$

Thus, by 6.1.20, we have both $\alpha \trianglelefteq \beta$ and $\beta \trianglelefteq \alpha$, which imply $\alpha = \beta$, and the proof is complete. □

We denote the multiplicities of the irreducible constituents of Young representations as follows:

$$\kappa_{\alpha\beta} := (IS_\alpha \uparrow S_n, [\beta]) = [\xi^\alpha \mid \zeta^\beta].$$

These numbers are called the *Kostka numbers* . With respect to the *reverse lexicographic order* on $P(n)$, the matrix consisting of these multiplicities is of the following form (cf. 6.1.13 and 6.1.20):

6.1.22 $$K_n := (\kappa_{\alpha\beta}) = \begin{pmatrix} 1 & & & & \\ \vdots & \ddots & & & \\ 1 & & 1 & & \\ \vdots & \star & & \ddots & \\ 1 & \dots & f^\beta & \dots & 1 \end{pmatrix},$$

where f^β denotes the dimension of $[\beta]$. The triangular form of this *Kostka matrix* K_n has several important consequences. We denote by ζ^α_β and ξ^α_β the value of ζ^α and that of ξ^α on the conjugacy class of elements with cycle partition β. Putting these values into matrices, say

$$Z_n := (\zeta^\alpha_\beta), \ \Xi_n := (\xi^\alpha_\beta),$$

we obtain, using the lexicographic order of the partitions:

6.1.23
$$\Xi_n = K_n Z_n, \text{ and hence } Z_n = K_n^{-1}\Xi_n.$$

As K_n^{-1} is a matrix over $\mathbb{Z}$ (see 6.1.22), this implies

6.1.24 Corollary *Each ordinary irreducible character ζ^α of S_n is a uniquely determined $\mathbb{Z}$–linear combination of the Young characters ξ^β, $\beta \vdash n$.*

6.1.24 shows the importance of the Kostka matrix K_n. It can be evaluated as soon as the numbers of double cosets are known:

6.1.25
$$\sum_{\gamma \vdash n} \kappa_{\alpha\gamma}\kappa_{\beta\gamma} = [\xi^\alpha \mid \xi^\beta] = |S_\alpha \backslash S_n / S_\beta|,$$

while these numbers can be obtained as follows:

6.1.26 Lemma *$|S_\alpha \backslash S_n / S_\beta|$ is the coefficient of the monomial*

$$x^\alpha y^\beta := x_0^{\alpha_0} x_1^{\alpha_1} \dots y_0^{\beta_0} y_1^{\beta_1} \dots$$

in the formal power series

$$\prod_{i,k\in\mathbb{N}} (1 - x_i y_k)^{-1} := \prod_{i,k} \Big(\sum_{\nu=0}^{\infty} x_i^\nu y_k^\nu \Big).$$

But 6.1.23 also shows that K_n^{-1} is even more important than K_n. We shall therefore give a better description of the inverse of the Kostka matrix later. It remains to describe how $\Xi_n = (\xi_\beta^\alpha)$ can be evaluated, so that we can derive the character table Z_n by an application of 6.1.23 and 6.1.26, say. The Young character ξ^α is the character of the action of S_n on the set $\alpha(n)$ of α–flags. A flag $n^\alpha = (n_0^\alpha, \dots, n_{h-1}^\alpha)$ can be illustrated by the corresponding *tabloid* that contains in its i–th row the elements of the i–th block n_i^α of the flag in their natural order:

$$\begin{array}{c} \overline{r_0 \cdot\cdot\cdot\cdot\cdot\cdot r_{\alpha_0 - 1}} \\ \cdot\cdot\cdot\cdot\cdot\cdot\cdot\cdot\cdot \\ \overline{t_0 \dots t_{\alpha_{h-1}-1}} \end{array},$$

if $n_0^\alpha = \{r_0 < \dots < r_{\alpha_0 - 1}\}, \dots, n_{h-1}^\alpha = \{t_0 < \dots < t_{\alpha_{h-1}-1}\}$. More explicitly we shall also call such a tabloid an α–*tabloid*. For example

$$\begin{array}{c} \overline{013} \\ \overline{24} \end{array}$$

is the (3,2)-tabloid representing the flag $5^{(3,2)} := \{\{0, 3, 1\}, \{4, 2\}\}$. It is obvious that such a tabloid is fixed under $\pi \in S_n$ if and only if, for each cyclic factor of π, the points contained in it belong to the same block, or, in other words, if and only if each block n_i^α is a union of orbits of π. This proves

6.1.27 Lemma *For each $\alpha = (\alpha_0, \dots, \alpha_{h-1}) \vdash n$ and any $\pi \in S_n$, the Young character value $\xi^\alpha(\pi)$ can be expressed in terms of multinomial coefficients as follows:*

$$\xi^\alpha(\pi) = \sum_{a^{(0)},\dots,a^{(h-1)}} \prod_{i=1}^n \binom{a_i(\pi)}{a_i^{(0)}, \dots, a_i^{(h-1)}}.$$

The sum has to be taken over the cycle types $a^{(0)} \dashv\vdash \alpha_0, \dots, a^{(h-1)} \dashv\vdash \alpha_{h-1}$ which sum up to $a(\pi)$ in the following sense:

$$\sum_i a^{(i)} := \Big(\sum_i a_1^{(i)}, \dots, \sum_i a_n^{(i)}\Big) = (a_1(\pi), \dots, a_n(\pi)).$$

For example:

6.1.28 $$\xi^{(n)}(\pi) = 1, \xi^{(n-1,1)}(\pi) = a_1(\pi), \xi^{(n-2,2)}(\pi) = \binom{a_1(\pi)}{2} + a_2(\pi).$$

Lemma 6.1.27 shows that each ξ^α and hence also each ζ^α is a polynomial function in the a_i. Let us consider the multinomial factors of its summands. As

$$\binom{a_i(\pi)}{a_i^{(0)}, \dots, a_i^{(h-1)}} := \frac{a_i(\pi)!}{a_i^{(0)}! \dots a_i^{(h-1)}!} = \binom{a_i(\pi)}{a_i^{(0)}} \binom{\sum_1^{h-1} a_i^{(j)}}{a_i^{(1)}, \dots, a_i^{(h-1)}},$$

we see that $\xi^\alpha(\pi)$ is the value, at $(x_1, \dots, x_n) := a(\pi)$, of the polynomial function $\Xi^{\hat\alpha}$, corresponding to the *truncated partition* $\hat\alpha := (\alpha_1, \alpha_2, \dots)$, that belongs to the polynomial

$$\Xi^{\hat\alpha} := \sum_{a^{(1)},\dots,a^{(h-1)}} \prod_{i=1}^n \binom{\sum_1^{h-1} a_i^{(j)}}{a_i^{(1)}, \dots, a_i^{(h-1)}} \binom{x_i}{\sum_1^{h-1} a_i^{(j)}} \in \mathbb{Q}[x_1, x_2, \dots].$$

The sum has to be taken over all the cycle types $a^{(1)} \dashv\vdash \alpha_1, \dots, a^{(h-1)} \dashv\vdash \alpha_{h-1}$, $\alpha_1, \dots, \alpha_{h-1}$ being the (nonzero) parts of the truncated partition $\hat\alpha$. Moreover, if $\hat\beta = \hat\alpha$, then $\Xi^{\hat\beta} = \Xi^{\hat\alpha}$, hence $\Xi^{\hat\alpha}$ gives an *infinite series* of Young characters, namely all the ξ^β for which $\beta = (\beta_0, \alpha_1, \alpha_2, \dots)$, via

$$\xi^\beta(\pi) = \Xi^{\hat\alpha}(a_1(\pi), \dots, a_m(\pi)), \text{ if } \beta \vdash m = \beta_0 + \sum_1^{h-1} \alpha_i.$$

Hence $\Xi^{\hat\alpha} \in \mathbb{Q}[x_1, x_2, \dots]$ has prescribed values at infinitely many points $a = (a_1, a_2, \dots)$, and at each such point only finitely many coordinates a_i are nonzero. This proves

6.1.29 Theorem *For each proper partition $\hat\alpha = (\alpha_1, \alpha_2, \dots, \alpha_{h-1}) \vdash n$ there exists the uniquely determined polynomial*

$$\Xi^{\hat{\alpha}} := \sum_{a^{(1)},\dots,a^{(h-1)}} \prod_{i=1}^{n} \binom{\sum_1^{h-1} a_i^{(j)}}{a_i^{(1)},\dots,a_i^{(h-1)}} \binom{x_i}{\sum_1^{h-1} a_i^{(j)}} \in \mathbb{Q}[x_1, x_2, \dots]$$

(where the sum is over all $a^{(1)} \vDash \alpha_1, \dots, a^{(h-1)} \vDash \alpha_{h-1}$, if $\alpha_1, \dots, \alpha_{h-1}$ are the nonzero parts of $\hat{\alpha}$) such that the corresponding polynomial function yields the values $\xi^\beta(\pi)$ of all the Young characters ξ^β, where $\beta = (\beta_0, \alpha_1, \dots, \alpha_{h-1})$ via

$$\xi_a^\beta = \Xi^{\hat{\alpha}}(a_1, \dots, a_m), \text{ if } \beta \vdash m = n + \beta_0.$$

We call these polynomials *Young polynomials*. Here are the smallest examples (cf. 6.1.28):

6.1.30 $$\Xi^{\emptyset} = 1,\ \Xi^{(1)} = x_1,\ \Xi^{(2)} = \binom{x_1}{2} + x_2,\ \Xi^{(1^2)} = x_1(x_1 - 1).$$

Many further examples of Young polynomials and the corresponding character polynomials which we shall introduce later, and which yield infinitely many irreducible characters, can be found in the appendix.

Exercises

Exercise 6.1.1 Evaluate K_3 and K_4, Z_3 and Z_4.

6.2 Tableaux and Matrices

The notion of a *tableau* is motivated by the equation $|M'_{\alpha\alpha'}| = 1$, which means that to an α–flag (n^α) and an α'–flag $(n^{\alpha'})$ there corresponds *exactly one* double coset $S(n^\alpha)\pi S(n^{\alpha'})$ with *trivial intersection property*

$$S(n^\alpha) \cap \pi S(n^{\alpha'})\pi^{-1} = \{1\}.$$

We want to construct a representative π of this particular double coset. In order to do this we replace the nodes in the i-th *row* of the Young diagram $[\alpha]$ by the elements of the block n_i^α and take for the k-th block of $\pi(n^{\alpha'})$ the resulting set of elements in the k–th *column*. Obviously

$$S(n^\alpha) \cap \pi S(n^{\alpha'})\pi^{-1} = S(n^\alpha) \cap S(\pi(n^{\alpha'})) = \{1\}.$$

The following example illustrates this. The flags

$$\left(n^\alpha\right) := \left(\{0, 2, 3\}, \{1, 4\}\right), \text{ and } \left(n^{\alpha'}\right) := \left(\{0, 1\}, \{2, 3\}, \{4\}\right)$$

yield the *tableau* (the exact definition will be given later)

$$\begin{matrix} 2 & 0 & 3 \\ 1 & 4 & \end{matrix},$$

if we choose the permutation

$$\pi = \begin{pmatrix} 0 & 1 & 2 & 3 & 4 \\ 1 & 2 & 0 & 4 & 3 \end{pmatrix},$$

which satisfies $S(n^\alpha) \cap \pi S(n^{\alpha'})\pi^{-1} = \{1\}$. More formally, and in fact more generally, we can describe this process of replacing the nodes of the diagram $[\alpha]$ by a mapping as follows. The Young diagram $[\alpha]$ can be considered as a subset of $\mathbb{N} \times \mathbb{N}$:

$$[\alpha] = \bigcup_i \{(i, 0), \ldots, (i, \alpha_i - 1)\},$$

where (i, j) is the coordinate pair of the node in the i–th row and the j–th column of $[\alpha]$. Let now $\mathcal{R}$ be a nonempty *range* set. Each mapping

$$T: [\alpha] \to \mathcal{R}: (i, j) \mapsto t_{ij}$$

is called an *α–tableau* over $\mathcal{R}$. The partition $\text{sh}(T) := \alpha$ is called the *shape* of T, and $t_{ij} = T((i, j))$ is called the *entry* of T at (i, j). Usually the range $\mathcal{R}$ will be a totally ordered set and we shall mostly take $\mathcal{R} = n$, if $\text{sh}(T) \vdash n$. In this case we put

$$c(T) := (\lambda_0, \lambda_1, \ldots), \text{ where } \lambda_i := |T^{-1}(i)|,$$

and call this sequence the *content* of T. It is an improper partition of n: $c(T) \models n$. Tableaux of content (1^n) will turn out to be of particular importance for us; they are called *Young tableaux* . The example given above is in fact a Young tableau. A tableau T over n will be called *row injective* if and only if T is injective on each of its *rows*. A tableau T is called *strictly (weakly) row monotone* if and only if, for each i, $t_{ij} < t_{i,j+1} (t_{ij} \leq t_{i,j+1})$, for each j. Monotony along the columns is defined in a similar way. T is called a *standard tableau* if and only if it is weakly monotone in its rows and strictly monotone in its columns. The set of α–tableaux over n, the set of standard α–tableaux over n, the set of α–tableaux and the set of all the standard α–tableaux of content λ will be denoted by

$$T^\alpha(n),\ ST^\alpha(n),\ T^\alpha(\lambda), \text{ and } ST^\alpha(\lambda),$$

respectively, while the set of all the Young tableaux of shape α and the set of all the standard Young tableaux of shape α will be abbreviated as follows:

$$T^\alpha := T^\alpha((1^n)), \text{ and } ST^\alpha := ST^\alpha((1^n)).$$

The definition of standard tableau immediately implies that we have, for each $\alpha \vdash n$ and any $\lambda \models n$, the implication

6.2.1 $$ST^\alpha(\lambda) \neq \emptyset \Longrightarrow \lambda \trianglelefteq \alpha,$$

where, for any $\lambda, \mu \models n$, the *dominance* of improper partitions is defined in the obvious way:

$$\lambda \trianglelefteq \mu :\Longleftrightarrow \forall\, i\colon \sum_0^i \lambda_\nu \leq \sum_0^i \mu_\nu .$$

We are now in a position to connect tableaux, Young characters and matrices over $\mathbb{N}$ with prescribed row and column sums. Our first step will be to establish bijections between $M_{\lambda\mu}$ and $M'_{\lambda\mu}$ and the sets

$$\bigcup_{\alpha \vdash n} ST^\alpha(\lambda) \times ST^\alpha(\mu) \text{ and } \bigcup_{\alpha \vdash n} ST^\alpha(\lambda) \times ST^{\alpha'}(\mu),$$

of *pairs* of standard tableaux, respectively. In order to show how this can be done, we map the matrix $Z := (z_{ij}) \in M_{\lambda\mu}$ onto the *double sequence*

$$z := \begin{pmatrix} z^0 \\ z^1 \end{pmatrix} := \begin{pmatrix} \underbrace{\begin{matrix} 0 \ldots 0 \\ 0 \ldots 0 \end{matrix}}_{z_{00}} \underbrace{\begin{matrix} 0 \ldots 0 \\ 1 \ldots 1 \end{matrix}}_{z_{01}} \begin{matrix} \ldots \\ \ldots \end{matrix} \underbrace{\begin{matrix} i \ldots i \\ j \ldots j \end{matrix}}_{z_{ij}} \begin{matrix} \ldots \\ \ldots \end{matrix} \end{pmatrix}.$$

For example

$$Z := \begin{pmatrix} 0 & 0 & 2 \\ 1 & 1 & 0 \\ 3 & 0 & 0 \end{pmatrix} \longmapsto z = \begin{pmatrix} 0 & 0 & 1 & 1 & 2 & 2 & 2 \\ 2 & 2 & 0 & 1 & 0 & 0 & 0 \end{pmatrix}.$$

We note that the content $c(z^0)$ of the row z^0 of z is λ, while $c(z^1) = \mu$. The double sequence z will now be mapped onto a pair of standard tableaux, a *standard bitableau.* We start with the lower row z^1 and apply to it the *Insertion Algorithm* which yields a standard tableau, the construction of which yields then a second tableau. This algorithm is due to D. E. Knuth ([83]) and it generalizes particular cases already described by G. de B. Robinson, G. Schensted ([136]) and M. P. Schützenberger ([140],[139],[138]).

6.2.2 The Insertion Algorithm *The following method associates with each matrix over $\mathbb{N}$, having row sums vector λ and column sums vector μ, where $\lambda, \mu \models n$, a standard tableau T with $c(T) = \mu$:*

Construct first the corresponding double sequence z and then from $z^1 = (z^1_0, z^1_1, \ldots)$ a sequence $T^0, T^1, \ldots, T^n$ of tableaux as follows:

- *$T^0 := \emptyset$, the empty tableau.*
- *For $1 \leq i \leq n$ the tableau T^i is obtained from the tableau T^{i-1} by insertion of $r := z^1_i$ according to the following rules:*

  1. *If the first row of T^{i-1} does not contain an element greater than r, put r at the end of that row.*

2. *Otherwise take the leftmost $s > r$ from the row with number 0 of T^{i-1}, replace it by r and call the result $(T^{i-1})'$. Compare the element s which was "bumped" by r with the elements in the row with number 1 of $(T^{i-1})'$. If there is no $t > s$, then put s at the end of this row, otherwise bump the leftmost $t > s$, obtain $(T^{i-1})''$ and compare t with the elements in the row with number 2 of $(T^{i-1})''$, and so on, until finally the bumped element is put at the end of the next row since this row does not contain any bigger entry.*

The example z shown above yields the sequence

$$T^0 = \emptyset \mapsto T^1 = 2 \mapsto T^2 = 2 \quad 2 \mapsto T^3 = \begin{matrix} 0 & 2 \\ 2 & \end{matrix} \mapsto T^4 = \begin{matrix} 0 & 1 \\ 2 & 2 \end{matrix}$$

$$\mapsto T^5 = \begin{matrix} 0 & 0 \\ 1 & 2 \\ 2 & \end{matrix} \mapsto T^6 = \begin{matrix} 0 & 0 & 0 \\ 1 & 2 & \\ 2 & & \end{matrix} \mapsto T^7 = \begin{matrix} 0 & 0 & 0 & 0 \\ 1 & 2 & & \\ 2 & & & \end{matrix} .$$

Proof of the theorem: We use induction. The tableau T^1 is obviously standard. Assume that T^{i-1} is standard. The insertion of z_i^1 yields T^i which is weakly monotone in its rows. But T^i is also strictly monotone in its columns, which follows easily by an indirect proof. □

We now put $T_1 := T^n$. The other tableau T_0 is obtained in parallel to the construction of T_1 by simply putting the element z_i^0 of z^0, the upper row of z, into the position which is filled when obtaining T^i from T^{i-1}. Our example gives

$$T_0 = \begin{matrix} 0 & 0 & 2 & 2 \\ 1 & 1 & & \\ 2 & & & \end{matrix} .$$

Thus we have obtained the standard bitableau

$$(T_0, T_1) = \begin{pmatrix} \begin{matrix} 0 & 0 & 2 & 2 \\ 1 & 1 & & \\ 2 & & & \end{matrix} , & \begin{matrix} 0 & 0 & 0 & 0 \\ 1 & 2 & & \\ 2 & & & \end{matrix} \end{pmatrix},$$

which is an element of $SBT^{(4,2,1)}((2,2,3),(4,1,2))$, if we use the abbreviation

$$SBT^{\alpha}(\lambda, \mu) := ST^{\alpha}(\lambda) \times ST^{\alpha}(\mu).$$

The construction of (T_0, T_1) shows that the resulting bitableau is a *standard bitableau*, i. e. both T_0 and T_1 are standard also in the general case. Furthermore it is obvious that T_0 and T_1 are of the same shape. The stated bijectivity between $M_{\lambda\mu}$ and $\cup_\alpha SBT^{\alpha}(\lambda, \mu)$ follows from the fact that the double sequence z, and hence also the matrix Z can be reconstructed from (T_0, T_1) by an application of

6.2.3 The Cancellation Algorithm *If an insertion procedure according to 6.2.2 amounts to filling the box at the end of the j-th row of T^{i-1} in order to get T^i, then we can determine which number r was inserted and also we can recover T^{i-1} as follows:*

- *In the case when $j = 0$ then $r = r_0$, the element at the end of the row with number 0 of T^i.*
- *In the case when $j > 0$, then we have to consider the entry r_j which is in the position at the end of the j-th row of T^i. In the $(j-1)$-th row there exists an entry less than r_j which is as far to the right as possible. We call it r_{j-1} and exchange it with r_j. If $j > 1$ we compare r_{j-1} with the elements in the $(j-2)$-th row, exchange it with the element less than r_{j-1} which is as far as possible to the right, which we denote by r_{j-2}, and so on. We do this until finally an element r_0 remains which originates from the row with number 0.*

The element $r := r_0$ is the element which was inserted.

An example is provided by the following tableau

$$\begin{matrix} 0 & 0 & 0 & 0 \\ 1 & 2 & & \\ 2 & & & \end{matrix} .$$

We wish to remove the element at the end of the lowest row. According to 6.2.3 the procedure is as follows

$$\begin{matrix} 0 & 0 & 0 & 0 \\ 1 & 2 & & \\ 2 & & & \end{matrix} \quad\mapsto\quad \begin{matrix} 0 & 0 & 0 & 0 \\ 2 & 2 & & \end{matrix} \quad\mapsto\quad \begin{matrix} 0 & 0 & 0 & 1 \\ 2 & 2 & & \end{matrix}$$

so the element removed is $r = 1$. In order to reconstruct z and Z from (T_0, T_1), this cancellation is applied in the following way:

6.2.4 Theorem *The sequence of the positions which have to be cancelled is read off from T_0 as follows: From right to left we look for places containing the biggest entries and cancel these positions one after the other from T_1. Then we take the positions with the second biggest entries, and so on. These positions have to be cancelled from T_1 according to 6.2.3. The resulting pairs of entries $(z_i^0 \in T_0, z_i^1 \in T_1)$ yield the double sequence z, but in reverse order.*

For our example the first two steps are trivial, since the rightmost entries 3 occur in the first row of T_0:

$$\left(\begin{matrix} 0 & 0 & 2 & 2 \\ 1 & 1 & & \\ 2 & & & \end{matrix} , \begin{matrix} 0 & 0 & 0 & 0 \\ 1 & 2 & & \\ 2 & & & \end{matrix}\right) \mapsto \left(\begin{matrix} 0 & 0 & 2 \\ 1 & 1 & \\ 2 & & \end{matrix} , \begin{matrix} 0 & 0 & 0 \\ 1 & 2 & \\ 2 & & \end{matrix}\right)$$

$$\mapsto \left(\begin{matrix} 0 & 0 \\ 1 & 1 \\ 2 & \end{matrix} , \begin{matrix} 0 & 0 \\ 1 & 2 \\ 2 & \end{matrix}\right) .$$

Hence these cancellations yield the pairs $(3, 1) = (z_6^0, z_6^1)$ and $(3, 1) = (z_5^0, z_5^1)$. In the next step the elements in the third row and first column have to be canceled. According to 6.2.4, we get

$$\left(\begin{matrix}0&0\\1&1\\2&\end{matrix}\;,\;\begin{matrix}0&0\\1&2\\2&\end{matrix}\right)\mapsto\left(\begin{matrix}0&0\\1&1\end{matrix}\;,\;\begin{matrix}0&1\\2&2\end{matrix}\right),$$

and so we obtain the pair $(3, 1) = (z_4^0, z_4^1)$. In the remaining pair the rightmost element 1 of T_0 is in the second row and second column, hence we proceed as follows:

$$\mapsto\left(\begin{matrix}0&0\\1&\end{matrix}\;,\;\begin{matrix}0&2\\2&\end{matrix}\right)\mapsto(0\;\;0\;\;,\;\;2\;\;2)$$

$$\mapsto(0\;\;,\;\;2)\mapsto(\emptyset\;\;,\;\;\emptyset).$$

This process yields the double sequence introduced above which in turn gives us the matrix Z. Hence (exercise 6.2.1) the mapping $Z \mapsto (T_0, T_1)$ is *injective*. It can be shown that, conversely, by this cancellation method we obtain from each standard bitableau $(T_0, T_1) \in ST^\alpha(\lambda) \times ST^\alpha(\mu)$ a matrix $A \in M_{\lambda\mu}$. This establishes the desired relation between matrices and bitableaux:

6.2.5 Corollary *For any $\lambda, \mu \models n$, the Insertion Algorithm establishes a bijection*

$$M_{\lambda\mu} \to \bigcup_{\alpha \vdash n} SBT^\alpha(\lambda, \mu).$$

In particular this gives the following identity, for $m_{\lambda\mu} := |M_{\lambda\mu}|$ and $st^\alpha(\lambda) := |ST^\alpha(\lambda)|$:

$$m_{\lambda\mu} = \sum_{\alpha \vdash n} st^\alpha(\lambda) st^\alpha(\mu).$$

It is surprising to see that this fact has the following remarkable consequence for the representation theory of symmetric groups:

6.2.6 Corollary *The Kostka numbers are equal to the numbers of standard tableaux:*

$$\kappa_{\alpha\beta} = st^\beta(\alpha);$$

thus Young characters decompose into irreducible characters as follows:

$$\xi^\alpha = \sum_\beta st^\beta(\alpha)\zeta^\beta.$$

In particular f^α, the dimension of $[\alpha]$, is equal to the number $t^\alpha = st^\alpha(1^n)$ of standard Young tableaux of shape α.

Proof: We note first that, with respect to the lexicographic order of the proper partitions, both the Kostka matrix Ξ_n and the matrix containing the number $st^\beta(\alpha)$ in the row with number α and in the column with number β have the same scalar products of rows: An application of 6.2.5 and of 6.1.9 gives:

$$\sum_\gamma st^\gamma(\alpha) st^\gamma(\beta) = m_{\alpha\beta} = \sum_\gamma \kappa_{\alpha\gamma}\kappa_{\beta\gamma}.$$

Furthermore both these matrices are lower triangular, by 6.1.22 and 6.2.1, and they have ones along their main diagonal. Thus, since their last rows are equal, an induction yields that they are elementwise equal. This also implies the statement on the dimensions (cf. 6.1.22). □

A direct implication is obtained from 6.1.14:

$$\epsilon\xi^{\alpha} = \sum_{\gamma} st^{\gamma}(\alpha)\zeta^{\gamma'} = \sum_{\gamma} st^{\gamma'}(\alpha)\zeta^{\gamma},$$

so that we obtain the mulitplicity

6.2.7 $$m'_{\alpha\beta} = [\xi^{\beta} \mid \epsilon\xi^{\alpha}] = \sum_{\gamma\vdash n} st^{\gamma}(\beta)st^{\gamma'}(\alpha).$$

With these results at hand it is now relatively easy to evaluate ordinary irreducible characters using Young characters. For example $ST^{\beta}((n))$ is nonempty if and only if $\beta = (n)$, in which case this set consists of the single tableau $0\ldots0$. Hence we have $1\cdot\zeta^{(n)} = \xi^{(n)}$, the identity character. This is of course obvious also from the definition of $[n]$. But what about $\zeta^{(n-r,r)}$, where $0 < r \leq n/2$? As $ST^{\beta}((n-r,r))$, if it is nonempty, consists of the elements

$$\begin{matrix} 0\ldots\ldots01\ldots1 \\ 1\ldots1 \end{matrix},$$

we derive from 6.2.6 the following decomposition:

6.2.8 $$\xi^{(n-r,r)} = \sum_{0\leq s\leq r} \zeta^{(n-s,s)}.$$

This, together with 6.1.28 gives

6.2.9 $$\zeta^{(n-1,1)}(\pi) = a_1(\pi) - 1,\ \zeta^{(n-2,2)} = \frac{a_1(\pi)(a_1(\pi)-3)}{2} + a_2(\pi).$$

There exist in fact polynomial functions which yield irreducible character values if we apply them to cycle types. They do in fact depend on the truncated partition $\hat{\alpha}$ only, so that we can denote them by $Z^{\hat{\alpha}}$. Here are the smallest examples (cf. 6.1.30):

6.2.10 $$Z^{\emptyset} = 1,\ Z^{(1)} = x_1 - 1,\ Z^{(2)} = \frac{x_1(x_1-3)}{2} + x_2.$$

We shall consider the general case later.

Exercises

Exercise 6.2.1 Give the remaining details of the proof that $Z \mapsto (T_0, T_1)$ is an injective mapping.

6.3 The Determinantal Form

In order to invert 6.2.6 by expressing the irreducible character ζ^{α} of S_n in terms of the Young characters ξ^{β}, we shall derive a basic combinatorial lemma which shows that $st^\alpha(\lambda)$ is an alternating sum of the numbers $m'_{\mu\nu}$. It should be mentioned already here that there exists an easier proof of this result. It is due to Littlewood (cf. his "University Algebra") and it uses Schur polynomials which we shall introduce later on. Moreover it is basically a comparison of coefficients between certain polynomials while the proof given here uses an inclusion–exclusion argument and bijections. It therefore yields in a sense more information.

For $\tau \in S_n$ and $\mu \models n$, put

$$\tau\mu := (\mu_0 + \tau 0 - 0, \mu_1 + \tau 1 - 1, \ldots, \mu_{n-1} + \tau(n-1) - (n-1)).$$

Using this notation we can consider, for each f from the set $IP(n)$ of improper partitions of n to the complex field $\mathbb{C}$, the expression

$$\sum_{\tau:\, \tau\lambda \models n} f(\tau\lambda),$$

where the sum has to be taken over all the $\tau \in S_n$ such that $\tau\lambda$ is an improper partition of n, i. e. for which each $\lambda_i + \tau i - i \geq 0$. The crucial result concerning expressions of this form will turn out to be

6.3.1 Doubilet's Lemma *For each $\alpha \vdash n$, $\lambda \models n$ the following equation holds:*

$$st^{\alpha'}(\lambda) = \sum_{\tau:\, \tau\alpha \models n} \epsilon(\tau) m'_{\tau\alpha,\lambda}.$$

Proof:

i) For $\tau \neq 1$ we describe a complete dissection of $M'_{\tau\alpha,\lambda}$ into pairwise disjoint subsets $C_\tau(t,i,j)$ and a subset B_τ. Then we show that the summand $\epsilon(\tau)c_\tau(t,i,j)$, where $c_\tau(t,i,j) := |C_\tau(t,i,j)|$, of $\epsilon(\tau)m'_{\tau\alpha,\lambda}$ cancels with $\epsilon(\tau(ij))c_{\tau(ij)}(t,i,j)$, the corresponding summand of $\epsilon(\tau(ij))m'_{\tau(ij)\alpha,\lambda}$.

a) The decomposition of $M'_{\tau\alpha,\lambda}$: For each matrix A we denote by $s(t,i,A)$ the sum of the first t elements in its i-th row and define

$$C_\tau := \{A \in M'_{\tau\alpha,\lambda} \mid \exists\, t,i,j\colon i < j, s(t,i,A) - \tau i = s(t,j,A) - \tau j\}.$$

Then

$$M'_{\tau\alpha,\lambda} = C_\tau \,\dot{\bigcup}\, B_\tau, \ \textit{where}\ B_\tau := M'_{\tau\alpha,\lambda} \backslash C_\tau,$$

and C_τ again can be decomposed into disjoint subsets $C_\tau(t,i,j)$, consisting of the $A \in C_\tau$ for which (t,i,j) is the lexicographically smallest triple such that

$$s(t,i,A) - \tau i = s(t,j,A) - \tau j. \qquad (*)$$

Hence

$$M'_{\tau\alpha,\lambda} = B_\tau \,\dot{\cup}\, \left(\dot{\cup}_{t,i,j} C_\tau(t,i,j)\right).$$

Of course some of the $C_\tau(t,i,j)$ may very well be empty.

b) A bijection between $C_\tau(t, i, j)$ and $C_{\tau(ij)}(t, i, j)$: Consider the matrix $A = (a_{rs})$ in $C_\tau(t, i, j)$. We exchange the first t elements of the i-th row of A with the elements in the j-th row and the same column. Let $A' := (a'_{rs})$ denote the resulting matrix. An easy check shows that A' is an element of $M'_{\tau(ij)\alpha,\lambda}$, while the minimality of (t, i, j) yields that A' is contained in $C_{\tau(ij)}(t, i, j)$. Furthermore $A \mapsto A'$ establishes a *bijection* between $C_\tau(t, i, j)$ and $C_{\tau(ij)}(t, i, j)$ since this operation is an involution.

c) We have thus obtained that each summand $\epsilon(\tau)c_\tau(t, i, j)$ of $\epsilon(\tau)m'_{\tau\alpha,\lambda}$ cancels with the corresponding summand:

$$\epsilon(\tau(ij))c_{\tau(ij)}(t, i, j) = -\epsilon(\tau)c_\tau(t, i, j),$$

which is a summand of $\epsilon(\sigma)m'_{\sigma\alpha,\lambda}$, where $\sigma := \tau(ij)$.

ii) Part i) has already shown that all the summands $\epsilon(\tau)|C_\tau|$ cancel so that it remains to examine the cases when $C_\tau \neq M'_{\tau\alpha,\lambda}$.

a) $\tau \neq 1$ implies that $C_\tau = M'_{\tau\alpha,\lambda}$: In order to prove this we have to verify that, for $\tau \neq 1$ and $A \in M'_{\tau\alpha,\lambda}$, there exists a triple (t, i, j) such that $(*)$ holds. But $\tau \neq 1$ implies the existence of $i < j$ such that $\tau i > \tau j$. Hence

$$s(0, i, A) - \tau i = -\tau i < -\tau j = s(0, j, A) - \tau j,$$

and also

$$s(\infty, i, A) - \tau i = \alpha_i - i > \alpha_j - j = s(\infty, j, A) - \tau j.$$

Now consider the sequences (x_t) and (y_t), where

$$x_t := s(t, i, A) - \tau i, \text{ and } y_t := s(t, j, A) - \tau j.$$

As A is a 0–1–matrix, these sequences increase by 0 or 1 only. Thus $x_0 < y_0$ and $x_\infty > y_\infty$ yield the existence of t such that $x_t = y_t$ and hence $A \in C_\tau$.

b) a) together with i) yields

$$\sum_{\tau\alpha \models n} \epsilon(\tau)m'_{\tau\alpha,\lambda} = |M'_{\alpha,\lambda} \backslash C_1| = |B_1|.$$

It therefore remains to show that B_1 and $ST^{\alpha'}(\lambda)$ are bijective.

c) Consider $A \in M'_{\alpha,\lambda} \backslash C_1$, so that in particular for each t and $i < j$ we have

$$s(t, i, A) - i \neq s(t, j, A) - j.$$

We again examine the sequences (x_t) and (y_t), for which we have in this case $\tau = 1$:

$$x_0 = s(0, i, A) - i > -j = s(0, j, A) - j = y_0.$$

Thus $A \notin C_1$ implies that for each t and $i < j$ we have

$$x_t = s(t, i, A) - i > s(t, j, A) - j = y_t. \qquad (**)$$

This will help us to show that the following mapping is a bijection between B_1 and $ST^{\alpha'}(\lambda)$. We map A onto T_A, the tableau which contains in its i-th column the j's (in their natural order) for which $a_{ij} = 1$, e. g.

$$\begin{pmatrix} 1 & 0 & 1 \\ 0 & 1 & 0 \\ 0 & 0 & 0 \end{pmatrix} \mapsto \begin{matrix} 0 & 1 \\ 2 & \end{matrix} \quad \textit{and} \quad \begin{pmatrix} 1 & 1 & 0 \\ 0 & 0 & 1 \\ 0 & 0 & 1 \end{pmatrix} \mapsto \begin{matrix} 0 & 2 & 2 \\ 1 & & \end{matrix}.$$

Now $A \in M'_{\alpha\lambda}$ yields that $\mathrm{sh}(T_A) = \alpha'$ and $c(T_A) = \lambda$. The strict monotony in the columns is trivial and hence it remains to check the weak monotony in the rows, which we prove indirectly, assuming that (i, j) is the lexicographically smallest pair for which this monotony is violated, i. e. for which we have $k := t_{i,j-1} > t_{i,j} =: l$. This assumption implies that $s(k, j-1, A) = s(l, j, A) = i$ and so we have, for $k > l$:

$$s(k, j-1, A) = s(l, j, A). \qquad (***)$$

But as $a_{j-1,k} = 1$, we have

$$s(k, j-1, A) > s(l, j-1, A) \geq_{(**)} s(l, j, A),$$

which contradicts $(***)$.

□

The main consequence of Doubilet's Lemma now easily follows. We use the lexicographic order on $P(n)$ so that, for each $\varphi\colon P(n) \times P(n) \to \mathbb{C}$, we obtain a matrix $(\varphi_{\alpha\beta})$ containing the value $\varphi_{\alpha\beta} := \varphi(\alpha, \beta)$ in its α-th row and β-th column. Using this abbreviation and the notation A^t for the transpose of A, we already know that

6.3.2 $$(m'_{\alpha\beta}) = \left(st^{\alpha}(\beta)\right)^t \left(st^{\alpha'}(\beta)\right).$$

Putting

$$\epsilon_{\alpha\beta} := \sum_{\tau:(\tau\alpha)^*=\beta} \epsilon(\tau),$$

where $(\tau\alpha)^*$ again denotes the *proper* partition obtained from $\tau\alpha$ by reordering this sequence in the case when it is an improper partition, Doubilet's Lemma together with 6.3.2 yields the crucial equation due to D. E. Littlewood:

6.3.3 The Standard Inversion $\left(\epsilon_{\alpha\beta}\right)^{-1} = \left(st^{\alpha}(\beta)\right)^t.$

Hence $(\epsilon_{\alpha\beta})$ is the inverse of the Kostka matrix and we can invert 6.2.6 obtaining

6.3.4 Corollary *For each $\alpha \vdash n$ is the unique $\mathbb{Z}$–linear combination of ζ^{α} in terms of Young characters is*

$$\zeta^{\alpha} = \sum_{\beta \vdash n} \epsilon_{\alpha\beta}\xi^{\beta} = \sum_{\beta \trianglerighteq \alpha} \epsilon_{\alpha\beta}\xi^{\beta},$$

(recall 6.1.24).

Another useful consequence of Doubilet's Lemma is

6.3.5 Corollary *For each $\alpha \vdash n$, $\lambda \models n$ and the corresponding proper partition λ^* obtained from λ by rearrangement, we have*

$$st^\alpha(\lambda) = st^\alpha(\lambda^*).$$

Our next aim is to determine a very explicit expression for the dimension f^α of $[\alpha]$. Corollary 6.3.4 has shown how its character can be expressed in determinantal form using the Young characters $\xi^{\tau\alpha}$. With the aid of the symbol # which we use for the outer tensor product and keeping in mind that $[m]$ is the identity representation of S_m, we see that, for $\tau\alpha \models n$, $\xi^{\tau\alpha}$ is the character of the representation

$$[\alpha_0 + \tau 0 - 0][\alpha_1 + \tau 1 - 1] \ldots := [\alpha_0 + \tau 0 - 0] \# [\alpha_1 + \tau 1 - 1] \# \ldots \uparrow S_n.$$

(More generally we put, for $\alpha \models m$, $\beta \models n$:

$$[\alpha][\beta] \; := \; [\alpha] \# [\beta] \uparrow S_{m+n},$$

where we mean by $[\alpha] \# [\beta]$ the irreducible representation of the subgroup $S_m \oplus S_n$ of S_{m+n}.) This notation allows us to rephrase the first part of 6.3.4 as follows:

6.3.6 The Determinantal Form *The irreducible representation $[\alpha]$, $\alpha \vdash n$ of S_n can be expressed as follows in terms of Young representations:*

$$[\alpha] = \det\big([\alpha_i + j - i]\big)_{i,j\in h}.$$

But this notation must be used with care, it must be interpreted in terms of generalized characters since it contains summands with negative signs. It describes, as it was already stated, ζ^α as a generalized character, and since we defined summands $\xi^{\tau\alpha}$ if and only if $\tau\alpha \models n$, we have to put $\xi^{\tau\alpha} = 0$ otherwise, and also we have to set $\xi^{(0)} := 1$. An example illustrates this:

$$[3, 1^2] = \det\begin{pmatrix} [3] & [4] & [5] \\ 1 & [1] & [2] \\ 0 & 1 & [1] \end{pmatrix} = [3][1][1] - [3][2] - [4][1] + [5].$$

This has to be considered as being a "virtual representation". It is preferable to consider the corresponding generalized character (cf. exercise 11.5.4):

$$\zeta^{(3,1^2)} = \xi^{(3,1,1)} - \xi^{(3,2)} - \xi^{(4,1)} + \xi^{(5)}.$$

From this determinantal form of $[\alpha]$ we can easily derive important results on the values of irreducible characters, e. g. the character of the n–cycle:

6.3.7 $$\zeta^\alpha\big((0,\ldots,n-1)\big) = \begin{cases} (-1)^r, & \text{if } \alpha = (n-r, 1^r), r \le n-1, \\ 0, & \text{otherwise.} \end{cases}$$

Proof: The only Young subgroup S_β containing n–cycles is $S_{(n)} = S_n$ itself. Thus $\zeta^\alpha((0,\ldots,n-1)) \neq 0$ implies that $\det([\alpha_i + j - i])$ must contain an entry equal to $[n]$, and hence

$$\det\big([\alpha_i + j - i]\big) = \det \begin{pmatrix} [\alpha_0] & \ldots & [n] \\ & \ddots & \vdots \\ \star & & [\alpha_{h-1}] \end{pmatrix},$$

which implies that $\alpha = (n-(h-1), 1^{h-1})$. Furthermore in this case we get

$$\zeta^\alpha\big((0,\ldots,n-1)\big) = (-1)^{h-1}\xi^{(n)}\big((0,\ldots,n-1)\big) = (-1)^{h-1}.$$

□

Thus $\zeta^\alpha((0,\ldots,n-1))$ is nonzero if and only if $[\alpha]$ is a diagram of shape:

$$[\alpha] = [n-r, 1^r] = \begin{matrix} \times & \times & \ldots & \times \\ \times & & & \\ \vdots & & & \\ \times & & & \end{matrix}.$$

Such diagrams are called *hooks* and they play a very important role in the representation theory of the symmetric groups. Hooks allow many results to be formulated in an easy manner. We define the (i,j)–*hook*

$$H^\alpha_{ij}$$

to be the following subset of $[\alpha]$: The node (i,j) itself, which we call the *corner* of this hook, together with its *arm* consisting of the nodes $(i,k), k > j$, and also the *leg* of the hook, consisting of the nodes $(l,j), l > i$.

$$\begin{matrix} & \times & \underbrace{\times \ldots \times}_{arm} \\ leg \left\{ \begin{matrix} \\ \\ \\ \end{matrix} \right. & \begin{matrix} \times \\ \vdots \\ \times \end{matrix} & \end{matrix}$$

The overall number h^α_{ij} of nodes in H^α_{ij} is called the *length* of the hook, while the *leg length* is denoted by l^α_{ij} :

$$h^\alpha_{ij} = \alpha_i - i + \alpha'_j - j - 1, \; l^\alpha_{ij} = \alpha'_j - i - 1.$$

It is important to realize that to each hook H^α_{ij} there corresponds a uniquely determined part

$$R^\alpha_{ij}$$

of the *rim* of $[\alpha]$ which also consists of h^α_{ij} nodes and which begins with the node (i, α_i) and ends with the node (α'_j, j). For example to $H^{(3,2,1^2)}_{00}$ there corresponds the part of the rim of $[3, 2, 1^2]$ which we indicated by encircled nodes:

$$\begin{array}{ccc} \times & \otimes & \otimes \\ \otimes & \otimes & \\ \otimes & & \\ \otimes & & \end{array}$$

The main fact is that the deletion of R^α_{ij} from $[\alpha]$ leaves a Young diagram, we denote it by

$$[\alpha]\backslash R^\alpha_{ij}.$$

For example: $[3, 2, 1^2]\backslash R^{(3,2,1^2)}_{00} = \times = [1]$.

Using this notation we can express the dimension f^α of $[\alpha]$ in terms of hook lengths as follows:

6.3.8 The Hook Formula *The dimension f^α of the ordinary irreducible representation $[\alpha]$ of S_n can be expressed as follows in terms of hook lengths:*

$$f^\alpha = \frac{n!}{\prod_{i,j} h^\alpha_{ij}}.$$

Proof: We apply the determinantal formula which yields

$$f^\alpha = n! \cdot \det\Big(\frac{1}{(\alpha_i + j - i)!}\Big)$$

with the convention that $(\alpha_i + j - i)!^{-1} := 0$, if $\alpha_i + j - i < 0$. Elementary transformations applied to this determinant yield

$$\begin{aligned} \det\Big(\frac{1}{(\alpha_i + j - i)!}\Big) &= \frac{1}{\prod_i h^\alpha_{i0}!} \det\Big(\frac{h^\alpha_{i0}!}{(\alpha_i + j - i)!}\Big) \\ &= \frac{1}{\prod_i h^\alpha_{i0}!} \det\Big(\prod_{r=1}^{(h-1)-j} (h^\alpha_{i0} + r + j - h)\Big). \end{aligned}$$

Further elementary transformations of the determinant on the right hand side of this equation yield the Vandermonde determinant

$$\det\Big((h^\alpha_{i0})^{h-j}\Big) = \prod_{i<j} (h^\alpha_{i0} - h^\alpha_{j0}).$$

This shows that f^α can be expressed in terms of the hooks H^α_{i0} which have their corner in the 0-th column. It also shows that we can complete the proof by showing that

$$\prod_{j=i+1}^{h} (h^\alpha_{i0} - h^\alpha_{j0}) \prod_{\nu=0}^{\alpha_i-1} h^\alpha_{i\nu} = h^\alpha_{i0}!. \tag{6.3.9}$$

The verification of this is left as an exercise. □

Another useful application of the determinantal form of $[\alpha]$ is the proof of the following theorem on the irreducible characters ζ^α which was already mentioned in connection with 6.1.29 when the analogous result for Young characters ξ^α was proved:

6.3.10 Specht's Theorem *For each partition $\hat{\alpha} := (\alpha_1, \ldots, \alpha_{h-1}) \vdash n$ there exists a uniquely determined polynomial*

$$Z^{\hat{\alpha}} = \sum_\sigma \epsilon(\sigma) \Xi^{(\alpha_1+\sigma(1)-1,\ldots,\alpha_{h-1}+\sigma(h-1)-h+1)} \in \mathbb{Q}\,[x_1, x_2, \ldots],$$

sum over all the $\sigma \in S_h$, that yields all the values of all the irreducible characters ζ^β for which $\beta = (\beta_0, \alpha_1, \ldots, \alpha_{h-1})$ by applying the corresponding polynomial function to the cycle type in question:

$$\zeta^\beta(\pi) = Z^{\hat{\alpha}}(a_1(\pi), \ldots, a_m(\pi)), m := n + \beta_0.$$

Proof: The determinantal form of $[\alpha]$, $\alpha = (\alpha_0, \alpha_1, \ldots, \alpha_{h-1}) \vdash n$, yields

$$\zeta^\alpha(\pi) = \sum_\sigma \epsilon(\sigma) \xi^{(\alpha_0+\sigma(0)-0, \alpha_1+\sigma(1)-1,\ldots,\alpha_{h-1}+\sigma(h-1)-(h-1))},$$

so that the statement follows by an applications of 6.1.29. □

Since the name "Specht polynomials" is already used for a different series of polynomials we call the $Z^{\hat{\alpha}}$ *character polynomials.* Here are two further examples ($Z^{\emptyset}$, $Z^{(1)}$ and $Z^{(2)}$ were already shown in 6.2.10):

$$Z^{(1^2)} = \binom{x_1}{2} - \binom{x_2}{1} - \binom{x_1}{1} + 1,$$

$$Z^{(3)} = \binom{x_3}{1} + \binom{x_1}{1}\binom{x_2}{1} + \binom{x_1}{3} - \binom{x_2}{1} - \binom{x_1}{2}.$$

(I did not simplify these expressions since 6.3.10 shows that each character polynomial is in fact a $\mathbb{Z}$–linear combination of products of binomials, see also the further character polynomials given in the appendix.)

6.4 Standard Bideterminants

We have used bijections between sets of matrices and sets of pairs of standard tableaux in order to derive the ordinary irreducible representations of symmetric groups. A further bijection leads from sets of pairs of tableaux to polynomials on which the symmetric group acts in a natural way and from which we shall obtain *bases* that yield the ordinary irreducible *matrix representations*, as will be described next. These polynomials came up in invariant theory first (see the book [156] by Turnbull and various papers by Rota et al., e. g. [39]). Their application to representation theory of symmetric groups is due to Clausen (e. g. [37],[35],[36]).

Consider the ring of polynomials over $\mathbb{Z}$ in the commuting indeterminates X_{ij}, where $i, j \in \mathbb{N}$:

$$\mathbb{Z}[X] := \mathbb{Z}[X_{ij} \mid i, j \in \mathbb{N}].$$

Since we assume the indeterminates to be commutative, we can decompose $\mathbb{Z}[X]$ into submodules according to degree. In order to do this we put

$$X^Z := \prod_{i,j} X_{ij}^{z_{ij}}, \text{ for each } Z \in M_{\lambda\mu},$$

and indicate by $\mathbb{Z}_{\lambda\mu}$, where $\lambda, \mu \models n$, the $\mathbb{Z}$–span of these monomials:

$$\mathbb{Z}_{\lambda\mu} := \langle X^Z \mid Z \in M_{\lambda\mu} \rangle_{\mathbb{Z}},$$

from which we obtain the obvious decomposition

6.4.1
$$\mathbb{Z}[X] = \bigoplus_n \bigoplus_{\lambda,\mu \models n} \mathbb{Z}_{\lambda\mu}.$$

Now we introduce the following abbreviation, for $i_\nu, j_\mu \in \mathbb{N}$:

$$\left(\begin{array}{c|c} i_0 & j_0 \\ \vdots & \vdots \\ i_{m-1} & j_{m-1} \end{array} \right) := \det \begin{pmatrix} X_{i_0 j_0} & \dots & X_{i_0 j_{m-1}} \\ \vdots & & \vdots \\ X_{i_{m-1} j_0} & \dots & X_{i_{m-1} j_{m-1}} \end{pmatrix} \in \mathbb{Z}[X].$$

For example

$$\left(\begin{array}{c|c} 1 & 4 \\ 3 & 8 \\ 3 & 2 \end{array} \right) = \det \begin{pmatrix} X_{14} & X_{18} & X_{12} \\ X_{34} & X_{38} & X_{32} \\ X_{34} & X_{38} & X_{32} \end{pmatrix} = 0.$$

And, for any bitableau (S, T), where S and T are tableaux of *the same shape* over $\mathbb{N}$, we denote by $(S \mid T)$ the product of the determinants arising from corresponding columns:

$$\left(\begin{array}{ccc|ccc} i_0 & k_0 & \dots & j_0 & l_0 & \dots \\ \vdots & \vdots & & \vdots & \vdots & \\ \vdots & k_{n-1} & & \vdots & l_{n-1} & \\ i_{m-1} & & & j_{m-1} & & \end{array} \right) = \det\big(X_{i_\nu j_\mu}\big) \cdot \det\big(X_{k_\nu l_\mu}\big) \cdot \dots,$$

in particular

$$(i_0 \dots k_0 \mid j_0 \dots l_0) = X_{i_0 j_0} \cdot \dots \cdot X_{k_0 l_0}.$$

There are natural actions of the symmetric group of $\mathbb{N}$ on $\mathbb{Z}[X]$ from the left and from the right:

$$\sigma f(\dots, X_{ij}, \dots) := f(\dots, X_{\sigma i, j}, \dots),$$

$$f(\ldots, X_{ij}, \ldots)\tau := f(\ldots, X_{i,\tau^{-1}j}, \ldots).$$

Clearly each of the submodules $\mathbb{Z}_{(1^n),\mu}$ is invariant under the restriction to S_n of the action from the left. The same holds for $\mathbb{Z}_{\lambda,(1^n)}$ and the action from the right. Furthermore, if $\sigma, \tau \in S_n$, then we have

6.4.2
$$\left(\begin{array}{c|c} i_{\sigma 0} & j_{\tau 0} \\ \vdots & \vdots \\ i_{\sigma(n-1)} & j_{\tau(n-1)} \end{array}\right) = \operatorname{sign}(\sigma\tau) \left(\begin{array}{c|c} i_0 & j_0 \\ \vdots & \vdots \\ i_{n-1} & j_{n-1} \end{array}\right).$$

Now we recall that a tableau S over $\mathbb{N}$ of shape α is a mapping. with domain

$$[\alpha] = \bigcup_i \{(i, 0), \ldots, (i, \alpha_i - 1)\},$$

and range $\mathbb{N}$, namely

$$S: [\alpha] \to \mathbb{N}: (i, j) \mapsto s_{ij}.$$

The content of the tableau S was defined to be the following sequence of multiplicities of values:

$$c(S) = \left(|S^{-1}(0)|, |S^{-1}(1)|, \ldots\right),$$

so that $c(S) \models n$, if the shape α of S is a partition of the natural number n. Now we note that there are two canonic Young subgroups of the symmetric group

$$S_{[\alpha]}$$

on the set of nodes $(i, j) \in [\alpha]$: The *horizontal group* H_α, consisting of the *horizontal* or *row permutations,* and the *vertical group* V_α consisting of the *vertical* or *column permutations*:

$$H_\alpha := \{\pi \in S_{[\alpha]} \mid \pi \text{ leaves each } (i, j) \text{ in its row}\},$$

$$V_\alpha := \{\pi \in S_{[\alpha]} \mid \pi \text{ leaves each } (i, j) \text{ in its column}\}.$$

These subgroups of $S_{[\alpha]}$ will help us easily to describe several polynomials associated with a bitableau (S, T) consisting of two tableaux S and T of shape α: The *natural monomial* is defined to be

$$\{S \mid T\} := \prod_{(i,j)\in[\alpha]} X_{s_{ij}, t_{ij}},$$

in terms of which the corresponding *bideterminant* now reads as follows:

6.4.3
$$(S \mid T) = \sum_{\sigma \in V_\alpha} \operatorname{sign}(\sigma)\{S \circ \sigma \mid T\} = \sum_{\tau \in V_\alpha} \operatorname{sign}(\tau)\{S \mid T \circ \tau\}.$$

The bideterminants that correspond to *standard* bitableaux are called *standard bideterminants,* and it is our main aim to show that they form a *basis* of $\mathbb{Z}[X]$. For example

$$(21 \mid 12) = (12 \mid 12) - \left(\begin{array}{c|c} 1 & 2 \\ 1 & 2 \end{array} \right).$$

In order to prove the independence of the standard bideterminants, we introduce an important result from multilinear algebra. It holds not only for diagrams $[\alpha]$ but also for arbitrary finite subsets A of $\mathbb{N} \times \mathbb{N}$, for which we can define tableaux, horizontal or vertical groups H_A or V_A, and bideterminants $(S \mid T)$ in an obvious way. For example

$$\left(\begin{array}{ccc|cc} & 3 & & 7 & \\ 2 & & & 6 & \\ 5 & & 1 & 8 & 1 \end{array} \right) := \left(\begin{array}{c|c} 2 & 6 \\ 5 & 8 \end{array} \right) \cdot \left(\begin{array}{c|c} 3 & 7 \\ 1 & 1 \end{array} \right).$$

6.4.4 Theorem (Clausen) *Assume that $A, B \subseteq \mathbb{N} \times \mathbb{N}$ are finite, that $\varphi \colon A \to B$ is a bijection, S a tableau with shape A and T a tableau with shape B. Put $\psi := \varphi^{-1}$, denote by $\mathcal{T}(\varphi)$ a transversal of*

$$\varphi V_A \varphi^{-1} / \varphi V_A \varphi^{-1} \cap V_B,$$

by $\mathcal{T}(\psi)$ a transversal of

$$\psi V_B \psi^{-1} / \psi V_B \psi^{-1} \cap V_A.$$

Then the following identity holds:

$$\sum_{\sigma \in \mathcal{T}(\psi)} \operatorname{sign}(\sigma)(S \circ \sigma \mid T \circ \varphi) = \sum_{\tau \in \mathcal{T}(\varphi)} \operatorname{sign}(\tau)(S \circ \psi \mid T \circ \tau).$$

Proof: We consider the sum of monomials

$$(S \mid T)_\varphi := \sum_{\sigma \in \psi V_B \psi^{-1} \cdot V_A} \operatorname{sign}(\sigma)\{S \circ \sigma \mid T \circ \varphi\},$$

which is, by 6.4.3, equal to

$$\sum_\sigma \operatorname{sign}(\sigma)\{S \mid T \circ \varphi \circ \sigma^{-1}\} = \sum_\sigma \{S \circ \varphi^{-1} \mid T \circ \varphi \circ \sigma^{-1} \circ \varphi^{-1}\}.$$

Now we note that $\sigma \mapsto \varphi \circ \sigma^{-1} \circ \varphi^{-1}$ is a bijection between $\psi V_B \psi^{-1} \cdot V_A$ and $\varphi V_A \varphi^{-1} \cdot V_B$, so that we can proceed as follows:

$$= \sum_{\tau \in \varphi V_A \varphi^{-1} \cdot V_B} \operatorname{sign}(\tau)\{S \circ \varphi^{-1} \mid T \circ \tau\} =: {}_\psi(S \mid T).$$

Moreover, as

$$\psi V_B \psi^{-1} \cdot V_A = \sum_{\eta \in \mathcal{T}(\psi)} \eta V_A,$$

we have that

$$(S \mid T)_\varphi = \sum_{\eta \in \mathcal{T}(\psi)} \operatorname{sign}(\eta) \sum_{\theta \in V_A} \operatorname{sign}(\theta)\{S \circ \eta \circ \theta \mid T \circ \varphi\}$$

$$= \sum_{\eta \in \mathcal{T}(\psi)} \operatorname{sign}(\eta)(S \circ \eta \mid T \circ \varphi),$$

which is the left hand side of the stated equality. Hence, by symmetry, ${}_\psi(S \mid T)$ is equal to the right hand side, and this completes the proof. □

6.4.5 Application (expansion of a determinant) Consider the tableaux

$$S := \begin{array}{cc} 0 & \\ \vdots & \\ k-1 & \\ & k \\ k+1 & \\ \vdots & \\ n-1 & \end{array}, \text{ and } T := \begin{array}{c} 0 \\ \vdots \\ k-1 \\ k \\ k+1 \\ \vdots \\ n-1 \end{array},$$

in which case $\varphi = \psi$ is the transposition $((k,0),(k,1))$, while

$$V_A = S_{\{(0,0),\dots,(k-1,0),(k+1,0),\dots,(n-1,0)\}} \oplus S_{\{(k,1)\}} \simeq S_{n-1},$$

$$V_B = S_{\{(0,0),\dots,(n-1,0)\}} \simeq S_n.$$

Hence

$$\mathcal{T}(\varphi) = \{1\},\ \mathcal{T}(\psi) = \{((j,0),(k,1)) \mid j \in n \backslash \{k\}\} \cup \{0\}.$$

Thus we obtain for the right hand side of the equation proved in 6.4.4:

$$\sum_{\tau \in \mathcal{T}(\psi)} \operatorname{sign}(\tau)(S \circ \psi \mid T \circ \tau) = (S \circ \psi \mid T \circ \tau) = \left(\begin{array}{c|c} 0 & 0 \\ \vdots & \vdots \\ n-1 & n-1 \end{array} \right).$$

The left hand side is

$$\sum_{\sigma \in \mathcal{T}(\psi)} \operatorname{sign}(\sigma)(S \circ \sigma \mid T \circ \varphi) = \sum_{\sigma \in \mathcal{T}(\psi)} -(S \circ \sigma \mid T \circ \varphi),$$

where, for $\sigma := ((j,0),(k,1))$, we have (if $j \leq k-1$):

$$(S\circ\sigma|T\circ\varphi)=\left(\begin{array}{c|c} 0 & 0 \\ \vdots & \vdots \\ j-1 & j-1 \\ k & j \\ j+1 & j+1 \\ \vdots & \vdots \\ k-1 & k-1 \\ j & k \\ k+1 & k+1 \\ \vdots & \vdots \\ n-1 & n-1 \end{array}\right)=(j|k)\left(\begin{array}{c|c} 0 & 0 \\ \vdots & \vdots \\ j-1 & j-1 \\ k & j \\ j+1 & j+1 \\ k-1 & k-1 \\ k+1 & k+1 \\ \vdots & \vdots \\ n-1 & n-1 \end{array}\right)$$

$$=(j\mid k)(-1)^{k-j-1}\left(\begin{array}{c|c} 0 & 0 \\ \vdots & \vdots \\ j-1 & k-1 \\ j+1 & k+1 \\ \vdots & \vdots \\ n-1 & n-1 \end{array}\right).$$

This proves the well known expansion of the determinant of (X_{jk}) along the k-th column:

$$\left(\begin{array}{c|c} 0 & 0 \\ \vdots & \vdots \\ n-1 & n-1 \end{array}\right)=\sum_j(-1)^{j+k}(j\mid k)\left(\begin{array}{c|c} 0 & 0 \\ \vdots & \vdots \\ j-1 & k-1 \\ j+1 & k+1 \\ \vdots & \vdots \\ n-1 & n-1 \end{array}\right).$$

It is clear that many other determinantal identities can also be obtained from the theorem proved above. ◇

Now we introduce the following total order $\leq$ on $\mathbb{N}\times\mathbb{N}$, since we want to define canonic transversals:

$$(i,j)\leq(k,l)\;:\Longleftrightarrow\;(j>l)\vee(j=l\wedge i\leq k).$$

The smallest elements with respect to $\leq$ in their left coset form *canonic transversals* $\mathcal{C}(\varphi)$ and $\mathcal{C}(\psi)$. In order to describe them more explicitly we recall the notion of tabloid, and we remember that a tabloid represents the lexicographically smallest element of its left coset with respect to the corresponding Young subgroup. Since here we are dealing with Young subgroups, too, namely with V_A, V_B, $\varphi V_A \varphi^{-1}$, $\psi V_B \psi^{-1}$ and their intersections, their smallest representatives are the permutations that are increasing (with respect to $\leq$) on each block. Denoting by A^j and B^j the columns of A and B, we therefore obtain the following result:

$$\mathcal{C}(\varphi) = \left\{ \pi \in S_B \mid \forall\, j, k \colon \pi[\varphi[A^j]] = \varphi[A^j], \pi \text{ increases on } \varphi[A^j] \cap B^k \right\},$$

$$\mathcal{C}(\psi) = \left\{ \pi \in S_A \mid \forall\, j, k \colon \pi[\psi[B^j]] = \psi[B^j], \pi \text{ increases on } \psi[B^j] \cap A^k \right\}.$$

These permutations are called *shuffles* with respect to φ and ψ, respectively.

6.4.6 Example Consider $A := [2^4, 1]$, $B := [3, 2, 1^4]$ and the following bijection $\varphi \colon [2^4, 1] \to [3, 2, 1^4]$, which we define by replacing $(i, j) \in [3, 2, 1^4]$ by its inverse image $(r, s) := \psi((i, j)) = \varphi^{-1}((i, j))$:

$$\psi := \begin{array}{lll} (0,1) & (2,1) & (0,0) \\ (1,1) & (3,1) & \\ (1,0) & & \\ (2,0) & & \\ (3,0) & & \\ (4,0) & & \end{array} .$$

Then $\psi V_B \psi^{-1} = V_{\psi[B]}$ is equal to

$$S_{\{(0,1),(1,1),(1,0),(2,0),(3,0),(4,0)\}} \oplus S_{\{(2,1),(3,1)\}} \oplus S_{\{(0,0)\}} \simeq S_{(6,2,1)},$$

and its intersection with V_A is

$$S_{\{(0,1),(1,1)\}} \oplus S_{\{(1,0),(2,0),(3,0),(4,0)\}} \oplus S_{\{(2,1),(3,1)\}} \oplus S_{\{(0,0)\}} \simeq S_{(2,4,2,1)}.$$

Thus $\mathcal{C}(\psi)$ is the set of elements of $\psi V_B \psi^{-1}$ that are increasing on each of the sets

$$\{(0,1),(1,1)\}, \{(1,0),(2,0),(3,0),(4,0)\}, \{(2,1),(3,1)\}, \{(0,0)\}.$$

We express this symbolically in the following way (abbreviating (i, k) by ik):

$$\mathcal{C}(\psi) = \left\| \begin{array}{c} 01\ 11 \\ \nearrow \end{array} \middle| \begin{array}{c} 10\ 20\ 30\ 40 \\ \nearrow \end{array} \right\| \begin{array}{c} 21\ 31 \\ \nearrow \end{array} \left\| \begin{array}{c} 00 \\ \nearrow \end{array} \right\| .$$

For example

$$\begin{pmatrix} 01 & 11 & 10 & 20 & 30 & 40 & 21 & 31 & 00 \\ 11 & 30 & 01 & 10 & 20 & 40 & 21 & 31 & 00 \end{pmatrix} \in \mathcal{C}(\psi),$$

and

$$|\mathcal{C}(\psi)| = \binom{6}{2,4} \cdot \binom{2}{2} \cdot \binom{1}{1} = 15.$$

◇

Now we introduce a total order on the set $BT(\lambda, \mu)$ of bitableaux: Assume that $(S, T) \in BT^{\alpha}(\lambda, \mu)$ and $(S', T') \in BT^{\beta}(\lambda, \mu)$, then we define

$$(S, T) \leq (S', T') \;:\Longleftrightarrow\; (\alpha' > \beta') \vee (\alpha' = \beta' \wedge S * T \leq S' * T'),$$

where $S*T$ means the concatenation of the words obtained from S and T by reading one column after the other from top to bottom and from left to right:

$$S * T = s_{00}s_{10} \ldots s_{\alpha_0'-1,0}s_{01}s_{11} \ldots t_{00}t_{10} \ldots,$$

and $S' * T'$ is defined correspondingly, while $S * T \leq S' * T'$ uses the lexicographic order of these words.

6.4.7 Theorem *If* $(U, V) \in BT(\lambda, \mu)$ *is not standard, then its bideterminant* $(U \mid V)$ *is a* $\mathbb{Z}$*–linear combination of bideterminants corresponding to bitableaux* $(S, T) \in BT(\lambda, \mu)$ *that are strictly smaller than* (U, V).

Proof: By induction using the total order introduced above.

i) If (U, V) consists of tableaux U, V with one column only, then, since (U, V) is not standard, we obtain a unique standard tableau (S, T) by reordering both U and V, hence $(S, T) < (U, V)$ and $(U \mid V) = \pm(S \mid T)$.

ii) Suppose that $(U, V) \in BT(\lambda, \mu)$ is not standard, that the shape α of U and V has more than one column, and that the statement is true for each nonstandard $(U', V') < (U, V)$. We note that we can make the following additional assumptions:

- Both U and V consist of injective columns, since otherwise $(U \mid V)$ is zero (in which case the statement holds).
- The columns are strictly increasing from top to bottom, since reordering within a column means only a change in sign.

Hence the first violation of the standardness looks as follows (assuming that these are the columns U^j and U^{j+1} of U):

$$
\begin{array}{ccc}
a_0 & \leq & b_0 \\
\wedge & & \wedge \\
\vdots & & \vdots \\
\wedge & & \wedge \\
a_{k-1} & \leq & b_{k-1} \\
\wedge & & \wedge \\
a_k & > & b_k \\
\wedge & & \wedge \\
\vdots & & \vdots \\
\vdots & & \wedge \\
\vdots & & b_{t-1} \\
\vdots & & \\
\wedge & & \\
a_{s-1} & &
\end{array}
\;.
$$

Correspondingly we have for the bideterminant that

$$(U \mid V) = (U^j \mid V^j)(U^{j+1} \mid V^{j+1}) \prod_{i:j \neq i \neq j+1} (U^i \mid V^i).$$

Let us consider the following bijection on the shape A of $U^j U^{j+1}$:

$$
\varphi = \psi^{-1} \colon A \to B, \text{ defined by }
\begin{array}{cc}
(0,1) & (0,0) \\
(1,1) & (1,0) \\
\vdots & \vdots \\
(k-1,1) & (k-1,0) \\
(k+1,0) & (k,0) \\
(k+2,0) & (0,2) \\
\vdots & \vdots \\
\vdots & (t-k-1,2) \\
(s,0) &
\end{array}
\;.
$$

Furthermore we put $S := U^j U^{j+1}$, and $T := (V^j V^{j+1}) \circ \psi$, obtaining from theorem 6.4.4 the following expression for $(S \mid T \circ \varphi)$:

$$- \sum_{1 \neq \sigma \in \mathcal{C}(\psi)} \operatorname{sign}(\sigma)(S \circ \sigma \mid T \circ \varphi) + \sum_{\tau \in \mathcal{C}(\varphi)} \operatorname{sign}(\tau)(S \circ \psi \mid T \circ \tau).$$

Let us consider each of the two sums on the right hand side separately.

a) We note that the transversal is

$$\mathcal{C}(\psi) = \left\| \begin{array}{c} 00 \dots k-1, 0 \\ \nearrow \end{array} \right\| \begin{array}{c} 01 \dots k1 \\ \nearrow \end{array} \left| \begin{array}{c} k0 \dots (s-1)0 \\ \nearrow \end{array} \right\| \begin{array}{c} k+1, 1 \dots (t-1)1 \\ \nearrow \end{array} \right\|.$$

Hence the word corresponding to $S \circ \sigma$ is of the form

$$a_0 \dots a_{k-1} w b_{k+1} \dots b_{t-1},$$

where w is a word that is a permutation of the chain

$$b_0 < \dots < b_k < a_k < \dots < a_{s-1}.$$

Thus $S \circ \sigma < S$, if $\sigma \neq 1$, and hence $S \circ \sigma * T \circ \varphi = U^j U^{j+1} * V^j V^{j+1}$ is lexicographically smaller than $S * T \circ \varphi$.

b) For the second sum in the above equation we need only note that the image B of A, the shape of $U^j U^{j+1}$, consists of *three* columns, so that the word $S \circ \psi * T \circ \tau$ is lexicographically smaller than the word $U^j U^{j+1} * V^j V^{j+1}$, too, and we are done.

□

In order to proceed with the proof of the independency of the standard bideterminants, we shall introduce linear operators on the set of polynomials

$$\mathbb{Z}[X]_n := \{f \in \mathbb{Z}[X] \mid f \text{ is homogeneous of degree } n\}.$$

For this purpose we construct, for each $Z = (z_{ij}) \in M_{\lambda\mu}, \lambda, \mu \models n$, two $\mathbb{Z}$–endomorphisms L_Z and R_Z of $\mathbb{Z}[X]_n$. They will be defined in terms of sets of the following type:

$$Sub_Z(U) := \left\{ S \in T^\alpha(\mathbb{N}) \mid \forall\, i, j : z_{ij} = |S^{-1}(i) \cap U^{-1}(j)| \right\},$$

corresponding to $U \in T^\alpha(\mathbb{N})$. Note that *$Sub_Z(U)$ is the set of all the α–tableaux S which arise from the α–tableau U over $\mathbb{N}$ by substituting z_{ij} entries j in U by i, for each $(i, j) \in [\alpha]$, simultaneously and independently.* Hence in particular $Sub_Z(U) = \emptyset$, if $U \notin T^\alpha(\mu)$. We put

$$L_Z\{U \mid V\} := \sum_{S \in Sub_Z(U)} \{S \mid V\},$$

and, correspondingly,

$$R_Z\{U \mid V\} := \sum_{T \in Sub_Z(V)} \{U \mid T\}.$$

It is not difficult to check that these mappings are well defined (exercise 6.4.1), so that we can linearly extend them to $\mathbb{Z}[X]_n$. The resulting $\mathbb{Z}$–endomorphisms L_Z

and R_Z are called the *left* and the *right substitution corresponding to* Z. From their definition we can easily derive that they satisfy

6.4.8 $$L_Z(U \mid V) = \sum_{S \in Sub_Z(U)} (S \mid V),\ R_Z(U \mid V) = \sum_{T \in Sub_Z(V)} (U \mid T).$$

Now we use that we can restrict the summation to the $S \in Sub_Z(U)$ that consist of injective columns only: $L_Z(U \mid V)$ is equal to

$$\sum_{c.inj.\ S \in Sub_Z(U)} (S \mid V),\ R_Z(U \mid V) = \sum_{c.inj.\ T \in Sub_Z(V)} (U \mid T).$$

Hence, in particular, $L_Z(U \mid V) = 0$, or $R_Z(U \mid V) = 0$, if $Sub_Z(U)$ or $Sub_Z(V)$ does not contain any column injective element.

Now we restrict attention to specific matrices $Z(S)$ arising from tableaux $S \in T^\alpha(n)$ by putting z_{ij}, $i, j \in n$, equal to the number of j occuring in the i-th row of S:

$$Z(S)_{ij} := |\{i\} \times \mathbb{N} \cap S^{-1}(j)|.$$

For example

$$Z\begin{pmatrix} 0 & 0 & 2 & 2 \\ 1 & 2 & 3 & \\ 3 & & & \end{pmatrix} = \begin{pmatrix} 2 & 0 & 2 & 0 \\ 0 & 1 & 1 & 1 \\ 0 & 0 & 0 & 1 \\ 0 & 0 & 0 & 0 \end{pmatrix}.$$

Using this notation we put

$$C_{ST} := L_{Z(S)} R_{Z(T)},$$

and call it the *Capelli operator associated with* (S, T). In order to reach a position where we can use these Capelli operators for a proof of the independency of the standard bideterminants, we need the following result:

6.4.9 The Sorting Lemma *Each tableau T on n, the columns of which are strictly increasing from top to bottom, yields a standard tableau*

$$T^s$$

when we reorder the elements of each of its rows in the natural way.

Proof: By induction on the maximal entry m of T.

i) If $m = 0$, then $T = 0 \ldots 0 = T^s$ is standard.

ii) Assume that $m > 0$. Each m is the bottom entry of its column. We order each set of columns of the same length lexicographically according to their bottom entries. The resulting diagram is called X, and obviously $X^s = T^s$. Moreover $X \mapsto X^s$ does not move any entry m, so that we can cancel m and apply the induction assumption to the remaining rest.

□

For example

$$T := \begin{array}{cccc} 0 & 1 & 1 & 6 \\ 2 & 2 & 8 & 7 \\ 4 & 3 & 9 & \\ 7 & 4 & & \end{array} \mapsto T^s = \begin{array}{cccc} 0 & 1 & 1 & 6 \\ 2 & 2 & 7 & 8 \\ 3 & 4 & 9 & \\ 4 & 7 & & \end{array}.$$

6.4.10 Lemma *Let S and T denote α–tableaux, $\alpha \vdash n$, with strictly increasing columns. Then we have, for*

$$T_\alpha := \begin{array}{ccccc} 0 & 0 & \dots & \dots & 0 \\ 1 & 1 & \dots & 1 & \\ \dots & \dots & \dots & & \\ \alpha'_0 - 1 & \dots & \alpha'_0 - 1 & & \end{array},$$

that

$$L_{Z(S)}(S \mid T) = (T_\alpha \mid T) \neq 0, \; R_{Z(S)}(S \mid T) = (S \mid T_\alpha) \neq 0,$$

and hence the Capelli operator corresponding to (S, T) satisfies

$$C_{ST}(S \mid T) = (T_\alpha \mid T_\alpha) \neq 0.$$

Proof: It clearly suffices to prove the first statement. We first remark that, for each $\sigma \in H_\alpha$, we have $Z(S) = Z(S \circ \sigma)$, and hence, since $(S \circ \sigma \mid T \circ \sigma) = (S \mid T)$,

$$L_{Z(S)}(S \mid T) = L_{Z(S \circ \sigma)}(S \circ \sigma \mid T \circ \sigma).$$

We can therefore assume without restriction that the maximal entries m of S occur in adjacent positions at the end of rows of equal length, and so, if S' denotes the remaining rest of S, we have that

$$S = \begin{array}{|llll} \hline S' & & & \cdot^{\cdot^{\cdot}} \\ & & m \dots m & \\ & & \cdot^{\cdot^{\cdot}} & \\ & m \dots m & & \\ \dots \;\; \cdot^{\cdot^{\cdot}} & & & \end{array}$$

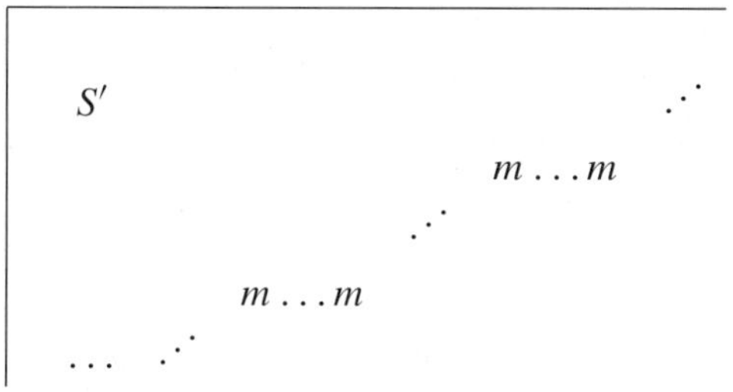

Moreover we can inductively assume that T_μ, μ the shape of S', is the only column injective element of $Sub_{Z(S')}(S')$. Hence each column injective element of $Sub_{Z(S)}(S)$ is obtained by replacing S' by T_μ and then replacing each entry m by the number of its row (recall the definition of $Sub_{Z(S)}(S)$!). Thus T_α is in fact the only column injective element of $Sub_{Z(S)}(S)$, and we are done. □

Our next aim is a necessary condition for pairs of bitableaux (S, T) and (U, V), for which $C_{ST}(U \mid V) \neq 0$, and where (S, T) is assumed to be a standard bitableau. We are going to show that such pairs are related by a generalization of the dominance order. A *standard* tableau $S \in ST^{\alpha}(\lambda)$ can be reconstructed from the following sequence of inverse images of subsets of entries under S:

$$\bar{S} := \left(S^{-1}(\{0\}), S^{-1}(\{0,1\}), \ldots, S^{-1}(\{0,\ldots,k-1\})\right),$$

where k is maximal such that $\lambda_{k-1} \neq 0$. The coimage $S^{-1}(\{0,\ldots,i-1\})$ is a subdiagram of $[\alpha]$ consisting of $\sum_0^{i-1} \lambda_j$ nodes. For example the tableau

$$S := \begin{matrix} 1 & 1 & 2 & 4 \\ 2 & 2 & 4 & \\ 4 & & & \end{matrix} \quad \in ST^{(4,3,1)}(0,2,3,0,3)$$

gives rise to the sequence

$$\bar{S} = \left(\emptyset,\ \begin{matrix} \times & \times \end{matrix},\ \begin{matrix} \times & \times & \times \\ \times & \times & \end{matrix},\ \begin{matrix} \times & \times & \times \\ \times & \times & \end{matrix},\ \begin{matrix} \times & \times & \times & \times \\ \times & \times & \times & \\ \times & & & \end{matrix} \right),$$

from which S easily can be reconstructed.

Now we use these sequences in order to define the announced *dominance order* on the set $ST(\lambda)$, by putting

$$S \trianglelefteq W :\Longleftrightarrow \bar{S} \trianglelefteq \bar{W} :\Longleftrightarrow \forall i \colon S^{-1}(\{0,\ldots,i-1\}) \trianglelefteq W^{-1}(\{0,\ldots,i-1\}).$$

Another description of this situation can be given in terms of the following matrix: For $S \in ST^{\alpha}(\lambda)$ we put

$$a_{pq}(S) := \left| \{(i,j) \in [\alpha] \mid s_{ij} \leq p, i \leq q\} \right|,$$

i. e. $a_{pq}(S)$ is *the number of entries* $\leq p$ *in the rows with numbers* $0, \ldots, q-1$ *of* S.

6.4.11 Lemma *For standard tableaux S and W of same content the following is true:*

$$S \trianglelefteq W \iff \forall\, p, q : a_{pq}(S) \leq a_{pq}(W).$$

Proof: Since S is assumed to be standard, the diagram $S^{-1}(\{0,\ldots,i-1\})$ is

$$[a_{i0}(S), a_{i1}(S) - a_{i0}(S), a_{i2}(S) - a_{i1}(S), \ldots].$$

Thus $S^{-1}(\{0,\ldots,i-1\}) \trianglelefteq W^{-1}(\{0,\ldots,i-1\})$ is equivalent to $a_{iq}(S) \leq a_{iq}(W)$, for each q, and the statement follows. □

This lemma shows that $\trianglelefteq$ is antisymmetric, and since it is obviously transitive, it is in fact a *partial order* on $ST(\lambda)$, we call it the *dominance order*. It can be refined as follows: We say that an A–tableau S and a B–tableau T, where $A, B \subseteq \mathbb{N} \times \mathbb{N}$, are *row–equivalent*, for short: $S \leftrightarrow T$, if they have the same column contents:

$$\forall\, i, j\colon \left|\{i\} \times \mathbb{N} \cap S^{-1}(\{j\})\right| = \left|\{i\} \times \mathbb{N} \cap T^{-1}(\{j\})\right|.$$

Correspondingly the *column equivalence* $S \updownarrow U$ is defined. And we put

$$S \leftrightarrow\!\!\!\!\updownarrow U :\Longleftrightarrow \exists\, T\colon S \leftrightarrow T \updownarrow U.$$

For example

$$\begin{array}{cccc} 0 & 0 & 1 & 1 \\ 2 & & & \end{array} \leftrightarrow\!\!\!\!\updownarrow \; 0 \;\; 0 \;\; 1 \;\; 1 \;\; 2$$

since

$$\begin{array}{cccc} 0 & 0 & 1 & 1 \\ 2 & & & \end{array} \leftrightarrow \begin{array}{cccc} & 0 & 0 & 1 & 1 \\ 2 & & & & \end{array} \updownarrow \; 0 \;\; 0 \;\; 1 \;\; 1 \;\; 2\,.$$

The next implication is now immediate from 6.4.11:

6.4.12 $$S \leftrightarrow\!\!\!\!\updownarrow U \Longrightarrow S \trianglelefteq U.$$

Hence $\leftrightarrow\!\!\!\!\updownarrow$ is antisymmetric, and the transitive closure of $\leftrightarrow\!\!\!\!\updownarrow$ is a partial order on $ST(\lambda)$, which we call the *hyperdominance order*, and we denote it by $\twoheadrightarrow\!\!\!\!\uparrow$. We are now in a position to derive the following necessary condition for $C_{ST}(U \mid V) \neq 0$:

6.4.13 Lemma *If both* (S, T) *and* (U, V) *are bitableaux of content* (λ, μ), *then*

$$C_{ST}(U \mid V) \neq 0 \Longrightarrow S \leftrightarrow\!\!\!\!\updownarrow U \text{ and } T \leftrightarrow\!\!\!\!\updownarrow V.$$

Proof: It clearly suffices to prove that $L_{Z(S)}(U \mid V) \neq 0$ implies $S \leftrightarrow\!\!\!\!\updownarrow U$. Assume that α is the shape of S and β the shape of U. An easy check yields that $Z(S) \in M_{\alpha\lambda}$. $L_{Z(S)}(U \mid V) \neq 0$ implies the existence of a $W \in Sub_{Z(S)}(U)$ which is column injective. Note that W arises from U by replacing, for each i and j, as many j by i as there are j's in the i-th row of S. But these j arising from collinear i in S, lie in different columns of W. Hence each j of S can be moved within its row to the column that it occupies in U, and the statement follows.

□

We shall indicate this necessary condition as follows:

$$(S, T) \leftrightarrow\!\!\!\!\updownarrow (U, V) :\Longleftrightarrow S \leftrightarrow\!\!\!\!\updownarrow U \text{ and } T \leftrightarrow\!\!\!\!\updownarrow V,$$

and call the resulting partial order on $SBT(\lambda, \mu)$ the *hyperbidominance* order. We shall also indicate it by

$$\twoheadrightarrow\!\!\!\!\uparrow\,.$$

Using this partial order we shall prove the following main result:

6.4.14 Theorem *For each pair $\lambda, \mu \models n$, the set $SBD(\lambda, \mu)$ of standard bideterminants $(S \mid T)$, where $(S \mid T) \in ST^\alpha(\lambda) \times ST^\alpha(\mu)$, for some $\alpha \vdash n$, forms a $\mathbb{Z}$–basis of $\mathbb{Z}_{\lambda\mu}$:*

$$\mathbb{Z}_{\lambda\mu} = \ll SBD(\lambda, \mu) \gg_{\mathbb{Z}} .$$

Proof: We first show that these standard bideterminants are linearly independent, Consider a linear relation

$$0 = \sum_{(U|V) \in SBD(\lambda,\mu)} a_{UV}(U \mid V),$$

and take a pair (S, T) that is maximal (with respect to hyperdominance) in the support

$$\{(U, V) \in SBD(\lambda, \mu) \mid a_{\mu\nu} \neq 0\}.$$

The corresponding Capelli operator C_{ST} yields, when applied to the linear relation, that

$$0 = a_{ST}(S \mid T) = a_{ST} \underbrace{(T_\alpha \mid T_\alpha)}_{\neq 0},$$

which implies $a_{ST} = 0$. Hence the standard bideterminants are linearly independent. Now we recall theorem 6.4.7. It says that each nonstandard tableau is a linear combination of bideterminants corresponding to smaller bitableaux. But the smallest bitableaux are standard, and so they are in fact linear combinations of standard bideterminants, which shows that $SBD(\lambda, \mu)$ does generate $\mathbb{Z}_{\lambda\mu}$, and we are done. □

Let us consider the following example in detail:

$$\mathbb{Z}_{(1^n)(1^n)}.$$

To each permutation $\pi \in S_n$ we can associate the mapping

$$(S \mid T) \mapsto (\pi S \mid T)$$

and we can linearly extend it to $\mathbb{Z}_{(1^n)(1^n)}$, which yields a representation of the symmetric group. Let us order the basis $SBD(1^n, 1^n)$ consisting of the standard bideterminants via the total order $\leq$ of the corresponding set of standard bitableaux $SBT(1^n, 1^n)$ which was defined above. This total order is compatible with the (partial) hyperbidominance. In this total order the elements of the same shape form intervals, and within such an interval we have the double lexicographic order with respect to the columns, so that the elements (S, T) with the same second component T for intervals, too. We therefore put these elements into sets according to the second diagram. For $n = 4$ we obtain the following sets:

$$M_0 := \left\{ \left(\begin{array}{c|c} 0 & 0 \\ 1 & 1 \\ 2 & 2 \\ 3 & 3 \end{array} \right) \right\},$$

$$M_1 := \left\{ \left(\begin{array}{c|c} 03 & 03 \\ 1 & 1 \\ 2 & 2 \end{array} \right), \left(\begin{array}{c|c} 02 & 03 \\ 1 & 1 \\ 3 & 2 \end{array} \right), \left(\begin{array}{c|c} 01 & 03 \\ 2 & 1 \\ 3 & 2 \end{array} \right) \right\},$$

$$M_2 := \left\{ \left(\begin{array}{c|c} 03 & 02 \\ 1 & 1 \\ 2 & 3 \end{array} \right), \left(\begin{array}{c|c} 02 & 02 \\ 1 & 1 \\ 3 & 3 \end{array} \right), \left(\begin{array}{c|c} 01 & 02 \\ 2 & 1 \\ 3 & 3 \end{array} \right) \right\},$$

$$M_3 := \left\{ \left(\begin{array}{c|c} 03 & 01 \\ 1 & 2 \\ 2 & 3 \end{array} \right), \left(\begin{array}{c|c} 02 & 01 \\ 1 & 2 \\ 3 & 3 \end{array} \right), \left(\begin{array}{c|c} 01 & 01 \\ 2 & 2 \\ 3 & 3 \end{array} \right) \right\},$$

$$M_4 := \left\{ \left(\begin{array}{c|c} 02 & 02 \\ 13 & 13 \end{array} \right), \left(\begin{array}{c|c} 01 & 01 \\ 23 & 23 \end{array} \right) \right\},$$

$$M_5 := \left\{ \left(\begin{array}{c|c} 02 & 01 \\ 13 & 23 \end{array} \right), \left(\begin{array}{c|c} 01 & 01 \\ 23 & 23 \end{array} \right) \right\},$$

$$M_6 := \left\{ \left(\begin{array}{c|c} 023 & 023 \\ 1 & 1 \end{array} \right), \left(\begin{array}{c|c} 013 & 023 \\ 2 & 1 \end{array} \right), \left(\begin{array}{c|c} 012 & 023 \\ 3 & 1 \end{array} \right) \right\},$$

$$M_7 := \left\{ \left(\begin{array}{c|c} 023 & 013 \\ 1 & 2 \end{array} \right), \left(\begin{array}{c|c} 013 & 013 \\ 2 & 2 \end{array} \right), \left(\begin{array}{c|c} 012 & 013 \\ 3 & 2 \end{array} \right) \right\},$$

$$M_8 := \left\{ \left(\begin{array}{c|c} 023 & 012 \\ 1 & 3 \end{array} \right), \left(\begin{array}{c|c} 013 & 012 \\ 2 & 3 \end{array} \right), \left(\begin{array}{c|c} 012 & 012 \\ 3 & 3 \end{array} \right) \right\},$$

$$M_9 := \{ (\, 0123 \mid 0123 \,) \} .$$

Now we denote the submodule generated by the standard bideterminants contained *in the first k of these sets M_i* by $U_k(\mathbb{Z})$,

$$U_k(\mathbb{Z}) := \ll M_0, \ldots, M_k \gg_{\mathbb{Z}},$$

obtaining the chain

$$U_0(\mathbb{Z}) \subseteq \ldots \subseteq U_{r-1}(\mathbb{Z}),$$

where r denotes the number of standard Young tableaux with n entries. It is easy to see that each of these sumodules is invariant under the linear extension of

$$\pi(S \mid T) := (\pi S \mid T),$$

and so each $U_k(\mathbb{Z})$ is an *invariant submodule* of $\mathbb{Z}_{(1^n)(1^n)}$ (cf. exercise 6.4.6). Moreover, if T_k denotes the right hand side standard tableau of the bitableaux in the k-th row, then the restriction of $R_{Z(T_k)}$ to $U_k(\mathbb{Z})$, which is in fact a S_n–homomorphism, maps $U_k(\mathbb{Z})$ onto the *Specht module*

$$\mathcal{S}_\alpha(\mathbb{Z}) := \ll (S \mid T_\alpha) \mid S \in ST^\alpha(1^n) \gg_{\mathbb{Z}},$$

where α denotes the shape of T_k. The submodule $U_{k-1}(\mathbb{Z})$ lies in the kernel of this restriction. Since the Specht module does only depend of the shape α, it occurs f^α–times in $\mathbb{Z}_{(1^n)(1^n)}$, and hence Maschke's theorem shows that the modules

$$\mathcal{S}_\alpha(\mathbb{Q}) := \mathbb{Q} \otimes_\mathbb{Z} \mathcal{S}_\alpha(\mathbb{Z})$$

form a complete system of ordinary irreducible representation modules for S_n. For example

$$\mathcal{S}_{(4)}(\mathbb{Q}) = \ll \left(\begin{array}{c|c} 0123 & 0000 \end{array} \right) \gg_\mathbb{Q},$$

$$\mathcal{S}_{(3,1)}(\mathbb{Q}) = \ll \left(\begin{array}{c|c} 023 & 000 \\ 1 & 1 \end{array} \right), \left(\begin{array}{c|c} 013 & 000 \\ 1 & 1 \end{array} \right), \left(\begin{array}{c|c} 012 & 000 \\ 3 & 1 \end{array} \right) \gg_\mathbb{Q},$$

$$\mathcal{S}_{(2^2)}(\mathbb{Q}) = \ll \left(\begin{array}{c|c} 02 & 00 \\ 13 & 11 \end{array} \right), \left(\begin{array}{c|c} 01 & 00 \\ 23 & 11 \end{array} \right) \gg_\mathbb{Q},$$

$$\mathcal{S}_{(2,1^2)}(\mathbb{Q}) = \ll \left(\begin{array}{c|c} 03 & 00 \\ 1 & 1 \\ 2 & 2 \end{array} \right), \left(\begin{array}{c|c} 02 & 00 \\ 1 & 1 \\ 3 & 2 \end{array} \right), \left(\begin{array}{c|c} 01 & 00 \\ 2 & 1 \\ 3 & 2 \end{array} \right) \gg_\mathbb{Q},$$

$$\mathcal{S}_{(1^4)}(\mathbb{Q}) = \ll \left(\begin{array}{c|c} 0 & 0 \\ 1 & 1 \\ 2 & 2 \\ 3 & 3 \end{array} \right) \gg_\mathbb{Q}.$$

Having these bases at hand we are now in a position to evaluate *matrix representations*. A detailed example may be in order, let us evaluate the representation afforded by $\mathcal{S}_{(3,1)}(\mathbb{Q})$. The basis elements are

$$\left(\begin{smallmatrix} 023 \\ 1 \end{smallmatrix} \middle| \begin{smallmatrix} 000 \\ 1 \end{smallmatrix} \right) = X_{00}X_{11}X_{20}X_{30} - X_{01}X_{10}X_{20}X_{30},$$

$$\left(\begin{smallmatrix} 013 \\ 2 \end{smallmatrix} \middle| \begin{smallmatrix} 000 \\ 1 \end{smallmatrix} \right) = X_{00}X_{10}X_{21}X_{30} - X_{01}X_{10}X_{20}X_{30},$$

$$\left(\begin{smallmatrix} 012 \\ 3 \end{smallmatrix} \middle| \begin{smallmatrix} 000 \\ 1 \end{smallmatrix} \right) = X_{00}X_{10}X_{20}X_{31} - X_{01}X_{10}X_{20}X_{30}.$$

The action of the transposition (01) on the basis elements is as follows:

$$(01) \left(\begin{smallmatrix} 023 \\ 1 \end{smallmatrix} \middle| \begin{smallmatrix} 000 \\ 1 \end{smallmatrix} \right) = \left(\begin{smallmatrix} 123 \\ 0 \end{smallmatrix} \middle| \begin{smallmatrix} 000 \\ 1 \end{smallmatrix} \right) = - \left(\begin{smallmatrix} 023 \\ 1 \end{smallmatrix} \middle| \begin{smallmatrix} 000 \\ 1 \end{smallmatrix} \right),$$

$$\begin{aligned} (01) \left(\begin{smallmatrix} 013 \\ 2 \end{smallmatrix} \middle| \begin{smallmatrix} 000 \\ 1 \end{smallmatrix} \right) = \left(\begin{smallmatrix} 103 \\ 2 \end{smallmatrix} \middle| \begin{smallmatrix} 000 \\ 1 \end{smallmatrix} \right) &= X_{00}X_{10}X_{21}X_{30} - X_{00}X_{11}X_{20}X_{30} \\ &= \left(\begin{smallmatrix} 013 \\ 2 \end{smallmatrix} \middle| \begin{smallmatrix} 000 \\ 1 \end{smallmatrix} \right) - \left(\begin{smallmatrix} 023 \\ 1 \end{smallmatrix} \middle| \begin{smallmatrix} 000 \\ 1 \end{smallmatrix} \right), \\ (01) \left(\begin{smallmatrix} 012 \\ 3 \end{smallmatrix} \middle| \begin{smallmatrix} 000 \\ 1 \end{smallmatrix} \right) = \left(\begin{smallmatrix} 102 \\ 3 \end{smallmatrix} \middle| \begin{smallmatrix} 000 \\ 1 \end{smallmatrix} \right) &= X_{00}X_{10}X_{20}X_{31} - X_{00}X_{11}X_{20}X_{30} \\ &= \left(\begin{smallmatrix} 012 \\ 3 \end{smallmatrix} \middle| \begin{smallmatrix} 000 \\ 1 \end{smallmatrix} \right) - \left(\begin{smallmatrix} 023 \\ 1 \end{smallmatrix} \middle| \begin{smallmatrix} 000 \\ 1 \end{smallmatrix} \right). \end{aligned}$$

Hence this action of (12) is represented by the matrix

$$\mathbf{D}((01)) = \begin{pmatrix} -1 & -1 & -1 \\ 0 & 1 & 0 \\ 0 & 0 & 1 \end{pmatrix}.$$

The matrices representing (12) and (23) are easier to evaluate since these transpositions just permute the basis elements:

$$\mathbf{D}((12)) = \begin{pmatrix} 0 & 1 & 0 \\ 1 & 0 & 0 \\ 0 & 0 & 1 \end{pmatrix}, \quad \mathbf{D}((23)) = \begin{pmatrix} 1 & 0 & 0 \\ 0 & 0 & 1 \\ 0 & 1 & 0 \end{pmatrix}.$$

In order to check these calculations we verify the relations in order to show that we have in fact evaluated matrices that define a representation of the symmetric group in question (the regularity of the evaluated matrices is obvious): In fact an easy check shows that

$$\mathbf{D}((01))^2 = \mathbf{D}((12))^2 = \mathbf{D}((23))^2$$

$$= \mathbf{D}((01)(12))^3 = \mathbf{D}((12)(23))^3 = \begin{pmatrix} 1 & 0 & 0 \\ 0 & 1 & 0 \\ 0 & 0 & 1 \end{pmatrix}.$$

Finally we identify this representation with one of the irreducible representations introduced above by calculating the character χ, the trace function. It has the values

$$\chi(1) = 3, \ \chi((12)) = 1, \ \chi((123)) = 0, \ \chi((12)(34)) = \chi((1234)) = -1,$$

so that $\mathbf{D}$ is a matrix representation corresponding to the irreducible representation $[3, 1]$, according to the character table shown in the appendix.

In case we want to evaluate bigger representations, we can organize the calculations as follows. We again number the elements $(S \mid T_\alpha)$, $S \in ST^\alpha(1^n)$, according to the column lexicographic order on $ST^\alpha(1^n)$, obtaining the sequence

$$(S_0 \mid T_\alpha) < \ldots < (S_{f^\alpha - 1} \mid T_\alpha).$$

Each $(\pi S_k \mid T_\alpha)$ is a $\mathbb{Z}$–linear combination of these standard bideterminants $(S_i \mid T_\alpha)$, say

$$(\pi S_k \mid T_\alpha) = \sum_{i=0}^{f^\alpha - 1} d^\alpha_{ik}(\pi)(S_i \mid T_\alpha),$$

where we have to evaluate the coefficients $d^\alpha_{ik}(\pi)$, since they form the desired representing matrix $\mathbf{D}^\alpha(\pi)$. Now for each $j \leq f^\alpha - 1$ there is the Capelli operator

$$C_i := C_{S_j T_\alpha},$$

and so

$$C_j(\pi S_k \mid T_\alpha) = \sum_i d^\alpha_{ik}(\pi) C_j(S_i \mid T_\alpha).$$

We recall that $C_j(S_i \mid T_\alpha) \neq 0$ implies that $i \leq j$, and $C_j(S_j \mid T_\alpha) = (T_\alpha \mid T_\alpha) \neq 0$. Hence the coefficient of $(S_i \mid T_\alpha)$ in $(\pi S_k \mid T_\alpha)$ is equal to the coefficient of the monomial $\{T_\alpha \mid T_\alpha\}$ in $C_i(\pi S_k \mid T_\alpha)$. We therefore indicate this coefficient as follows:

$$b_{ij} := \text{coefficient of } \{T_\alpha \mid T_\alpha\} \text{ in } C_j(S_i \mid T_\alpha).$$

The matrix consisting of these coefficients b_{ij} is therefore upper triangular, and it has ones along its main diagonal. But the main point is the equation

$$(b_{ij}) \begin{pmatrix} d^\alpha_{0,k}(\pi) \\ \vdots \\ d^\alpha_{f^\alpha-1,k}(\pi) \end{pmatrix} = \begin{pmatrix} b_0 \\ \vdots \\ b_{f^\alpha-1} \end{pmatrix},$$

where

$$b_i := \text{coefficient of } \{T_\alpha \mid T_\alpha\} \text{ in } C_i(\pi S_k \mid T_\alpha).$$

Exercises

Exercise 6.4.1 Prove that $L_Z\{U \mid V\}$ is in fact well defined.

Exercise 6.4.2 Verify 6.4.8.

Exercise 6.4.3 Evaluate the order diagram of $ST(2^2, 1)$ with respect to hyperdominance.

Exercise 6.4.4 Assume that S is a standard tableau that is hyperdominated by a tableau U which is strictly increasing down each of its columns. Show that there exist *standard* tableaux $S_0, \ldots, S_{r-1}$ such that

$$S = S_0 \leftrightarrow\!\!\!\!\!\updownarrow S_1 \leftrightarrow\!\!\!\!\!\updownarrow \ldots \leftrightarrow\!\!\!\!\!\updownarrow S_{r-1} \leftrightarrow\!\!\!\!\!\updownarrow U.$$

Exercise 6.4.5 Prove that a standard bitableau (S, T) is hyperbidominated by (U, V) with strictly increasing columns if and only if there exist standard bitableaux $(S_0, T_0), \ldots, (S_{r-1}, T_{r-1})$ such that

$$S = S_0 \leftrightarrow\!\!\!\!\!\updownarrow T_0 \leftrightarrow\!\!\!\!\!\updownarrow \ldots \leftrightarrow\!\!\!\!\!\updownarrow S_{r-1} \leftrightarrow\!\!\!\!\!\updownarrow U, \textit{ and } T = T_0 \leftrightarrow\!\!\!\!\!\updownarrow \ldots T_{r-1} \leftrightarrow\!\!\!\!\!\updownarrow U.$$

Exercise 6.4.6 Prove that each of the submodules $U_k(\mathbb{Z})$ introduced above is in fact an invariant submodule for the representation defined by $\pi(S \mid T) := (\pi S \mid T)$. Show that U_k is the image of $R_{Z(T_k)}$, while U_{k-1} is the kernel of this map, when it is applied to U_k.

7. Further Applications

We shall now refine our results on the enumeration of symmetry classes of mappings by using linear representations of finite groups, in particular of symmetric groups. Conversely, we shall apply to representation theory what we know about combinatorial enumeration. A first striking example is the theory of Schur polynomials. These symmetric polynomials will be defined here with the aid of irreducible characters of the symmetric group, and later on we shall show what these polynomials count. We shall discuss the diagram lattice, using representation theory first, and afterwards we can show its significance for unimodality questions about generating functions of enumeration theory.

Then we will see that symmetrization and permutrization of representations are natural generalizations of situations which we met already in connection with the enumeration of symmetry classes of mappings. Another striking example is the plethysm of representations, the corresponding permutation representation case we have already seen. Two very important theorems of representation theory of symmetric groups will be derived, the Littlewood–Richardson Rule and the Murnaghan–Nakayama Rule. They show how certain decompositions of induced representations can be evaluated, and how we can recursively calculate the ordinary irreducible characters. These results can be used in connection with representations, too.

7.1 Schur Polynomials

We now refine enumeration under finite group action with the aid of the preceding results on linear representations. A first remark is a rephrasing of the Cauchy-Frobenius Lemma 2.1.1 and 2.1.13 in terms of inner products of characters:

7.1.1 Corollary *For any finite action ${}_GX$, the corresponding permutation character $\chi\colon g \mapsto |X_g|$, the identity character $\iota\colon g \mapsto 1$ of G and the alternating character $\epsilon\colon g \mapsto \epsilon(\bar{g})$ of G, we have:*

- *The number of orbits of G on X is equal to $[\chi \mid \iota]$.*
- *If $G \neq G^+ := \{g \in G \mid \epsilon(\bar{g}) = 1\}$, then the number of orbits which split over G^+ is equal to $[\chi \mid \epsilon]$, while the number of G–orbits which are also G^+–orbits is equal to $[\chi \mid \iota] - [\chi \mid \epsilon]$.*

The main applications are those which take place in the enumeration of symmetry classes of mappings by weight. In order to describe some of them we recall the definition of the cycle indicator polynomial:

$$C(G,X)=\frac{1}{|G|}\sum_{g\in G}\prod_{i=1}^{|X|} z_i^{a_i(\bar{g})}=\frac{1}{|\bar{G}|}\sum_{\bar{g}\in\bar{G}}\prod_{i=1}^{|X|} z_i^{a_i(\bar{g})}.$$

We notice that it is a sum of monomials $\prod_i z_i^{a_i(\bar{g})}$ which depend only on the cycle type $a(\bar{g})=(a_1(\bar{g}),\ldots,a_{|X|}(\bar{g}))$ of $\bar{g}$. Hence we can rewrite $C(G,X)$ using the notation C^a for the conjugacy class of S_X consisting of the elements of cycle type $a \vdash\!\!\dashv |X|$:

7.1.2 $$C(G,X)=\frac{1}{|X|!}\sum_{a\vdash\!\dashv|X|}\frac{|X|!\,|C^a\cap\bar{G}|}{|\bar{G}|\quad |C^a|}|C^a|\prod_{i=1}^{|X|} z_i^{a_i}.$$

In terms of the character $\iota_{\bar{G}}\uparrow S_X$ induced by the identity character $\iota_{\bar{G}}$ of $\bar{G}$ and its value $(\iota_{\bar{G}}\uparrow S_X)_a$ on the class of elements of type a, this is

7.1.3 $$=\frac{1}{|X|!}\sum_{a\vdash\!\dashv|X|}(\iota_{\bar{G}}\uparrow S_X)_a|C^a|\prod_i z_i^{a_i}.$$

Hence we can use the *generalized* cycle indicator polynomial 3.2.13 corresponding to ${}_GX$ and *the representation D of G* with character χ^D:

$$C(G,X,D):=C(G,X,\chi^D)=\frac{1}{|G|}\sum_{g\in G}\chi^D(g^{-1})\prod_{i=1}^{|X|} z_i^{a_i(\bar{g})}.$$

Using this generalization we obtain an expression for $C(G,X)$ in terms of generalized cycle indicator polynomials of the natural action of the symmetric S_X:

7.1.4 Foulkes' Lemma *For each finite action ${}_GX$ we have:*

$$C(G,X)=\sum_{\alpha\vdash|X|}(I\bar{G}\uparrow S_X,[\alpha])C(S_X,X,[\alpha]).$$

This together with Pólya's Theorem gives the following form of the generating function for the enumeration of G–classes on Y^X by weight

$$\sum_{\alpha\vdash|X|}(I\bar{G}\uparrow S_X,[\alpha])C\big(S_X,X,[\alpha]\,\big|\sum_{y\in Y}y\big).$$

We therefore introduce the abbreviation, for $\alpha\vdash n$ and any set Y of indeterminates,

$$\{\alpha,Y\}:=C\big(S_n,n,[\alpha]\,\big|\sum_{y\in Y}y\big)=\frac{1}{n!}\sum_{a\vdash\!\dashv n}\zeta_a^\alpha|C^a|\prod_{i=1}^{n}\Big(\sum_{y\in Y}y^i\Big)^{a_i}.$$

This polynomial, for which we sometimes simply write

$$\{\alpha\},$$

is called the *Schur polynomial* corresponding to α in the indeterminates $y \in Y$. Using this notation and the *group reduction function*

$$\operatorname{Grf}(G, Y^X) := C\big(G, X \mid \sum y\big),$$

as well as the corresponding generalization

$$\operatorname{Grf}(G, Y^X, D) := C\big(G, X, D \mid \sum y\big),$$

Pólya's Theorem now reads as follows:

7.1.5 Corollary *The generating function for the enumeration of G–classes on Y^X by weight is equal to the group reduction function*

$$\operatorname{Grf}(G, Y^X) = \sum_{\alpha \vdash |X|} (I\bar{G} \uparrow S_X, [\alpha])\{\alpha, Y\}.$$

A particular example is the group reduction function corresponding to the natural action of the alternating group A_X on X:

7.1.6 $$\operatorname{Grf}(A_X, Y^X) = \{(|X|), Y\} + \{(1^{|X|}), Y\},$$

since clearly $IA_n \uparrow S_n = [n] + [1^n]$. This implies (recall 7.1.1):

7.1.7 Corollary *The generating function for the enumeration of injective S_X–classes by weight is equal to the Schur polynomial*

$$\{(1^{|X|}), Y\}.$$

7.1.8 Application (molecular libraries by weight) We recall the molecular libraries mentioned in the section on applications to chemistry. One of them arose from the xanthene as central molecule:

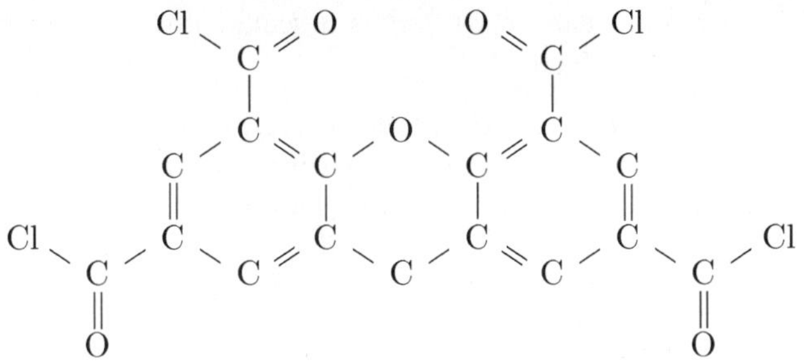

To this central molecule we attached amino acids, assuming that the symmetry group is the cyclic group of order 2. Hence the generating function for the enumeration of the elements in the library by weight is the group reduction function of the symmetry group C_2,

$$\mathrm{Grf}\,(C_2, 4) = \{4\} + \{3, 1\} + 2 \cdot \{2^2\} + \{2, 1^2\}.$$

In the case when we allow at most 3 different amino acids to be attached, the group reduction function in 3 indeterminates a, b, c, say, turns out to be

$$c^4 + 2bc^3 + 4b^2c^2 + 2b^3c + b^4 + 2ac^3 + 6abc^2 + 6ab^2c$$

$$+2ab^3 + 4a^2c^2 + 6a^2bc + 4a^2b^2 + 2a^3c + 2a^3b + a^4$$

◇

These examples may suffice for the moment. We recall that *using Schur polynomials we get a decomposition of the generating function and therefore of the enumeration of G–classes on Y^X*, so that it remains to show what is counted by Schur polynomials. It will turn out that this can be formulated in terms of standard tableaux.

Consider, for $\alpha \vdash n$, the natural action of S_α on Y^n. The corresponding group reduction function is (use 6.2.6)

7.1.9 $$\mathrm{Grf}\,(S_\alpha, Y^n) = \sum_{\beta \vdash n} st^\beta(\alpha)\{\beta, Y\}.$$

Another expression for this group reduction function can be obtained f using Breaking of Symmetry as described in 1.2.11. Since S_X is transitive on the set of elements with the same content as a fixed $f \in Y^X$ we have

7.1.10 Corollary (Ruch) *If $_GX$ is a finite action and $\bar{G}$ the corresponding subgroup of S_X, then, for a fixed $f \in Y^X$ and its stabilizer $(S_X)_f$, the mapping*

$$G \backslash\!\!\backslash_f Y^X \to \bar{G}\backslash S_X/(S_X)_f\colon\ G(f \circ \pi^{-1}) \mapsto \bar{G}\pi(S_X)_f$$

is a bijection between the set $G \backslash\!\!\backslash_f Y^X$ of orbits of G on Y^X consisting of mappings with the same content $c(f, -)$ as f (recall 3.1.5) and the set of $\bar{G}\backslash S_X/(S_X)_f$ of $(\bar{G}, (S_X)_f)$-double cosets in the symmetric group S_X.

The order of $\bar{G}\backslash S_X/(S_X)_f$ can be written as a scalar product of characters (see exercise 11.5.8), and so we obtain

7.1.11 Corollary *If $_GX$ is a finite action, $\bar{G}$ the corresponding subgroup of S_X, $f \in Y^X$, then the number of orbits of G on Y^X consisting of elements of content $c(f, -)$ is , for finite Y, the scalar product of the characters of the induced representations $I\bar{G} \uparrow S_X$ and $I(S_X)_f \uparrow S_X$:*

$$\left|G \backslash\!\!\backslash_f Y^X\right| = \left[\chi^{I\bar{G}\uparrow S_X} \mid \chi^{I(S_X)_f\uparrow S_X}\right].$$

Using an ordered set $Y := \{y_0, \ldots, y_{m-1}\}$ and the abbreviation

$$Y^\lambda := y_0^{\lambda_0} \cdots y_{m-1}^{\lambda_{m-1}},$$

we obtain, for example,

7.1.12 Corollary *The group reduction function for the natural action of the Young subgroup S_α on the set of mappings m^n can be expressed as follows in terms of double cosets, of characters and of numbers of standard tableaux:*

$$\begin{aligned}\mathrm{Grf}\,(S_\alpha, m^n) &= \sum_{(\lambda_0,\dots,\lambda_{m-1})\models n} |S_\lambda \backslash S_n / S_\alpha| Y^\lambda \\ &= \sum_{(\lambda_0,\dots,\lambda_{m-1})\models n} [\xi^\lambda \mid \xi^\alpha] Y^\lambda \\ &= \sum_{(\lambda_0,\dots,\lambda_{m-1})\models n} \sum_{\beta\vdash n} st^\beta(\lambda) st^\beta(\alpha) Y^\lambda.\end{aligned}$$

Comparing this equation with 7.1.9 we finally obtain

7.1.13 Corollary *For each finite set Y of indeterminates and any proper partition $\alpha \vdash n$ we have*

$$\{\alpha, Y\} = \sum_{(\lambda_0,\dots,\lambda_{|Y|-1})\models n} st^\alpha(\lambda) Y^\lambda.$$

In other words: Schur polynomials generate numbers of standard tableaux by weight.

We recall that $ST^\alpha(\lambda) \neq \emptyset$ implies that the number of nonzero λ_i must be at least the number of nonzero parts α_i of α. Hence we obtain

7.1.14 Corollary *For each proper partition α the Schur polynomial $\{\alpha, Y\}$ is the zero polynomial if the number of indeterminates $y \in Y$ is smaller than the number of nonzero parts of α.*

A further consequence of 7.1.13 is

7.1.15 Lemma *The Gaussian polynomials (cf. 5.2.7) are specializations of the Schur polynomials corresponding to one rowed partitions:*

$$\begin{bmatrix} m+n \\ n \end{bmatrix} = \{m\}(1, x, \dots, x^n),$$

if $\{m\}$ is taken over the set $Y := \{y_0, \dots, y_n\}$ of indeterminates.

Proof: We obtain from 7.1.13 that

$$\{m\}(1, x, \dots, x^n) = \sum_{(\lambda_0,\dots,\lambda_n)\models m} st^{(m)}(\lambda) x^{\Sigma_i i\cdot\lambda_i}.$$

We recall that the Gaussian polynomial counts (see 5.2.7) partitions that fit into a certain rectangle. It therefore suffices to establish a bijection

$$\bigcup_{(\lambda_0,\dots,\lambda_n)\models m} ST^{(m)}(\lambda) \longrightarrow \{\alpha \mid \exists\, k \le m\cdot n\colon \alpha \vdash k, \alpha_0' \le m, \alpha_0 \le n\}.$$

Such a bijection is provided by

$$T := t_0 \dots t_{m-1} \mapsto \alpha(T) := (t_{m-1}, \dots, t_0) \vdash (\Sigma_i t_i) = \sum_i i\cdot\lambda_i \le m\cdot n.$$

□

Another immediate corollary is 6.3.5. Since $\{\alpha, Y\}$ is symmetric, we obtain again the interesting equality $st^\alpha(\lambda) = st^\alpha(\lambda^*)$, for any improper partition $\lambda \models n$. We consider an example of a group reduction function. The generating function for the enumeration of graphs on 4 vertices is

$$C\left(S_4^{[2]}, \binom{4}{2} \Big| y_0 + y_1\right) = \sum_{\alpha \vdash 6} (IS_4^{[2]} \uparrow S_6, [\alpha])\{\alpha\} = \{6\} + \{4,2\} + \{3^2\}.$$

This follows from 7.1.14 together with an application of character theory which gives

$$IS_4^{[2]} \uparrow S_6 = [6] + [4,2] + [3^2] + [2^3] + [2^2, 1^2] + [1^6].$$

Hence 7.1.13 shows that the group reduction function for graphs on 4 vertices is equal to the sum of monomials corresponding to the standard tableaux

$$\begin{matrix}0&0&0&0&0&0\end{matrix}, \quad \begin{matrix}0&0&0&0&0&1\end{matrix}, \quad \begin{matrix}0&0&0&0&1&1\end{matrix},$$

$$\begin{matrix}0&0&0&1&1&1\end{matrix}, \quad \begin{matrix}0&0&1&1&1&1\end{matrix}, \quad \begin{matrix}0&1&1&1&1&1\end{matrix},$$

$$\begin{matrix}1&1&1&1&1&1\end{matrix}, \quad \begin{matrix}0&0&0&0\\1&1\end{matrix}, \quad \begin{matrix}0&0&0&1\\1&1\end{matrix},$$

$$\begin{matrix}0&0&1&1\\1&1\end{matrix}, \quad \begin{matrix}0&0&0\\1&1&1\end{matrix}.$$

But until now there is no natural bijection known between the set of graphs on v vertices and the corresponding set of standard tableaux.

Another interesting consequence of the bijection from double cosets and orbits allows to reformulate Redfield's cap product of cycle indicator polynomials in terms of characters (use 3.6.2 and exercise 11.5.8 again):

7.1.16 Corollary *For any two actions of finite groups G and H on the set n, the corresponding cap product of the cycle indicator polynomials is equal to the inner product of the corresponding characters of S_n induced by the identity characters:*

$$C(G,n) \cap C(H,n) = \left[\chi^{I\bar{G}\uparrow S_n} \mid \chi^{I\bar{H}\uparrow S_n}\right].$$

Correspondingly we have, for more than 2 factors, the identity

$$C(G,X) \cap \ldots \cap C(K,X) = \left[\chi^{I\bar{G}\uparrow S_n} \cdots \chi^{I\bar{K}\uparrow S_n} \mid \chi^{IS_n}\right].$$

Hence the cap product of cycle indices can be used in order to evaluate the number $|\bar{G}\backslash S_X/\bar{H}|$ of double cosets, in the symmetric group S_X, of finite permutation representations $\bar{G}$ and $\bar{H}$ of abstract groups G and H, say, on a finite set X. We obtain from 7.1.16 the following equation:

7.1.17 $$|\bar{G}\backslash S_X/\bar{H}| = C(G,X) \cap C(H,X).$$

This is implemented in SYMMETRICA and turned out to be efficient.

Exercises

Exercise 7.1.1 Evaluate the generating function — as a sum of Schur polynomials — for the molecular library with benzene triacid chlorine as central molecule and amino acids as building blocks (see exercise 3.4.1).

7.2 Symmetric Polynomials

We now consider Schur polynomials in a more general context. There is a natural action of S_m on the set $\mathbb{Q}\,[Y] := \mathbb{Q}\,[y_0, \ldots, y_{m-1}]$ of all the polynomials in the set $Y := \{y_0, \ldots, y_{m-1}\}$ of commuting indeterminates over $\mathbb{Q}$,

$$S_m \times \mathbb{Q}\,[Y] \to \mathbb{Q}\,[Y] : (\pi, f) \mapsto f(x_{\pi^{-1}(0)}, \ldots, x_{\pi^{-1}(m-1)}).$$

The invariants of this action, i. e. the elements of

$$\mathbb{Q}\,[y_0, \ldots, y_{m-1}]_{S_m},$$

are called *symmetric polynomials* (over $\mathbb{Q}$). Consider the $\mathbb{Q}$–vector space

$$\mathcal{HS}_n[Y]$$

consisting of all the symmetric polynomials that are homogeneous and of degree $n \leq m$ in the indeterminates $y_i \in Y$, together with the zero polynomial. We shall establish an isometry between $\mathcal{HS}_n[Y]$ and the vector space

$$CF(S_n, \mathbb{Q})$$

consisting of the $\mathbb{Q}$–valued class functions on S_n. In order to do this we introduce various series of symmetric polynomials. For $p \geq 1$ we set

$$e_p := \sum_{i_j < i_{j+1}} y_{i_0} \cdots y_{i_{p-1}},$$

($e_0 := 1$) and note that $e_p = 0$ if $p > m$. This polynomial is in fact an *orbit sum*, the sum of the elements of the orbit $S_m(y_0 \cdots y_{p-1})$,

$$e_p = \underline{S_m(y_0 \cdots y_{p-1})}.$$

Now, for each $\alpha \vdash n$, introduce

$$e_\alpha := e_{\alpha_0} e_{\alpha_1} \cdots$$

and call these the *elementary symmetric* polynomials. Furthermore we define:

$$h_p := \sum_{(i_0, \ldots, i_{m-1}) \models p} y_0^{i_0} \cdots y_{m-1}^{i_{m-1}},\ h_0 := 1,\ h_\alpha := h_{\alpha_0} h_{\alpha_1} \cdots,$$

the *complete homogeneous symmetric* polynomials. Another series of symmetric polynomials is formed by the

$$s_\alpha := s_{\alpha_0} s_{\alpha_1} \cdots, \text{ where } s_p := \sum_i y_i^p = \underline{S_m(y_0^p)},\ s_0 := 1.$$

These s_p are called the *symmetric power sums*. Finally put

$$k_\alpha := \sum y_{i_0}^{\alpha_0} y_{i_1}^{\alpha_1} \cdots = \underline{S_m(y_0^{\alpha_0} y_1^{\alpha_1} \cdots y_{m-1}^{\alpha_{m-1}})},$$

the *monomial symmetric* polynomials, where the sum has to be taken over all the *different* monomials $y_{i_0}^{\alpha_0} y_{i_1}^{\alpha_1} \cdots$ with *pairwise different* indices $i_j \in m$, thus $k_\alpha = 0$, if $\alpha_0' > m$. We note in passing that k_α is an orbit sum. The crucial connections between these series of symmetric polynomials and the Schur polynomials are the following ones

7.2.1 Lemma *For each proper partition* α *of* n *we have:*

$$h_\alpha = \sum_{\beta \vdash n} m_{\alpha\beta} k_\beta,\; e_\alpha = \sum_{\beta \vdash n} m'_{\alpha\beta} k_\beta,\; k_\alpha = \sum_{\lambda \models n:\, \lambda^* = \alpha} y_0^{\lambda_0} y_1^{\lambda_1} \cdots,$$

and

$$\{\alpha\} = \frac{1}{n!} \sum_{\beta \vdash n} |C^\beta| \zeta_\beta^\alpha s_\beta = \sum_{\beta \vdash n} \kappa_{\beta\alpha} k_\beta,$$

so that, by inversion, we obtain

$$k_\alpha = \sum_{\beta \vdash n} \epsilon_{\beta\alpha} \{\beta\},\; h_\alpha = \sum_{\beta \vdash n} \kappa_{\alpha\beta} \{\beta\},\; \{\alpha\} = \sum_{\beta \vdash n} \epsilon_{\alpha\beta} h_\beta.$$

Proof: It follows from the definition of h_α that

$$h_\alpha = \sum_{(i_0,\ldots,i_{m-1}) \models \alpha_0} y_0^{i_0} y_1^{i_1} \cdots y_{m-1}^{i_{m-1}} \sum_{(j_0,\ldots,j_{m-1}) \models \alpha_1} y_0^{j_0} y_1^{j_1} \cdots y_{m-1}^{j_{m-1}} \sum \cdots .$$

Each summand of this expression corresponds to a matrix

$$M := \begin{pmatrix} i_0 & i_1 & \ldots & i_{m-1} \\ j_0 & j_1 & \ldots & j_{m-1} \\ \ldots & \ldots & \ldots & \ldots \end{pmatrix}$$

with row sums α_i. If $M \in M_{\alpha\lambda}$, then $\lambda \models n$ and the corresponding summand is $y_0^{\lambda_0} y_1^{\lambda_1} \cdots$, and so we can proceed as follows:

$$= \sum_{\lambda \models n} m_{\alpha\lambda} y_0^{\lambda_0} y_1^{\lambda_1} \cdots = \sum_{\beta \vdash n} m_{\alpha\beta} k_\beta,$$

since obviously $k_\beta = \sum_{\lambda^* = \beta} y_0^{\lambda_0} y_1^{\lambda_1} \cdots$. This proves the first statement, while the second is obtained analogously, the matrices M now being elements of $M'_{\alpha\lambda}$, $\lambda \models n$. The expression of k_α in terms of the monomials Y^λ is obvious. The linear combinations of $\{\alpha\}$ in terms of the s_β and of the k_α is clear from the definition of Schur polynomials and the fact that the Kostka numbers are numbers of standard tableaux, for which we know that $st^\alpha(\lambda)$ is equal to $st^\alpha(\lambda^*)$. The final row of equations is obtained by inversion. □

It is trivial that $\{k_\alpha \mid \alpha \vdash n\}$ is a $\mathbb{Q}$–basis of $\mathcal{HS}_n[Y]$, and hence we obtain from 7.2.1:

7.2.2 Corollary *The $s_\alpha, e_\alpha, h_\alpha, k_\alpha, \{\alpha\}$ form $\mathbb{Q}$–bases of $\mathcal{HS}_n[Y]$, if α runs through the proper partitions $(\alpha_0, \ldots, \alpha_{m-1}) \vdash n$, where $m := |Y|$. Some of the intertwining matrices which lead from one basis to the others were shown in 7.2.1.*

The following example, taken from the enumeration of symmetry classes, exhibits the relationship between the e_α and the s_α.

7.2.3 Example We recall from 7.1.7 that the generating function for the enumeration of injective S_n–classes on Y^n is the Schur polynomial

$$\{1^n\} = \frac{1}{n!}\sum_{a \dashv\vdash n}(-1)^{\sum(i-1)\cdot a_i}|C^a|\prod_i s_i^{a_i} = \sum_a \frac{(-1)^{a_2+a_4+\ldots}}{\prod_i i^{a_i}a_i!}\prod_i s_i^{a_i}.$$

On the other hand it is obvious from the definition of e_n that it is the generating function for the numbers of injective S_n–classes. Thus we have proved the identity

7.2.4
$$e_n = \sum_{a \dashv\vdash n}(-1)^{a_2+a_4+\ldots}\prod_i \frac{1}{a_i!}\left(\frac{s_i}{i}\right)^{a_i}.$$

From this identity also a recursion for the e_n follows (cf. exercise 7.2.1). ◇

We now introduce a bilinear form $\langle -, - \rangle$ on $\mathcal{HS}_n[Y]$, *Hall's inner product*, which is defined as the bilinear extension of

$$\langle h_\alpha, k_\beta \rangle := \delta_{\alpha\beta}.$$

7.2.5 Lemma *$\langle -, - \rangle$ is symmetric, and $\langle h_\alpha, h_\beta \rangle = m_{\alpha\beta}$, while $\langle h_\alpha, e_\beta \rangle = m'_{\alpha\beta}$.*

Proof: Using 7.2.1 we obtain

$$\langle h_\alpha, h_\beta \rangle = \langle h_\alpha, \sum_\gamma m_{\beta\gamma}k_\gamma \rangle = m_{\beta\alpha} = m_{\alpha\beta} = \langle h_\beta, h_\alpha \rangle,$$

which yields the first two statements. The third statement also follows by an application of 7.2.1. □

In seeking for an orthonormal basis with respect to Hall's inner product we note

7.2.6 Theorem *The $\{\alpha, Y\}$, $\alpha \vdash n$, form an orthonormal basis of $\mathcal{HS}_n[Y]$ which is interrelated with $\{k_\alpha \mid \alpha \vdash n\}$, $\{h_\alpha \mid \alpha \vdash n\}$, and $\{e_\alpha \mid \alpha \vdash n\}$, by matrices over $\mathbb{Z}$.*

Proof: The orthonormality follows from an application of 7.2.5:

$$\langle \{\alpha\}, \{\beta\} \rangle = \sum_{\gamma,\delta \vdash n} \epsilon_{\alpha\gamma}\epsilon_{\beta\delta}\langle h_\gamma, h_\delta \rangle = \delta_{\alpha\beta},$$

where the final equation is clear from the proof of 7.2.5. The other statements follow from 7.2.2. □

The only orthogonal matrices over $\mathbb{Z}$, which are invertible over $\mathbb{Z}$, are the matrices which contain in each row and in each column exactly one nonzero entry, and this entry is ± 1. Hence 7.2.6 yields that up to factors ± 1 the Schur polynomials are the only orthonormal basis of $\mathcal{HS}_n[Y]$:

7.2.7 Corollary *The Schur polynomials* $\{\alpha, Y\}, \alpha \vdash n$, *form essentially the only orthonormal basis of* $\mathcal{HS}_n[Y]$.

We are now in a position to compare the vector space $CF(S_n, \mathbb{Q})$, and its scalar product $[- \mid -]$, with $\mathcal{HS}_n[Y]$, equipped with Hall's inner product $\langle - \mid - \rangle$. Consider the *Frobenius mapping*

$$F: CF(S_n, \mathbb{Q}) \to \mathcal{HS}_n[Y]: \psi \mapsto \frac{1}{n!} \sum_{\alpha \vdash n} \psi_\alpha |C^\alpha| s_\alpha,$$

where ψ_α denotes the value of ψ on the conjugacy class C^α. Hence, for example,

7.2.8 $$F(\zeta^\alpha) = \{\alpha\}.$$

Now we point to the connection between induced characters and group reduction functions, expressed in terms of Frobenius' mapping:

7.2.9 Lemma *For each representation* D *of a subgroup* $U \leq S_n$ *we have:*

$$F(\chi^D \uparrow S_n) = \mathrm{Grf}\,(S_n, Y^n, D \uparrow S_n) = \mathrm{Grf}\,(U, Y^n, D).$$

In particular, for each $\alpha \vdash n$ *and the corresponding Young character* ξ^α, *we have*

$$F(\xi^\alpha) = F(\xi^{\alpha_0}) F(\xi^{\alpha_1}) \cdots = h_\alpha = e_{\alpha'},$$

and in particular

$$h_p = \mathrm{Grf}\,(S_p, Y^p).$$

Proof: The first statement is clear. An application of 3.2.7 gives

$$F(\xi^\alpha) = \mathrm{Grf}\,(S_\alpha, Y^n) = \prod_i \mathrm{Grf}\,(S_{\alpha_i}, Y^{\alpha_i}) = \prod_i F(\xi^{\alpha_i}).$$

It therefore suffices to prove that $F(\xi^{(p)}) = h_p = e_{(1^p)} = e_1 e_1 \cdots$. But $h_p = e_{(1^p)}$ is clear by definition. Furthermore h_p is the sum of all the monomials of degree p while $F(\xi^{(p)}) = F(\zeta^{(p)}) = \mathrm{Grf}\,(S_p, Y^p)$, which counts the mappings $f \in Y^p$ by content, and the orbits of S_p on this set are characterized by the weights $y_0^{i_0} \cdots y_{m-1}^{i_{m-1}}$. □

We are now in a position to prove the main result of this section:

7.2.10 Theorem *The Frobenius map* F *is an isometry between the vector spaces* $CF(S_n, \mathbb{Q})$ *and* $\mathcal{HS}_n[Y]$.

Proof: The Young characters ξ^α form a basis of $CF(S_n, \mathbb{Q})$, so, by 7.2.9, F is an isomorphism. Furthermore

$$[\xi^\alpha \mid \xi^\beta] = m_{\alpha\beta} = \langle h_\alpha, h_\beta \rangle = \langle F(\xi^\alpha), F(\xi^\beta) \rangle,$$

and so F is even an isometry. □

It is important to realize that 7.2.10 means that we either can use characters or we can use symmetric polynomials in order to do representation theory of the symmetric group!

Exercises

Exercise 7.2.1 Derive from 7.2.4 *Newton's identity*,

$$\sum_{l=0}^{n-1}(-1)^l e_l s_{n-l} + (-1)^n n e_n = 0.$$

Exercise 7.2.2 What is the inverse image of s_α under the Frobenius isometry? Use this in order to express $|C^\alpha|$ in terms of $\langle s_\alpha, s_\alpha\rangle$.

7.3 The Diagram Lattice

Lemma 7.1.4 implies

7.3.1 Corollary *The group reduction function of a finite action ${}_GX$ can be expressed as follows:*

$$\begin{aligned}\mathrm{Grf}\,(G, Y^X) &= \sum_{\alpha\vdash|X|}(I\bar{G}\uparrow S_X, [\alpha]) \sum_{(\beta_0,\ldots,\beta_{|Y|-1})\vdash|X|} \kappa_{\beta\alpha}k_\beta \\ &= \sum_{\alpha,\beta}(I\bar{G}\uparrow S_X, [\alpha])st^\alpha(\beta)k_\beta.\end{aligned}$$

In order to compare the coefficients $\kappa_{\beta\alpha} = st^\alpha(\beta)$, we first derive the following result which will turn out to be crucial for the development of the representation theory of symmetric groups:

7.3.2 Theorem *For $\alpha, \beta \vdash n$ there exists a character $\chi_{\alpha\beta}$ of S_n such that*

$$\xi^\beta + \chi_{\alpha\beta} = \xi^\alpha,$$

if and only if $\alpha \lhd \beta$.

Proof: The existence of such a $\chi_{\alpha\beta}$ implies that $\alpha \neq \beta$ and that

$$1 = [\xi^\beta \mid \zeta^\beta] \leq [\xi^\alpha \mid \zeta^\beta].$$

Hence, by 6.1.20, $\alpha \lhd \beta$ must hold. If on the other hand $\alpha \lhd \beta$, then there exists a sequence of partitions $\alpha^i \vdash n$ such that

$$\alpha = \alpha^0 \lhd \alpha^1 \lhd \ldots \lhd \alpha^r = \beta,$$

where α^i covers α^{i-1}. Thus we need only prove the statement in the case when β covers α. We can therefore assume the existence of i and j which satisfy

$$0 \le i < j,\ \beta_i = \alpha_i + 1,\ \beta_j = \alpha_j - 1,\ \forall\, k \neq i, j \colon \alpha_k = \beta_k.$$

We shall express $\chi_{\alpha\beta}$ in terms of these indices i and j. Consider the representation with character ξ^α,

$$[\alpha_0] \,\#\, [\alpha_1] \,\#\, \ldots \uparrow S_n = \big(([\alpha_i] \,\#\, [\alpha_j]) \,\#\, (\,\#_{\,k \neq i,j}\, [\alpha_k])\big) \uparrow S_n.$$

The Determinantal Form 6.3.6 yields

$$[\alpha_i] \,\#\, [\alpha_j] = [\alpha_i, \alpha_j] + [\alpha_i + 1][\alpha_j - 1],$$

thus we can proceed as follows (since induction is transitive):

$$[\alpha_0] \,\#\, [\alpha_1] \,\#\, \ldots \uparrow S_n = \big(([\alpha_i, \alpha_j] + [\alpha_i + 1][\alpha_j - 1]) \,\#\, (\,\#\, [\alpha_k])\big) \uparrow S_n$$

$$= \big([\alpha_i, \alpha_j] \,\#\, (\,\#_{\,k \neq i,j}\, [\alpha_k])\big) \uparrow S_n + [\beta_0] \,\#\, [\beta_1] \,\#\, \ldots \uparrow S_n,$$

and hence the character $\chi_{\alpha\beta}$ of the representation

$$\big([\alpha_i, \alpha_j] \,\#\, (\,\#_{\,k \neq i,j}\, [\alpha_k])\big) \uparrow S_n$$

is equal to $\xi^\alpha - \xi^\beta$. □

An immediate consequence of this theorem, 6.2.1 and 6.3.5 is

7.3.3 Corollary *For each $\alpha \vdash n$ and $\lambda \models n$ we have*

$$ST^\alpha(\lambda) \neq \emptyset \iff \lambda^* \trianglelefteq \alpha.$$

Moreover $st^\alpha(\alpha) = 1$ and, if also $\mu \models n$, then

7.3.4 $$\lambda^* \triangleleft \mu^* \trianglelefteq \alpha \Longrightarrow st^\alpha(\lambda) > st^\alpha(\mu) > 0.$$

In order to use this for a proof of an existence theorem on 0-1–matrices, we consider the identity

$$m'_{\alpha\beta} = [\xi^\alpha \mid \epsilon\xi^\beta].$$

Since, by 6.1.14, we have

$$\epsilon \cdot \xi^\beta = \epsilon \cdot \sum_\gamma st^\gamma(\beta)\zeta^\gamma = \sum_\gamma st^{\gamma'}(\beta)\zeta^\gamma,$$

we can deduce the identity

7.3.5 $$m'_{\alpha\beta} = \sum_\gamma st^\gamma(\alpha) st^{\gamma'}(\beta).$$

It shows that $M'_{\alpha\beta}$ is not empty if and only if a $\gamma \vdash n$ exists such that $\alpha \trianglelefteq \gamma$ and $\gamma \trianglelefteq \beta'$, i. e. if and only if $\alpha \trianglelefteq \beta'$. An easy generalization gives the following important existence theorem:

7.3.6 The Gale–Ryser Theorem *For any* $\lambda, \mu \models n$ *we have*

$$M'_{\lambda,\mu} \neq \emptyset \iff \lambda^* \trianglelefteq (\mu^*)'.$$

In terms of Young characters this reads as follows:

7.3.7 The Ruch–Schönhofer Theorem *For any* $\alpha, \beta \vdash n$ *and the corresponding Young characters of* S_n *we have*

$$[\xi^\alpha \mid \epsilon\xi^\beta] > 0 \iff \alpha \trianglelefteq \beta'.$$

These characterizations of dominance in terms of standard tableaux, 0-1–matrices and common irreducible constituents show the importance of the diagram lattice $(P(n), \trianglelefteq)$. The following subsets I_α and F_α of this lattice are called the *principal ideal* and the *principal filter* corresponding to α:

$$I_\alpha := \{\beta \vdash n \mid \beta \trianglelefteq \alpha\},\ F_\alpha := \{\gamma \vdash n \mid \alpha \trianglelefteq \gamma\}.$$

In terms of these the above characterizations of dominance read as follows:

7.3.8 Corollary *For each* $\alpha \vdash n$,

$$I_\alpha = \{\beta \vdash n \mid ST^\alpha(\beta) \neq \emptyset\} = \{\beta \vdash n \mid M'_{\alpha'\beta} \neq \emptyset\}$$

$$= \{\beta \vdash n \mid [\xi^\beta \mid \zeta^\alpha] > 0\} = \{\beta \vdash n \mid [\xi^{\alpha'} \mid \epsilon\xi^\beta] > 0\},$$

while

$$F_\alpha = \{\beta \vdash n \mid ST^{\alpha'}(\beta') \neq \emptyset\} = \{\beta \vdash n \mid M'_{\alpha\beta'} \neq \emptyset\}$$

$$= \{\beta \vdash n \mid [\xi^\alpha \mid \zeta^\beta] > 0\} = \{\beta \vdash n \mid [\xi^\alpha \mid \epsilon\xi^{\beta'}] > 0\}.$$

There are various other sets of partitions which are unions of principal ideals of filters in diagram lattices.

7.3.9 Application (graphical partitions) The *edge degree* of a vertex in a labeled graph Γ is the number of edges which meet in this vertex. The edge degrees of all the v vertices of a labeled graph with e edges form, after reordering (if necessary), a nonincreasing sequence $(\alpha_0, \ldots, \alpha_{v-1})$ which gives a partition $\alpha := (\alpha_0, \alpha_1, \ldots) \vdash 2e$ after adding further $\alpha_i := 0, i \geq v$. We call this sequence the *edge degree sequence* of Γ and of the labelded graphs which are represented by Γ. We ask for a characterization of *graphical partitions*, i. e. of partitions which occur as edge degree sequences. (Recall the basic problem of combinatorial enumeration mentioned above, the construction of chemical isomers, which amounts to the construction of connected multigraphs with a given edge degree sequence!)

Assume that $\alpha \vdash n = 2e$ is an edge degree sequence, and that Γ has its i-th vertex of degree α_i. The labeled graph Γ can be described by a tableau $T(\Gamma)$ which contains in its i-th row the numbers of the vertices which are connected with the i-th vertex. In order to have $T(\Gamma)$ uniquely determined, we assume that the row entries are put in increasing order. An example is shown in Figure 7.1. We note that both

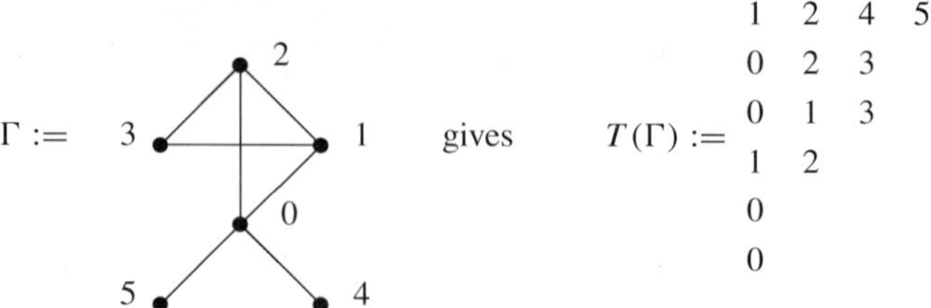

Fig. 7.1. A labeled graph Γ and its tableau $T(\Gamma)$

the shape and the content of $T(\Gamma)$ is equal to α. Furthermore it is clear that

7.3.10
$$\sum_{i\in j}(\alpha'_i - \alpha_i) \geq j,\ j \in \underline{d(\alpha)},$$

if $d(\alpha)$ denotes the length of the main diagonal of the diagram $[\alpha]$, i. e. if

$$d(\alpha) - 1 := \max\{i \mid \alpha_i > i\} = \max\{j \mid \alpha'_j > j\}.$$

The square subdiagram

$$[d(\alpha)^{d(\alpha)}] \subseteq [\alpha]$$

occurs very often in the combinatorial theory of partitions, it is called the *Durfee square* of $[\alpha]$.

7.3.11 Lemma *The partitions $\alpha \vdash n = 2e$ with the property*

$$\sum_{i\in j}(\alpha'_i - \alpha_i) = j,\ j \in \underline{d(\alpha)},$$

are graphical partitions.

Proof: The tableau

$$\begin{array}{llllllllll}
1 & 2 & 3 & 4 & 5 & \dots & \dots & \dots & \dots & \dots\ \alpha'_0 - 1 \\
0 & 2 & 3 & 4 & 5 & \dots & \dots & \dots\ \alpha'_1 - 1 & & \\
0 & 1 & 3 & 4 & 5 & \dots\ \alpha'_2 - 1 & & & & \\
0 & 1 & 2 & 4 & 5 & \dots & & & & \\
\vdots & \vdots & \vdots & & & & & & & \\
0 & 1 & & & & & & & & \\
\vdots & & & & & & & & & \\
0 & & & & & & & & &
\end{array}$$

which is of shape and content α shows the existence of a labeled graph with edge degree sequence α. □

We shall show that these particular partitions form an antichain in $P(n)$. In fact the following more general result is true:

7.3.12 Lemma *For each sequence* $z = (z_0, \ldots, z_{d(\alpha)-1})$ *of integers* z_i*, the set*

$$P(n)_z := \{\alpha \vdash n \mid \forall\, j \in \underline{d(\alpha)} : \Sigma_{i \in j}(\alpha'_i - \alpha_i) = z_j\}$$

is an antichain, i. e. the elements of this set are pairwise incomparable with respect to the dominance order.

Proof: Using an indirect proof we assume that $\alpha,\ \beta \in P(n)_z$ are comparable, say $\alpha \trianglelefteq \beta$, and we shall show that this implies $\alpha = \beta$. The assumption $\alpha \trianglelefteq \beta$ yields $\alpha_0 \leq \beta_0$, hence also $\alpha_0 + z_0 \leq \beta_0 + z_0$, i. e. $\alpha'_0 \leq \beta'_0$. But $\alpha \trianglelefteq \beta$ is the same as $\beta' \trianglelefteq \alpha'$, so that also $\beta'_0 \leq \alpha'_0$ must hold, and we obtain that $\alpha'_0 = \beta'_0$, which gives $\alpha_0 = \beta_0$. Using this we can derive in the same way that also $\alpha_1 = \beta_1, \ \ldots, \alpha_{d(\alpha)-1} = \beta_{d(\alpha)-1}$. □

Hence in particular the set $P(n)_{(1,2,\ldots)}$, which consists, if n is even, of graphical partitions (cf. 7.3.11), is an antichain. In order to show that the union of the principal ideals of the elements in $P(n)_{(1,2,\ldots)}$ is the complete set of graphical partitions, we need one further result:

7.3.13 Lemma *If the graphical partition* α *covers* β*, then* β *is also graphical.*

Proof: As α covers β, there exist $i < j$ such that $\alpha_i = \beta_i + 1, \alpha_j = \beta_j - 1$, and $\alpha_k = \beta_k$, for each $k \neq i, j$. Consider a tableau $T(\Gamma)$ that describes a labeled graph with edge degree sequence α. The inequality $\beta_i \geq \beta_j$ implies that $\alpha_i \geq \alpha_j + 2$, so that there exists a vertex of Γ that is connected with the i-th vertex but not with the j-th vertex, let k be its label. The labeled graph Γ' which arises from Γ by erasing the edge that joins the i-th and the k-th vertex and adding an edge that connects the k-th and the j-th vertex obviously has edge degree sequence β, which proves the theorem. □

Thus, by 7.3.11, $P(n)_{(1,2,\ldots)}$ consists of graphical partitions, if n is even. Furthermore, the ideals of the elements of $P(n)_{(1,2,\ldots)}$ consist of graphical partitions, by 7.3.13. Since 7.3.10 together with 7.3.11 shows that a graphical partition α is dominated by the element $\beta \in P(n)_{(1,2,\ldots)}$, defined by

$$\beta_0 := \alpha_0, \ldots, \beta_{d(\alpha)-1} := \alpha_{d(\alpha)-1}, \text{ and } \beta'_0 := \alpha_0 + 1, \ldots, \beta'_{d(\alpha)-1} := \alpha_{d(\alpha)-1} + 1,$$

we have proved the following:

7.3.14 Theorem (Gutman–Ruch) *For even n, the set*

$$G(n) := \{\alpha \in P(n) \mid \exists\, \beta \in P(n)_{(1,2,\ldots)} \colon \alpha \trianglelefteq \beta\},$$

i. e. the union of the ideals generated by elments of $P(n)_{(1,2,\ldots)}$*, is the set of graphical partitions of n, i. e. the set of edge degree sequences of graphs with* $n/2$ *edges.*

For example the set of graphical partitions of $n = 6$ is

$$I_{(2^3)} \cup I_{(3,1^3)} = \{(2^3), (3, 1^3), (2^2, 1^2), (2, 1^4), (1^6)\}.$$

◇

The theorem of Gutman and Ruch suggests to compare the *branching* of graphs. They say that a graph with edge degree sequence α is *more branched* than a graph with edge degree sequence β if and only if $\alpha \rhd \beta$. In this sense the graphs with edge degree sequence $\alpha \in P(2e)_{(1,2,\dots)}$ are the *maximally branched* graphs with e edges. For example $\alpha := (e, 1^e)$, the edge degree sequence of the *star* with e edges, gives a maximally branched graph. In order to enumerate the elements of $P(2e)_{(1,2,\dots)}$, we note that $\alpha \in P(2e)_{(1,2,\dots)}$ is uniquely determined by the sequence $(\alpha_0, \alpha_1 - 1, \dots, \alpha_{d(\alpha)-1} - (d(\alpha) - 1))$ which is strictly decreasing. Furthermore each strictly decreasing sequence of natural numbers $(n_0, \dots, n_{r-1})$ yields an $\alpha \in P(n)_{(1,2,\dots)}$, n even, by putting $\alpha_0 := n_0, \alpha_1 := n_1 + 1, \dots, \alpha_{d(\alpha)-1} := n_{d(\alpha)-1} + (d(\alpha) - 1)$, where $d(\alpha) := r$, and $\alpha'_i := \alpha_i + 1$, for $0 \leq d(\alpha) - 1$. Hence $P(2e)_{(1,2,\dots)}$ is of the same cardinality as the set of strictly decreasing partitions of e, which proves

7.3.15 Corollary *The number of elements in $P(2e)_{(1,2,\dots)}$, i. e. the number of edge degree sequences that correspond to maximally branched graphs with e edges, is the coefficient c_e in the formal power series*

$$\sum_{e\in\mathbb{N}} c_e x^e := \prod_{n\in\mathbb{N}^*} (1 + x^n) = 1 + x + x^2 + 2x^3 + 2x^4 + 3x^5 + \dots .$$

7.4 Unimodality

In an earlier chapter we have seen that the numbers of graphs on 4 vertices which have 0,1,2,3,4,5,6 edges are 1,1,2,3,2,1,1. For the graphs on 5 vertices we obtain the sequence 1,1,2,4,6,6,6,4,2,1,1 according to the number of edges. Both these sequences weakly increase up to their middle term and then they weakly decrease, i. e. they both are *unimodal* sequences. Furthermore we note that, if $g(v, e)$ denotes the number of graphs with v vertices and e edges, at least for the cases $v := 4, 5$, we have $g(v, e) = g(v, \binom{v}{2} - e)$, i. e. corresponding terms are equal and hence these two sequences are also *reciprocal*. Thus the sequences of numbers of graphs on v vertices according to their number of edges are reciprocal and unimodal, at least for the cases $v = 4$ and $v = 5$. It is our aim to show that this is true for *each* v, in fact we shall prove a much more general result on unimodal and reciprocal sequences.

The reciprocity of the sequence $g(v, e)$, for each v, is trivial by complementation of graphs:

7.4.1 $$\forall\, v \in \mathbb{N}\colon\ g(v, e) = g\left(v, \binom{v}{2} - e\right), \text{ for each } e.$$

In order to prove the unimodality of this sequence is we recall 7.3.1 which yields

7.4.2
$$\mathrm{Grf}(G, Y^X) = \sum_{(\beta_0,\ldots,\beta_{|Y|-1})\vdash n} [\chi^{I\bar{G}\uparrow S_X} \mid \xi^\beta] k_\beta.$$

Since $k_\beta = \sum_{\lambda^*=\beta} Y^\lambda$, this gives

7.4.3 Corollary *The number of G–classes on Y^X, the elements of which are of content $\lambda \models |X|$, is equal to the following inner product of characters:*

$$[\chi^{I\bar{G}\uparrow S_X} \mid \xi^\lambda] = [\chi^{I\bar{G}\uparrow S_X} \mid \xi^{\lambda^*}].$$

An application of 7.3.2 to this result proves

7.4.4 Theorem *If $\lambda, \mu \models n$ and $\lambda^* \lhd \mu^*$, then the number of G–classes of content λ on Y^X is greater than or equal to the number of G–classes of content μ.*

Hence in particular each sequence of numbers of G–classes on Y^X by content is monotone as soon as the partitions λ^* corresponding to the contents λ form a chain. An example of a chain of partitions is formed by the two–rowed partitions:

$$(|X|) \rhd (|X|-1, 1) \rhd (|X|-2, 2) \rhd \ldots \rhd (|X| - \lfloor |X|/2 \rfloor, \lfloor |X|/2 \rfloor).$$

This gives

7.4.5 Corollary *In the case when $|Y| = 2$, the sequence of numbers of G–classes on Y^X by content is unimodal and reciprocal.*

In terms of the group reduction function this means that the sequence of the coefficients of the monomials in $C(G, X \mid y_0 + y_1)$ is unimodal. The coefficient of $y_0^{|X|-k} y_1^k$ in this polynomial is equal to the coefficient of y^k in $C(G, X \mid 1 + y)$. We therefore can also use $C(G, X \mid 1 + y)$ which is easier to write down. Let us now generalize our definitions of unimodality and reciprocity to *polynomials* in $\mathbb{R}[y]$. Consider $p = b_0 + b_1 y + \ldots + b_k y^k \in \mathbb{R}[y]$. If $p = b_i y^i + b_{i+1} y^{i+1} + \ldots + b_j y^j$, where $b_i \neq 0 \neq b_j$, then we call $m(p) := (i + j)/2$ the *middle* of p, while $d(p) := i + j$. Now we say that p is a *unimodal polynomial* if and only if its coefficients weakly increase up to the middle while they weakly decrease afterwards:

$$b_0 \leq b_1 \leq \ldots \leq b_{\lfloor m(p) \rfloor} \geq b_{\lfloor m(p) \rfloor + 1} \geq \ldots \geq b_{d(p)}.$$

Thus, for example, $1+y+2y^2+2y^3+y^4+y^5$ and y^3+y^4 are unimodal. Furthermore we call p a *reciprocal polynomial* if and only if, for each $i \leq d(p)$, we have that

$$b_i = b_{d(p)-i}.$$

Hence 7.4.5 can be rephrased as follows:

7.4.6 Corollary *For each finite action ${}_G X$ the polynomial $C(G, X \mid 1 + y)$ is unimodal and reciprocal.*

A trivial example is formed by the identity subgroup of S_n which gives

$$C(\{1\}, n \mid 1+y) = (1+y)^n = \sum_{k=0}^{n} \binom{n}{k} y^k.$$

Hence, as it is well known, the sequence of binomial coefficients is unimodal and reciprocal. A less trivial example follows from 2.2.8/3.2.9. If $p(k; m, n)$ denotes the number of partitions $\alpha \vdash k$ such that $\alpha_0' \leq m$ and $\alpha_0 \leq n$, then we have, since the actions of $S_m \wr S_n$ on $m \times n$ and on mn are similar (recall 3.2.8):

7.4.7 $$\begin{bmatrix} m+n \\ n \end{bmatrix} = C(S_m \odot S_n, mn \mid 1+y) = \sum_{k=0}^{mn} p(k; m, n) y^k,$$

and so the sequence of numbers $p(k; m, n)$ and the Gaussian polynomial $\left[{m+n \atop n}\right]$ are reciprocal and unimodal. If we compare this result with 7.1.15, we obtain the interesting equation

7.4.8 $$C(S_m \odot S_n, mn \mid 1+y) = \{m\}(1, y, \ldots, y^n).$$

We next aim to show that even $C(G, X \mid 1 + y + y^2 + \ldots y^k)$ is reciprocal and unimodal. In order to prove this we note (exercise 7.4.1)

7.4.9 Lemma *For the polynomials in a single indeterminate y over the field $\mathbb{R}$ of real numbers we have:*

- *$p \in \mathbb{R}[y]$ is reciprocal if and only if $p = y^{d(p)} \cdot p(y^{-1})$.*
- *Products of reciprocal polynomials are reciprocal.*
- *The sum of reciprocal unimodal polynomials p and q with the same middle $m(p) = m(q)$ is reciprocal and unimodal.*
- *Multiplying any reciprocal and unimodal polynomial by a nonegative real number we obtain a reciprocal and unimodal polynomial.*
- *Products of reciprocal and unimodal polynomials are reciprocal and unimodal.*

Proof: The first statement is obviously valid and it implies the second, while the third and the fourth statement are trivial. In order to prove the fifth statement, we consider $p, q \in \mathbb{R}[y]$, both being unimodal and reciprocal. We can rewrite $p = b_0 + b_1 y + \ldots + b_k y^k$ in the following ways:

$$\begin{aligned} p &= b_0 + b_1 y + \ldots + b_1 y^{d(p)-1} + b_0 y^{d(p)} \\ &= b_0(1 + y + \ldots + y^{d(p)}) + (b_1 - b_0)(y + \ldots + y^{d(p)-1}) + \ldots . \end{aligned}$$

Doing the same with $q = c_0 + \ldots + c_l y^l$, we get that pq is equal to

$$\sum_{i,j} (a_{i+1} - a_i)(b_{j+1} - b_j)(y^{i+1} + \ldots + y^{d(p)-(i+1)})(y^{j+1} + \ldots + y^{d(q)-(j+1)}).$$

Each one of these summands is reciprocal by part one, and it is not difficult to check that they are also unimodal. Since $(d(p) + d(q))/2$ is the middle of each of them, also $p \cdot q$, their sum, must be unimodal. $\square$

We now remark that 7.1.13 yields

7.4.10 Corollary *For* $Y = \{y_0, y_1\}$ *we have that*

$$\{\alpha, Y\} = y_0^{\alpha_0} y_1^{\alpha_1} + y_0^{\alpha_0 - 1} y_1^{\alpha_1 + 1} + \ldots + y_0^{\alpha_1} y_1^{\alpha_0},$$

if α *has at most two rows and* $\{\alpha, Y\} = 0$ *otherwise. Hence in particular*

$$C(S_n, n, [\alpha] \mid 1 + y) = \begin{cases} y^{\alpha_1} + y^{\alpha_1 + 1} + \ldots + y^{\alpha_0}, & \text{if } \alpha_2 = 0, \\ 0, & \text{otherwise.} \end{cases}$$

Thus $C(S_n, n, [\alpha] \mid 1 + y)$ *is in any case reciprocal and unimodal and its coefficients are natural numbers.*

In order to generalize this to $C(S_n, n, [\alpha] \mid 1 + \ldots + y^k)$, we shall express this polynomial in terms of plethysm of cycle indicator polynomials. Let us therefore introduce the plethysm of polynomials first: We define $p(y_1, \ldots) \odot q(z_1, \ldots)$, where we assume that $p(0, \ldots, 0) = 0$, i. e. we assume that p has no constant term, to be

$$q(p(y_{1\cdot 1}, y_{1\cdot 2}, \ldots), \ldots, p(y_{n\cdot 1}, y_{n\cdot 2}, \ldots)).$$

We note that this definition is in accordance with 3.2.9, and we also note that in order to introduce this definition, we have to *renumber the indeterminates*. Using this generalization of plethysm of cycle indicator polynomials we can state that the following identity holds:

7.4.11 $C(S_n, n, [\alpha] \mid 1 + y + \ldots + y^k) = C(S_k, k \mid 1 + y) \odot C(S_n, n, [\alpha])$.

Proof: 3.2.9 implies that $C(S_k, k \mid y_1 + y_2) \odot C(S_n, n, [\alpha])$ is equal to

$$\frac{1}{n!} \sum_{\rho \in S_n} \zeta^\alpha(\rho^{-1}) \prod_{i=1}^{n} \underbrace{\Big(\frac{1}{k!} \sum_{\pi \in S_k} \prod_{j=1}^{k} (y_1^{i\cdot j} + y_2^{i \cdot j})^{a_j(\pi)} \Big)}_{= y_1^{i\cdot k} + y_1^{i\cdot(k-1)} y_2^{i} + \ldots + y_2^{i \cdot k}, \text{ by 7.4.10.}}^{a_i(\rho)},$$

which is the same as $C(S_n, n, [\alpha] \mid y_1^k + y_1^{k-1} y_2 + \ldots + y_2^k)$, and so we obtain the statement by putting $y_1 := 1$, $y_2 := y$. □

Now $C(S_k, k \mid y_1 + y_2) \odot C(S_n, n, [\alpha])$ belongs to $\mathcal{HS}_{k \cdot n}[Y]$, and it has integral coefficients, hence it is a $\mathbb{Z}$–linear combination of Schur polynomials in y_1 and y_2. Thus $C(S_k, k \mid 1 + y) \odot C(S_n, n, [\alpha])$ is a sum of Schur polynomials $\{\alpha, Y\}$ taken at $y_1 := 1$, $y_2 := y$, and hence, by 7.4.10, we obtain

7.4.12 Corollary $C(S_n, n, [\alpha] \mid 1 + y + \ldots + y^k)$ *is reciprocal and unimodal and it has natural coefficients.*

In order to complete the proof of the general theorem on unimodal and reciprocal polynomials we need to prove two further results on Schur polynomials:

7.4.13 Lemma (Cauchy) *If X and Y are two disjoint finite sets of indeterminates, then*

$$\prod_{x\in X, y\in Y}(1-xy)^{-1} = \sum_{n\in\mathbb{N}}\sum_{\alpha\vdash n} h_\alpha(X)k_\alpha(Y) = \sum_{n,\alpha}\{\alpha, X\}\{\alpha, Y\},$$

and

$$\prod_{x\in X, y\in Y}(1+xy) = \sum_{n\in\mathbb{N}}\sum_{\alpha\vdash n} e_\alpha(X)k_\alpha(Y) = \sum_{n,\alpha}\{\alpha', X\}\{\alpha, Y\}.$$

Proof: $\prod(1-xy)^{-1}$ is a formal power series, defined by

$$\prod_{x,y}(1-xy)^{-1} = \prod_{x,y}(1+xy+x^2y^2+\ldots) = \sum_{n\in\mathbb{N}}\sum_{\lambda,\mu\models n} m_{\lambda\mu}X^\lambda Y^\mu$$

$$= \sum_n\sum_{\alpha\vdash n}\sum_{\lambda\models n} st^\alpha(\lambda)X^\lambda \sum_{\mu\models n} st^\alpha(\mu)Y^\mu = \sum_{n,\alpha}\{\alpha, X\}\{\alpha, Y\}.$$

The second statement follows analogously. □

The next equation we would like to derive uses the notion of *skew representation* $[\alpha\backslash\beta]$, which is defined as follows. For $\alpha\vdash n$ and $\beta\vdash m\leq n$ we put

7.4.14
$$[\alpha\backslash\beta] := \sum_{\gamma\vdash n-m}([\beta][\gamma],[\alpha])[\gamma].$$

The corresponding *skew Schur polynomials* have the following property:

7.4.15 Lemma *If X and Y denote two disjoint finite sets of indeterminates, then*

$$\{\alpha, X\cup Y\} = \sum_{m<n}\sum_{\beta\vdash m}\{\alpha\backslash\beta, X\}\{\beta, Y\}.$$

Proof: For $m<n$ and a third finite set Z of indeterminates, disjoint with X and Y, we have, by definition of $\{\alpha\backslash\beta\}$:

$$\sum_{\alpha\vdash n}\sum_{m<n}\sum_{\beta\vdash m}\{\alpha\backslash\beta, X\}\{\alpha, Z\}\{\beta, Y\} =$$

$$= \sum_{\alpha,\beta}\Big(\sum_{\gamma\vdash n-m}([\beta][\gamma],[\alpha])\{\gamma, X\}\Big)\{\alpha, Z\}\{\beta, Y\}$$

$$= \sum_{\beta,\gamma}\{\beta, Y\}\{\gamma, X\}\sum_\alpha([\beta][\gamma],[\alpha])\{\alpha, Z\}$$

$$= \sum_{\beta,\gamma}\{\beta, Y\}\{\gamma, X\}\{\beta, Z\}\{\gamma, Z\}$$

$$= \prod_{y,z}(1-yz)^{-1}\prod_{x,z}(1-xz)^{-1} = \sum_\alpha\{\alpha, X\cup Y\}\{\alpha, Z\}.$$

This proves the lemma since the $\{\alpha, Z\}$ form a basis. □

We are now in the position to prove the theorem that the Pólya insertion of any reciprocal and unimodal polynomial with natural coefficients into a cycle indicator of a finite action gives a reciprocal and unimodal polynomial. But instead of formally proving this, we demonstrate it by an example which shows that all this works in the general case, too.

Consider $u(y) := 1 + 2y + 2y^2 + y^3$, which is a unimodal and reciprocal polynomial with coefficients in $\mathbb{N}$. By Foulkes' Lemma $C(G, X \mid u(y))$ is reciprocal and unimodal if this is also true for $C(S_n, n, [\alpha] \mid u(y))$. In order to prove that the latter polynomials are in fact reciprocal and unimodal, we use 7.4.13 which implies the last of the following equations: $C(S_n, n, [\alpha] \mid 1 + 2y + 2y^2 + y^3)$ is equal to

$$C(S_n, n, [\alpha] \mid 1 + y + y^2 + y^3 + z + z^2)|_{z:=y}$$

$$= \{\alpha, \{y_1, y_2, y_3, y_4\} \cup \{z_1, z_2\}\}|_{y_i:=y^{i-1}, z_i:=y^i}$$

$$= \sum_\beta C(S_{n-m}, n-m, [\alpha \backslash \beta] \mid 1 + y + y^2 + y^3) C(S_m, m, [\beta] \mid z + z^2)|_{z:=y}.$$

We now remark that

$$C(S_m, m, [\beta] \mid z + z^2) = z^m C(S_m, m, [\beta] \mid 1 + z),$$

which is reciprocal and unimodal by 7.4.9/ 7.4.10. Moreover

$$C(S_{n-m}, n-m, [\alpha \backslash \beta] \mid 1 + y + y^2 + y^3)$$

$$= \sum_\gamma ([\beta][\gamma], [\alpha]) C(S_{n-m}, n-m, [\gamma] \mid 1 + y + y^2 + y^3)$$

is also reciprocal and unimodal since each of the summands has this property and all the summands have the same middle. Thus, by 7.4.13, each $C(S_n, n, [\alpha] \mid 1 + 2y + 2y^2 + y^3)$ and hence also each $C(G, X \mid 1 + 2y + 2y^2 + y^3)$ is reciprocal and unimodal. In the same way we can prove

7.4.16 Theorem *Pólya insertion of a unimodal and reciprocal polynomial $u(y)$ with natural coefficients into a cycle indicator of a finite action ${}_GX$ gives the polynomial $C(G, X \mid u(y))$ which is reciprocal and unimodal.*

Note that the above argument needs in fact that $u(y)$ has all its coefficients in $\mathbb{N}$. But more than that: 7.4.16 *does not hold* if we allow $u(y)$ to have rational coefficients:

$$C\left(S_2, 2 \,\middle|\, \frac{1}{2} + \frac{1}{2}y\right) = \frac{3}{8} + \frac{2}{8}y + \frac{3}{8}y^2.$$

We are now going to determine which numbers are generated by these unimodal and reciprocal polynomials $C(G, X \mid 1 + y + y^2 + y^3)$. From the preceding sections it should in fact be clear that this polynomial generates the numbers of G–classes on $Y^X := (k+1)^X$ by weight, if we take for the weight the function

$$w \colon Y^X \to \mathbb{R} \colon f \mapsto y^{\Sigma f(x)},$$

so that, for example, $C(S_v, \binom{v}{2} \mid 1 + y + \ldots + y^k)$ generates the numbers of k–graphs on v vertices with respect to the number of edges, the coefficient of y^e in this polynomial is in fact equal to the number of k–graphs on v vertices which have a total of e edges.

Finally it should be mentioned that unimodality considerations suggest the question for *bijective* proofs. For example, the unimodality of the Gaussian polynomial

$$G_{mn} = \sum_{k=0}^{mn} p(k; m, n)x^k = \begin{bmatrix} m+n \\ n \end{bmatrix}$$

suggests a proof by constructing a natural *embedding* of a vector space of dimension $p(k-1; m, n)$ into a space of dimension $p(k; m, n)$, for $1 < k \leq mn/2$. (There are other proofs, due to O'Hara, using an embedding of the set $P(k-1; m, n)$ of partitions, which will be introduced below, into $P(k; m, n)$, see [60],[114] and the papers of Zeilberger on this result.)

A particularly easy proof was given in the dissertation by H. von Koch (see [85] and [86]). In order to describe this proof (in fact there are three further and very different proofs) we recall that the coefficient $p(k; m, n)$ is the number of proper partitions $\alpha = (\alpha_0, \ldots) \vdash k$ such that $\alpha_0 \leq n$ and $\alpha'_0 \leq m$. The set of these partitions α that fit into an $(m \times n)$-rectangle,

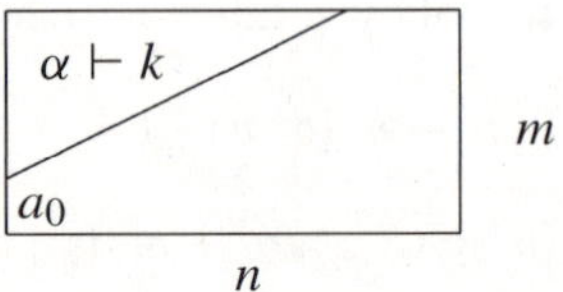

will be indicated as follows:

$$P(k; m, n) := \{\alpha \vdash k \mid k \leq m \cdot n, \alpha_0 \leq n, \alpha'_0 \leq m\}.$$

It is helpful to rewrite these partitions in terms of corresponding *extended cycle types*

$$a(\alpha) := (a_0(\alpha), a_1(\alpha), \ldots, a_n(\alpha)),$$

where $a_i(\alpha), i > 0$, denotes the number of parts α_i of length i, while $a_0(\alpha) := m - \alpha'_0$. The set of all these extended cycle types, corresponding to partitions of k contained in the rectangle, will be denoted as follows:

$$A(k; m, n) := \left\{ a := (a_0, a_1, \ldots, a_n) \in \mathbb{N}^{n+1} \mid \sum_{i>0} i a_i = k, a_0 = m - \sum_{i>0} a_i \right\}.$$

The mapping

$$P(k; m, n) \ni \alpha \mapsto a(\alpha) \in A(k; m, n)$$

is an embedding of $P(k; m, n)$ into the set

$$Z(m, n) := \left\{ b := (b_0, b_1, \ldots, b_n) \in \mathbb{Z}^{n+1} \mid \sum_{i \geq 0} b_i = m \right\}.$$

Now we are going to consider the vector space

$$V(m,n) := \mathbb{Q}^{Z(m,n)} = \Big\{ \sum_{b\in Z(m,n)} q_b b \;\Big|\; q_b \in \mathbb{Q},\ \text{only finitely many } q_b \neq 0 \Big\}$$

which has $Z(m,n)$ as basis. We shall use linear mappings on $V(m,n)$ which are defined as linear extensions of

$$\varphi_i\colon (b_0,\ldots,b_{i-1},b_i,\ldots,b_n) \mapsto (b_0,\ldots,b_{i-1}+1,b_i-1,\ldots,b_n),$$

$$\gamma_i\colon (b_0,\ldots,b_{i-1},b_i,\ldots,b_n) \mapsto (b_0,\ldots,b_{i-1}-1,b_i+1,\ldots,b_n).$$

Using these linear mappings on $V(m,n)$ we define the linear mappings F and G by putting

$$F(b) := \sum_{i>0} i b_i \varphi_i(b),\quad G(b) := \sum_{i>0} (n+1-i) b_{i-1} \gamma_i(b)$$

and extending linearly.

We shall now restrict attention to the subspace of $V(m,n)$ which is generated by $A(k;m,n)$. An easy calculation, using the commutativity $\gamma_i\varphi_j = \varphi_j\gamma_i$ as well as $\varphi_i\gamma_i = \mathrm{id}$, shows that, for each $b \in A(k;m,n)$,

$$\begin{aligned}(GF-FG)(b) &= \sum_{i=1}^{n} (n+1-i)i(b_i - b_{i-1})\varphi_i\gamma_i(b)\\ &= \Big(\sum_{i=0}^{n} (2i-n)b_i\Big)b\\ &= (2k-mn)b.\end{aligned}$$

H. von Koch proved

7.4.17 Theorem *For $1 \le k \le m \cdot n/2$, the restriction of G to the subspace of $V(m,n)$ that is generated by $A(k-1;m,n)$ is injective.*

Proof: To begin with we show, by induction on t, that for $0 \le t \le k$ and a vector

$$v \in \langle A(k-1;m,n)\rangle \cap ker(G)$$

we have

7.4.18 $$G^t F^t(v) = q_t \cdot v,$$

with a rational coefficient $q_t \neq 0$. The proof is by induction on t. As v is supposed to lie in the kernel of G, we have

$$GF^t(v) = (GF^t - F^tG)(v).$$

Telescoping yields

$$GF^t - F^tG = \sum_{i=1}^{t} (F^{i-1}GF^{t-i+1} - F^iGF^{t-i}) = \sum_{i=1}^{t} F^{i-1}(GF-FG)F^{t-i},$$

while $F^{t-i}(v) \in \langle A(k-1-(t-i); m, n)\rangle$ and $(GF - FG)(b) = (2k - mn)b$, for $b \in A(k; m, n)$, give

$$\begin{aligned} GF^t(v) &= \sum_{i=1}^{t} F^{i-1}(GF - FG)F^{t-i}(v) \\ &= \sum_{i=1}^{t} F^{i-1}(2(k-1-(t-i)) - mn)F^{t-i}(v) \\ &= \underbrace{\left(\sum_{i=1}^{t}(2(k-1-t+i) - mn)\right)}_{\neq 0} F^{t-1}(v). \end{aligned}$$

This proves 7.4.18 since we can apply the mapping G^{t-1}, which is injective by the induction hypothesis, to both sides of the last equation.

Thus, in particular, $G^k F^k(v) = q_k \cdot v$, with a nonzero scalar factor q_k. But $F^k(v) = 0$, as $v \in A(k-1; m, n)$, and so we can conclude that $v = 0$. This proves the injectivity of G. □

The injectivity of the mapping $G\colon \mathbb{Q}^{A(k-1;m,n)} \longrightarrow \mathbb{Q}^{A(k;m,n)}$, for $1 \leq k \leq mn/2$, gives the desired inequality for the corresponding dimensions:

$$p(k-1; m, n) = |A(k-1; m, n)| \leq |A(k; m, n)| = p(k; m, n),$$

which is another proof of the unimodality of the Gaussian polynomials. The complementation of the partitions inside of the rectangle, which is

$$c\colon A(k; m, n) \longrightarrow A(mn-k; m, n)\colon (a_0, \ldots, a_k) \mapsto (a_k, \ldots, a_1, a_0),$$

verifies the reciprocity of these polynomials: $|A(k; m, n)| = |A(mn-k; m, n)|$. These two arguments together give the unimodality (and reciprocity) of the Gaussian polynomials.

Exercises

Exercise 7.4.1 Prove lemma 7.4.9.

7.5 The Littlewood–Richardson Rule

If ${}_GX$ is an action then we call $x \in X$ a *G–invariant* or simply an *invariant* if and only if $gx = x$, for each $g \in G$. If X is not only a set but also a vector space and the action ${}_GX$ a representation D of G on X, then we can generalize this definition of G–invariant to that of a *relative G–invariant* by which we now mean an $x \in X$ such that, for a suitable 1–dimensional character χ of G, we have

$$\forall\, g \in G\colon\ gx := D(g)x = \chi(g)x.$$

(Thus G–invariants which are sometimes called *absolute* G–invariants are the relative G–invariants corresponding to the identity character.) In other words, a relative G–invariant is an element of a vector space affording a 1–dimensional representation of G.

The symmetric polynomials over $\mathbb{Q}$ and in the set of indeterminates $Y = \{y_0, \ldots, y_{m-1}\}$ were defined to be the absolute invariants of the natural action of S_m on $\mathbb{Q}[Y]$. Since ι and ϵ are the only onedimensional characters of S_m, the relative S_m–invariants of $\mathbb{Q}[Y]$ which are not absolute invariants are the polynomials p which satisfy

$$\forall\, \pi \in S_m\colon\ p(y_{\pi^{-1}0}, \ldots, y_{\pi^{-1}(m-1)}) = \epsilon(\pi) p(y_0, \ldots, y_{m-1}).$$

We call them the *alternating polynomials* in the $y_i \in Y$ over $\mathbb{Q}$. Clearly the following is true (exercise 7.5.1):

7.5.1 Lemma *The polynomial $p \in \mathbb{Q}[Y]$ is alternating if and only if it is a $\mathbb{Q}$–linear combination of the polynomials*

$$\sum_{\pi \in S_m} \epsilon(\pi) y_{\pi 0}^{d_0} \cdots y_{\pi(m-1)}^{d_{m-1}} = \det(y_i^{d_k})_{i,k \in m},$$

where $d_0 > \ldots > d_{m-1} \geq 0$.

As each such d_i is greater than or equal to $m - 1 - i$, we can replace $d = (d_0, \ldots, d_{m-1})$ by $\alpha := (\alpha_0, \ldots, \alpha_{m-1})$, where $\alpha_i := d_i - m + 1 + i$, obtaining in this way a proper partition α of $\sum d_i - \binom{m}{2}$. Thus 7.5.1 can be rephrased as follows:

7.5.2 Corollary *$p \in \mathbb{Q}[Y]$ is alternating if and only if it is a $\mathbb{Q}$–linear combination of the polynomials*

$$\Delta_\alpha := \sum_{\pi \in S_m} \epsilon(\pi) y_{\pi 0}^{\alpha_0+m-1} \cdots y_{\pi(m-1)}^{\alpha_{m-1}+m-m} = \det\left(y_i^{\alpha_k+m-k-1}\right)_{i,k \in m},$$

where $\alpha := (\alpha_0, \ldots, \alpha_{m-1})$ is a proper partition of some $n \in \mathbb{N}$ such that $\alpha'_0 \leq m$.

The Δ_α are obviously linearly independent, we therefore denote by

$$\mathcal{HA}_n[Y] = \ll \Delta_\alpha \mid \alpha \vdash n, \alpha'_0 \leq m \gg_{\mathbb{Q}}$$

the subspace of $\mathbb{Q}[Y]$ consisting of the *homogeneous* alternating polynomials of degree $n + \binom{m}{2}$ which has the Δ_α as $\mathbb{Q}$–basis. The vector space $\mathcal{HA}[Y]$ containing *all* the alternating polynomials therefore satisfies

7.5.3 $$\mathcal{HA}[Y] = \bigoplus_{n \in \mathbb{N}} \mathrm{A}_n[Y].$$

We now consider the particular alternating polynomial corresponding to the zero partition $\alpha := (0, \ldots)$:

$$\Delta_0 = \det(y_i^{m-k-1}) = \prod_{i<j}(y_i - y_j),$$

the well known *Vandermonde determinant*. It yields an interesting endomorphism of $\mathbb{Q}[Y]$ via left multiplication:

$$\Delta_0 : \mathbb{Q}[Y] \to \mathbb{Q}[Y] : p \mapsto \Delta_0 \cdot p.$$

This mapping is obviously linear, injective, and it maps $\mathcal{HS}_n[Y]$ into $\mathcal{HA}_n[Y]$. Moreover, it establishes a $\mathbb{Q}$– isomorphism between these two vector spaces:

7.5.4 $$\Delta_0 : \mathcal{HS}_n[Y] \simeq \mathcal{HA}_n[Y].$$

Proof: As Δ_0 is injective and the Δ_α are a basis of $\mathcal{HA}_n[Y]$, we need only show that they occur in the image which again means that they should be divisible by Δ_0. Now Δ_α can be considered as a polynomial in the single indeterminate y_j by taking $\mathbb{Q}[Y\backslash\{y_j\}]$ as a ring of coefficients, i. e. we consider Δ_α as an element of the polynomial ring $\mathbb{Q}[Y\backslash\{y_j\}][y_j]$. But as Δ_α is alternating, we get zero if we replace the indeterminate y_j by the element $y_i \in \mathbb{Q}[Y\backslash\{y_j\}]$, and so Δ_α must be divisible by $(y_i - y_j)$ (recall that $\mathbb{Q}[Y\backslash\{y_j\}][y_j]$ is also Gaussian, and $y_i - y_j$ is an irreducible element therein) and hence also by the Vandermonde determinant $\prod(y_i - y_j)$. □

We should like to show that the inverse image Δ_α/Δ_0 of Δ_α is the Schur polynomial $\{\alpha\}$. The first step to prove this is to examine

$$\begin{aligned}
\Delta_\alpha \cdot e_p &= \det(y_i^{\alpha_k+m-1-k}) \sum_{M\in\binom{m}{p}} \prod_{i\in M} y_i \\
&= \sum_{\pi\in S_m} \epsilon(\pi) \prod_{k\in m} y_{\pi k}^{\alpha_k+m-1-k} \sum_{M\in\binom{m}{p}} \prod_{k\in M} y_{\pi k} \\
&= \sum_M \sum_\pi \epsilon(\pi) \prod_k y_{\pi k}^{\alpha_k+m-1-k+I_M(k)},
\end{aligned}$$

where I_M denotes the *characteristic function* of $M \subseteq m$:

$$I_M(k) := \begin{cases} 1, & \text{if } k \in M, \\ 0, & \text{otherwise.} \end{cases}$$

Thus we have obtained that

7.5.5 $$\Delta_\alpha \cdot e_p = \sum_{M\in\binom{m}{p}} \det\left(y_i^{\alpha_k+m-1-k+I_M(k)}\right)_{i,k\in m}.$$

In order to simplify the right hand side of this equation we notice that such a determinantal summand is zero if and only if, for some $k < l$, we have $\alpha_k - k - 1 + I_M(k) = \alpha_l - l - 1 + I_M(l)$. Since $\alpha_k \geq \alpha_l$ and $I_M(k), I_M(l) \in \{0, 1\}$ this equation implies $l = k+1, \alpha_k = \alpha_k + 1$ and $I_M(k) = 0, I_M(k+1) = 1$. Thus the nonzero summands correspond to p–subsets M of m, for which $\beta := (\ldots, \alpha_k + I_M(k), \ldots)$ is a *proper partition* of $m + p$. Hence we have proved

7.5.6 Corollary *For each $\alpha \vdash n$ and any $p \in \mathbb{N}$ we have*

$$\Delta_\alpha \cdot e_p = \sum \Delta_\beta,$$

if the sum is taken over all the $\beta \vdash n+p$ the diagrams $[\beta]$ of which arise from $[\alpha]$ by adding p nodes in p different rows.

Iteration of this yields, for $\alpha \vdash n$:

$$\Delta_0 \cdot e_\alpha = \sum_{(\beta_0,\ldots,\beta_{m-1})\vdash n} st^{\beta'}(\alpha)\Delta_\beta,$$

or, equivalently:

$$e_\alpha = \sum_{(\beta_0,\ldots,\beta_{m-1})\vdash n} st^{\beta'}(\alpha)\frac{\Delta_\beta}{\Delta_0}.$$

On the other hand, 7.2.1 gives

$$e_\alpha = \sum_\gamma m'_{\alpha\gamma}k_\gamma = \sum_{\gamma,\beta} m'_{\alpha\gamma}\epsilon_{\beta\gamma}\{\beta\} = \sum_{\beta,\gamma}\epsilon_{\beta\gamma}m'_{\gamma\alpha}\{\beta\} = \sum_\beta st^{\beta'}(\alpha)\{\beta\}.$$

Comparing these two expressions of e_α we finally obtain

7.5.7 Corollary *For $\alpha = (\alpha_0, \ldots, \alpha_{m-1}) \vdash n$ and $Y = \{y_0, \ldots, y_{m-1}\}$ we have*

$$\frac{\Delta_\alpha(Y)}{\Delta_0(Y)} = \{\alpha, Y\}.$$

Besides the representation theoretical consequences which will be discussed later, this result yields, for combinatorial purposes, a useful method for decomposing a given symmetric polynomial into a linear combination of Schur polynomials.

7.5.8 Corollary *The coefficient of $\{\alpha, Y\}$ in $p \in \mathcal{HS}_n[Y]$ is equal to the coefficient of $\prod y_i^{\alpha_i+m-1-i}$ in $\Delta_0(Y) \cdot p$.*

Proof: The statement follows immediately from 7.5.4 since $\Delta_0(Y) \cdot p \in \mathcal{HA}_n[Y]$ and the $\Delta_\alpha(Y)$ form a basis of $\mathcal{HA}_n[Y]$ while $\prod y_i^{\alpha_i+m-i-1}$ is the leading term in $\Delta_\alpha(Y)$. □

The identification 7.5.7 allows to rephrase 7.5.6 in terms of representations. For 7.5.6 says that $\{\alpha\} \cdot e_p = \sum\{\beta\}$, where the $[\beta]$ arise from $[\alpha]$ by adding p nodes in p different rows, and e_p is the symmetric polynomial corresponding to the alternating representation of S_p (cf. 7.2.4), which is $[1^p]$. This yields

7.5.9 Corollary *The diagrams $[\beta]$ of the constituents of $[\alpha][1^p]$ are obtained by adding p nodes to pairwise different rows of $[\alpha]$ or to the empty row following the last nonempty one of $[\alpha]$.*

As multiplication by the alternating character means taking instead of α, (1^p) or β its associate α', (p) or β', we also get

7.5.10 Young's Rule *The diagrams $[\beta]$ of the constituents of $[\alpha][p]$ are obtained by adding p nodes to pairwise different columns of $[\alpha]$ or to the empty column following the last nonempty one of $[\alpha]$.*

For example, if $\alpha := (3, 2^2)$ and $p := 2$ we obtain, indicating the added nodes by 0, the following diagrams according to 7.5.10:

$$\begin{array}{ccccc} \times & \times & \times & 0 & 0 \\ \times & \times & & & \\ \times & \times & & & \end{array}\;, \quad \begin{array}{cccc} \times & \times & \times & 0 \\ \times & \times & 0 & \\ \times & \times & & \end{array}\;, \quad \begin{array}{cccc} \times & \times & \times & 0 \\ \times & \times & & \\ \times & \times & & \\ 0 & & & \end{array}\;,$$

$$\begin{array}{ccc} \times & \times & \times \\ \times & \times & 0 \\ \times & \times & \\ 0 & & \end{array}\;, \quad \begin{array}{ccc} \times & \times & \times \\ \times & \times & \\ \times & \times & \\ 0 & 0 & \end{array}\;.$$

which yields the decomposition

$$[3, 2^2][2] = [5, 2^2] + [4, 3, 2] + [4, 2^2, 1] + [3^2, 2, 1] + [3, 2^3].$$

We can apply this process repeatedly in order to get the constituents of $IS_\alpha \uparrow S_n$. For example if we want to decompose $[3][2][1]$, we first evaluate $[3][2]$ obtaining

$$\begin{array}{ccccc} \times & \times & \times & 0 & 0 \end{array} + \begin{array}{cccc} \times & \times & \times & 0 \\ 0 & & & \end{array} + \begin{array}{ccc} \times & \times & \times \\ 0 & 0 & \end{array}.$$

Thus $[3][2] = [5] + [4, 1] + [3, 2]$, which yields the decomposition of $[3][2][1]$:

$$\begin{array}{cccccc} \times & \times & \times & 0 & 0 & 1 \end{array} + \begin{array}{ccccc} \times & \times & \times & 0 & 0 \\ 1 & & & & \end{array} + \begin{array}{ccccc} \times & \times & \times & 0 & 1 \\ 0 & & & & \end{array}$$

$$+ \begin{array}{cccc} \times & \times & \times & 0 \\ 0 & 1 & & \end{array} + \begin{array}{cccc} \times & \times & \times & 0 \\ 0 & & & \\ 1 & & & \end{array}$$

$$+ \begin{array}{cccc} \times & \times & \times & 1 \\ 0 & 0 & & \end{array} + \begin{array}{ccc} \times & \times & \times \\ 0 & 0 & 1 \end{array} + \begin{array}{ccc} \times & \times & \times \\ 0 & 0 & \\ 1 & & \end{array}$$

so that we have received the following decomposition:

$$[3][2][1] = [6] + 2 \cdot [5, 1] + 2 \cdot [4, 2] + [4, 1^2] + [3^2] + [3, 2, 1].$$

More generally the constituents $[\alpha]$, $\alpha \vdash m + n$, of the representation

$$[\beta][\delta_0][\delta_1] \ldots = ([\beta] \# IS_\delta \uparrow S_n) \uparrow S_{m+n}\,,\ \beta \vdash m\,,\ \delta \models n,$$

are obtained by first adding to the diagram $[\beta]$ δ_0 nodes 0 in all the admissible ways (see 7.5.10), then to each one of the resulting diagrams add δ_1 further nodes 1 in all the admissible ways, and so on. This procedure motivates the introduction of *skew diagrams* $[\alpha \backslash \beta]$ arising from $[\alpha]$ by deleting the subdiagram $[\beta]$, for example,

$$[8, 7^2, 5\backslash 4, 3^2] = \begin{array}{cccccccc} & & & & \times & \times & \times & \times \\ & & & \times & \times & \times & \times & \\ & & & \times & \times & \times & \times & \\ \times & \times & \times & \times & \times & & & \end{array} .$$

The reader notices that the resulting diagram does *not* determine $[\alpha]$ or $[\beta]$, e. g.

$$[8, 7^2, 5\backslash 4, 3^2] = [9, 8^2, 6\backslash 5, 4^2, 1].$$

To each skew diagram there correspond *skew tableaux* and *standard skew tableaux* in the obvious way. Let

$$st^{\alpha\backslash\beta}(\lambda)$$

denote the number of standard tableaux of shape $\alpha\backslash\beta$ and content λ. If we are given a skew tableau, say

$$\begin{array}{ccc} & 0 & 0 \\ & 1 & \\ 1 & 2 & \end{array} \in ST^{(3,2^2)\backslash(1^2)}(2, 2, 1),$$

then we can read its entries from *right to left* in each row and one row after the other, *downwards*, obtaining a sequence of natural numbers. The given example yields 00121. Such a sequence is called a *lattice permutation* if the following holds: For each place j of the sequence the number of i's which occur among the j leftmost elements is greater than or equal to the number of $(i + 1)$'s, for each i and any j. Hence 00121 is a lattice permutation, while 00211 is not.

For the sake of simplicity we now allow the nodes of a diagram or a skew diagram to be replaced by elements of an arbitrary totally ordered set. We prove a combinatorial lemma which will turn out to be crucial:

7.5.11 Lemma *Assume that $\alpha \vdash n_0 + n_1 + b$, where $n_i, b \in \mathbb{N}$ and $\beta \vdash n_1, \gamma \vdash n_0$. Then the number of standard tableaux of shape $\alpha\backslash\gamma$ which contain β_i numbers $i, i \in \mathbb{N}$, together with b symbols x such that the elements i form a lattice partition, is the same if the order is defined by*

$$x < 0 < 1 < \ldots < \beta_0' - 1,$$

as if it is defined by

$$0 < 1 < \ldots < \beta_0' - 1 < x.$$

Proof: We denote by

$$M_k$$

the set of standard tableaux of shape $\alpha\backslash\gamma$ which contain β_i numbers i, b symbols x, and where *standard* is defined with respect to the order

$$0 < \ldots < k < x < k + 1 < \ldots < \beta_0' - 1.$$

For example, M_{-1} is the set of standard tableaux with this content and with respect to the order $x < 0 < \ldots$, while M_0 is defined with respect to $0 < x < \ldots$, and so on, until $M_{\beta'_0-1}$, where standard refers to $0 < \ldots < \beta'_0 - 1 < x$. The subset of M_k, consisting of the tableaux the entries i of which yield lattice permutations, will be indicated by

$$L_k.$$

Our method of proof is to construct bijections

$$\varphi_k \colon M_k \to M_{k-1}, 0 \leq k \leq \beta'_0 - 1,$$

which satisfy

$$\varphi_0\varphi_1 \ldots \varphi_{\beta'_0-1}[L_{\beta'_0-1}] = L_{-1},$$

so that the statement immediately follows by definition of $L_{\beta'_0}$ and L_0 from the fact that the φ_k are bijections.

i) *The bijections*: We define φ_k on M_k. If $T \in M_k$, then $\varphi_k(T)$ arises from T in the following way:

a) In each column of T which contains both x and k, we interchange x and k (remember that T is standard, so that each column of T contains both x and k at most once and next to each other), and afterwards
b) in each row containing both x's and k's which were left fixed under a), we shift the x's to the left of the k's.

Obviously $\varphi_k(T) \in M_{k-1}$, so that $\varphi_k \colon M_k \to M_{k-1}$. Furthermore φ_k is a bijection, since it is inverted by $\psi_k \colon M_{k-1} \to M_k$, which is defined by its operation on an element of M_k as follows.

a′) interchange x and k when they occur in the same column, and then
b′) in each row which contains both x's and k's fixed under a′), shift the x's to the right of the k's.

ii) $\varphi_0 \ldots \varphi_{\beta'_0-1}[L_{\beta'_0-1}] \subseteq L_{-1}$: Take a $T_{\beta'_0-1} \in L_{\beta'_0-1}$. Suppose, inductively, that the i's in each of the following tableaux form a lattice permutation:

$$T_{\beta'_0-1}, \varphi_{\beta'_0-1}(T_{\beta'_0-1}), \ldots, \varphi_{k+1} \ldots \varphi_{\beta'_0-1}(T_{\beta'_0-1}) =: T_k.$$

We have to show that this is also true for

$$T_{k-1} := \varphi_k(T_k) = \varphi_k \ldots \varphi_{\beta'_0-1}(T_{\beta'_0-1}).$$

(Note that we do *not* prove that $\varphi_k[L_k] \subseteq L_{k-1}$, which is in fact false.) The definition of φ_k shows that the lattice property holds for all the numbers in $\varphi_k(T_k)$, except, perhaps, between the numbers k and $k+1$. In particular $\varphi_{\beta'_0-1}(T_{\beta'_0-1}) \in L_{\beta'_0-1}$. We may therefore assume $k < \beta'_0 - 1$. Suppose now that there is a $k+1$ in the i-th row of T_k. Then the required lattice property in $\varphi_k(T_k)$ is true for $k+1$ unless there is an x to the left of it in the same row of T_k and this x lies below a k. We must examine the latter case more closely:

$$\begin{array}{cc|cccc} & <k & k\dots k & k\dots k & x\dots x & x\dots \\ \hline & \dots x & \underbrace{x\dots x}_{>0} & \underbrace{k+1\dots k+1}_{a} & \underbrace{k+1\dots k+1}_{b} & \multicolumn{1}{|l}{>k+1 \quad \leftarrow i} \end{array}$$

Every $k+1$ in the i-th row of T_k has a k or an x immediately above it, since $k < x < k+1$, and there is an x in the i-th row with a k above it. Let a (≥ 0) be the number of $(k+1)$'s in this row which lie below an x. Suppose that the last k in the $(i-1)$-th row of T_k lies in the j-th column. Since $T_k = \varphi_{k+1}(T_{k+1}) \in L_k$, we have

$$b \leq |\{(i', j') \mid k \text{ in } (i', j') \text{ of } T_k, i' < i-1, \text{or } i' = i-1 \text{ and } j' > j\}|$$

$$-|\{(i', j') \mid k+1 \text{ in } (i', j') \text{ of } T_k, i' < i-1 \text{ or } i' = i-1 \text{ and } j' > j\}|\,.$$

Thus, in $\varphi_k(T_k)$, the number of k's in the first $i-1$ rows minus the number of $(k+1)$'s in the first $i-1$ rows is at least $a+b$. This shows that the lattice property holds for all the $(k+1)$'s in the i-th row of $\varphi_k(T_k)$, and completes the proof of ii).

iii) The inclusion $\psi_{\beta'_0} \dots \psi_0[L_{-1}] \subseteq L_{\beta'_0-1}$ follows quite analogously. □

7.5.12 Corollary *The multiplicity* $([\beta][\delta_0][\delta_1]\dots, [\gamma]), \beta \vdash m, \delta \vdash n, \gamma \vdash m+n,$ *is equal to the number of standard tableaux of shape γ which contain β_i symbols i, δ_j symbols $\overline{j}$, $0 \leq i \leq \beta'_0 - 1, 0 \leq j \leq \delta'_0 - 1$, subject to the ordering $\overline{0} < \dots < \overline{\delta'_0 - 1} < 0 < \dots < \beta'_0 - 1$ and such that the i's yield a lattice permutation when we read the rows from the right to the left and downwards.*

Proof: The considerations above have shown that

$$([\beta][\delta_0][\delta_1]\dots, [\gamma]) = st^{\gamma\backslash\beta}(\delta),$$

which is the number of standard tableaux of shape $\gamma\backslash\beta$ and of content δ. Now the tableau

$$\begin{array}{ccccc} 0 & \dots & \dots & \dots & 0 \\ 1 & \dots & \dots & 1 & \\ \dots & \dots & \dots & & \\ \beta'_0 - 1 & \dots & \beta'_0 - 1 & & \end{array}$$

is the only standard tableau of shape and content β, its entries form a lattice permutation. Thus $st^{\gamma\backslash\beta}(\delta)$ is equal to the number of standard tableaux of shape γ which contain β_i i's and δ_j $\overline{j}$'s, subject to the order $0 < \dots < \beta'_0 - 1 < \overline{0} < \dots < \overline{\delta'_0 - 1}$ and such that the i's form a lattice permutation. Successive applications of 7.5.11 to $x := \overline{0}, \overline{1}, \dots, \overline{\delta'_0 - 1}$ yield the statement. □

We are now in a position to prove the main theorem:

7.5.13 The Littlewood–Richardson Rule *The multiplicity*

$$([\alpha][\beta], [\gamma]), \alpha \vdash m, \beta \vdash n, \gamma \vdash m+n,$$

is equal to the number of standard tableaux of shape $\gamma\backslash\alpha$ and content β which yield lattice permutations when we read their entries from the right to the left and downwards.

Proof: Let $g^{\gamma}_{\alpha\beta}$ denote this number of standard tableaux. 7.5.12 yields

$$([\delta_0][\delta_1]\dots[\beta],[\gamma]) = \sum_{\alpha\vdash m} st^{\alpha}(\delta) g^{\gamma}_{\alpha\beta},$$

so that, for each $\delta \vdash m$:

$$\sum_{\alpha} st^{\alpha}(\delta) g^{\gamma}_{\alpha\beta} = ([\delta_0][\delta_1]\dots[\beta],[\gamma]) = \sum_{\alpha} \kappa_{\delta\alpha}([\alpha][\beta],[\gamma]).$$

The statement now follows from $st^{\alpha}(\delta) = \kappa_{\delta\alpha}$ and the regularity of the Kostka matrix. □

In terms of Schur polynomials and in terms of ordinary irreducible representations this reads as follows:

7.5.14 Corollary *For each $\alpha \vdash m$, $\beta \vdash n$ we have:*

$$\{\alpha\}\{\beta\} = \sum_{\gamma\vdash m+n} g^{\gamma}_{\alpha\beta}\{\gamma\}, \text{ and } [\alpha][\beta] = \sum_{\gamma\vdash m+n} g^{\gamma}_{\alpha\beta}[\gamma].$$

Moreover

$$\{\gamma\backslash\alpha\} = \sum g^{\gamma}_{\alpha\beta}\{\beta\}, \text{ and } [\gamma\backslash\alpha] = \sum g^{\gamma}_{\alpha\beta}[\beta],$$

in accordance with the definition of skew representation, so that skew representations correspond to skew diagrams and skew tableaux.

An application of this result yields all the irreducible constituents of the restriction $[\alpha] \downarrow S_{n-1}$. In order to describe this *branching* of the irreducible representations of symmetric groups, we introduce the following notation. Let α denote a partition of n, and put

$$\alpha^{i\pm} := (\alpha_0, \dots, \alpha_{i-1}, \alpha_i \pm 1, \alpha_{i+1}, \dots).$$

In terms of this we get (already from 7.5.10):

7.5.15 The Branching Theorem *If $\alpha \vdash n$, then we have, for the corresponding irreducible representation $[\alpha]$ of S_n and its restriction to S_{n-1}, the stabilizer of the subset $n-1 \subseteq n$:*

$$[\alpha\backslash 1] = [\alpha] \downarrow S_{n-1} = \sum_{i:\ \alpha_i > \alpha_{i+1}} [\alpha^{i-}],$$

which means that the irreducible components of this restriction correspond to the diagrams $[\beta]$ that can be obtained from $[\alpha]$ by deleting one node, wherever this is possible. Conversely, if S_n denotes the stabilizer of $n \subseteq n+1$, then the representation of S_{n+1} induced by $[\alpha]$, has the following decomposition into irreducible constituents:

$$[\alpha] \uparrow S_{n+1} = \sum_{i:\ \alpha_i < \alpha_{i-1}} [\alpha^{i+}].$$

This means that the irreducible components of this representation correspond to the diagrams $[\beta]$ that can be obtained from $[\alpha]$ by adding just one node.

For example

$$[3, 2, 1^2] \downarrow S_6 = [2^2, 1^2] + [3, 1^3] + [3, 2, 1],$$

while

$$[3, 2, 1^2] \uparrow S_8 = [4, 2, 1^2] + [3^2, 1^2] + [3, 2^2, 1] + [3, 2, 1^3].$$

An iterative application of the Branching Theorem shows that, for $\alpha \vdash n, m \in \mathbb{N}$ and $\beta \vdash m + n$, we have the implication

7.5.16 $$([\alpha] \uparrow S_{m+n}, [\beta]) > 0 \iff \alpha \subseteq \beta,$$

where $\alpha \subseteq \beta$ means that $\alpha_i \leq \beta_i$, for each i. We therefore consider the set

$$P(\mathbb{N}) := \bigcup_{n \in \mathbb{N}} P(n) = \{\alpha \mid \exists\, n \in \mathbb{N}\colon\ \alpha \vdash n\}, \text{ where } P(0) := \{\emptyset\},$$

consisting of *all* the proper partitions and partially ordered by *inclusion*:

$$\alpha \subseteq \beta \iff \forall\, i\colon \alpha_i \leq \beta_i.$$

This poset $(P(\mathbb{N}), \subseteq)$ has the following canonical lattice structure:

$$\alpha \wedge \beta = (\min\{\alpha_0, \beta_0\}, \min\{\alpha_1, \beta_1\}, \ldots),$$

$$\alpha \vee \beta = (\max\{\alpha_0, \beta_0\}, \max\{\alpha_1, \beta_1\}, \ldots).$$

We call this lattice $(P(\mathbb{N}), \wedge, \vee)$ *Young's Lattice*. The first part of its order diagram is shown in figure 7.2. Figure 7.3 again shows this lower part of Young's lattice, but the diagrams are replaced by bullets, and furthermore in each of its levels the dominance order is indicated.

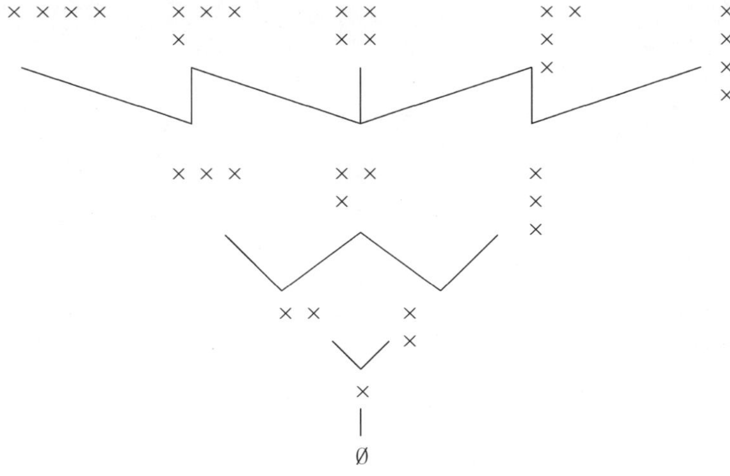

Fig. 7.2. Young's Lattice, lower part

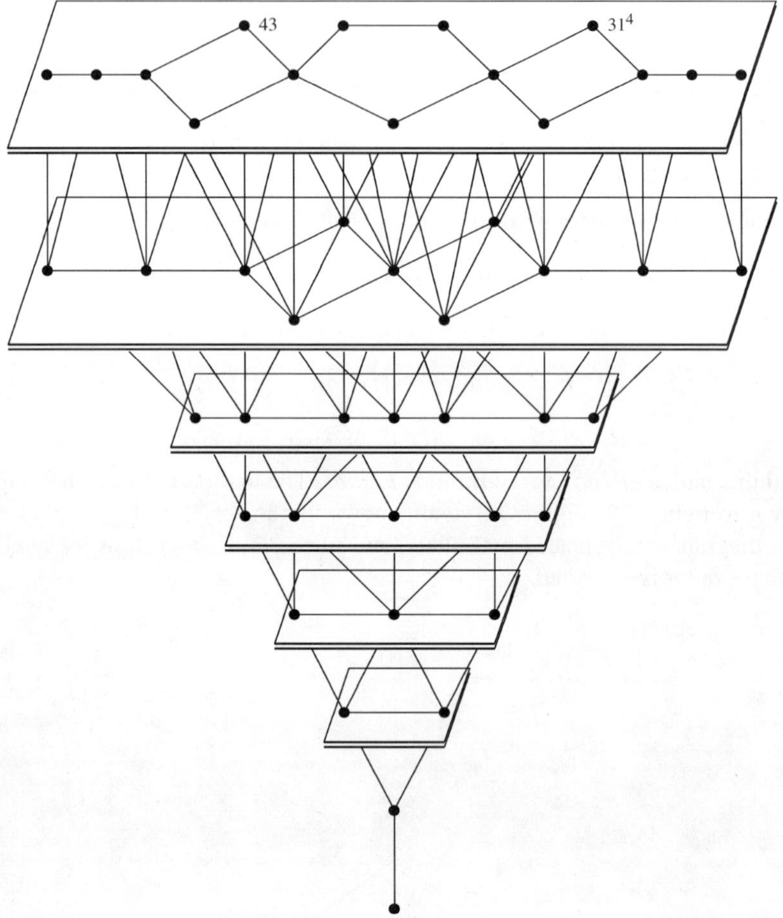

Fig. 7.3. Young's Lattice, lower part, with dominance and branching

7.5.17 Corollary *Young's Lattice visualizes the branching of the ordinary irreducible representations of symmetric groups.*

Exercises

Exercise 7.5.1 Prove 7.5.1.

Exercise 7.5.2 Use 7.2.1 and 7.5.7 in order to give another proof of the Standard Inversion Theorem 6.3.3. (Hint: Multiply both sides of $k_\alpha = \sum_\beta \epsilon_{\beta\alpha}\{\beta\}$ by Δ_0, use 7.5.7 and compare coefficients.)

7.6 The Murnaghan–Nakayama Rule

Besides the direct methods using character polynomials there are recursion formulae for the evaluation of characters of symmetric groups. Suppose we want to know the value ζ^γ_a, where $\gamma \vdash n$, $a \mathrel{\vdash\!\!\!\dashv} n$ and $a_k > 0$. We can use the fact that $\zeta^\gamma_a = \zeta^\gamma(\pi)$, where $\pi = \rho \cdot \sigma$, $\rho \in S_{n-k}$ and σ is the k–cycle $(n-k, \ldots, n-1) \in S_{n\backslash n-k}$. Thus, by restriction to the subgroup $S_{n-k} \oplus S_{n\backslash n-k}$, we obtain

$$\zeta^\gamma_a = \zeta^\gamma(\pi) = \sum_{\alpha\vdash n-k,\beta\vdash k} g^\gamma_{\alpha\beta}\zeta^\alpha(\rho)\zeta^\beta(\sigma),$$

where again $g^\gamma_{\alpha\beta}$ denotes the multiplicity of $[\alpha]$ # $[\beta]$ in the restriction of $[\gamma]$ to the subgroup $S_{n-k} \oplus S_{n\backslash n-k}$. But we need not take the sum over *all* the $\beta \vdash k$, since we have seen in 6.3.7 that $\zeta^\beta(\sigma) \neq 0$ only if $[\beta]$ is a hook, i. e. $[\beta] = [k-r, 1^r]$,for a suitable r, where $0 \leq r \leq k-1$. Hence

7.6.1
$$\zeta^\gamma_a = \sum_{\alpha\vdash n-k} \zeta^\alpha(\rho) \sum_{r=0}^{k-1} (-1)^r g^\gamma_{\alpha,(k-r,1^r)},$$

where, by Frobenius' reciprocity law 11.5.11,

$$g^\gamma_{\alpha,(k-r,1^r)} := \big([\gamma] \downarrow S_{n-k} \oplus S_{n\backslash n-k}, [\alpha] \# [k-r, 1^r]\big) = \big([\gamma], [\alpha][k-r, 1^r]\big).$$

We need to evaluate the alternating sum of these $g^\gamma_{\alpha,(k-r,1^r)}$.

7.6.2 Lemma *The sum* $\sum_r (-1)^r g^\gamma_{\alpha,(k-r,1^r)}$ *is* $(-1)^s$ *if* $[\gamma\backslash\alpha]$ *is a rim hook of leg length* s *in* $[\gamma]$*, otherwise this sum is zero.*

Proof: First of all we note that the Littlewood–Richardson Rule shows that $[\gamma]$ can occur in $[\alpha][k-r, 1^r]$ only if $[\gamma\backslash\alpha]$ is part of the rim of $[\gamma]$. Furthermore we recall that each irreducible constituent $[\gamma]$ of $[\alpha][k-r, 1^r]$ arises from $[\alpha]$ by adding $k-r$ symbols 0 and the symbols $1, \ldots, r$ to $[\alpha]$ according to the Littlewood–Richardson Rule. This means that each *connected* part of the arising skew tableau of shape $[\gamma\backslash\alpha]$ which is added is either of the form

$$\begin{array}{ccccccc} & & & & & 0 & \dots 0 \\ & & & 0 & \dots & 0 & 1 \\ & \dots & 0 & 2 & & & \\ \cdot^{\cdot^{\cdot}} & & & & & & \end{array}$$

if this is the highest part which is added, otherwise it is of one of the following forms:

$$\begin{array}{cccccccc} & & & & 0 & \dots & 0 & i \\ & & 0 & \dots & 0 & i+1 & & \\ \dots & i+2 & & & & & & \\ \cdot^{\cdot^{\cdot}} & & & & & & & \end{array},$$

or

$$\begin{array}{cccccccc} & & & & 0 & \dots & 0 \\ & & 0 & \dots & 0 & j & \\ \dots & j+1 & & & & & \\ \cdot^{\cdot^{\cdot}} & & & & & & \end{array}.$$

We now distinguish the cases if this rim part $[\gamma\backslash\alpha]$ is connected or not.

i) In the case when $[\gamma\backslash\alpha]$ is connected, there is a unique skew tableau that fits into $[\gamma\backslash\alpha]$ according to the Littlewood–Richardson Rule and hence also a unique r such that $g^{\gamma}_{\alpha,(k-r,1^r)} \neq 0$, and for this r we have $g^{\gamma}_{\alpha,(k-r,1^r)} = 1$, which proves the statement for this particular case.

ii) If $[\gamma\backslash\alpha]$ is a disconnected part of the rim of $[\gamma]$ then everything cancels as we shall show next. Recall that the second highest connected component either starts with a row of the form $0\dots0$ or with a row of the form $0\dots0j$, where the entry j is defined by the highest connected component. Hence each skew tableau, the second component of which starts with the row $0\dots0$, corresponds to a *unique* skew tableau arising by replacing the last 0 of this row by the unique admissible j and corresponding renumbering of the other entries > 0, and conversely. If the first of these tableaux, with first row $0\dots0$ in its second highest connected component, arises by adding the hook $[k-r, 1^r]$ according to the Littlewood–Richardson Rule, then the second arises by adding the hook $[k-r-1, 1^{r+1}]$, and hence the corresponding contributions to $(-1)^r g^{\gamma}_{\alpha,(k-r,1^r)}$ cancel. □

This shows that the α occuring on the right hand side of 7.6.1 are the partitions of $n-k$ the diagrams of which can be obtained from $[\gamma]$ by erasing a (connected) part R^{γ}_{ij} of its rim, in short: replace γ by $\gamma\backslash R^{\gamma}_{ij}$. In terms of the corresponding hook H^{γ}_{ij} we can express k as follows: $k = b^{\gamma}_{ij} := \gamma'_j - i$, the *leg length* of this hook. This yields the desired recursion formula for the irreducible characters of S_n:

7.6.3 The Murnaghan–Nakayama Formula *If $a \vdash\!\!\dashv n$, $a_k > 0$, and $a' := (a_1, \dots, a_{k-1}, a_k - 1, a_{k+1}, \dots, a_n) \vdash\!\!\dashv n-k$, then we have, for the value ζ^{γ}_a of $\zeta^{\gamma}, \gamma \vdash n$, on the class of elements of cycle type a in S_n the following recursion:*

$$\zeta_a^\gamma = \sum_{i,j:\, h_{ij}^\gamma = k} (-1)^{b_{ij}^\gamma} \zeta_{a'}^{\gamma \backslash R_{ij}^\gamma},$$

if we put $\zeta_0^\emptyset := 1$.

A particular example is $\zeta^{(3^3)}((0123)(45)(678))$, which satisfies

$$= \zeta^{(2^3)}((0123)(45)) - \zeta^{(3,2,1)}((0123)(45)) + \zeta^{(3^2)}((0123)(45))$$

$$= -\zeta^{(2,1^2)}((0123)) + \zeta^{(2^2)}((0123)) - \zeta^{(2^2)}((0123)) + \zeta^{(3,1)}((0123))$$

$$= -1 - 1 = -2.$$

The Murnaghan–Nakayama Formula means that the value ζ_a^γ can be obtained by summing all the values ± 1 which we obtain while completely erasing the diagram $[\gamma]$ by erasing hooks. The lengths of the hooks in question are the lengths of the cyclic factors of an element of type a in some fixed order, i. e. these hook lengths form an improper partition $\lambda \models n$, and $[\gamma]$ has to be erased in all possible ways by successively erasing hooks of lengths λ_0, λ_1, and so on. Such a complete erasement of nodes will be called an *$(\gamma, \lambda, \emptyset)$–path*. Notice that this is a *directed* path. For the above example we used $\lambda := (3, 2, 4)$, and a picture of the corresponding destruction process is shown in figure 7.4. The *weight of the edge*, defined to be $(-1)^{b_{ij}}$,

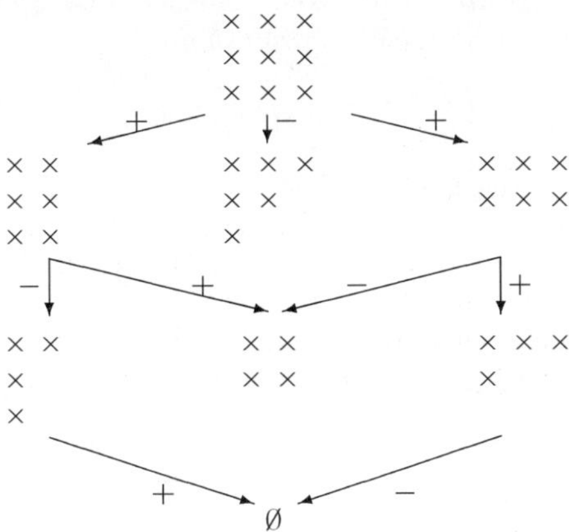

Fig. 7.4. The destruction process

where b_{ij} is the leg length of the hook erased, is indicated for each edge. Now we define the *weight of a directed path* to be the *product* of the edges which form this path. Using this notion we can reformulate the Murnaghan–Nakayama Formula as follows:

7.6.4 Corollary *For each* $\lambda \models n$ *such that* $\lambda^* = \beta$, *the character value* ζ^γ_β *is equal to the sum of weights of all the* $(\gamma, \lambda, \emptyset)$*–paths. In particular* $\zeta^\gamma_\beta = 0$, *if there exists no such path.*

An application of the Murnaghan–Nakayama Rule is (exercise 7.6.1):

7.6.5 Corollary *For* $\gamma \vdash n$ *and the partition* $\beta := (h^\gamma_{00}, h^\gamma_{11}, \ldots) \vdash n$, *consisting of the lengths of the hooks with corner in the main diagonal of* $[\gamma]$, *we have that*

$$\zeta^\gamma_\beta = (-1)^{\sum_i (\gamma'_i - i - 1)}.$$

Furthermore, for any $\delta \vdash n$, *the following holds:*

$$\zeta^\gamma_\delta \neq 0 \Longrightarrow \delta \trianglelefteq \beta.$$

The Murnaghan–Nakayama formula can be generalized considerably as we shall show next. In order to do this we recall the *skew representation* $[\gamma \backslash \alpha]$ which was introduced in 7.4.14: For $\gamma \vdash m + n$ and each $\alpha \vdash m$ where $\alpha_i \leq \gamma_i$ we have

7.6.6
$$[\gamma \backslash \alpha] = \sum_{\beta \vdash n} g^\gamma_{\alpha\beta} [\beta],$$

and it is our aim to show that this skew *representation* is in fact closely related to the skew *diagram* $[\gamma \backslash \alpha]$. A first indication of this close relationship shows

7.6.7 Lemma *The dimension of the skew representation* $[\gamma \backslash \alpha]$ *is the number of standard Young tableaux with skew diagram shape* $[\gamma \backslash \alpha]$*:*

$$f^{\gamma \backslash \alpha} = st^{\gamma \backslash \alpha}(1^n).$$

Proof: Young's Rule shows that

$$st^{\gamma \backslash \alpha}(1^n) = ([\alpha][1] \cdots [1], [\gamma]).$$

Since $[1] \cdots [1]$ is the regular representation $\sum_\beta f^\beta [\beta]$, an application of the Littlewood–Richardson Rule completes the proof as follows:

$$= \sum_{\beta \vdash n} f^\beta ([\alpha][\beta], [\gamma]) = \sum_{\beta \vdash n} f^\beta g^\gamma_{\alpha\beta}.$$

□

But the relationship between skew representations and skew diagrams is much closer as we shall show by deriving the corresponding determinantal form and a Murnaghan–Nakayama formula. In order to do this we need a generalization of the notion of proper and improper partition of n. A *composition* λ of n is a sequence $(\lambda_0, \lambda_1, \ldots)$ with the following properties:

$$\lambda_i \in \mathbb{Z}, \text{ only finitely many } \lambda_i \neq 0, \text{ and } \sum \lambda_i = n.$$

We now slightly generalize the definition of ξ^λ to compositions λ of n by putting

$$\xi^\lambda := \begin{cases} \text{the character of } IS_\lambda \uparrow S_n, & \text{if } \lambda \models n, \\ \text{the zero mapping on } S_n, & \text{otherwise.} \end{cases}$$

Now we note that S_n can be considered as the subgroup of the *restricted symmetric group on* $\mathbb{N}$

$$\mathcal{S} := \{\pi : \mathbb{N} \to \mathbb{N} \mid \pi \text{ bijective, moves only finitely many points}\}$$

which consists of the $\pi \in \mathcal{S}$ that fix each $i \in \mathbb{N} \backslash n$. Using pointwise addition on the set of mappings from $\mathbb{N}$ to $\mathbb{Z}$, so that e. g.

$$\lambda - \mathrm{id} + \pi = (\lambda_0 - 0 + \pi 0, \lambda_1 - 1 + \pi 1, \ldots),$$

we can introduce the generalized character

$$\chi^\lambda := \sum_{\pi \in \mathcal{S}} \mathrm{sign}(\pi) \xi^{\lambda - id + \pi},$$

(Note that $\mathrm{sign}(\pi)$ is well defined, and that this sum is a finite sum.) Since there exists $h \in \mathbb{N}$ such that $\lambda_i = 0$ for each $i > h$ we have, for each $i \geq h$, that

7.6.8
$$\chi^\lambda = \sum_\pi \mathrm{sign}(\pi) \xi^{\lambda - id + \pi},$$

where the sum is taken over the $\pi \in S_i$. Moreover the determinantal form of ζ^α shows that, for each proper partition α, we have

7.6.9
$$\chi^\alpha = \zeta^\alpha.$$

It will turn out crucial for the following derivations to note that for each composition λ and the related sequence

$$\mu := (\lambda_0, \ldots, \lambda_{i-1}, \lambda_{i+1} - 1, \lambda_i + 1, \lambda_{i+2}, \ldots)$$

we have the folowing important identity

7.6.10
$$\chi^\mu = -\chi^\lambda.$$

Proof: Consider the transposition $\tau := (i, i+1) \in \mathcal{S}$. It satisfies $\xi^{\mu - id + \pi\tau} = \xi^{\lambda - id + \pi}$, and so

$$\chi^\lambda = \sum_\pi \mathrm{sign}(\pi) \xi^{\lambda - id + \pi} = \sum_\pi \mathrm{sign}(-\pi\tau) \xi^{\mu - id + \pi\tau} = -\chi^\mu.$$

□

Now we consider restrictions to certain Young subgroups:

7.6.11 Lemma *If λ is a composition of $m+k$, then*

$$\xi^\lambda \downarrow S_{(m,k)} = \sum_{\mu \models k} \xi^{\lambda-\mu} \# \xi^\mu,$$

and

$$\chi^\lambda \downarrow S_{(m,k)} = \sum_{\mu \models k} \chi^{\lambda-\mu} \# \xi^\mu.$$

Proof: In order to prove the first statement we note that both sides of the equation are zero if λ has negative parts λ_i. We can therefore assume that $\lambda \models m+k$. Since ξ^λ is zero or a permutation character, its restriction to the Young subgroup in question satifies

$$\xi^\lambda \downarrow S_{(m,k)} = \sum_{S_{(m,k)}\pi S_\lambda} I(S_{(m,k)} \cap \pi S_\lambda \pi^{-1}) \uparrow S_{(m,k)}.$$

But the double coset symbol associated with $S_{(m,k)}\pi S_\lambda$ (recall 6.1.8) is the matrix

$$\begin{pmatrix} \dots & |m \cap \pi[n_j^\lambda]| & \dots \\ \dots & |n \backslash m \cap \pi[n_j^\lambda]| & \dots \end{pmatrix}, \ j \in n.$$

This matrix is uniquely determined by its second row from which we obtain the improper partition

$$\mu := (\dots, |n\backslash m \cap \pi[n_j^\lambda]|, \dots) \models k.$$

Moreover the intersection $S_{(m,k)} \cap \pi S_\lambda \pi^{-1}$ is conjugate to the Young subgroup $S_{\lambda-\mu} \oplus S_\mu$, so that we finally obtain

$$I(S_{(m,k)} \cap \pi S_\lambda \pi^{-1}) \uparrow S_{(m,k)} = I(S_{\lambda-\mu} \oplus S_\mu) \uparrow S_{(m,k)}$$

$$= (IS_{\lambda-\mu} \uparrow S_m) \# (IS_\mu \uparrow S_k) = \xi^{\lambda-\mu} \# \xi^\mu,$$

which completes the proof of the first statement. In order to prove the second statement we consider the definition of χ^λ together with an application of the first statement which yield that

$$\chi^\lambda \downarrow S_{(m,k)} = \sum_\pi \operatorname{sign}(\pi) \xi^{\lambda-id+\pi} \downarrow S_{(m,k)}$$

$$= \sum_{\mu \models k} \sum_\pi \operatorname{sign}(\pi) \xi^{\lambda-id+\pi-\mu} \# \xi^\mu = \sum_{\mu \models k} \chi^{\lambda-\mu} \# \xi^\mu.$$

□

Now we introduce another generalized character by putting

$$\chi^{\lambda\backslash\mu} := \sum_{\pi \in S} \operatorname{sign}(\pi) \xi^{\lambda-id-(\mu-id)\circ\pi}.$$

It is the aim to show that this is in fact the character of the skew representation $[\gamma\backslash\alpha]$. In order to prove this we first state

7.6.12 Lemma *For each composition λ of $m+k$ we have that*

$$\chi^\lambda \downarrow S_{(m,k)} = \sum_{\pi \in \mathcal{S}} \sum_{\beta \vdash k} \chi^{\lambda\setminus\beta} \# \chi^\beta.$$

Proof: From the proof of lemma 7.6.11 we derive that

$$\chi^\lambda \downarrow S_{(m,k)} = \sum_{\pi \in \mathcal{S}} \mathrm{sign}(\pi) \sum_{\mu \models k} \xi^{\lambda - id + \pi - \mu} \# \xi^\mu,$$

where we can replace μ by $\mu \circ \pi$ obtaining

$$\sum_\pi \mathrm{sign}(\pi) \sum_\mu \xi^{\lambda - id - (\mu - id)\circ\pi} \# \xi^\mu.$$

Now, for each μ, μ_i is eventually zero. Moreover we can assume that μ − id does not contain equal parts, since otherwise $\chi^{\lambda\setminus\mu} = 0$. Thus for which

$$\mu - \mathrm{id} = (\beta - \mathrm{id}) \circ \sigma,$$

and

$$\beta_0 - 0 \geq \beta_1 - 1 \geq \ldots.$$

We can continue as follows:

$$= \sum_\pi \mathrm{sign}(\pi) \sum_{\beta:} \sum_{\sigma \in S^r_{\mathbb{N}*}} \xi^{\lambda - id - (\beta - id)\circ\sigma\circ\pi} \# \xi^{(\beta - id)\circ\sigma + id}$$

$$= \sum_{\pi,\sigma} \mathrm{sign}(\pi\sigma)\mathrm{sign}(\sigma) \sum_\beta \xi^{\lambda - id - (\beta - id)\circ\sigma\circ\pi} \# \sum_\sigma \mathrm{sign}(\sigma) \xi^{(\beta - id) + \sigma^{-1}}$$

$$= \sum_\beta \sum_\rho \mathrm{sign}(\rho) \xi^{\lambda - id - (\beta - id)\rho} \# \chi^\beta = \sum_\beta \chi^{\lambda\setminus\beta} \# \chi^\beta = \sum_{\beta \vdash k} \chi^{\lambda\setminus\beta} \# \chi^\beta.$$

The last equation holds since $\chi^\beta = 0$, if $\beta_i - i = \beta_{i+1} - (i+1)$, by 7.6.10.

□

In the next step we show that $\chi^{\gamma\setminus\alpha}$ is in fact the character of the skew representation $[\gamma\setminus\alpha]$:

7.6.13 Theorem *If $\gamma \vdash m+n$, $\alpha \vdash n$, $\alpha_i \leq \gamma_i$, then*

$$\chi^{\gamma\setminus\alpha} = \sum_\pi \mathrm{sign}(\pi) \xi^{\gamma - id - (\alpha - id)\circ\pi}$$

is in fact the character of

$$[\gamma\setminus\alpha] = \sum_\beta g^\gamma_{\alpha\beta} [\beta].$$

Proof: The definition of $[\gamma\backslash\alpha]$ shows that we have to prove the following identity

$$[\chi^{\gamma\backslash\alpha} \mid \zeta^\beta] = [\zeta^\gamma \downarrow S_{(m,n)} \mid \zeta^\alpha\zeta^\beta],$$

or, in more detail, that

$$\frac{1}{n!}\sum_{\pi\in S_n}\chi^{\gamma\backslash\alpha}(\pi)\zeta^\beta(\pi) = \frac{1}{m!n!}\sum_{(\rho,\sigma)\in S_m\times S_n}\zeta^\gamma(\rho,\sigma)\zeta^\alpha(\rho)\zeta^\beta(\sigma).$$

The right hand side of this equation is, by 7.6.12, equal to

$$\frac{1}{m!n!}\sum_{(\rho,\sigma)}\sum_{\epsilon\vdash n}\chi^{\gamma\backslash\epsilon}(\sigma)\zeta^\epsilon(\rho)\zeta^\alpha(\rho)\zeta^\beta(\sigma)$$

$$= \frac{1}{n!}\sum_\sigma\chi^{\gamma\backslash\epsilon}(\sigma)\zeta^\beta(\sigma)\underbrace{\frac{1}{m!}\sum_\rho\zeta^\epsilon(\rho)\zeta^\alpha(\rho)}_{\delta_{\epsilon,\alpha}} = \frac{1}{n!}\sum_{\sigma\in S_n}\chi^{\gamma\backslash\alpha}(\sigma)\zeta^\beta(\sigma),$$

as we have to show. □

This yields the desired determinantal form:

7.6.14 Corollary *The skew representation $[\gamma\backslash\alpha]$ (to be exact: its character) can be expressed as follows in determinantal form:*

$$[\gamma\backslash\alpha] = \det\big([\gamma_i - \alpha_j + j - i]\big),$$

if we agree upon putting $[0] := 1$ and $[r] := 0$, if $r < 0$.

In order to derive also a Murnaghan–Nakayama formula, we prove

7.6.15 Lemma *If λ is a composition of $n+r$, and ν a composition of r and $m+k = n$, then we have that*

$$\chi^{\lambda\backslash\nu}\downarrow S_{(m,k)} = \sum_{\mu\models k}\chi^{(\lambda-\mu)\backslash\nu}\,\#\,\xi^\mu.$$

Moreover, if $\pi\in S_n$ contains a k–cycle, and if

$$a' := (a_1(\pi),\ldots,a_{k-1}(\pi),a_k(\pi)-1,a_{k+1}(\pi),\ldots),$$

while $\rho\in S_m$ is of type a', then

$$\chi^{\lambda\backslash\nu}(\pi) = \sum_{i=1}^{\infty}\chi^{(\lambda_0,\ldots,\lambda_i-k,\ldots)\backslash\nu}(\rho).$$

Proof: The definition of $\chi^{\lambda\backslash\nu}$ shows that

$$\chi^{\lambda\backslash\nu}\downarrow S_{(m,k)} = \sum_\pi \mathrm{sign}(\pi)\xi^{\lambda-id-(\nu-id)\circ\pi}\downarrow S_{(m,k)},$$

which, by 7.6.11, is equal to

$$\sum_{\pi} \operatorname{sign}(\pi) \sum_{\mu \models k} \xi^{\lambda-\mu-id-(\nu-id)\circ\pi} \# \xi^{\mu}$$

$$= \sum_{\mu \models k} \underbrace{\left(\sum_{\pi} \operatorname{sign}(\pi) \xi^{\lambda-\mu-id-(\nu-id)\circ\pi} \right)}_{=\chi^{(\lambda-\mu)\backslash\nu}} \# \xi^{\mu},$$

as it is claimed. An element $\pi \in S_n$ containing a k–cycle can be considered as being an element of the Young subgroup $S_{(m,k)}$, for charcter theoretic purposes. Hence, by an application of the statement just proved, we obtain

$$\chi^{\lambda\backslash\nu}(\pi) = \chi^{\lambda\backslash\nu} \downarrow S_{(m,k)}(\pi) = \sum_{\mu \models k} \chi^{(\lambda-\mu)\backslash\nu}(\rho) \xi^{\mu}_{(k)}.$$

but $\xi^{\mu}_{(k)}$ is nonzero only if $\mu = (0, \ldots, 0, k, 0, \ldots)$, in which case ξ^{μ} is the identity character. This completes the proof. □

As we can delete from $[\gamma\backslash\alpha]$ in an obvious way all the parts of its rim which correspond to hooks that contain no node of the subdiagram $[\alpha]$, we finally get

7.6.16 The Murnaghan–Nakayama Formula for skew representations

$$\chi^{\gamma\backslash\alpha}(\pi) = \sum_{i,j: h^{\gamma}_{ij}=k} (-1)^{b^{\gamma}_{ij}} \chi^{(\gamma\backslash\alpha)\backslash R^{\gamma}_{ij}}(\rho).$$

Exercises

Exercise 7.6.1 Prove 7.6.5.

7.7 Symmetrization and Permutrization

Having all these concepts and results on ordinary representations and their characters at hand we are now in a position to refine the permutation theoretical considerations of actions on Y^X to representation theoretical ones. In order to do this we simply take Y^X as a basis by forming the free vector space over $\mathbb{C}$ with Y^X as basic set, i. e. we consider

$$\mathbb{C}^{(Y^X)} = \ll f \in Y^X \gg_{\mathbb{C}} .$$

Assume now a representation D of H on $\mathbb{C}^Y$ (which replaces the action ${}_HY$) and an action ${}_GX$. Denoting by $\mathbf{D}$ the corresponding matrix representation with respect to the basis Y of $\mathbb{C}^Y$, we obtain a matrix representation $\widetilde{\mathbf{D}}$ of $H \wr_X G$ on $\mathbb{C}^{(Y^X)}$ in a canonic way by defining the (linear) action of $(\psi, g) \in H \wr_X G$ on an element f of the basis as follows:

7.7.1
$$(\psi, g)f := \sum_{f' \in Y^X} d_{f', f\circ\bar{g}^{-1}}(\psi) f',$$

where

$$d_{f', f\circ\bar{g}^{-1}}(\psi) := \prod_x d_{f'(x), f(g^{-1}x)}(\psi(x)).$$

It is not difficult to check that this is in fact a representation, and it is obvious that it generalizes the permutation representation of $H \wr_X G$ on Y^X which we introduced in 1.3.2 for enumerative purposes. The permutation character with its values (cf. 2.2.12)

$$\chi((\psi, g)) = \prod_{\nu \in c(\bar{g})} a_1(h_\nu(\psi, g))$$

is now replaced by

7.7.2
$$\chi^{\widetilde{D}}((\psi, g)) = \prod_\nu \chi^D(h_\nu(\psi, g)),$$

as can be shown in an analogous way. We now restrict this representation to the subgroup $H^* \cdot G' \simeq H \times G$, i. e. to the elements $(\psi, g) \in H \wr_X G$, where ψ is constant. We can identify such a (ψ, g) with $(h, g) \in H \times G$, h being the value of ψ, and get

7.7.3
$$\chi^{\widetilde{D}}((h, g)) = \prod_{i=1}^{|X|} \chi^D(h^i)^{a_i(\bar{g})},$$

a character of $H \times G$ arising from D of H and the action ${}_GX$ in a natural way. A further restriction to the subgroup $G' \simeq G$ gives

7.7.4
$$\chi^{\widetilde{D}}(g) = |Y|^{c(\bar{g})} = (\dim(D))^{c(\bar{g})}.$$

The restriction to the diagonal subgroup of H^* which is isomorphic to H gives

7.7.5
$$\chi^{\widetilde{D}}(h) = (\chi^D(h))^{|X|}.$$

The representation $\widetilde{D} \downarrow H^* \cdot G'$ of $H \times G$, the character of which is given in 7.7.3, gives rise to a few more general remarks concerning representations of direct products of groups. They will turn out to be useful since they show that there is a close connection between the decomposition of $\widetilde{D} \downarrow H^*$ and the decomposition of $\widetilde{D} \downarrow G'$.

Assume that we are given a vector space V affording a representation $\widetilde{D}$ of $H \times G$ and define D^0 and D^1 by $D^0(h) := \widetilde{D}(h, 1)$ and $D^1(g) := \widetilde{D}(1, g)$. Then we first note that

7.7.6
$$\forall\, (h, g) \in H \times G\colon\ D^0(h)D^1(g) = D^1(g)D^0(h).$$

Assuming that G is finite, since D^1 of G is completely reducible, we may consider a basis of V whith respect to which D^1 decomposes as follows

$$\mathbf{D}^1(g) = \begin{pmatrix} \mathbf{D}_0^1(g) & & & & & & \\ & \ddots & & & & 0 & \\ & & \mathbf{D}_0^1(g) & & & & \\ & & & \ddots & & & \\ & & & & \mathbf{D}_{r-1}^1(g) & & \\ & 0 & & & & \ddots & \\ & & & & & & \mathbf{D}_{r-1}^1(g) \end{pmatrix}$$

$$= \sum_{i \in r} n_i \mathbf{D}_i^1(g).$$

The direct summand of V which affords n_i times the irreducible constituent $\mathbf{D}_i^1$ is called the *homogeneous component of type* $\mathbf{D}_i^1$ of V (as representation space for G).

With respect to the same basis the matrix representation $\mathbf{D}^0$ of H takes the following form:

$$\mathbf{D}^0(h) = \begin{pmatrix} \mathbf{D}_{0,0}^{0,0}(h) & \dots & \mathbf{D}_{0,n_0-1}^{0,0}(h) & \mathbf{D}_{0,0}^{0,1}(h) & \dots & \mathbf{D}_{0,n_1-1}^{0,1}(h) & \dots \\ \vdots & & \vdots & \vdots & & \vdots & \\ \mathbf{D}_{n_0-1,0}^{0,0}(h) & \dots & \mathbf{D}_{n_0-1,n_0-1}^{0,0} & \mathbf{D}_{0,0}^{0,1}(h) & \dots & \mathbf{D}_{0,n_0-1}^{0,1}(h) & \dots \\ \vdots & & \vdots & \vdots & & \vdots & \end{pmatrix}$$

The submatrices satisfy (cf. 7.7.6), for each h, g, i, j, k and l:

$$\mathbf{D}_{kl}^{ij}(h)\mathbf{D}_j^1(g) = \mathbf{D}_i^1(g)\mathbf{D}_{kl}^{ij}(h),$$

and so we obtain, by an application of Schur's Lemma:

$$\mathbf{D}_{kl}^{ij}(h) = \begin{cases} 0\text{–matrix}, & \text{if } i \neq j, \\ d_{kl}^i(h) \cdot \mathbf{I}, & \text{otherwise,} \end{cases}$$

where $\mathbf{I}$ denotes the identity matrix and $d_{kl}^i(h) \in \mathbb{C}$ is a numerical factor. Using these numbers, we define

$$\mathbf{D}_i^0(h) := (d_{kl}^i(h)),\, k, l \in n_i,\, i \in r.$$

Hence there exists a permutation matrix P such that, for each $h \in H$, $P\mathbf{D}^1(h)P^{-1}$ is equal to

$$\begin{pmatrix} \mathbf{D}_0^0(h) & & & & & & \\ & \ddots & & & & 0 & \\ & & \mathbf{D}_0^0(h) & & & & \\ & & & \ddots & & & \\ & & & & \mathbf{D}_{r-1}^0(h) & & \\ & 0 & & & & \ddots & \\ & & & & & & \mathbf{D}_{r-1}^0(h) \end{pmatrix} = \sum_{i \in r} f_i \mathbf{D}_i^0(h).$$

We notice in particular that the multiplicity n_i of $\mathbf{D}_i^1$ is equal to the dimension of $\mathbf{D}_i^0$, while the dimension f_i of $\mathbf{D}_i^1$ is equal to the multiplicity of $\mathbf{D}_i^0$ in $\mathbf{D}^1$. An application of this general argument to the particular case $D^0(h) := \widetilde{D}(h, 1)$ and $D^1(g) := \widetilde{D}(1, g)$ gives (since, by 7.7.5, D^0 is equivalent to $\otimes^{|X|} D$, the $|X|$–fold inner tensor power):

7.7.7 Theorem *Let H denote a group, D an ordinary representation of H on the vector space $\mathbb{C}^Y$, ${}_GX$ a finite action, D_i an ordinary irreducible representation of G, f_i its dimension. Then the homogeneous component of type D_i of $\mathbb{C}^{(Y^X)}$ affords, if it is nonzero, f_i–times a representation of H (which need not be irreducible). We denote this representation by*

$$D \boxdot D_i$$

and call it the $|X|$–fold symmetrization *of D by D_i corresponding to the action ${}_GX$. Its dimension is equal to the multiplicity $n_i := (P_Y^X, D_i)$ of D_i in P_Y^X, the permutation representation of G afforded by $\mathbb{C}^{(Y^X)}$. $D \boxdot D_i$ satisfies, for each $h \in H$, $g \in G$, and with respect to a suitable basis of $\mathbb{C}^{(Y^X)}$ the equation*

$$\widetilde{\mathbf{D}}((h, g)) = \otimes^{|X|}\mathbf{D}(h) \cdot \mathbf{P}_Y^X(g) = \sum_i (\mathbf{D} \boxdot \mathbf{D}_i)(h) \otimes \mathbf{D}_i(g).$$

(We allow $D \boxdot D_i$ to denote the zero "representation" of H if D_i is not an irreducible constituent of P_Y^X, and so the corresponding homogeneous component of type D_i is the zero subspace. In this sense $D \boxdot D_i$ is defined for each irreducible D_i of G, but the corresponding summand in the equation above may have to be neglected.)

An important particular case is that when ${}_GX$ is the natural action of S_X on X, in which we meet the symmetrization $D \boxdot [\alpha]$, $\alpha \vdash |X|$, obtaining the following decomposition of the n–fold tensor power of D:

7.7.8
$$\otimes^{|X|} D = \sum_{\alpha \vdash |X|} f^\alpha (D \boxdot [\alpha]),$$

for example

$$D \otimes D \otimes D = D \boxdot [3] + 2(D \boxdot [2, 1]) + D \boxdot [1^3].$$

Returning to the general case we note that the character of $D \boxdot D_i$ satisfies (cf. 7.7.7 and use 7.7.3):

$$\forall\, h \in H, g \in G\colon \prod_{k=1}^{|X|} \chi^D(h^k)^{a_k(\bar{g})} = \sum_i \chi^{D \boxdot D_i}(h) \zeta^i(g).$$

An application of the orthogonality relations now gives

7.7.9
$$\chi^{D \boxdot D_i}(h) = \frac{1}{|G|} \sum_g \zeta^i(g^{-1}) \prod_{k=1}^{|X|} \chi^D(h^k)^{a_k(\bar{g})}.$$

For the particular case where ${}_GX$ is the natural action of S_X on X, this right hand side takes the form

$$\frac{1}{|X|!} \sum_{\pi \in S_X} \zeta^\alpha(\pi^{-1}) \prod_{k=1}^{|X|} \chi^D(h^k)^{a_k(\pi)},$$

and so we have proved

7.7.10 $$\chi^{D \boxdot [\alpha]}(h) = \{\alpha, Y\}(\epsilon_0, \ldots, \epsilon_{m-1}),$$

which means that we obtain the character value of $D \boxdot [\alpha]$ at h by taking the *Schur function* $\{\alpha, Y\}$, i. e. the polynomial *function* corresponding to the *polynomial* $\{\alpha, Y\}$ and applying it to the eigenvalues $\epsilon_0, \ldots, \epsilon_{m-1}$ of $D(h)$. In particular we obtain for the dimension of this representation:

7.7.11 $$f^{D \boxdot [\alpha]} = \sum_{(\lambda_0, \ldots, \lambda_{m-1}) \models |X|} \kappa_{\lambda^*, \alpha},$$

which implies

7.7.12 Corollary *$D \boxdot [\alpha]$ is nonzero if and only if the number α'_0 of parts of α is less than or equal to the dimension of D.*

The same argument used in proving theorem 7.7.7 on symmetrization gives another construction which is, in a certain sense, its dual:

7.7.13 Theorem *Let H denote a finite group, D an ordinary representation of H on the vector space $\mathbb{C}^Y$ and assume ${}_GX$ to be a finite action. If D_i denotes an ordinary irreducible representation of H and f_i its dimension, then the homogeneous component of $\mathbb{C}^{(Y^X)}$ (as representation space for H, affording $\otimes^{|X|} D$) gives, if it is nonzero, f_i–times a representation of G (which may be reducible). We denote it by $D \vartriangle_{|X|} D_i$ and call it the $|X|$–fold* permutrization *of D by D_i corresponding to the action ${}_GX$. Its dimension is equal to the multiplicity $(\otimes^{|X|} D, D_i)$, and it satisfies, for each $h \in H$, $g \in G$, and with respect to a suitable basis of $\mathbb{C}^{(Y^X)}$ the equation*

$$\widetilde{\mathbf{D}}(h, g) = \otimes^{|X|} \mathbf{D}(h) \cdot \mathbf{P}_Y^X(g) = \sum_i \mathbf{D}_i(h) \otimes (\mathbf{D} \vartriangle \mathbf{D}_i)(g).$$

(We allow $D \vartriangle_{|X|} D_i$ to denote the zero "representation" of G if D_i is not an irreducible constituent of $\otimes^{|X|} D$, and so the corresponding homogeneous component of type D_i is the zero subspace. In this sense $D \vartriangle_{|X|} D_i$ is defined for each irreducible D_i of H, but the corresponding summand in the equation above may have to be neglected.)

To 7.7.8 there corresponds the following equation:

7.7.14 $$P_Y^X = \sum_i f^i (D \vartriangle_{|X|} D_i),$$

while we have for the character

7.7.15 $$\chi^{D\triangle_{|X|}D_i}(g) = \frac{1}{|H|}\sum_{h\in H}\zeta^i(h^{-1})\prod_{k=1}^{|X|}\chi^D(h^k)^{a_k(\bar{g})}.$$

These equations, in particular the last, have important consequences. First of all there is a close relationship between symmetrization and permutrization expressed in the following equality of multiplicities which is immediate from 7.7.15 and 7.7.9:

7.7.16 The Duality Theorem *For each ordinary irreducible representation D_i of H and D_k of G we have*

$$(D\triangle_{|X|}D_i, D_k) = (D\boxdot D_k, D_i).$$

Various other results follow from the consideration of particular cases of 7.7.15. Interpreting $D\triangle_{|X|}D_i$ as a representation of $\bar{G}\leq S_X$, we see that it can be extended to S_X since ${}_GX$ is a subaction of the natural action of S_X. Thus we obtain

7.7.17 Corollary *If H denotes a finite group with an ordinary character χ and an ordinary irreducible character ζ^i, then, for each $a_1,\dots,a_n\in\mathbb{N}$, the expression*

$$\frac{1}{|H|}\sum_{h\in H}\zeta^i(h^{-1})\prod_k\chi(h^k)^{a_k}$$

is the value of a character of S_n, where $n:=\sum_i i\cdot a_i$, at an element π of cycle type $(a_1,\dots,a_n)$:

$$\chi^{D\triangle_n D_i}(\pi) = \frac{1}{|H|}\sum_{h\in H}\zeta^i(h^{-1})\prod_k\chi(h^k)^{a_k(\pi)}.$$

Hence the right hand side is an integer, and so in particular the mapping

$$\chi_{(k)}\colon H\to\mathbb{C}\colon h\mapsto\chi(h^k)$$

is a generalized character, for each χ and any $k\in\mathbb{N}$.

Exercises

Exercise 7.7.1 Show that the dimensions $f^{D\boxdot[|X|]}$ and $f^{D\boxdot[1^{|X|}]}$ are binomial coefficients in terms of $|X|$ and $|Y|$.

7.8 Plethysm of Representations

Recall the representation $\widetilde{D}$ of $H\wr_X G$ defined in 7.7.1: For (ψ, g) in $H\wr_X G$, $f\in Y^X$, and the matrix representation $\mathbf{D}$ of H on $\mathbb{C}^Y$, we have put

$$(\psi,g)f := \sum_{f'\in Y^X} d_{f',f\circ\bar{g}^{-1}}(\psi)f',\ d_{f',f\circ\bar{g}^{-1}}(\psi) := \prod_x d_{f'(x),f(g^{-1}x)}(\psi(x)).$$

We shall multiply this representation with $\mathbf{E}'$, another matrix representation of $H \wr_X G$, arising from the matrix representation $\mathbf{E}$ of G on $\mathbb{C}^X$, and defined by

$$\mathbf{E}'(\psi, g) := \mathbf{E}(g).$$

The inner tensor product $\widetilde{\mathbf{D}} \otimes \mathbf{E}'$ of these two representations is a matrix representation of $H \wr_X G$ on the vector space $\mathbb{C}^{(Y^X)}$, and its character has the values (cf. 7.7.2)

7.8.1
$$\chi^{\widetilde{D}\otimes E'}(\psi, g) = \chi^E(g) \prod_\nu \chi^D(h_\nu(\psi, g)).$$

In the case when $H \wr_X G = S_m \wr S_n$, $D := [\alpha]$, $\alpha \vdash m$, and $E := [\beta]$, $\beta \vdash n$, we write

$$(\alpha; \beta) := \widetilde{[\alpha]} \otimes [\beta]',$$

and as $S_m \wr S_n$ embeds into S_{mn} in a natural way (cf. 2.2.1), this representation induces in a canonical way a representation of S_{mn} which we call the *plethysm* of $[\alpha]$ and $[\beta]$ and which we indicate as follows:

$$[\alpha] \odot [\beta] := (\alpha; \beta) \uparrow S_{mn}.$$

An important particular case is the plethysm of identity representations

$$[m] \odot [n] = (m; n) \uparrow S_{mn} = IS_m \odot S_n \uparrow S_{mn},$$

which gives, according to Foulkes' Lemma, the following helpful expression for the *plethysm* of cycle indicator polynomials (cf. 3.2.9):

7.8.2
$$C(S_m \odot S_n, mn) = \sum_{\alpha \vdash mn} ([m] \odot [n], [\alpha]) C(S_{mn}, mn, [\alpha]).$$

But there are only very few results known on the decomposition of plethysm. One of them is

7.8.3 Theorem *For each $n \in \mathbb{N}^*$ we have the following decomposition of plethysm of identity representations of symmetric groups:*

$$[2] \odot [n] = \sum_{\alpha \vdash n} [2\alpha],$$

if $2\alpha := (2\alpha_0, 2\alpha_1, \ldots)$.

Proof: By induction on n. The case $n = 1$ is obvious: $[2] \odot [1] = [2]$. If $n > 1$, the induction hypothesis together with (exercise 7.8.1)

7.8.4
$$[2] \odot [n] \downarrow S_{2n-1} = [2] \odot [n-1] \uparrow S_{2n-1}$$

and the branching theorem yields that $[2] \odot [n] \downarrow S_{2n-1}$ is equal to $\sum [\beta]$, where β runs through the partitions of $2n - 1$ which contain exactly one odd part. Another application of the branching rule shows that for these β the following is true:

$$\sum_{\beta}[\beta] = \sum_{\alpha\vdash n}[2\alpha] \downarrow S_{2n-1}.$$

It therefore suffices to prove that $[2] \odot [n]$ cannot contain any constituent $[\gamma]$ with odd parts γ_i. Since for each such γ every constituent of $[\gamma] \downarrow S_{2n-1}$ has exactly one odd part, γ must be of the form $\gamma = (\gamma_1, \gamma_2)$, γ_i odd. Hence it remains to show that no $\gamma = (\gamma_1, \gamma_2)$, γ_i odd, can be a constituent of $[2] \odot [n]$.

In order to prove this we notice that by 7.8.4 together with the induction hypothesis, $[2] \odot [n]$ must be multiplicity–free, i. e. it contains no constituent more than once. Now $[2n]$ is a constituent of $[2] \odot [n]$ since this is a permutation representation. We would like to check that $[2n-1, 1]$ cannot occur. As $[2n-1, 1] \downarrow S_{2n-1} = [2n-2, 1] + [2n-1]$, the occurrence of both $[2n]$ and $[2n-1, 1]$ in $[2] \odot [n]$ would imply that $[2n-1]$ occurs *twice* in $[2] \odot [n-1] \uparrow S_{2n-1}$, which is in fact multiplicity–free. Thus $[2n]$ is a constituent of $[2] \odot [n]$, while $[2n-1, 1]$ is not.

In order to prove that $[2n-2, 2]$ is a constituent of $[2] \odot [n]$, we consider $[2n-2, 1]$, a constituent of $[2] \odot [n-1] \uparrow S_{2n-1}$. It satisfies

$$[2n-2, 1] \uparrow S_{2n} = [2n-1, 1] + [2n-2, 2] + [2n-2, 1^2].$$

Hence $[2n-2, 1]$ arises from the restriction of $[2n-2, 2]$, for we have seen already that neither $[2n-1, 1]$ nor $[2n-2, 1^2]$ occur in $[2] \odot [n]$.

This gives us an idea how to proceed inductively. Let us assume that we have shown that $[2n-k, k]$, k even, is a constituent of $[2] \odot [n]$. We have to check that $[2n-k-1, k+1]$ does not occur, while $[2n-k-2, k+2]$ does. As

$$[2n-k-1, k+1] \downarrow S_{2n-1} = [2n-k-2, k+1] + [2n-k-1, k],$$

$$[2n-k, k] \downarrow S_{2n-1} = [2n-k-1, k] + [2n-k, k-1],$$

$[2n-k-1, k+1]$ cannot occur, for otherwise $[2] \odot [n-1] \uparrow S_{2n-1}$ would not be multiplicity–free. On the other hand

$$([2] \odot [n-1] \uparrow S_{2n-1}, [2n-k-2, k+1]) > 0$$

and $[2n-k-2, k+1] \uparrow S_{2n}$ is equal to

$$[2n-k-1, k+1] + [2n-k-2, k+2] + [2n-k-2, k+1, 1],$$

so that $[2n-k-2, k+1]$ stems from the restriction of $[2n-k-2, k+2]$, which therefore must occur in $[2] \odot [n]$. This completes the proof.

□

Hence, for example,

$$[2] \odot [4] = [8] + [6, 2] + [4^2] + [4, 2^2] + [2^4].$$

These particular plethysms, arising from the hyperoctahedral group $S_2 \wr S_m$, can be used in order to construct a natural *model* for the symmetric group S_n, i. e. a representation that decomposes into $\sum_\alpha [\alpha]$, which means that it contains each irreducible representation of S_n *exactly once* as an irreducible constituent. How this can be done will be described next. Later on we shall see that the character of the model can be used in order to count square roots in symmetric groups.

The result 7.8.3 together with the special case 7.5.9 of the Littlewood–Richardson Rule shows that $([2] \odot [m])[1^r]$ decomposes into $\sum[\beta]$, where β runs through the partitions of $n := 2m + r$ which contain exactly r odd parts β_i. This implies

7.8.5 Theorem *For each $n > 2$ we have the decomposition*

$$\sum_{0 \le m \le n/2} ([2] \odot [m])[1^{n-2m}] = \sum_{\alpha \vdash n} [\alpha] = \sum_{0 \le m \le n/2} ([1^2] \odot [m])[n - 2m],$$

i. e. the left and the right hand side is a model for the symmetric group S_n.

Now we shall show that this model of S_n is contained in the *conjugation representation* κ of S_n, by which we mean the linear representation of S_n arising from the permutation representation of S_n on itself via conjugation:

$$S_n \times S_n \to S_n \colon (\pi, \rho) \mapsto \pi \rho \pi^{-1}.$$

It is obvious that the character χ^κ of κ at π is the order of the centralizer of π:

7.8.6
$$\chi^\kappa(\pi) = |C_S(\pi)|,$$

and it is easy to check that the multiplicity of $[\alpha]$ in κ is the sum of the corresponding row of the character table:

7.8.7
$$[\chi^\kappa \mid \zeta^\alpha] = \sum_{a \dashv\vdash n} \zeta^\alpha_a.$$

On the other hand the conjugation action of S_n on itself has the conjugacy classes of elements as orbits. Hence, if $\mathcal{C}$ denotes a transversal of the conjugacy classes of S_n, κ is the following sum of transitive permutation representations:

7.8.8
$$\kappa = \sum_{\pi \in \mathcal{C}} IC_S(\pi) \uparrow S_n.$$

Hence, in order to show that κ contains the model, it suffices to prove that the model is contained in a suitable partial sum of the right hand side of 7.8.8. We shall do this in three steps.

7.8.9 Lemma *If $\pi \in S_n$, $n = 2m + r$, r odd, consists of an r–cycle together with m 2–cycles, then $IC_S(\pi) \uparrow S_n$ contains* $([2] \odot [m])[1^r]$.

Proof: The centralizer of π is similar to $(S_2 \odot S_m) \oplus C_r$. Hence

$$IC_S(\pi) \uparrow S_n = (IS_2 \odot S_m \uparrow S_{2m} \# IC_r \uparrow S_r) \uparrow S_n.$$

But $(IC_r \uparrow S_r, [1^r]) = ([1^r] \downarrow C_r, IC_r) = 1$, as r is odd. This yields the statement.

□

7.8.10 Lemma *If $\pi \in S_n$, $n = 2m + r$, r even and $\neq 2$, consists of a 1–cycle, an $(r-1)$–cycle and m 2–cycles, then $IC_S(\pi) \uparrow S_n$ contains $([2] \odot [m])[1^r]$.*

Proof: The proof of 7.8.9 shows, since $r-1$ is odd, that $IC_S(\pi) \uparrow S_n$ contains

$$([2] \odot [m])([1^{r-1}][1]) = ([2] \odot [m])([1^r] + [2, 1^{r-1}]),$$

which implies the statement. □

It remains to consider the case when $r = 2$.

7.8.11 Lemma *Assume that $n = 2m + 2 > 2$ and $\alpha(\pi) = (1, 2^{m-1}, 3)$. Then the induced representation $IC_S(\pi) \uparrow S_n$ contains each $[\beta]$, where $\beta \vdash n$ has exactly two odd parts.*

Proof: The induced representation $IC_S(\pi) \uparrow S_n$ is equal to

$$([2] \odot [m-1])(IC_3 \uparrow S_3)[1] = ([2] \odot [m-1])([4] + [3, 1] + [2, 1^2] + [1^4]).$$

Assume that $\alpha \vdash n$ contains exactly two odd parts, we have to show that $[\alpha]$ is contained in the right hand side.

There exists $\beta \vdash n - 4$ such that each β_i is even and $\leq \alpha_i$. The representation $[\beta]$ is a constituent of $[2] \odot [m-1]$, and, for each i, we have that $\alpha_i - \beta_i \leq 3$. The sequence of nonzero differences $\alpha_i - \beta_i$, ordered according to i, is therefore one of the following five sequences:

$$(2, 1, 1), (1, 2, 1), (1, 1, 2), (3, 1), (1, 3).$$

The Littlewood–Richardson Rule now shows, that $[\alpha]$ is contained in $[\beta][2, 1^2]$ or in $[\beta][3, 1]$. This implies the statement.

□

Summarizing we obtain

7.8.12 Corollary *For $n \neq 2$ the conjugation representation of S_n contains each $[\alpha]$, where $\alpha \vdash n$, at least once, or, equivalently, the sum of each row in the character table of S_n is strictly positive, if $n > 2$.*

An application of plethysm is provided by a proof of certain results on the *Gaussian coefficient* $p(k; m, n)$ which, for $k, m, n \in \mathbb{N}$, is defined to be the number of partitions of k into $\leq m$ parts, each one being $\leq n$. The corresponding polynomial

$$G_{mn}(x) = \sum_{k=0}^{mn} p(k; m, n)x^k$$

was already introduced (cf. 7.4.7) and we called it a *Gaussian polynomial*. We have seen that the sequence of its coefficients is *unimodal*, i. e. that

$$p(0; m, n) \leq p(1; m, n) \leq \ldots \leq p(\lfloor mn/2 \rfloor; m, n)$$

$$= p(\lfloor mn/2+1\rfloor; m, n) \geq \ldots \geq p(mn; m, n).$$

We are now in a position to express these Gaussian coefficients as multiplicities. An application of 7.1.11 gives

$$p(k; m, n) = ([mn-k][k], [m] \odot [n]) = \sum_{j=0}^{k} ([mn-j, j], [m] \odot [n]).$$

Conversely we can use this in order to derive a result on plethysms:

$$([mn-k, k], [n] \odot [m]) = p(k; m, n) - p(k-1; m, n)$$

$$= p(k; n, m) - p(k-1; n, m) = ([mn-k, k], [m] \odot [n]).$$

7.8.13 Corollary *For two–rowed diagrams* $[\alpha]$, $\alpha \vdash mn$, *we have the following equation for multiplicities:*

$$([n] \odot [m], [\alpha]) = ([m] \odot [n], [\alpha]).$$

Still open is the following generalization which is called *Foulkes' Conjecture*:

7.8.14 $$\forall\, m \leq n, \alpha \vdash mn: \ ([m] \odot [n], [\alpha]) \geq ([n] \odot [m], [\alpha]).$$

There are various helpful rules for calculating with plethysms. First of all plethysm is distributive on the right hand side:

7.8.15 $$[\alpha] \odot ([\beta] + [\gamma]) = [\alpha] \odot [\beta] + [\alpha] \odot [\gamma].$$

Since the embedding of $S_m \wr (S_r \oplus S_s)$ into $S_{m(r+s)}$ yields a subgroup which is conjugate to $(S_m \wr S_r) \oplus (S_m \wr S_s)$, we have

7.8.16 $$[\alpha] \odot ([\beta][\gamma]) = ([\alpha] \odot [\beta])([\alpha] \odot [\gamma]).$$

Hence there is also a *determinantal form* for plethysm:

7.8.17 $$[\alpha] \odot [\beta] = \det([\alpha] \odot [\beta_i + j - i]).$$

This reduces the problem of decomposing $[\alpha] \odot [\beta]$ to the problem of decomposing plethysms of the special form $[\alpha] \odot [m]$. But still there is no explicit formula for the decomposition of $[\alpha] \odot [\beta]$, this problem is one of the big open problems in representation theory of symmetric groups.

One of the most interesting rules known for plethysms is

$$([\lambda] + [\mu]) \odot [\nu] = \sum_{(r,s): \nu \vdash r+s} \ \sum_{\alpha \vdash r, \beta \vdash s} ([\nu], [\alpha][\beta])([\lambda \odot [\alpha])([\mu] \odot [\beta]).$$

The definition 7.4.14 allows to rephrase it as follows:

$$([\lambda] + [\mu]) \odot [\nu] = \sum_{r,s,\alpha,\beta} ([\nu \backslash \alpha], [\beta])([\lambda] \odot [\alpha])([\mu] \odot [\beta])$$

$$= \sum_{r,s,\alpha} ([\lambda] \odot [\alpha])([\mu] \odot [\nu\backslash\alpha]),$$

and so we obtain

7.8.18 $$([\lambda] + [\mu]) \odot [\nu] = \sum_{0 \leq r \leq |\nu|} \sum_{\alpha \vdash r} ([\lambda] \odot [\alpha])([\mu] \odot [\nu\backslash\alpha]).$$

Plethysm allows to generalize the plethysm of cycle indicator polynomials as follows:

7.8.19 Lemma *For $\alpha \vdash m$ and $\beta \vdash n$ the following is true:*

$$C(S_{mn}, mn, [\alpha] \odot [\beta]) = C(S_m, m, [\alpha]) \odot C(S_n, n, [\beta]),$$

i. e. in terms of $(\alpha; \beta)$ we have:

$$\frac{1}{m!^n n!} \sum_{(\varphi,\rho) \in S_m \wr S_n} \chi^{(\alpha;\beta)}((\varphi, \rho)^{-1}) \prod_{k=1}^{mn} x_k^{a_k((\bar{\varphi},\rho))}$$

$$= \frac{1}{n!} \sum_{\rho \in S_n} \zeta^{\beta}(\rho^{-1}) \prod_{k=1}^{n} \Big(\frac{1}{m!} \sum_{\pi \in S_m} \zeta^{\alpha}(\pi^{-1}) \prod_{i=1}^{m} x_{ik}^{a_i(\pi)} \Big)^{a_k(\rho)}.$$

Proof: First of all we rewrite the right hand side obtaining

$$\frac{1}{m!^n n!} \sum_{\rho} \zeta^{\beta}(\rho^{-1}) m!^{n-c(\rho)} \sum_{(\pi_0,\ldots,\pi_{c(\rho)-1})} \zeta^{\alpha}(\pi_0^{-1}) \cdots \zeta^{\alpha}(\pi_{c(\rho)-1}^{-1}) \prod_k \prod_i x_{i,l_k(\rho)}^{a_i(\pi_k)},$$

where $l_k(\rho)$ denotes the length of the k-th cycle of $\rho \in S_n$, with respect to the standard cycle notation

$$\prod_{k \in c(\rho)} \Big(j_k \ldots \rho^{l_k(\rho)-1}(j_k) \Big),$$

where j_k is the smallest entry in the corresponding cycle. Now, to $\psi: n \to S_m$ and $\widetilde{\varphi}: n\backslash\{j_0, \ldots, j_{c(\rho)-1}\} \to S_m$ there exists exactly one $\varphi: n \to S_m$ that extends $\widetilde{\varphi}$ and for which

$$\psi(j_k) \cdots \psi(\rho^{l_k(\rho)-1}(j_k)) = \varphi(j_k) \cdots \varphi(\rho^{l_k(\rho)-1}(j_k)), \text{ for each } k \in c(\rho).$$

But these elements are just the cycle products $h_k(\psi, \rho)$ and $h_k(\varphi, \rho)$. Thus, to a fixed $c(\rho)$–tuple $(\pi_0, \ldots, \pi_{c(\rho)-1})$ of elements of S_m, there exist exactly $m!^{n-c(\rho)}$ mappings $\varphi: n \to S_m$ such that

$$\pi_k = h_k(\varphi, \rho), \text{ for each } k \in c(\rho).$$

This shows that the above expression is equal to

$$\frac{1}{m!^n n!} \sum_{(\varphi,\rho) \in S_m \wr S_n} \chi^{(\alpha;\beta)}((\varphi, \rho)^{-1}) \prod_k \prod_i x_{i,l_k(\rho)}^{a_i(h_k(\varphi,\rho))}.$$

□

Exercises

Exercise 7.8.1 Prove 7.8.4.

7.9 Actions on Chains

We recall that an action ${}_G X$ on a poset $(X, \leq)$ is called a *poset action* if and only if the following condition is satisfied:

$$x \leq x' \iff gx \leq gx'.$$

If this holds, then *chains are mapped onto chains:*

7.9.1 $$g : x_0 < x_1 < \ldots < x_l \longmapsto gx_0 < gx_1 < \ldots < gx_l.$$

We therefore introduce the set of all the chains on X:

$$\mathrm{Ch}(X) := \{x_0 < \ldots < x_l \mid l \in \mathbb{N}, x_i \in X\}.$$

The subset consisting of all the chains of the same *length* l is indicated as follows:

$$\mathrm{Ch}(X, l) := \{x_0 < \ldots < x_l \mid x_j \in X\}.$$

For example, $\mathrm{Ch}(X, 0) = X$. We shall also use the notion of the *empty chain* $\emptyset$, a chain which we consider to be of length -1, so that $\mathrm{Ch}(X, -1) = \{\emptyset\}$. Moreover, for technical reasons, we put $\mathrm{Ch}(X, l) := \emptyset$, for $-1 > l \in \mathbb{Z}$. We also note in passing that $\mathrm{Ch}(X, l) = \emptyset$, as soon as l is bigger than the maximal chain length in X. Thus, $\mathrm{Ch}(X, l)$ is now defined for every $l \in \mathbb{Z}$. The set $\mathrm{Ch}(X, l)$ is a basis of the free vector space

$$\mathrm{Ch}(X, l, \mathbb{C}) := \mathbb{C}^{Ch(X,l)}, l \in \mathbb{Z},$$

on which G acts in a natural way: The linear mapping induced by $g \in G$ on $\mathrm{Ch}(X, l, \mathbb{C})$ is denoted by $g^{(l)}$, it is the linear extension of the action of g on the basis:

$$g^{(l)}: \mathrm{Ch}(X, l, \mathbb{C}) \to \mathrm{Ch}(X, l, \mathbb{C}): x_0 < \ldots < x_l \mapsto g^{-1}x_0 < \ldots < g^{-1}x_l.$$

over $\mathbb{C}$ with the sets of chains $\mathrm{Ch}(X, l)$ as bases. For example, $\mathrm{Ch}(X, -1, \mathbb{C})$ is onedimensional, while $\mathrm{Ch}(X, l, \mathbb{C})$ is the zero space if $l < -1$ or l is greater than the maximal chain length in X. We introduce the following mapping called *differential:*

$$d_l: \mathrm{Ch}(X, l, \mathbb{C}) \to \mathrm{Ch}(X, l-1, \mathbb{C}), l \in \mathbb{Z},$$

defined to be the linear extension of

$$x_0 < \ldots < x_l \mapsto \sum_{i=0}^{l} (-1)^i x_0 < \ldots < \hat{x}_i < \ldots < x_l,$$

where $\hat{x}_i$ means, as usual, that this element is left out. An easy check (exercise 7.9.1) shows the following:

7.9.2 Lemma *The differentials, the actions of the group G on the kernels and the images of them have the following properties:*

- $d_l \circ d_{l+1} = 0$,
- $d_l \circ g^{(l)} = g^{(l-1)} \circ d_l$,
- $\mathrm{im}(d_{l+1}) \subseteq \ker(d_l)$,
- $\ker(d_{l+1})$ *and* $\mathrm{im}(d_l)$ *are* G*-invariant.*

We therefore introduce the factor space

$$H_l(X, \mathbb{C}) := \ker(d_l)/\mathrm{im}(d_{l+1}),$$

the l-th *homology group* of the *chain complex*

$$\cdots \xrightarrow{d_{l+1}} \mathrm{Ch}(X, l, \mathbb{C}) \xrightarrow{d_l} \mathrm{Ch}(X, l-1, \mathbb{C}) \xrightarrow{d_{l-1}} \cdots$$

Lemma 7.9.2 shows that to each $g \in G$ there corresponds an endomorphism of this complex, which can be visualized by the following commutative diagram:

$$\begin{array}{ccccccc} \cdots \xrightarrow{d_{l+1}} & \mathrm{Ch}(X, l, \mathbb{C}) & \xrightarrow{d_l} & \mathrm{Ch}(X, l-1, \mathbb{C}) & \xrightarrow{d_{l-1}} \cdots \\ & \downarrow g^{(l)} & & \downarrow g^{(l-1)} & \\ \cdots \xrightarrow[d_{l+1}]{} & \mathrm{Ch}(X, l, \mathbb{C}) & \xrightarrow[d_l]{} & \mathrm{Ch}(X, l-1, \mathbb{C}) & \xrightarrow[d_{l-1}]{} \cdots \end{array}$$

If we allow zero representations for sake of convenience, we obtain a natural linear representation of G on $H_l(X, \mathbb{C})$ as follows: We denote the linear operator on $H_l(X, \mathbb{C})$ which corresponds to $g \in G$ by g_l, and define it as follows:

$$g_l \colon H_l(X, \mathbb{C}) \to H_l(X, \mathbb{C}) \colon x + \mathrm{im}(d_{l+1}) \mapsto g^{(l)}x + \mathrm{im}(d_{l+1}).$$

The alternating sum of the traces of these mappings,

$$\Lambda_X(g) := \sum_{l \in \mathbb{Z}} (-1)^l \mathrm{tr}\,(g_l),$$

is called the *Lefschetz number* of g on X. We shall relate it to the *Euler characteristic* of X:

$$\chi(X) := \sum_{l \in \mathbb{Z}} (-1)^l |\mathrm{Ch}(X, l)|.$$

The Euler characterstic can be expressed in terms of numbers of chains as follows:

7.9.3 The Euler–Poincaré Formula

$$\chi(X) = \sum_{l=0}^{r+1} (-1)^l \dim(H_l(X, \mathbb{C})).$$

Proof: From the definition of the l-th homology group we obtain its dimension:

$$\dim(H_l(X,\mathbb{C})) = |\mathrm{Ch}(X,l)| - \dim(\mathrm{im}(d_l)) - \dim(\mathrm{im}(d_{l+1})),$$

and so the statement follows by forming the alternating sum of the right hand sides and cancelling. □

Now we relate the Lefschetz number to the Euler characteristic of the fixed point set:

7.9.4 The Hopf–Lefschetz Fixed Point Theorem

$$\Lambda_X(g) = \chi(X_g).$$

Proof: We use that $H_l(X,\mathbb{C})$, $\mathrm{Ch}(X,l,\mathbb{C})$, $\ker(d_l)$ and $\mathrm{im}(d_{l+1})$ are invariant under g, so that we can decompose the trace of g_l in the following way:

$$\begin{aligned}\mathrm{tr}\,(g_l) &= \mathrm{tr}\,(g \downarrow H_l(X,\mathbb{C})) = \mathrm{tr}\,(g \downarrow \ker(d_l)) - \mathrm{tr}\,(g \downarrow \mathrm{im}(d_{l+1}))\\ &= \mathrm{tr}\,(g \downarrow \mathrm{Ch}(X,l,\mathbb{C})) - \mathrm{tr}\,(g \downarrow \mathrm{im}(d_l)) - \mathrm{tr}\,(g \downarrow \mathrm{im}(d_{l+1})).\end{aligned}$$

Hence we obtain, by summing up, that

$$\begin{aligned}\Lambda_X(g) &= \sum_{l=0}^{r+1}(-1)^l\mathrm{tr}\,(g_l) = \sum_{l=0}^{r+1}(-1)^l\mathrm{tr}\,(g \downarrow \mathrm{Ch}(X,l,\mathbb{C}))\\ &= \sum_{l=0}^{r+1}(-1)^l|\mathrm{Ch}(X_g,l)| = \chi(X_g).\end{aligned}$$

□

This shows that we obtain from an action of G on X in a canonical way permutation representations of G on the vector spaces $\mathrm{Ch}(X,l,\mathbb{C})$ and representations on the homology groups $H_l(X,\mathbb{C})$ as well as on the spaces $\ker(d_l)$ and $\mathrm{im}(d_l)$. Further representations can be obtained if we bring the *rank* into the game. In order to do this we assume that X contains a *smallest* element $\hat{0}$ and a *biggest* element $\hat{1}$ (if this is not true then we can add such elements, since both these elements must be fixed points of each $g \in G$). Moreover, we suppose that all the maximal chains (i. e. the chains in X which neither can be refined nor prolongued) have the same length r, and so we can call X a *graded* poset of *rank* r.

The assumption that X is graded of rank r implies the *Jordan–Dedekind* condition: *The chains between two given elements which cannot be refined are all of the same length.* Hence there exists a *rank function* $\rho\colon X \mapsto \mathbb{N}$ with the following properties: $\rho(\hat{0}) = 0$, $\rho(\hat{1}) = r$, and

7.9.5 If y covers x, then $\rho(y) = \rho(x) + 1$.

With the aid of this rank function we can pick subsets, on which G acts. The reason is that the following holds:

7.9.6 $$\forall\, g \in G, x \in X\colon r(gx) = r(x).$$

The proof is left as exercise 7.9.2. Now, if $C\colon x_0 < \ldots < x_l$ is a chain, then

$$R(C) := \{\rho(x_0), \ldots, \rho(x_l)\}\backslash\{0, r\}$$

is called the *rank set* of C, it is a subset of $\underline{n-1} = \{1, \ldots, n-1\}$. For a given set R of ranks we introduce the following subset of X:

$$X_R := \{x \in X \mid \rho(x) \in R\} \cup \{\hat{0}, \hat{1}\}.$$

In particular

7.9.7 $$X_{\underline{r-1}} = X,\ X_{\{i\}} = \{x \in X \mid \rho(x) = i\} \cup \{\hat{0}, \hat{1}\}.$$

It is crucial to note that, by 7.9.6, all these *rank selected subsets* X_R are unions of orbits, and so G acts on each X_R. We therefore introduce the corresponding sets

$$\mathrm{MCh}(X, R) := \{C \in \mathrm{Ch}(X) \mid R(C) = R,\ C \text{ maximal }\}$$

of *rank selected maximal chains*, for each subset $R \subseteq \underline{r-1}$. The character of the corresponding permutation representation on

$$\mathrm{MCh}(X, R, \mathbb{C}) := \mathbb{C}^{MCh(X,R)}$$

is indicated as follows:

7.9.8 $$\kappa_R(g) := \mathrm{tr}\,(g \downarrow \mathrm{MCh}(X, R, \mathbb{C})) = |\mathrm{MCh}(X_g, R)|.$$

Moreover, we introduce the characters

$$\gamma_{R,i}(g) := tr(g \downarrow H_i(X_R, \mathbb{C})),$$

together with the *Lefschetz character*

$$\nu_R(g) := \sum_i (-1)^{|R|-i} \gamma_{R,i}(g).$$

An interesting connection between these characters is described in

7.9.9 Theorem *The characters κ_R and the Lefschetz characters ν_R are related as follows:*

$$\kappa_R = \sum_{T \subseteq R} \nu_T,$$

or, equivalently, by

$$\nu_R = \sum_{T \subseteq R} (-1)^{|R\backslash T|} \kappa_T.$$

Proof: We shall compare two expressions for the Lefschetz number of g on X_g. The first one uses the Fixed Point Theorem 7.9.4 which gives

$$\Lambda_{X_R}(g) = \chi((X_R)_g) = \chi((X_g)_R) = \sum_i (-1)^i |\mathrm{MCh}((X_g)_R, i)|$$

$$= \sum_i (-1)^i \sum_{T \subseteq R : |T| = i} |\mathrm{MCh}(X_g, T)| = \sum_{T \subseteq R} (-1)^{|T|} \kappa_T(g).$$

The second expression uses the definition of the Lefschetz number which implies that $\Lambda_{X_R}(g)$ is

$$= \sum_i (-1)^i tr(g \downarrow H_i(X_R, \mathbb{C})) = \sum_i (-1)^i \gamma_{R,i}(g) = (-1)^{|R|} \nu_R(g).$$

A comparison of these two expressions yields the statement. □

Let us consider a few examples, where $G = S_n$, the symmetric group, and let us try to decompose the corresponding characters.

7.9.10 Example The first example is the natural action of the symmetric group S_n on the power set of n, which is the *Boolen algebra* of rank n. It can easily be identified with the set of mappings 2^n, where $f \in 2^n$ is identified with the set $\{i \in n \mid f(i) = 1\} \subseteq n$. This set of subsets is partially ordered by inclusion:

$$f \le f' \iff \forall\, i\colon\ f(i) \le f'(i).$$

The rank function is $\rho(f) := \sum f(i)$, and it is a graded poset of rank n. If now $R := \{n_0, \ldots, n_{k-1}\} \subseteq \underline{n}$, $n_i < n_{i+1}$, denotes a rank set, then the action of S_n on $\mathrm{MCh}(2^n, R)$ is transitive, and hence the corresponding permutation representation is induced by the identity representation of the stabilizer of any chain with rank set R. The chain

$$\emptyset = \hat{0} < f_0 < \ldots < f_{k-1} < \hat{1} = n$$

has rank set R if and only if

$$|f_i| = |\{j \mid f_i(j) = 1\}| = n_i,$$

for each $i \in k$. Its stabilizer is the set of permutations which leave each f_i fixed, and hence this stabilizer is

$$\bigcap_{i \in k} \left(S_{\{j \mid f_i(j)=1\}} \oplus S_{\{j \mid f_i(j)=0\}}\right)$$

which is, as $f_i < f_{i+1}$,

$$= S_{\{j \mid f_0(j)=1\}} \oplus S_{\{j \mid f_0(j)=0, f_1(j)=1\}} \oplus S_{\{j \mid f_1(j)=0, f_2(j)=1\}} \oplus \ldots .$$

This is a *Young subgroup* and isomorphic to S_λ, where λ denotes the improper partition

$$\lambda := (\lambda_0, \lambda_1, \ldots) := (n_0, n_1 - n_0, \ldots, n_{k-1} - n_{k-2}, n - n_{k-1}) \models n.$$

7.9.11 Corollary *The action of the symmetric group S_n on the Boolean algebra 2^n of rank n yields the following rank selected characters κ_R, for the rank set $R := \{n_0, \ldots, n_{k-1}\}$, where $n_i < n_{i+1}$, and $\lambda := (n_0, n_1 - n_0, \ldots, n_{k-1} - n_{k-2}, n - n_{k-1}) \models n$:*

$$\kappa_R = \xi^\lambda.$$

Using the decomposition into irreducibles

$$\xi^\lambda = \sum_{\alpha \vdash n} st^\alpha(\lambda)\zeta^\alpha,$$

we can decompose the character ν_R in the following way. We map the set of standard Young tableaux $ST^\alpha(\lambda)$ onto a subset of $ST^\alpha(1^n)$ by replacing the λ_0 entries 0 by entries $0, 1, \ldots, \lambda_0 - 1$ (from left to right), the entries 2 by $\lambda_0, \ldots, \lambda_0 + \lambda_1 - 1$, and so on. Let us denote this mapping by E. For example

$$E\colon \begin{matrix} 0 & 0 & 0 & 2 \\ 1 & 1 & 2 & \\ 2 & 3 & & \end{matrix} \quad \mapsto \begin{matrix} 0 & 1 & 2 & 7 \\ 3 & 4 & 6 & \\ 5 & 8 & & \end{matrix} \quad .$$

(Note that we *do not replace the entries rowwise but columnwise from left to right!*) This mapping E is clearly injective, since it can easily be inverted, if λ is given. Hence it remains to characterize the image of $ST^\alpha(\lambda)$, which, in general, will be a proper subset of $ST^\alpha(1^n)$ In order to do this we consider a standard tableau $T \in E(ST^\alpha(\lambda))$, where $\lambda \vdash n$. Since its inverse image $U := E^{-1}(T)$ contains all its entries 0 in its 0-th row, the 0-th row of T begins with the sequence $0, 1, \ldots, \lambda_0 - 1$. The entries 1 of U are contained in the *two* highest rows, and hence U contains an entry 1 in the row with the number 1 if and only if the entry λ_0 occurs in the row with the number 1 of T. If this holds, we call λ_0 a *descent* of T. The set of *all* the descents of T is called its *descent set*, it is a subset of $\Lambda := \{\lambda_0, \lambda_1, \ldots\}$. We therefore introduce a notation for the set of standard Young tableaux with descent set $D \subseteq n - 1$:

$$ST^\alpha_D(1^n).$$

Here is the above Young tableau, but now with its descents emphasized:

$$\begin{matrix} 0 & 1 & \boxed{2} & \boxed{7} \\ 3 & \boxed{4} & 6 & \\ 5 & 8 & & \end{matrix} \quad .$$

The following is clear from the foregoing:

7.9.12 Theorem *For each $\lambda \models n$ the following holds:*

$$E(ST^\alpha(\lambda)) = \bigcup_D ST^\alpha_D(1^n),$$

if the sum is taken over all the subsets

$$D \subseteq \Lambda = \{\lambda_0, \lambda_1, \ldots\}.$$

This result will help us to decompose the Lefschetz characters. In order to prepare this decomposition we deduce from 7.9.12 that

$$\kappa_R = \sum_{\alpha} \sum_{T \subseteq R} st_T^{\alpha}(1^n)\zeta^{\alpha},$$

and hence, by inversion, that

$$\sum_{\alpha} st_R^{\alpha}(1^n)\zeta^{\alpha} = \sum_{T \subseteq R} (-1)^{|R \setminus T|}\kappa_T,$$

so that we finally obtain:

7.9.13 Corollary *The characters κ_R and ν_R which correspond to the action of S_n on the Boolean algebra 2^n and subsets R of the rank set $n-1$, have the following decompositions into irreducibles:*

$$\kappa_R = \sum_{\alpha \vdash n} st^{\alpha}(\lambda)\zeta^{\alpha}, \ \nu_R = \sum_{\alpha \vdash n} st_R^{\alpha}(1^n)\zeta^{\alpha},$$

where

$$\lambda := (n_0, n_1 - n_0, \ldots, n_{k-1} - n_{k-2}, n - n_{k-1}),$$

if $R = \{n_0, \ldots, n_{k-1}\}, n_i < n_{i+1}, i \in k-1$.

Moreover, these decompositions show in which cases these characters are irreducible, namely, if and only if $st^{\alpha}(\lambda)$ or $st_R^{\alpha}(1^n)$ are zero except for one case, where they have to be 1. Cleary κ_R is irreducible if and only if $\lambda = (n)$, in which case κ_R is the identity character. The Lefschetz character ν_R is irreducible if and only if R is of the form$\{k, k+1, \ldots, n-2\}$, in which case it is a hook character:

7.9.14 Corollary *If S_n acts on the Boolean algebra of rank n, then the only irreducible κ_R is the identity character ($R = \emptyset$). The Lefschetz character ν_R is irreducible if and only if $R = \{k, k+1, \ldots, n-2\}$, in which case*

$$\nu_{\{k,\ldots,n-2\}} = \zeta^{(n-k+1,1^{k-1})}.$$

This completes our discussion of the action of the symmetric group on the Boolean algebra 2^n.

◇

A natural generalization of this example is the action of S_n on the set $m^n, m > 2$. In this case the chain characters κ_R turn out to be *sums of Young characters.* Another interesting example is the action of S_n on the set partitions of n. The corresponding characters κ_R are induced from products of plethysms $[r] \odot [s]$ (exercise 7.9.3). A quite different case is

7.9.15 Example The action of $GL_n(q)$ on the lattice $\mathcal{L}(n,q)$ of subspaces of $GF(q)^n$. This lattice is a graded poset, the rank $\rho(U)$ of a subspace U is its dimension. ◇

There is, of course, much more known about actions of groups on chains. The reader interested in more details is referred in particular to the corresponding papers by Stanley, Bjoerner, Welker, Sundaram.

Exercises

Exercise 7.9.1 Prove 7.9.2.

Exercise 7.9.2 Check 7.9.6.

Exercise 7.9.3 Evaluate the rank selected chain characters κ_R for the canonical action of the symmetric group S_n on the poset of set partitions of n.

8. Permutations

We begin with a consideration of multiply transitive actions. Numbers will be derived that allow directly to see from the cycle structure of the group elements if the action is multiply transitive or not. Afterwards we shall enumerate permutations with prescribed algebraic and combinatorial properties. We consider roots in finite groups, which means that we take a fixed natural number k and ask for the number of group elements x, the k-th power of which is equal to a given element g of the group G, $x^k = g$. The case when $g = 1$ is of particular interest. Then we restrict attention to the symmetric group, in order to derive expressions for the number of roots in terms of characters and to show how permutrizations can be applied. It will be shown that the function which maps a permutation onto the number of its k-th roots is in fact a proper character of the symmetric group in question.

We shall also examine combinatorial properties, in particular rises and falls of permutations, and we will introduce Foulkes characters for this purpose. It is in fact true that the enumeration theory of permutations by rises and falls is a very good link between combinatorics and representation theory. The character of the action of S_n on m^n will be decomposed into Foulkes characters, the dimensions of these Foulkes characters are the well known Eulerian numbers.

The final section of this chapter contains an introduction of a class of polynomials that correspond to permutations. These Schubert polynomials form a very important $\mathbb{Z}$–basis of the union of the polynomial rings $\mathbb{Z}[x_0, \ldots, x_{n-1}]$, they generalize the Schur polynomials, and they are the main tool used in the computer algebra system SYMMETRICA[1] for the examination and application of representation theory, invariant theory and combinatorics of symmetric groups.

8.1 Multiply Transitive Groups

Let P denote a subgroup of S_n. We derive first a few characterizations of multiple transitivity of P in terms of the cycle structure of its elements. Recall that, for a natural $k \leq n$, we denoted by $\binom{n}{k}$ the set of all the *k–subsets* of n. P acts canonically on this set as well as on n^k, the set of k–tuples, and on the set n^k_{inj}, which consists of all the injective k–tuples over n. The corresponding permutation groups on $\binom{n}{k}$, n^k, and n^k_{inj} will be denoted by $P^{[k]}$, $P^{(k)}$, and $P^{\langle k\rangle}$, respectively, and they

[1] available from http://www.mathe2.uni-bayreuth.de

will be called the *k–subsets group*, the *k–tuples group*, and the *injective k–tuples group*. The permutations corresponding to $\pi \in P$ will be indicated by $\pi^{[k]}$, $\pi^{(k)}$ and $\pi^{\langle k \rangle}$, respectively. The following result is very easy to check:

8.1.1 Lemma *The numbers of fixed points of $\pi \in S_n$ on $\binom{n}{k}$, n^k and n^k_{inj}, respectively, are*

$$a_1(\pi^{[k]}) = \sum_{b \dashv\vdash k} \prod_{i=1}^{k} \binom{a_i(\pi)}{b_i}, \; a_1(\pi^{(k)}) = a_1(\pi)^k, \; a_1(\pi^{\langle k \rangle}) = [a_1(\pi)]_k.$$

The permutation group P is, by definition, k–fold transitive if and only if $P^{\langle k \rangle}$ is transitive, and so in particular k–fold transitivity implies $(k-1)$–fold transitivity. The Cauchy–Frobenius Lemma yields the first characterization of k–fold transitivity:

8.1.2 Corollary *A subgroup $P \leq S_n$ is k–fold transitive if and only if*

$$\frac{1}{|P|} \sum_{\pi \in P} [a_1(\pi)]_k = 1.$$

This is a characterization in terms of the character of $P^{\langle k \rangle}$. The second will be in terms of the character of $P^{(k)}$, and it will be obtained by comparing the actions of $P^{(k)}$ and of ${S_n}^{(k)}$.

S_n is k–fold transitive, i. e. ${S_n}^{\langle k \rangle}$ is transitive. ${S_n}^{(k)}$ is intransitive if $k > 1$. Furthermore the k-tuples $(i_0, \ldots, i_{k-1}), (j_0, \ldots, j_{k-1}) \in n^k$ are in the same orbit of ${S_n}^{(k)}$ if and only if the following holds, for each μ and ν:

8.1.3 $$i_\mu = i_\nu \iff j_\mu = j_\nu.$$

Therefore the following is true:

8.1.4 Corollary *The group ${S_n}^{(k)}$ possesses as many orbits as there are partitions of the set k.*

Hence, in terms of the Bell numbers (cf. 2.4.17), we obtain:

8.1.5 $$|{S_n}^{(k)} \backslash\!\backslash n^k| = |S_n \backslash\!\backslash n^k| = B_k.$$

Thus $B_k = |P \backslash\!\backslash n^k|$ is equivalent to the fact that P has the *same* orbits on n^k as S_n. But if P has the same orbits as S_n on n^k, it has the same orbits on the subset n^k_{inj} as well. The converse is also true, as it is not difficult to see.

8.1.6 Corollary *A subgroup $P \leq S_n$ is k–fold transitive if and only if*

$$\frac{1}{|P|} \sum_{\pi \in P} a_1(\pi)^k = B_k.$$

The third characterization uses the character of $P^{[k]}$:

8.1.7 Lemma *A subgroup $P \leq S_n$ is k–fold transitive if and only if, for every choice of $b_1, \ldots, b_k \in \mathbb{N}$, we have*

$$\sum_i ib_i \leq k \Longrightarrow \frac{1}{|P|}\sum_{\pi\in P}\prod_i \binom{a_i(\pi)}{b_i} = \frac{1}{\prod_i i^{b_i} b_i!}.$$

Proof: i) If each such equation holds, then in particular

$$\frac{1}{|P|}\sum_{\pi\in P}\binom{a_1(\pi)}{k} = \frac{1}{k!},$$

so that P is k–fold transitive by 8.1.2.
ii) Now let P be k–fold transitive, and $b_i \in \mathbb{N}$ such that $\sum ib_i = k$. The expression

$$\sum_{\pi\in P}\binom{a_1(\pi)}{b_1}\cdots\binom{a_k(\pi)}{b_k}$$

is equal to the number of ways of picking from the elements $\pi \in P$ just b_1 1–cycles, b_2 2–cycles, ..., b_k k–cycles. Each such choice

$$\{(i_1), \ldots, (i_{b_1})\}, \{(i_{b_1+1}, i_{b_1+2}), \ldots, (i_{b_1+2b_2-1}, i_{b_1+2b_2})\}, \ldots$$

yields a k–tuple $(i_1, \ldots, i_k)$, and the expression

8.1.8 $$\Big(\prod_{i=1}^{k} i^{b_i} b_i!\Big)\Big(\sum_{\pi\in P}\binom{a_1(\pi)}{b_1}\cdots\binom{a_k(\pi)}{b_k}\Big)$$

is equal to the number of k–tuples which arise in this way, if each k–tuple is counted with its multiplicity. (Notice that, in order to form $(i_1, \ldots, i_k)$, we take first the chosen 1–cycles, then the chosen 2–cycles, and so on, respecting the order of the choices of 1–cycles, 2–cycles, etc., while from each chosen i–cycle we obtain i different i–tuples by cyclically permuting the points.)

For a given $(i_1, \ldots, i_k)$ there always exists a permutation $\pi \in P$ from which it arises by a suitable choice. If $(i_1, \ldots, i_k)$ arises from $\pi \in P$, then it arises exactly from the elements ρ in the left coset

$$\pi P_{\{i_1,\ldots,i_k\}}$$

of the stabilizer of the points $i_1, \ldots, i_k$. Hence $(i_1, \ldots, i_k)$ occurs $|P_{\{i_1,\ldots,i_k\}}|$ times. But all these stabilizers are conjugate subgroups, since P is k–fold transitive, so each k–tuple arises with the same multiplicity $|P_{\{i_1,\ldots,i_k\}}|$. Furthermore, by the k–fold transitivity of P, there are exactly $|P/P_{\{i_1,\ldots,i_k\}}|$ pairwise different k–tuples, and hence if each of them is counted with its multiplicity, there are $|P|$ of them. Thus 8.1.8 is equal to $|P|$, and this completes the proof, for k–fold transitivity implies $(k-1)$–fold transitivity. □

Here are a few easy cases:

8.1.9 Examples If P is 2–fold transitive, then

$$\frac{1}{|P|}\sum_{\pi\in P} a_2(\pi) = \frac{1}{2}.$$

If P is 3–fold transitive, then

$$\frac{1}{|P|}\sum_{\pi\in P} a_1(\pi)a_2(\pi) = \frac{1}{2}.$$

If P is 4–fold transitive, then both

$$\frac{1}{|P|}\sum_{\pi\in P} a_2(\pi)^2 = \frac{3}{4} \text{ and } \frac{1}{|P|}\sum a_1(\pi)^2 a_2(\pi) = 1.$$

◇

These examples show how we can get results on expressions of the form

8.1.10
$$\frac{1}{|P|}\sum_{\pi\in P} a_1(\pi)^{b_1}\cdots a_k(\pi)^{b_k}$$

recursively from 8.1.7 once P is $(\Sigma_i ib_i)$–fold transitive. In order to provide a direct approach, we shall define a matrix of combinatorial numbers in terms of which we can formulate *all* the results of this form. In order to do this we introduce, for each $i, k \in \mathbb{N}^*$, the number t_{ik} defined by

8.1.11
$$t_{ik} := \frac{i^k}{(i\cdot k)!}\sum_{\pi\in S_{i\cdot k}} a_i(\pi)^k,$$

and form the matrix

$$T := (t_{ik}).$$

This is a matrix with infinitely many rows and columns. Later on we shall prove that P is k–fold transitive if and only if $\sum ib_i \leq k$ implies that 8.1.10 is equal to the following expression:

8.1.12
$$\prod_{i:\, b_i>0} \frac{t_{ib_i}}{i^{b_i}}$$

But let us show first how T can be evaluated and that it is a matrix over $\mathbb{N}^*$. In order to do this we use exercise 2.4.7 which implies

8.1.13
$$\frac{1}{|P|}\sum_{\pi\in P} a_i(\pi)^k = \sum_{j=0}^{k} S(k,j)\frac{1}{|P|}\sum_{\pi\in P}[a_i(\pi)]_j.$$

8.1.14 Lemma *For each $i, k \in \mathbb{N}^*$ we have*

$$\frac{1}{(i\cdot k)!}\sum_{\pi\in S_{i\cdot k}} a_i(\pi)^k = \sum_{j=0}^{k}\frac{S(k,j)}{i^j},$$

so that in particular the following is true:

$$t_{ik} = \sum_{j=0}^{k} S(k, j) \cdot i^{k-j} \in \mathbb{N}^*.$$

Moreover, these numbers satisfy the recursion relation

$$t_{i,k+1} = \sum_{j=0}^{k} \binom{k}{j} i^{k-j} t_{ij}, \text{ if } t_{i0} := 1.$$

Proof: For $j \leq k$ the symmetric group $S_{i \cdot k}$ is $(i \cdot j)$–fold transitive, so that, by 8.1.7, we obtain

$$\frac{1}{(i \cdot k)!} \sum_{\pi \in S_{i \cdot k}} \binom{a_i(\pi)}{j} = \frac{1}{i^j j!},$$

which is in fact the same as

$$\frac{1}{(i \cdot k)!} \sum_{\pi} [a_i(\pi)]_j = \frac{1}{i^j}.$$

The first two statements now follow from 8.1.13. The recursion relation is easily obtained from the recursion for Stirling numbers

$$S(k+1, j) = \sum_{l=0}^{k} \binom{k}{l} S(l, j-1),$$

a recursion that is obvious from the definition of Stirling numbers of the second kind (exercise 2.4.8). □

This result shows how we can evaluate the coefficients of T. The upper left–hand corner of this matrix is

8.1.15
$$T = \begin{pmatrix} 1 & 2 & 5 & 15 & 52 & 203 & \dots \\ 1 & 3 & 11 & 49 & 257 & 1539 & \dots \\ 1 & 4 & 19 & 109 & 742 & 5815 & \dots \\ 1 & 5 & 29 & 201 & 1657 & 15821 & \dots \\ 1 & 6 & 41 & 331 & 3176 & 35451 & \dots \\ 1 & 7 & 55 & 505 & 5497 & 69823 & \dots \\ \vdots & \vdots & \vdots & \vdots & \vdots & \vdots & \end{pmatrix}$$

We notice that the first row of T contains the sequence of Bell numbers. Furthermore we are now in the position to prove the desired theorem which characterizes multiple transitivity in terms of the entries of T.

8.1.16 Theorem *A subgroup $P \leq S_n$ is k–fold transitive if and only if, for every choice of natural numbers b_i, the following holds:*

$$\sum i b_i \leq k \Longrightarrow \frac{1}{|P|} \sum_{\pi \in P} \prod_{i=1}^{k} a_i(\pi)^{b_i} = \prod_{i:\, b_i > 0} \frac{t_{i b_i}}{i^{b_i}}.$$

Proof: i) If P is k–fold transitive and $\sum i b_i \leq k$, then from 8.1.13 we get

$$\frac{1}{|P|} \sum_{\pi} \prod_{i=1}^{k} a_i(\pi)^{b_i} = \frac{1}{|P|} \sum_{\pi} \prod_{i=1}^{k} \sum_{j_i=0}^{b_i} S(b_i, j_i)[a_i(\pi)]_{j_i}$$

$$= \sum_{j_1,\ldots,j_k=0}^{b_1,\ldots,b_k} \left(\prod_{i=1}^{k} S(b_i, j_i) \right) \frac{1}{|P|} \sum_{\pi} \prod_{i} [a_i(\pi)]_{j_i},$$

which is (use 8.1.7) equal to

$$\sum_{j_1,\ldots,j_k} \left(\prod_{i} S(b_i, j_i) \right) \prod_{i} i^{-j_i} = \prod_{i} \sum_{j_i=0}^{b_i} \frac{S(b_i, j_i)}{i^{j_i}} = \prod_{i} \frac{t_{i b_i}}{i^{b_i}}.$$

ii) Conversely, suppose that $\sum i b_i \leq k$ implies that

$$\frac{1}{|P|} \sum_{\pi} \prod_{i} a_i(\pi)^{b_i} = \prod_{i} \frac{t_{i b_i}}{i^{b_i}}.$$

Then in particular

$$\frac{1}{|P|} \sum a_1(\pi)^k = t_{1k} = B_k,$$

and hence P is k–fold transitive by 8.1.6. □

The fact that the t_{ik} are nonnegative integers suggests to show that they can be interpreted as cardinalities, we are going to do this next. If $n \geq i \cdot k$, then S_n is k-fold transitive and so, by 8.1.16,

8.1.17
$$t_{ik} = \frac{1}{n!} \sum_{\pi \in S_n} (i \cdot a_i(\pi))^k.$$

This suggests, at first glance, to look for a combinatorial situation, where this number t_{ik} turns out to be the number of orbits of S_n. We shall construct such a situation, although a second glance shows that $\pi \mapsto (i \cdot a_i(\pi))^k$ is *not* the corresponding permutation character, so that the desired result will not be a direct application of the Cauchy–Frobenius Lemma. In fact this mapping *cannot* be a permutation character in general, which is easily seen by an application of the orthogonality relations: If we put $\chi(\pi) := n \cdot a_n(\pi)$, and write it as a linear combination of irreducible characters ζ^λ of the symmetric group S_n, say $\chi = \sum m_\lambda \zeta^\lambda$, then a multiplication of

$\chi_\alpha = \sum m_\lambda \zeta_\alpha^\lambda$, the value of χ on the conjugacy class C^α, by $|C^\alpha|\zeta_\alpha^\mu$ and summation over all the α gives the following result:

8.1.18
$$\chi = n \cdot a_n = \sum_i (-1)^i \zeta^{(n-i,1^i)}.$$

Thus at least $\pi \mapsto n \cdot a_n(\pi)$ is not in general a permutation character.

But we can use another clever argument (due to M. Klemm and B. Wagner) to show that the t_{ik} are equal to certain cardinalities. Consider r-tuples $b = (b_1, \ldots, b_r) \in \mathbb{N}^r$, a subgroup $P \leq S_n$, and the corresponding number

$$t_b(P) := \frac{1}{|P|} \sum_{\pi \in P} \prod_{i=1}^r (i \cdot a_i(\pi))^{b_i},$$

so that in particular

8.1.19
$$t_{ik} = t_b(S_n), \text{ for } b := (0, \ldots, 0, \underbrace{k}_{i-th\ entry}, 0, \ldots, 0).$$

Putting $s := \sum_{i=1}^r i \cdot b_i$ we can write the elements $f \in n^s$ as r-tuples $f = (f_1, \ldots, f_r)$ of $(i \times b_i)$-matrices f_i which contain the values of f in their columns of lengths $i, i \in \underline{r}$:

$$f_i := (f_{jk}^i) := \begin{pmatrix} f(\sum_{\nu=1}^{i-1} \nu \cdot b_\nu + 1) & f(\sum_{\nu=1}^{i-1} \nu \cdot b_\nu + i + 1) & \ldots \\ f(\sum_{\nu=1}^{i-1} \nu \cdot b_\nu + 2) & f(\sum_{\nu=1}^{i-1} \nu \cdot b_\nu + i + 2) & \ldots \\ \vdots & \vdots & \\ f(\sum_{\nu=1}^{i-1} \nu \cdot b_\nu + i) & f(\sum_{\nu=1}^{i-1} \nu \cdot b_\nu + 2i) & \ldots \end{pmatrix}.$$

Using this notation for the elements of n^s we build the cartesian product $P \times n^s$ and consider its subset

$$M_b(P) := \{(\pi, f) \mid \text{ columns of each } f_i \text{ are } i\text{-cycles of } \pi \}.$$

The projection onto the second component

$$M'_b(P) := \{f \in n^s \mid \exists \pi \in P\colon (\pi, f) \in M_b(P)\}$$

is a set on which $P \leq S_n$ acts in a natural way:

$$P \times M'_b(P) \to M'_b\colon (\pi, f) \mapsto \pi \circ f = (\pi f_1, \ldots, \pi f_r), \text{ where } \pi f_i := (\pi f_{jk}^i).$$

8.1.20 Theorem (Klemm/Wagner) *For each $P \leq S_n$ we have*

- $t_b(P) = |P \backslash\backslash M'_b(P)|$, *and*
- *if in addition P is s-fold transitive, $s = \sum i \cdot b_i$, then*

$$t_b(P) = \prod_{i=1}^r t_{i,b_i}.$$

Proof: We count the elements of $M_b(P)$ in two different ways: First of all we note that clearly

$$|M_b(P)| = \sum_{\pi \in P} \prod_{i=1}^{r} (i \cdot a_i(\pi))^{b_i}.$$

Secondly we note that $f \in n^s$ contributes to $M_b(P)$ either 0 pairs (π, f) or exactly $|P_f|$ (=order of stabilizer) pairs. Thus, if $f, \tilde{f} \in M_b'(P)$ belong to the same orbit, then their contributions to $M_b(P)$ are of the same size: $|P_f| = |P_{\tilde{f}}|$. Hence we obtain

$$|M_b(P)| = \sum_{f \in T \in \mathcal{T}(P \backslash\backslash M_b'(P))} |P_f| \cdot |P(f)| = |P| \cdot |P \backslash\backslash M_b'(P)|,$$

which proves the first statement. The second statement follows directly from 8.1.16. □

As $t_{ik} = t_b(S_n)$, for $b := (0, \ldots, 0, k, 0, \ldots, 0)$, we obtain the consequence

8.1.21 $$t_{ik} = |S_n \backslash\backslash M_b'(S_n)|, \text{ if } b = (0, \ldots, 0, \underbrace{k}_{i-th\ entry}, 0, \ldots, 0).$$

But two elements $f, \tilde{f} \in M_b'(S_n)$ belong to the same orbit of S_n if and only if they are constant on the same subsets of $\underline{i} \times k$. Since these matrices have i-cycles of permutations as columns, we obtain

8.1.22 Corollary *The natural number t_{ik} is equal to the number of equivalence relations $R \subseteq \underline{i} \times k$ such that $(j, k)R(j', k')$ implies the existence of a permutation $\rho \in \langle(1 \ldots i)\rangle$ such that for each $j_1, j_2 \in \underline{i}$ we have*

$$(j_1, k)R(j_2, k') \iff j_2 = \rho j_1.$$

It is a reasonable guess that some of these results can be reformulated in terms of characters, in order to emphasize the representation theoretical aspect. The permutation group $S_n{}^{\langle k \rangle}$ which is induced by S_n on the set n_{inj}^k of injective k–tuples over n is a transitive permutation representation of S_n, and hence it is induced from the identity representation of the stabilizer of any such tuple. The stabilizer of the particular k–tuple $(n - k + 1, \ldots, n)$, for example, is the Young subgroup

$$S_{(n-k,1^k)} := S_{n-k} \oplus S_1 \oplus \ldots \oplus S_1$$

of S_n. Hence $S_n{}^{\langle k \rangle}$ has the same character as

$$IS_{(n-k,1^k)} \uparrow S_n = [n-k][1] \cdots [1] =: [n-k][1]^k.$$

Thus $P^{\langle k \rangle}$ has the same character as has the following representation of P:

$$IS_{(n-k,1^k)} \uparrow S_n \downarrow P = [n-k][1]^k \downarrow P.$$

Since the number of orbits of a permutation representation equals the multiplicity of the identity representation in that permutation representation, we have the following equivalences:

8.1.23 Lemma *The subgroup $P \leq S_n$ is k–fold transitive if and only if*

$$1 = (IS_{(n-k,1^k)} \uparrow S_n \downarrow P, IP),$$

i. e. if and only if

$$1 = (IS_{(n-k,1^k)} \uparrow S_n, IP \uparrow S_n),$$

or, equivalently, if and only if $IS_n = [n]$ is the only irreducible constituent which $IS_{(n-k,1^k)} \uparrow S_n$ and $IP \uparrow S_n$ have in common.

The last of these characterizations leads us to the concept of *depth* of the partition $\alpha \vdash n$, defined by

$$d_\alpha := n - \alpha_0.$$

The following lemma gives an estimate for the depth of the constituents of $IP \uparrow S_n$, P being multiply transitive.

8.1.24 Lemma *A subgroup $P \leq S_n$ is k–fold transitive if and only if $[n]$ is the only constituent $[\alpha]$ of depth $d_\alpha \leq k$ which is contained in $IP \uparrow S_n$.*

Proof: The Littlewood–Richardson Rule shows that the irreducible constituents of $[n-k][1^k]$ are just the $[\alpha]$ of depth $d_\alpha \leq k$. □

Since the Kostka matrix K_n of S_n is upper triangular (cf. 6.1.22), the Young characters ξ^α of depth $d_\alpha \leq k$ form a basis of the space generated by the ζ^α with this same estimate (exercise 8.1.2):

8.1.25 $$V_{n,k} := \ll \zeta^\alpha \mid d_\alpha \leq k \gg_{\mathbb{C}} = \ll \xi^\alpha \mid d_\alpha \leq k \gg_{\mathbb{C}} .$$

Thus by linearity we get

8.1.26 Corollary *A subgroup $P \leq S_n$ is k–fold transitive if and only if, for each $\chi \in V_{n,k}$, we have*

$$[\chi^{IP\uparrow S_n} \mid \chi] = [\xi^{(n)} \mid \chi],$$

or, equivalently, if and only if, for each $\alpha \vdash n$ such that $d_\alpha \leq k$,

$$[\chi^{IP\uparrow S_n} \mid \xi^\alpha] = [\xi^{(n)} \mid \xi^\alpha] = 1.$$

Exercises

Exercise 8.1.1 Use 8.1.1 in order to derive a fromula for the number of graphs on v vertices.

Exercise 8.1.2 Prove 8.1.25.

8.2 Root Number Functions

We want to enumerate roots in finite groups, later on we shall restrict attention to roots of permutations. In order to derive a few results concerning this problem we indicate the set and the number of k-th *roots* of $g \in G$ as follows:

$$R_k(g) := \{x \in G \mid x^k = g\},\ r_k(g) := |R_k(g)|.$$

The corresponding power map

$$p_k\colon G \to G\colon g \mapsto g^k$$

induces the following set partition of G which consists of the inverse images:

$$G = \bigcup_{g \in G} R_k(g) = \bigcup_{g \in G} p_k^{-1}(g).$$

The mapping $r_k\colon G \to \mathbb{N}\colon g \mapsto r_k(g)$, let us call it the k-th *root number function*, is clearly a class function, moreover the following holds:

8.2.1 Lemma *The k-th root number function r_k is a $\mathbb{Z}$–linear combination of the irreducible characters of G, and hence a generalized character:*

$$r_k = \sum_i c_{i,k}\zeta^i, \ \text{and}\ c_{i,k} = \frac{1}{|G|}\sum_{g \in G}\zeta^i(g^k) = \chi^{D_i \Delta_k D_0}((0,\ldots,k-1)) \in \mathbb{Z}.$$

Proof: As

$$\sum_i c_{i,k}\zeta^i(g) = \frac{1}{|G|}\sum_{x \in G}\sum_i \zeta^i(x^k)\zeta^i(g),$$

an application of the second orthogonality relation 11.5.7 yields that this is just $r_k(g)$, the rest of the statement follows from 7.7.17. □

In order to examine the case $k = 2$ more closely we derive

8.2.2 Lemma *For each finite group G and any i we have that*

$$c_i := c_{i,2} = \frac{1}{|G|}\sum_{g \in G}\zeta^i(g^2) \in \{0, 1, -1\},$$

and $c_i \neq 0$ if and only if the corresponding character ζ^i is real–valued.

Proof: $D_i \Delta_2 D_0$ is either zero, in which case $c_{i,2} = 0$, or it is a onedimensional representation of S_2 and $c_{i,2} \in \{1, -1\}$. Moreover, $D_i \Delta_2 D_0$ is nonzero if and only if $\otimes^2 D_i$ contains the identity representation D_0, i. e. if and only if $1 = [\zeta^i \mid \overline{\zeta^i}]$, which completes the proof. □

This yields, for numbers of square roots of elements in G:

8.2.3 $$r_2(g) = \sum_{c_i=1} \zeta^i(g) - \sum_{c_j=-1} \zeta^j(g), \text{ and } r_2(1) = \sum_{c_i=1} f^i - \sum_{c_j=-1} f^j.$$

In fact it can be shown that $c_{i,2} = 1$ if and only if D_i can be realized over $\mathbb{R}$. Hence, in the case when all the ordinary irreducible representations of G can be realized over $\mathbb{R}$, then $r_2(g) = \sum_i \zeta^i(g)$, the sum of the elements in the column of the character table of G corresponding to the conjugacy class of g. In other words, r_2 is the *model character* in this particular case. An example is the symmetric group. We have not shown yet that each of its irreducible representation is realizable over $\mathbb{R}$, but we can obtain that each $c_{\alpha,2} = 1$ also from the Robinson–Schensted Construction in connection with 8.2.3. The Robinson–Schensted construction, which maps the permutation $\pi \in S_n$ onto the pair (T_0, T_1) of standard Young tableaux consisting of n entries, has the property that π^{-1} is mapped onto the pair (T_1, T_0). This implies that the number of elements of order ≤ 2 in S_n is equal to the total number of standard Young tableaux consisting of n entries, and hence, by 8.2.3, each coefficient $c_{\alpha,2}$ of an irreducible dimension f^α (which is the number of standard Young tableaux with diagram $[\alpha]$) must be positive. Putting, for $G := S_n$ and $\alpha \vdash n$,

$$r_k(\alpha) := \text{no. of } k\text{-th roots of an element with cycle partition } \alpha,$$

we have obtained:

8.2.4 Corollary *The number of square roots is the model character of the symmetric group S_n:*

$$r_2(\alpha) = \sum_{\beta \vdash n} \zeta^\beta_\alpha, \text{ in short: } r_2 = \sum_{\beta \vdash n} \zeta^\beta.$$

(Cf. 7.8.5 for the corresponding representation.)

This result shows in particular that r_2 is a *proper character* of the symmetric group S_n, which means that r_2 is an $\mathbb{N}$–linear combination of the irreducible characters $\zeta^\alpha, \alpha \vdash n$. But there are groups for which r_2 is *not* proper. Equation 8.2.3 shows that r_2 is not proper if and only if there exist irreducible characters ζ^i for which $c_{i,2} = -1$, and it was mentioned before that this holds if and only if ζ^i is real–valued, while D_i cannot be realized over $\mathbb{R}$. Here is another sufficient condition which has the advantage neither to use characters nor matrix representations:

8.2.5 Lemma *If the finite group G contains elements g for which $r_2(g) > r_2(1)$, then r_2 is not a proper character, or, equivalently, there exist ordinary irreducible characters ζ^i of G, for which $c_{i,2} = -1$.*

Proof: Indirectly. If $r_2 = \sum c_{i,2}\zeta^i$ is proper, then we obtain, using the first item of 11.5.1, that for each $g \in G$

$$r_2(g) \leq \sum_i c_{i,2}|\zeta^i(g)| \leq \sum_i c_{i,2} f^i = r_2(1).$$

□

The quaternion group $Q_8 = \{\pm 1, \pm j, \pm k, \pm l\}$ is an example since -1 has the six square roots $\pm j$, $\pm k$ and $\pm l$, while the identity element has the two square roots ± 1 only. Various calculations have shown that there exist many pairs (k, n) such that $k > 2$ and r_k is a proper character of S_n. It therefore was a reasonable guess that this holds for *each pair* (k, n) of natural numbers. This conjecture was proved by Th. Scharf, as it will be described next.

We introduce particular characters of centralizers. Recall from 2.2.3 that the centralizer of an element $\sigma \in S_n$ of type a is conjugate to a direct sum of plethysms. We abbreviate the centralizer of an element σ by $C(\sigma)$ and the conjugate direct sum of plethysms by $C(a)$, where a denotes the cycle type of σ, so that 2.2.3 gives

8.2.6 $$C(\sigma) \simeq C(a) := \oplus_i (C_i \odot S_{a_i}) \simeq \times_i \left(C_i \wr S_{a_i}\right).$$

The permutation σ is a k-th root of the identity element if and only if each cycle length of σ divides k. We therefore abbreviate cycle types of this particular form by writing

$$a \mathrel{\vdash\!\!\!\dashv}_k n :\Longleftrightarrow [a_i > 0 \Rightarrow i \mid k].$$

Now we pick a primitive k-th root of unity ϵ and map the cyclic factors σ_i of $\sigma = \sigma_0 \ldots \sigma_{s-1}$ onto powers of ϵ as follows:

8.2.7 $$\chi_\sigma \colon \sigma_i \mapsto \epsilon^{k/l_i},$$

where $l_i = |\langle \sigma_i \rangle|$, the length of the cyclic factor. This mapping trivially extends to a onedimensional character χ_σ of the centralizer of σ. Now we assume that we have chosen, for each cycle type $a \mathrel{\vdash\!\!\!\dashv}_k n$, a representative $\sigma(a)$ of the conjugacy class C^a, and that $\chi_{\sigma(a)}$ is constructed according to 8.2.7. Consider the induced characters

$$\chi^{(a)} := \chi_{\sigma(a)} \uparrow S_n,$$

and take the sum of all of them:

$$\chi_n^k := \sum_{a \mathrel{\vdash\!\!\!\dashv}_k n} \chi^{(a)}.$$

It is our aim to show that in fact $\chi_n^k = r_k$, the k-th root number function.

8.2.8 Lemma *If we denote by C_n^k the union of the conjugacy classes consisting of k-th roots of the identity element of S_n:*

$$C_n^k := \bigcup_{a \mathrel{\vdash\!\!\!\dashv}_k n} C^a,$$

then we have, for each $\pi \in S_n$, that

$$\chi_n^k(\pi) = \sum_\tau \chi_\tau(\pi),$$

if the sum is taken over all the $\tau \in C_n^k$ which are contained in the centralizer of π.

Proof: Consider the chosen representative $\sigma(a) \in C^a$, $a \vdash_k n$, and the decomposition of S_n into the left cosets of its centralizer $C_n(\sigma(a))$:

$$S_n = \bigcup_i \tau_i C_n(\sigma(a)).$$

By definition of $\chi^{(a)}$ we have (cf. 11.5.8)

$$\chi^{(a)}(\pi) = \sum_i \dot{\chi}_{\sigma(a)}(\tau_i^{-1}\pi\tau_i) = \sum_i \chi_{\sigma(a)}(\tau_i^{-1}\pi\tau_i),$$

if the last sum is taken over all the i such that $\tau_i^{-1}\pi\tau_i \in C_n(\sigma(a))$ or, equivalently, $\pi \in C_n(\tau_i\sigma(a)\tau_i^{-1})$. Now we obtain from the definition of the τ_i that to each $\rho \in C^a \ni \sigma(a)$ there corresponds a unique index $i = i(\rho)$ such that $\rho = \tau_{i(\rho)}\sigma(a)\tau_{i(\rho)}^{-1}$. Hence we can replace the above sum over certain i by the sum over the elements $\rho \in C^a$, for which $\pi \in C_n(\rho)$, obtaining in this way that

$$\chi^{(a)}(\pi) = \sum_\rho \chi_{\sigma(a)}(\tau_{i(\rho)}^{-1}\pi\tau_{i(\rho)}) = \sum_\rho \chi_\rho(\pi).$$

The last equation follows from the fact that $\pi \mapsto \tau_{i(\rho)}^{-1}\pi\tau_{i(\rho)}$ is a bijection between the centralizer of $\tau_{i(\rho)}\sigma(a)\tau_{i(\rho)}^{-1}$ and the centralizer of π, that $\sigma(a)$ is mapped onto ρ by conjugation with $\tau_{i(\rho)}$, and from the definition of $\chi_{\sigma(a)}$. This completes the proof by taking the sum over all the conjugacy classes C^a contained in C_n^k. □

Our next aim is to show that we may restrict attention to conjugacy classes corresponding to rectangular partitions $\alpha = (d^r)$. In order to prove this we notice that the following holds:

8.2.9 Lemma *Assume that $\pi \in S_n$ contains exactly r d–cycles which form the factor π_0 of $\pi = \pi_0\pi_1$. Then*

$$\chi_n^k(\pi) = \chi_{rd}^k(\pi_0)\chi_{n-rd}^k(\pi_1).$$

Proof: Without loss of generality we can assume that

$$\pi_0 = (0, \ldots, d-1)(d, \ldots, 2d-1)\cdots((r-1)d, \ldots, rd-1).$$

From 8.2.6 we obtain for the centralizer of π that

$$C(\pi) = (C_d \odot S_r) \oplus \ldots,$$

so that each $\tau \in C(\pi)$ is of the form $\tau = \tau_0\tau_1$, where the subset $n \backslash rd$ consists of fixed points of τ_0. Hence, as $\tau^k = 1$ if and only if each $\tau_i^k = 1$,

$$\chi_n^k(\pi) = \sum_\tau \chi_\tau(\pi) = \sum_{(\tau_0,\tau_1)} \chi_{\tau_0}(\pi_0)\chi_{\tau_1}(\pi_1)$$

$$= \sum_{\tau_0} \chi_{\tau_0}(\pi_0) \sum_{\tau_1} \chi_{\tau_1}(\pi_1) = \chi_{rd}^k(\pi_0)\chi_{n-rd}^k(\pi_1).$$

□

This multiplicativity of χ_n^k allows us to restrict attention to elements π with rectangular cycle partition $\alpha(\pi) = (d^r)$, assuming now without loss of generality that

$$\pi = \underbrace{(0, \ldots, d-1)}_{=:\rho_0} \underbrace{(d, \ldots, 2d-1)}_{=:\rho_1} \cdots \underbrace{((r-1)d, \ldots, rd-1)}_{=:\rho_{r-1}}.$$

Consider the canonical epimorphism Φ from $C(\pi) = C_d \odot S_r \simeq C_d \wr S_r$ onto S_r which maps τ onto the permutation $\hat{\tau}$, defined by

8.2.10
$$\tau \rho_i \tau^{-1} = \rho_{\hat{\tau} i}.$$

Since $\tau \mapsto \hat{\tau}$ is homomorphic, $\tau \in C_n^k$ implies $\hat{\tau} \in C_r^k$. The surjectivity of this map yields that for each $\sigma \in C_r^k$ there exist $\tau \in C(\pi)$ such that $\hat{\tau} = \sigma$. Moreover, by homomorphy, we can assume that τ lies in C_n^k. This gives the following decomposition into inverse images:

8.2.11
$$C_n^k \cap C(\pi) = \bigcup_{b \vdash\!\dashv_k r} \bigcup_{\sigma \in C^b} C_n^k \cap \Phi^{-1}(\sigma).$$

Now we note that the inverse images of conjugate σ' of σ are bijective: For σ, σ' in C^b there exist $\rho \in S_r$ such that $\rho \sigma' \rho^{-1} = \sigma = \hat{\tau}$, and each such ρ has an inverse image $\tilde{\rho}$ in $C(\pi)$, so that the following mapping is a bijection:

$$\phi : C_n^k \cap \Phi^{-1}(\sigma) \to C_n^k \cap \Phi^{-1}(\sigma') : \tau \mapsto \tilde{\rho}^{-1} \tau \tilde{\rho}.$$

Since ϕ means conjugation by an element in the centralizer $C(\pi)$, we have that $\chi_{\phi(\tau)}(\pi) = \chi_\tau(\pi)$, and hence we can derive from 8.2.11:

8.2.12 Corollary *The character χ_n^k satisfies the equations*

$$\chi_n^k(\pi) = \sum_\tau \chi_\tau(\pi) = \sum_{b \vdash\!\dashv_k r} |C^b| \sum_{\tau :\, \hat{\tau} = \hat{\tau}(b)} \chi_\tau(\pi)$$

$$= \sum_{b \vdash\!\dashv_k r} \frac{r!}{\prod_i i^{b_i} b_i!} \sum_{\tau :\, \hat{\tau} = \hat{\tau}(b)} \chi_{\hat{\tau}(b)}(\pi),$$

where $\hat{\tau} = \hat{\tau}(b)$, the representative of the conjugacy class C^b of S_r.

Let us simplify the last sum by considering the $\tau \in C_n^k \cap C(\pi)$. From the cycle decomposition $\prod_j \hat{\tau}(b)_j$ of $\hat{\tau}(b)$ we obtain the orbits of $\langle \tau, \pi \rangle$ on n, since the elements of the cycle $\hat{\tau}(b)_j$ are the *numbers* i of the cyclic factors ρ_i of π that form an orbit under τ via conjugation (cf. 8.2.10). Let us denote by ω_j this orbit which is associated with $\hat{\tau}(b)_j$, and let us indicate the restrictions accordingly:

$$\tau_j := \tau \downarrow \omega_j, \ \pi_j := \pi \downarrow \omega_j.$$

As π consists of d–cycles only, also π_j is a product of cyles of this length. Moreover to the restrictions τ_j to subsets of n there correspond the restrictions $\hat{\tau}_j$ of $\hat{\tau}$ to the according subsets of r. Hence we introduce

$$M_j := \{\tau_j \in C^k_{|\omega_j|} \cap C(\pi_j) \mid \hat{\tau}_j = \hat{\tau}(b)_j\}.$$

It is clear that

$$\{\tau \mid \hat{\tau} = \hat{\tau}(b)\} \to \times_j M_j \colon \tau \mapsto (\dots, \tau_j, \dots)$$

is a bijection and that $\chi_\tau(\pi) = \prod_j \chi_{\tau_j}(\pi_j)$. This implies

8.2.13 $$\sum_{\tau\colon \hat{\tau}=\hat{\tau}(b)} \chi_\tau(\pi) = \sum_{(\tau_j)\in\times_j M_j} \prod_j \chi_{\tau_j}(\pi_j) = \prod_j \sum_{\tau_j \in M_j} \chi_{\tau_j}(\pi_j).$$

We want to replace the last sum. Without loss of generality we can assume that π_j consists of the first t_j cyclic factors of π, i. e. $\pi_j = \rho_0 \cdots \rho_{t_j-1}$. Moreover we can assume that the representative $\hat{\tau}(b)$ of C^b was chosen in such a way that its factor $\hat{\tau}(b)_j$ is equal to the cyclic permutation $(0, \dots, t_j - 1)$. Hence M_j can be rewritten in the following way:

$$M_j := \{\tau_j \in C^k_{t_j d} \cap C(\pi_j) \mid \forall\, i < t_j - 1\colon\ \tau_j \rho_i \tau_j^{-1} = \rho_{i+1}\}.$$

This set can easily be decomposed by picking elements $z_0, \dots, z_{t_j-1}$ from the cyclic factors $\rho_0, \dots, \rho_{t_j-1}$ and putting

$$M_j(z_0, \dots, z_{t_j-1}) := \{\tau_j \in M_j \mid \forall\, i < t_j - 1\colon\ \tau_j z_i = z_{i+1}\}.$$

Since $\rho_0 = (0, \dots, d-1)$, we obtain a set partition of M_j for each $z \in d$, arising from the possible choices of the elements $z_1, \dots, z_{t_j-1}$:

$$M_j = \bigcup_{(z_1,\dots,z_{t_j-1})} M_j(z, z_1, \dots, z_{t_j-1}).$$

Moreover, the blocks $M_j(z, z_1, \dots, z_{t_j-1})$ are conjugates in $C_n(\pi_j)$, since, for each $(z_0, \dots, z_{t_j-1})$, there exist $\sigma \in C(\pi_j)$ such that

$$M_j(0, d, \dots, (t_j - 1)d) = M_j(\sigma z_0, \dots, \sigma z_{t_j-1}) = \sigma M_j(z_0, \dots, z_{t_j-1})\sigma^{-1}.$$

We have therefore proved

8.2.14 $$\sum_{\tau_j \in M_j} \chi_{\tau_j}(\pi) = d^{t_j-1} \sum_{\tau_j \in M_j(0,\dots,(t_j-1)d)} \chi_{\tau_j}(\pi_j).$$

This, together with 8.2.12 and 8.2.13, yields

8.2.15 Corollary *The character χ^k_n has the following values:*

$$\chi^k_n(\pi) = \sum_{b \dashv\!\dashv_k r} \frac{r!}{\prod_i i^{b_i} b_i!} d^{r-c(b)} \prod_j \sum_{\tau_j \in M_j(0,d,\dots,(t_j-1)d)} \chi_{\tau_j}(\pi_j),$$

where $c(b) := \sum b_i$.

Recall the definition of $M_j(0, d, \ldots, (t_j - 1)d)$. Each of its elements is uniquely determined by the image of the point $(t_j - 1)d$, say $\tau_j((t_j - 1)d) = \pi^i 0$. We want to characterize these exponents. In terms of wreath product notation, this particular τ_j is of the form (ψ, σ), where σ is the t_j–cycle $(0, d, \ldots, (t_j - 1)d)$, and $\psi(0)$ is the i-th power of a d–cycle of π. Thus, by 11.2.12 and 2.2.7, we obtain that τ_j consists of $\gcd(i, d)$ cyclic factors, each of which is of length $t_j d / \gcd(i, d)$. Furthermore τ_j is a k–th root of unity, so that the lengths of its cyclic factors divide k or, equivalently (as t_j divides k, since $|\omega_j| = t_j d$),

8.2.16 $$d / \gcd(i, d) \text{ divides } k / t_j.$$

If we put $d = d_0 d_1$, where $d_1 = \gcd(d, k/t_j)$, then 8.2.16 is equivalent to

$$i = i_0 d_0 \text{ and } d_1 / \gcd(d_1, i_0) \text{ divides } k / t_j,$$

which yields the equivalence

8.2.17 $$\tau_j \in M_j \Longleftrightarrow i = i_0 d_0, \text{ where } i_0 \in d_1.$$

In order to evaluate the desired sum $\sum_{\tau_j \in M_j} \chi_{\tau_j}(\pi_j)$, we fix $\tau_j \in M_j$, which in turn is characterized by a certain i_0 according to the equation

$$\tau_j((t_j - 1)d) = \pi_j^{i_0 d_0} 0.$$

In order to apply the definition of χ_{τ_j} we have to find a decomposition $\pi_j = \rho_j \sigma_j$ such that ρ_j is a product of certain powers of the cyclic factors of τ_j and σ_j lies in the complement of this base group.

In order to visualize this decomposition of τ_j we consider the following array which contains in its rows the cyclic factors of π_j. Without loss of generality we can assume that it looks as follows:

0	$\pi_j 0 = 1$	$\ldots$	$\pi_j^{d-1} 0 = d - 1$
d	$d + 1$	$\ldots$	$2d - 1$
$\vdots$			$\vdots$
$(t_j - 1)d$	$(t_j - 1)d + 1$	$\ldots$	$t_j d - 1$

Note that τ_j consists of cyclic factors of the same length and that each of them contains a union of columns in the above array: The order in which these columns have to be put together can be obtained from the wreath product notation which was already mentioned: $\tau_j = (\psi, \sigma)$ and $(\psi, \sigma) = (\psi(0), \ldots, \psi(t_j - 1); \sigma)$ where the $\psi(k)$ and σ are as follows:

$$(\underbrace{\psi_j^i, 1, \ldots, 1}_{\psi}; \underbrace{(0, \ldots, t_j - 1)}_{\sigma}).$$

If $\tau_{j,0}$, the 0-th cyclic factor of τ_j is the one containing the 0-th column, say, then it is therefore of the form

$$\tau_{j,0} = (\underbrace{0, d, \ldots, (t_j - 1)d}_{\text{0-th column}}, \underbrace{\pi_j^i 0, d + \pi_j^i 0, \ldots, (t_j - 1)d + \pi_j^i 0}_{\text{1-th column}}, \ldots).$$

Thus the order in which the columns are put together is determined by the numbers $n_l := l \cdot i = l \cdot i_0 \cdot d_0$ (and their residue classes modulo d). If there exist further cyclic factors of τ_j, i. e. if $i' = \gcd(i, d) = d_0 \gcd(i_0, d_1) > 1$, then they are obtained from $\tau_{j,0}$ via conjugation by powers of π_j. The next cyclic factor $\tau_{j,1}$, for example, is

$$\tau_{j,1} = (\underbrace{1, d+1, \ldots, (t_j - 1)d + 1}_{\text{0-th column}}, \underbrace{\pi_j^i 1, d + \pi_j^i 1, \ldots, (t_j - 1)d + \pi_j^i 1}_{\text{next column}}, \ldots).$$

Moreover, the desired σ_j is the permutation which performs this conjugation operation. This shows that σ_j is the following product of cyclic factors:

$$(0, 1, \ldots, i' - 1)(i', i' + 1, \ldots, 2i' - 1) \ldots (\ldots, d - 1)$$
$$\cdot (d, \ldots, d + i' - 1)(d + i', d + i' + 1, \ldots, d + 2i' - 1) \ldots (\ldots, 2d - 1).$$

This allows us to evaluate $\rho_j = \pi_j \sigma_j^{-1}$. For σ_j is defined as follows (note that, as d divides n and i' divides d, each element of n has a unique representation $fd+gi'+h$ with $0 \le h \le i' - 1, 0 \le g \le d/i' - 1, 0 \le f \le t_j - 1$):

$$\sigma_j(fd + gi' + h) = \begin{cases} fd + gi' + h + 1, & \text{if } h < i' - 1, \\ fd + gi', & \text{else,} \end{cases}$$

from which it follows that $\sigma(fd + gi' + h) = \pi(fd + gi' + h)$, if $h < i' - 1$. Thus the points $fd + gi' + h, h > 0$, are fixed points of ρ_j. For $fd + gi'$ we obtain

$$\rho_j(fd + gi') = \begin{cases} fd + gi' + i', & \text{if } gi' + i' < d - 1, \\ fd, & \text{else.} \end{cases}$$

This means that, for each f, the points $fd + gi'$ belong to one d/i'-cycle

$$(fd, fd + i', \ldots, (fd + (d - 1)i'),$$

and, omitting fixed points, ρ_j is the product of these t_j cycles:

$$(0, i', \ldots, (d-1)i')(d, d+i', \ldots, d+(d-1)i')(2d, 2d+i', \ldots, 2d+(d-1)i') \cdots$$

We have to rewrite ρ_j as a power of τ_j in order to evaluate $\chi_{\tau_j}(\pi_j) = \chi_{\tau_j}(\rho_j)$. To make things clear it is useful to consider an example:

8.2.18 Example Let $\pi_j := (1, 2, 3, 4, 5, 6)(7, 8, 9, 10, 11, 12)$. The corresponding picture is

0	1	2	3	4	5
6	7	8	9	10	11

We choose $\tau_j = (0, 6, 4, 10, 2, 8)(1, 7, 5, 11, 3, 9)$. It is immediate that

$$\sigma_j = (0, 1)(2, 3)(4, 5)(6, 7)(8, 9)(10, 11), \ \rho_j = (0, 2, 4)(6, 8, 10).$$

ρ_j contains only the entries of the cyclic factor $(0, 6, 4, 10, 2, 8)$ of τ_j. If we take the t_j-th power of this cycle, the resulting permutation acts only on the rows: $(0, 6, 4, 10, 2, 8)^2 = (0, 4, 2)(6, 10, 8)$. Note that this permutation and also ρ_j are elements of the same order in the cyclic group generated by π_j^2 and thus ρ_j is a certain power (in this case the square) of $(0, 4, 2)(6, 10, 8)$. ◇

Indeed, in the general case, ρ_j is a certain power of $\tau_{j,0}$ of order pt_j , where $p \in d_1/i'_0$ is the unique solution of the equation

$$p \cdot (i_0/i'_0) \equiv 1 \text{ modulo } (d_1/i'_0).$$

Note that by definition of i'_0 we have $\gcd(d_1/i'_0, i_0/i'_0) = 1$, thus a bijection from $\{j \in d_1 \mid \gcd(j, d_1) = i'_0\}$ to the set of units of the multiplicative group of $\mathbb{Z}/(d_1/i'_0)\mathbb{Z}$. From this it follows that, if i_0/i'_0 runs through the the elements of the set of units, the corresponding solution p also does. And, applying the bijection in the other direction, the set of the elements $i'_0 p$ is precisely $\{j \in d_1 \mid \gcd(j, d_1) = i'_0\}$. We can therefore deduce that

$$\chi_{\tau_j}(\pi_j) = \chi_{\tau_j}(\rho_j)\chi_{\tau_j}(\sigma_j) = \chi_{\tau_j}(\tau_{j,1})^{pt_j} = \epsilon^{(k/(t_j d/(i'_0 d_0)))pt_j} = (\epsilon^{k/d_1})^{i'_0 p},$$

obtaining

$$\sum_{\tau_j \in M_j} \chi_{\tau_j}(\pi_j) = \sum_{i_0=1}^{d_1} (\epsilon^{k/d_1})^{i_0}.$$

As ϵ^{k/d_2} is a primitive root of unity, this sum is nonzero if and only if $\epsilon^{k/d_1} = 1$, which means $d_1 = 1$, and in this case the sum is 1.

This allows us to give an explicit version of 8.2.15. Note that for a $b \vdash_k r$ the corresponding t_j give:

$$b_l = |\{j \mid t_j = l\}|$$

so that we conclude

8.2.19 Corollary *For $\pi \in S_n$ with cycle partition (d^r) we have*

$$\chi_n^k(\pi) = \sum_{b \vdash_k r} \frac{r!}{\prod_i i^{b_i} b_i!} d^{r-c(b)}$$

where the sum is over all $b \vdash_k r$ with the property that each l with $b_l > 0$ obeys $\gcd(d, k/l) = d_1 = 1$.

In the next section it will be shown that this number χ_n^k is the desired root number function!

8.3 Equations in Groups

The k-th roots of $g \in G$ are the solutions of the equation $x^k = g$ in G. Hence the enumeration of roots is a particular case of the enumeration of solutions of *equations in finite groups*. A few remarks concerning this more general problem are therefore in order. A quite general result in this field of problems is

8.3.1 Theorem *Let G denote a finite group and $C_0, \ldots, C_{k-1}$ some of its conjugacy classes (which need not be pairwise different) while m_j and n_j are given natural numbers, for $j \in k$. Then, for a fixed $g \in G$, the number of tuples $(g_0, \ldots, g_{k-1}) \in G^k$ such that*

$$g_0^{n_0} \cdots g_{k-1}^{n_{k-1}} = g \text{ and } g_j^{m_j} \in C_j,\ j \in k,$$

is equal to

$$\sum_i \frac{f^i}{|G|} \zeta^i(g^{-1}) \prod_j \Big(\frac{1}{f^i} \sum_{g_j:\, g_j^{m_j} \in C_j} \zeta^i(g_j^{n_j}) \Big),$$

where the ζ^i are the ordinary irreducible characters of G and the f^i their dimensions.

Proof: Let $\mathbf{D}_i$ denote a matrix representation corresponding to D_i. As the set

$$\{\, g_j^{n_j} \mid g_j \in G, g_j^{m_j} \in C_j \,\}$$

is a union of conjugacy classes of G, Schur's Lemma gives

$$\sum_{g_j : g_j^{m_j} \in C_j} \mathbf{D}_i(g_j^{n_j}) = \frac{1}{f^i} \sum_{g_j:\, g_j^{m_j} \in C_j} \zeta^i(g_j^{n_j}) \cdot \mathbf{I},$$

where $\mathbf{I}$ denotes the f^i–dimensional identity matrix. Multiplying k such equations we obtain

$$\sum \mathbf{D}_i(g_0^{n_0} \cdots g_{k-1}^{n_{k-1}} g^{-1}) = \Big(\frac{1}{f^i}\Big)^k \prod_j \sum_{g_j:\, g_j^{m_j} \in C_j} \zeta^i(g_j^{n_j}) \mathbf{D}_i(g^{-1}),$$

where the sum on the left hand side has to be taken over all the $(g_0, \ldots, g_{k-1}) \in G^k$ such that each $g_j^{m_j} \in C_j$. Taking trace on both sides and dividing by $|G|$, we get

$$\frac{1}{|G|} \sum f^i \zeta^i(g_0^{n_0} \cdots g_{k-1}^{n_{k-1}} g^{-1}) = \frac{1}{|G|} \Big(\frac{1}{f^i}\Big)^{k-1} \zeta^i(g^{-1}) \prod_j \sum \zeta^i(g_j^{n_j}).$$

The left hand side of this equation is equal to

$$\frac{1}{|G|} \sum \chi^R(g_0^{n_0} \cdots g_{k-1}^{n_{k-1}} g^{-1}),$$

(χ^R the character of the regular representation) which is just the desired number of solutions. □

A special case of 8.3.1 is the number of ways to express $\pi \in S_n$ as a product $\sigma_0 \cdots \sigma_{k-1}$ of k n–cycles $\sigma_i \in C^{(n)} \subseteq S_n$, the class of n–cycles. The theorem shows that this number is equal to

$$\sum_{\alpha \vdash n} \frac{f^\alpha}{n!} \zeta^\alpha(\pi^{-1}) \prod_j \Big(\frac{1}{f^\alpha} \sum_{\sigma_j \in C^{(n)}} \zeta^\alpha(\sigma_j) \Big).$$

We now remember that $\zeta^\alpha(\sigma_j)$, $\sigma_j \in C^{(n)}$, is nonzero if and only if α is a hook, say $\alpha = (n-r, 1^r)$, in which case $\zeta^\alpha(\sigma_j) = (-1)^r$ and $f^\alpha = \binom{n-1}{r}$. This gives

8.3.2 Corollary *The number of ways to express $\pi \in S_n$ as a product $\sigma_0 \cdots \sigma_{k-1}$ of n–cycles σ_i is equal to*

$$\frac{1}{n} \sum_{r=0}^{n-1} (r!(n-r-1)!)^{k-1} (-1)^{rk} \zeta^{(n-r,1^r)}(\pi).$$

Particular cases are described in

8.3.3 Corollary

- *The number of ways to write $1 \in S_n$ as a product of k n–cycles is equal to*

$$\frac{1}{n} \sum_{r=0}^{n-1} (r!(n-r-1)!)^{k-2} (-1)^{rk} (n-1)!.$$

- *The number of ways to express an n–cycle as a product of k n–cycles is equal to*

$$\frac{1}{n} \sum_{r=0}^{n-1} (r!(n-r-1)!)^{k-1} (-1)^{r(k+1)}.$$

The proof of theorem 8.3.1 allows a generalization. Instead of conjugacy classes $C_0, \ldots, C_{k-1}$ we can consider subsets $V_0, \ldots, V_{k-1}$ which are unions of conjugacy classes, for example $V_0 = \ldots = V_{k-1} = G$:

8.3.4 Corollary *Let G denote a finite group, and consider, for $0 \leq j \leq k-1$, natural numbers n_j. Then, for a given element $h \in G$ the number of solutions $(g_0, \ldots, g_{k-1}) \in G^k$ such that $g_0^{n_0} \cdots g_{k-1}^{n_{k-1}} = h$ is equal to*

$$\sum_i \Big(\frac{|G|}{f^i}\Big)^{k-1} \Big(\prod_j c_{i,n_j}\Big) \zeta^i(h^{-1}).$$

Since, according to 7.7.17, the c_{i,n_j} are integers, we also obtain

8.3.5 Corollary *The number*

$$\left|\{(g_0, \ldots, g_{k-1}) \mid g_0^{n_0} \cdots g_{k-1}^{n_{k-1}} = g\}\right|$$

is, for each element $g \in G$, divisible by the greatest common divisor

$$\gcd\Big\{ \Big(\frac{|G|}{f^i}\Big)^{k-1} \Big| \zeta^i \textit{ irreducible} \Big\}.$$

Now we would like to take a closer look at numbers of roots in symmetric groups. Recall that

$$r_m(\pi) := |\{\rho \in S_n \mid \rho^m = \pi\}| =: r_m(\alpha), \text{ if } \alpha = \alpha(\pi).$$

Aiming at a recursion we use the following multiplicativity property. If $\pi = \rho^m$ and $\rho = \rho_0 \cdots \rho_{r-1}$ is the cycle decomposition of ρ, then $\rho^m = \rho_0^m \cdots \rho_{r-1}^m = \pi$. Hence $r_m(\alpha)$, the number of m-th roots of any element with cycle partition α, is multiplicative in the following sense:

8.3.6 $$r_m(n^{a_n}, \ldots, 1^{a_1}) = \prod_i r_m(i^{a_i}).$$

It therefore remains to evaluate the numbers $r_m(i^{a_i})$.

8.3.7 Lemma *If $\pi \in S_n$ consists of k–cycles only, i. e. if $a_k(\pi) \cdot k = n$, then*

$$r_m(\pi) = r_m(k^{a_k}) = \sum_{j \in \underline{a_k}:\ gcd(k,m/j)=1} [a_k - 1]_{j-1} k^{j-1} r_m(k^{a_k - j}).$$

In particular the following holds:

$$r_m(1^n) = \sum_{j \in \underline{n}:\ j \mid m} [n-1]_{j-1} r_m(1^{n-j}),$$

and,

$$\forall\, n \geq 2\colon r_2(1^n) = r_2(1^{n-1}) + (n-1) r_2(1^{n-2}).$$

In terms of dimensions of the irreducibles, i. e. in terms of numbers of standard Young tableaux, this reads as follows:

$$\forall\, n \geq 2\colon \sum_{\alpha \vdash n} f^\alpha = \sum_{\beta \vdash n-1} f^\beta + (n-1) \sum_{\gamma \vdash n-2} f^\gamma.$$

Proof: We consider $\rho \in R_m(\pi)$, where $\alpha(\pi) = (k^{a_k})$, and $a_k(\pi) > 0$. Assume that $(j_0 \ldots j_{k-1})$, a cyclic factor of π, arises from the cyclic factor $(i_0 \ldots i_{r-1})$ of ρ. Then, by 11.2.12, it is not only $(j_0 \ldots j_{k-1})$ which comes from $(i_0 \ldots i_{r-1})$, but altogether $j := \gcd(r, m)$ k–cycles of π, and furthermore we have $k = r/j$. Canceling this cycle $(i_0 \ldots i_{r-1})$ from ρ, we obtain ρ^*, a permutation of degree $n - r = n - jk$, the m-th power of which has the cycle partition $(k^{a_k - j})$. It remains to count the number of different $(i_0 \ldots i_{r-1})$ which contribute $(j_0 \ldots j_{k-1})$, together with $j - 1$ further k–cycles of π. As we can cyclically permute the points without changing the cyclic permutation, we may assume that $j_{k-1} = i_{r-1}$, which fixes the places of the j_ν in this cycle:

$$(i_0 \ldots i_{r-1}) = (\ldots j_0 \ldots j_1 \ldots j_{k-1}).$$

Now the k points contained in one of the remaining $j - 1$ k–cycles $(j'_0 \cdots j'_{k-1})$ of π which also arise from $(i_0 \ldots i_{r-1})$ have also to be shuffled into this cycle. $(j'_0 \cdots j'_{k-1})$ Fixing the place of j'_{k-1} also fixes the places of the other j'_ν:

$$(\ldots j_0 \ldots j_1 \ldots j'_{k-1} \cdots j_{k-1}),$$

and hence there are exactly so many admissible ways:

$$[a_k(\pi) - 1]_{j-1} = (a_k(\pi) - 1)(a_k(\pi) - 2) \ldots (a_k(\pi) - j + 1).$$

This completes the proof, since $[a_k(\pi) - 1]_{j-1} = 0$, if $j > a_k(\pi)$. □

Besides the multiplicativity of r_m, we can use the following closed form for root numbers: If G denotes a finite group, $g \in G$, and $m \in \mathbb{N}$, then (exercise 8.3.1)

8.3.8 $$r_m(g) = |C_G(g)| \sum_x{}' |C_G(x)|^{-1},$$

if the sum $\sum'$ is taken over a system of representatives x of such conjugacy classes of G that contain m-th roots of g. In order to apply this to the symmetric group case, we have to characterize the conjugacy classes which contain m-th roots of an element of cycle partition (k^{a_k}). Each such conjugacy class corresponds to a cycle type $c \vdash\!\!\dashv k \cdot a_k$, where $c_i > 0$ implies that k divides i, say $i = jk$. We can therefore replace $c_i = c_{j \cdot k}$ by b_j, obtaining in this way a cycle type $b \vdash\!\!\dashv a_k$. Notice that $b_j > 0$ means that there is a $j \cdot k$–cycle, in the root, the m-th power of which consists of k–cycles. Thus, by 11.2.12, $j = \gcd(m, j \cdot k)$, so that the cycle types $b \vdash\!\!\dashv a_k$ in question are just the cycle types, the nonzero elements b_j of which satisfy the conditions

$$j | m \wedge \gcd(m/j, k) = 1.$$

This leads us to the following result:

8.3.9 Theorem *The number of m-th roots of an element consisting of a_k cycles of length k satisfies the equation*

$$r_m(k^{a_k}) = \sum_{b \vdash\!\dashv a_k, b_i \neq 0 \Rightarrow [i|m \wedge gcd(m/i,k)=1]} k^{a_k - \Sigma b_i} \frac{a_k!}{\prod_i i^{b_i} b_i!}.$$

The number of m-th roots of an element $\pi \in S_n$ of cycle type a is

$$r_m(\pi) = \prod_{k:a_k>0} k^{a_k} a_k! \sum_{b \vdash\!\dashv a_k, b_i \neq 0 \Rightarrow [i|m \wedge gcd(m/i,k)=1]} k^{a_k - \Sigma b_i} \frac{a_k!}{\prod_i i^{b_i} b_i!}.$$

Hence, in particular, $r_m = \chi_n^m$, and so, r_m, the root number function, is a proper character.

Exercises

Exercise 8.3.1 Prove that

$$r_2((2i)^{a_{2i}}) = \begin{cases} 0, & \text{if } a_{2i} \text{ is odd,} \\ (2i)^m (2m)!(2^m m!)^{-1}, & \text{if } a_{2i} = 2m. \end{cases}$$

Exercise 8.3.2 Derive the following in two ways, first using the multiplicativity of the root number function, and then by using character theory only:

$$r_2(\alpha) = 0, \text{ if an } a_{2i} \text{ is odd.}$$

8.4 Up–Down Sequences

Recall the method of displaying a permutation $\pi \in S_n$ by putting down the *list* of its values:

$$\pi = [\pi 0 \dots \pi(n-1)].$$

The i-th position, $i < n-1$, of this list is called an *up* if $\pi i < \pi(i+1)$, otherwise it is called a *down*. Replacing an up by "+", a down by "−", we obtain the *up–down sequence* $U\pi$ of π, a sequence of lenght $n-1$, for example

$$U[02137654] = (+ - + + - - -).$$

Before we enumerate permutations with given up–down sequence, we take a look at permutations with prescribed *number* of ups. The *Eulerian number* $E(n,k)$ is defined to be the number of $\pi \in S_n$ such that π contains exactly k ups. Hence in particular the following holds:

8.4.1
$$\sum_{k=0}^{n-1} E(n,k) = n!.$$

H. O. Foulkes was the first to notice that in order to examine Eulerian numbers we can associate with the up–down sequences *rims of Young diagrams*. In order to describe this we take a node "×" and we successively add further nodes to the left or downwards according to $U\pi$ and the following rule:

8.4.2
$$\begin{array}{cc} + \leftarrow & \times \\ & \downarrow - \end{array}$$

This means that to an entry + of U there corresponds a node × which has to be added at the left of the last node added, and in the same row. Correspondingly to an entry − of U there corresponds a node that has to be added just below the last node. For example the sequence $(+ - + + - - -)$ mentioned above gives

$$\begin{array}{cccc} & & \times & \times \\ \times & \times & \times & \\ \times & & & \\ \times & & & \\ \times & & & \end{array} \quad \text{according to} \quad \begin{array}{cccc} & & + & \times \\ + & + & - & \\ - & & & \\ - & & & \\ - & & & \end{array} .$$

We consider the resulting skew diagram as the rim R^{α}_{00} of a Young diagram $[\alpha]$ the shape α of which we denote by $\alpha(U)$, while we indicate the rim hook of $[\alpha(U)]$ as follows:

$$R(U) := R^{\alpha(U)}_{00}.$$

The partition $\alpha := \alpha(U)$ is uniquely determined, and $\alpha_0 = k+1$, k being the number of ups, while $\alpha'_0 = n-k$, one more than the number of downs. For our example we obtain

$$[\alpha(+-++---)] = \begin{matrix} \times & \times & \times & \times \\ \times & \times & \times & \\ \times & & & \\ \times & & & \\ \times & & & \end{matrix} = [4,3,1^3].$$

If we want to erase the rim R^{α}_{00} from $[\alpha(U)]$ by successively removing nodes in such a way that each step leaves a Young diagram, then we usually have many possibilities to do this. Each of these possibilities can be described by replacing the nodes of the rim by numbers $0, \ldots, n-1$, according to the sequence of removals. The above example offers, among others, the following two ways of removing the rim of $[4, 3, 1^3]$:

$$\begin{matrix} & & 7 & 6 \\ 5 & 4 & 3 & \\ 2 & & & \\ 1 & & & \\ 0 & & & \end{matrix} \quad , \quad \begin{matrix} & & 7 & 6 \\ 5 & 1 & 0 & \\ 4 & & & \\ 3 & & & \\ 2 & & & \end{matrix} \quad .$$

Reading these numbers row by row from top to bottom and in the rows from right to left, we obtain lists of permutations, for example

$$[67345210], \ [67015432].$$

The rule 8.4.2 implies that they all have the same up–down sequence, namely the up–down sequence which lead to $[\alpha]$. Hence the number of permutations with prescribed up–down sequence U is equal to the number of ways to remove the rim $R(U)$ from $[\alpha(U)]$ subject to the condition that in each step we remove just one node and the remaining rest is still a Young diagram. Replacing the entry i by $n-1-i$, for each $i \in n$, we clearly obtain the standard Young tableaux of shape $[R(U)]$, and hence, according to 7.6.7 or the Murnaghan Nakayama formula for skew representations, this is just the dimension $f^{R(U)}$ of the corresponding skew representation $[R(U)]$. This yields the following result on Eulerian numbers, since the leg length of $R(U)$ is equal to $\alpha'_0 - 1 = n - k - 1$:

8.4.3 Corollary *The Eulerian numbers satisfy the following identities:*

$$E(n,k) = \sum_{\alpha} f^{R^{\alpha}_{00}} = \sum_{U} f^{R(U)},$$

if the first sum is taken over all the proper partitions α such that $\alpha_0 = k+1$ and $\alpha'_0 = n-k$, while the second sum is taken over all the up–down sequences U containing exactly k entries $+$.

This looks circumstantial at first glance, but it allows to express $E(n,k)$ as a sum of dimensions of irreducible representations of S_n, i. e. $E(n,k)$ is a sum of numbers of standard Young tableaux. Moreover it opens a natural way of generalizing the Eulerian numbers by replacing the dimensions by the characters, this will be described later.

The summand $f(U)$ of $E(n, k)$ in 8.4.3, i. e. the dimension of the skew representation $[R(U)]$, is the sum of the dimensions of its irreducible components. The Murnaghan–Nakayama formula shows that it is the number of $(\alpha, (1^n), \alpha \backslash R^\alpha_{00})$–paths in Young's lattice, i. e. it is the number of ways to erase R^α_{00} from $[\alpha]$, as it was described above. The crucial point is now to construct, according to R^α_{00} or to the up–down sequence U which lead to α, the Young diagrams $[\beta]$ (together with the corresponding multiplicities) which form the irreducible constituents. In order to do this we start with a node "×" together with an up–down sequence U and add further nodes according to U and the following rule:

8.4.4
$$\begin{array}{ccc} & \times & \nearrow + \\ \swarrow - & & \end{array}$$

By this pictorial description I mean that to an entry $+$ of U there corresponds a node $\times$ which has to be added to the right of the last node, maybe in a higher row, while to an entry $-$ there corresponds a node added to the left of the last node or in a lower row. Consider once more the example $U = (+ - + + - - -)$. We start with a node $\times$, and the first entry of U is a $+$, so the corresponding node has to be added, according to 8.4.4, to the right of the starting node, i. e. we obtain the diagram $\times\otimes$, where the last node added is encircled. Now the second entry of U is a minus sign, hence the corresponding addition of a node is again uniquely determined, and we get the diagram

$$\begin{array}{cc} \times & \times \\ \otimes & \end{array} .$$

The next entry of U is a plus sign, so that there are two places open for an additional node which are to the right of the node which was added last time:

$$\begin{array}{ccc} \times & \times & \otimes \\ \times & & \end{array} \quad \text{and} \quad \begin{array}{cc} \times & \times \\ \times & \otimes \end{array} .$$

The next steps yield the following cascade of diagrams:

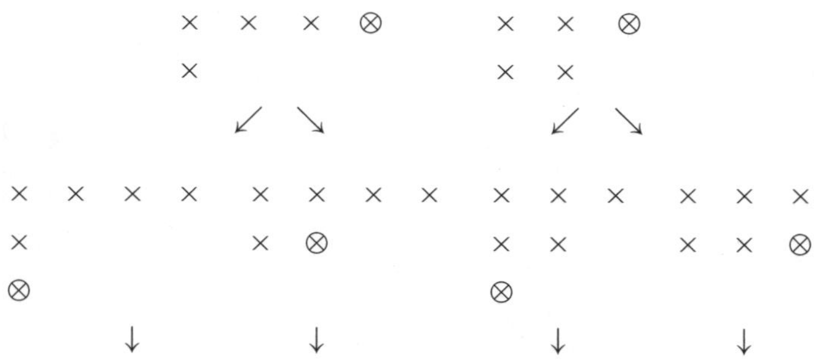

$$\begin{array}{cccc} \times & \times & \times & \otimes \\ \times & & & \end{array} \qquad \begin{array}{ccc} \times & \times & \otimes \\ \times & \times & \end{array}$$

$$\swarrow \quad \searrow \qquad\qquad \swarrow \quad \searrow$$

$$\begin{array}{cccc} \times & \times & \times & \times \\ \times & & & \\ \otimes & & & \end{array} \quad \begin{array}{cccc} \times & \times & \times & \times \\ \times & \otimes & & \end{array} \quad \begin{array}{ccc} \times & \times & \times \\ \times & \times & \\ \otimes & & \end{array} \quad \begin{array}{ccc} \times & \times & \times \\ \times & \times & \otimes \end{array}$$

$$\downarrow \qquad\qquad \downarrow \qquad\qquad \downarrow \qquad\qquad \downarrow$$

$$\begin{array}{llll} \times & \times & \times & \times \\ \times & & & \\ \times & & & \\ \times & & & \\ \times & & & \end{array} \qquad \begin{array}{llll} \times & \times & \times & \times \\ \times & \times & & \\ \times & & & \\ \times & & & \end{array} \qquad \begin{array}{lll} \times & \times & \times \\ \times & \times & \\ \times & & \\ \times & & \\ \times & & \end{array} \qquad \begin{array}{lll} \times & \times & \times \\ \times & \times & \times \\ \times & & \\ \times & & \end{array}$$

Hence from $U = (+ - + + - - -)$ we obtain the diagrams

$$[4, 1^4], [4, 2, 1^2], [3, 2, 1^3], [3^2, 1^2],$$

and each one of them exactly once. It is the aim to show that this is in fact the decomposition of $[R(U)]$ into its irreducible constituents. We therefore denote by

$$[\widetilde{R}(U)]$$

the sum of the representations $[\beta]$ obtained from U by 8.4.2, where each of them occurs with the multiplicity by which the diagram $[\beta]$ shows up. We have to prove the equation $[R(U)] = [\widetilde{R}(U)]$, and the following lemma will turn out to be the crucial step towards this:

8.4.5 Lemma *For each k and any up–down sequence U we have:*

$$[k+1][\widetilde{R}(U)] = [\widetilde{R}(U + + \ldots +)] + [\widetilde{R}(U - + \ldots +)].$$

Proof: Young's rule shows that the irreducible constituents of $[k+1][\widetilde{R}(U)]$ arise from the $[\mu]$ of $[\widetilde{R}(U)]$ by adding nodes in $k{+}1$ *different columns*. Now we consider, how these $[\nu]$ can be reached from $[\emptyset]$ in Young's lattice. $[\mu]$ comes from 8.4.2 by working through U, so that $[\nu]$ is obtained by working through $U + \ldots +$ or through $U - + \ldots +$, depending on the lowest node of $[\nu \backslash \mu]$. This proves the statement. □

We are now in a position to prove the desired equality $[R(U)] = [\widetilde{R}(U)]$, but let us first work out an example. Since

$$R(+ - - + + - - + ++) = [7, 6^2, 4^2 \backslash 5^2, 3^2],$$

the determinantal form for skew representations yields that

$$[R(+ - - + + - - + ++)] = \det \begin{pmatrix} [2] & [3] & [6] & [7] & [1] \\ 1 & [1] & [4] & [5] & [9] \\ 0 & 1 & [3] & [4] & [8] \\ 0 & 0 & 1 & [1] & [5] \\ 0 & 0 & 0 & 1 & [4] \end{pmatrix},$$

which, according to its last row, is equal to

$$[4]\det\begin{pmatrix}[2] & [3] & [6] & [7]\\ 1 & [1] & [4] & [5]\\ 0 & 1 & [3] & [4]\\ 0 & 0 & 1 & [1]\end{pmatrix} - \det\begin{pmatrix}[2] & [3] & [6] & [11]\\ 1 & [1] & [4] & [9]\\ 0 & 1 & [3] & [8]\\ 0 & 0 & 1 & [5]\end{pmatrix}$$

$$= [4][7,6^2,4\backslash 5^2,3^2] - [8,7^2,5\backslash 6^2,4]$$

$$= [4][R(+--++-)] - [R((+--++-+++))].$$

We note that $(+--++-)$ is shorter than the original sequence $(+--++--+++)$, while the other sequence $(+--++-+++)$ is of the same length, but it has *one down less*. This allows to prove the statement via induction on the length $n-1$ of U together with an induction on the number of downs (using that clearly $[\widetilde{R}(+\dots+)] = [R(+\dots+)] = [n]$) inside the induction on the length of U. But this same double induction procedure applies in the general case, too, as we can assume without loss of generality that U ends with a $+$. For otherwise we can use that for the *complementary* up–down sequence U', which arises from U by changing each $+$ into a $-$ and each $-$ into $+$, we have both

8.4.6 $$[\widetilde{R}(U)] = [\widetilde{R}(U')] \otimes [1^n], \text{ and } [R(U)] = [R(U')] \otimes [1^n].$$

In fact, if $U = (u_0, \dots, u_{n-2})$ closes with a down and then k ups, the determinantal form gives that $[R(U)]$ is equal to

$$[k+1][R(u_0,\dots,u_{n-k-3})] - [R(u_0,\dots,u_{n-k-3},+,u_{n-k-1},\dots,u_{n-2})].$$

This shows that the same argumentation goes through in the general case as well, and we have therefore proved

8.4.7 Theorem *For each up down sequence U the rule 8.4.4 yields the decomposition of $[R(U)]$ into its irreducible constituents, or, more formally:*

$$[R(U)] = [\widetilde{R}(U)].$$

Therefore the number of permutations with sequence $(+-++---)$ is equal to

$$f^{(4,1^4)} + f^{(4,2,1^2)} + f^{(3,2,1^3)} + f^{(3^2,1^2)} = 35 + 90 + 64 + 56 = 245.$$

Summarizing we obtain the following result about Eulerian numbers:

8.4.8 Theorem *The Eulerian number $E(n,k)$, i. e. the number of elements $\pi \in S_n$ containing exactly k ups, satisfies the equation*

$$E(n,k) = \sum_U f^{R(U)},$$

if the sum is taken over the $\binom{n-1}{k}$ up–down sequences U of length $n-1$ which contain exactly k entries $+$, and where $f^{R(U)}$ denotes the dimension of the skew representation $[R(U)]$. The summand $f^{R(U)}$ therefore satisfies the equation

$$f^{R(U)} = \sum_{\beta}([R(U)], [\beta]) f^{\beta},$$

where f^{β} *denotes the dimension of* $[\beta]$. *The multiplicity* $([R(U)], [\beta])$ *of the irreducible constituent* $[\beta]$ *can be obtained by carrying out the procedure indicated by 8.4.4 which yields the decomposition of* $[R(U)]$. *Hence the Eulerian number* $E(n, k)$ *is the following linear combination of numbers of standard Young tableaux, the coefficients of which can be obtained via 8.4.4:*

$$E(n, k) = \sum_{U:\, k \text{ ups}} \sum_{\beta}([R(U)], [\beta]) f^{\beta}.$$

Moreover we have obtained a recursion for $f^{R(U)}$ by the number of downs in U:

8.4.9 Corollary *Let* U *denote an up–down sequence containing at least one entry* $-$, *say*

$$U = (u_0 \dots u_{n-2}) = (\underbrace{u_0 \dots u_{n-k-3}}_{=:\widetilde{U}} - \underbrace{+ \dots +}_{k}),$$

then we have the following recursion (on the number of entries $-$ *of* U*):*

$$f(U) = \binom{n}{k+1} f(\widetilde{U}) - f(\widetilde{U} + u_{n-k-1} \dots u_{n-2}),$$

where $f(+ \dots +) = 1$, *and* $f(\emptyset) = 1$.

For example $f(+) = 1$ yields that

$$f(-) = \binom{2}{1} f(\emptyset) - f(+) = 2 \cdot 1 - 1 = 1,$$

so that, using $f(++) = 1$, we obtain

$$f(+-) = \binom{3}{1} f(+) - f(++) = 3 \cdot 1 - 1 = 2,$$

$$f(-+) = \binom{3}{2} f(\emptyset) - f(++) = 3 \cdot 1 - 1 = 2,$$

$$f(--) = \binom{3}{1} f(-) - f(-+) = 3 \cdot 1 - 2 = 1,$$

and so on. Up–down sequences of particular interest are the *alternating* sequences, i. e. the sequences of the form $(- + - + \dots)$, or $(+ - + - \dots)$. Again we may assume that they end with an entry $+$, and we put

$$t_n := \left|\{\pi \in S_n \mid U\pi = (\dots +) \text{ is alternating}\}\right|.$$

These numbers satisfy the following recursions (exercise 8.4.1):

8.4.10 Corollary *For odd numbers $n = 2k + 1$ we have*

$$t_n = (-1)^k + \sum_{i=1}^{k} (-1)^{i-1} t_{n-2i},$$

while, for even $n = 2k$, the following is true:

$$t_n = (-1)^{k-1} + \sum_{i=1}^{k-1} (-1)^{i-1} \binom{n}{2i} t_{n-2i}.$$

The starting values for these recursions are $t_1 = t_2 = 1$.

These numbers are often called the *Euler numbers* . They are in fact the coefficients in the tangens (for even n) and in the secans series (for odd n). The smallest of these numbers are shown in the following table:

$t_1 = 1$	$t_{10} = 50521$
$t_2 = 1$	$t_{11} = 353792$
$t_3 = 2$	$t_{12} = 2702765$
$t_4 = 5$	$t_{13} = 22368256$
$t_5 = 16$	$t_{14} = 199360981$
$t_6 = 61$	$t_{15} = 1903757312$
$t_7 = 272$	$t_{16} = 19391512145$
$t_8 = 1385$	$t_{17} = 209865342976$
$t_9 = 7936$	$t_{18} = 2404879675441$

Exercises

Exercise 8.4.1 Check the recursions of 8.4.9.

8.5 Foulkes Characters

The above considerations and results have shown that a natural generalization of the Eulerian numbers arises when we replace the *number* $E(n, k)$ by the *character*

$$\chi^{n,k} := \sum_{U:\, k\ ups} \chi^{R(U)},$$

where the sum is taken over all the up–down sequences U of length $n - 1$ which contain exactly k entries $+$. Thus, first of all, the dimensions of these characters are the Eulerian numbers:

8.5.1 $$\chi^{n,k}(1) = E(n, k).$$

I call these characters the *Foulkes characters* since they were apparently discovered by H. O. Foulkes. According to 8.4.4 we have, for example, that

8.5.2 $$\chi^{n,0} = \zeta^{(1^n)}, \ \chi^{n,n-1} = \zeta^{(n)}, \ \chi^{n,k} = \chi^{n,n-k-1} \otimes \zeta^{(1^n)}.$$

The most important property of these characters is the fact that their value on $\pi \in S_n$ *does only depend on the number of cyclic factors of* π:

8.5.3 Theorem *If the elements $\pi, \rho \in S_n$ consist of the same number of cyclic factors, i. e. if $c(\pi) = c(\rho)$, then, for each k, we have*

$$\chi^{n,k}(\pi) = \chi^{n,k}(\rho).$$

Proof: In order to prove this by induction on n we use that

$$\chi^{R(U)}(\pi) = \text{ sum of the weights of all } (\alpha(U), \mu, \alpha(U)\backslash R(U))\text{–chains},$$

where μ is a fixed improper partition, the summands μ_i of which are the lengths of the cyclic factors of π.

We have to take the sum over all these expressions, where U runs through all the $\binom{n-1}{k}$ up–down sequences containing k ups. From the corresponding rims $R(U)$ we obtain, by canceling the starting node, $\binom{n-1}{k}$ skew diagrams consisting of $n-1$ nodes. Each of those either belongs to the $\binom{n-2}{k}$ up–down sequences of length $n-2$ which still contain k ups or to the $\binom{n-2}{k-1}$ up–down sequences of length $n-2$ with $k-1$ ups, depending on the form of the rim which is either of the form

$$R_- = \begin{matrix} & & \times \\ & & \times \\ & \cdot^{\cdot^{\cdot}} & \end{matrix} \quad \text{or of the form } R_+ = \begin{matrix} & \times & \times \\ \cdot^{\cdot^{\cdot}} & & \end{matrix}.$$

Denoting this shorter sequence of length $n-2$ by U^*, we see that the proof amounts to show that there is a close connection between the

$$(\alpha(U), \mu, \alpha(U)\backslash R(U))\text{–chains}$$

and the

$$(\alpha(U^*), \mu^*, \alpha(U^*)\backslash R(U^*))\text{–chains}.$$

We therefore consider a fixed $(\alpha(U), \mu, \alpha(U)\backslash R(U))$– chain, assuming that the starting node is canceled together with a skew hook of length μ_i. This determines a $(\alpha(U^*), \mu^*, \alpha(U^*)\backslash R(U^*))$–chain, where

$$\mu^* := (\mu_0, \ldots, \mu_{i-2}, \mu_{i-1} - 1, \mu_i, \ldots).$$

In the case when $R(U)$ is of the form R_+, then the resulting shorter chain has the *same weight*, in case R_- the new weight is the opposite of the old one. This proves the helpful recursion

8.5.4 $$\chi^{n,k}_\mu = \underbrace{\chi^{n-1,k-1}_{\mu^*}}_{R_+} - \underbrace{\chi^{n-1,k}_{\mu^*}}_{R_-}.$$

Now the induction hypothesis yields, that the summands on the right hand side of this equation only depend on the number of cyclic factors, and this completes the proof. □

This theorem allows us to introduce the following notation:

$$\chi^{n,k}_j := \chi^{n,k}(\pi), \text{ if } c(\pi) = j, j \in \underline{n},$$

and to define the *Foulkes table* F_n of S_n by

$$F_n := (\chi^{n,k}_j)_{k \in n, j \in \underline{n}} = \begin{array}{c|cccc} k\backslash j & n & n-1 & \dots & 1 \\ \hline 0 & \chi^{n,0}_n & \chi^{n,0}_{n-1} & \cdots & \chi^{n,0}_1 \\ \vdots & \vdots & \vdots & & \vdots \\ n-1 & \chi^{n,n-1}_n & \chi^{n,n-1}_{n-1} & \cdots & \chi^{n,n-1}_1 \end{array}.$$

This table is a square table consisting of n rows and columns, and we shall show in a minute that it is invertible. Let us evaluate, for example, the fourth row of F_5, which contains the character $\chi^{5,3}$. The up–down sequences of length $5 - 1 = 4$ containing 3 ups are the sequences

$$(+++-), (++-+), (+-++), (-+++).$$

The corresponding Young diagrams (use 8.4.4) show that

8.5.5 $$\chi^{5,3} = 4\zeta^{(4,1)} + 2\zeta^{(3,2)}.$$

We therefore obtain from the character table of S_5 that

$$\chi^{5,3}_5 = 26,\ \chi^{5,3}_4 = 10,\ \chi^{5,3}_3 = 2,\ \chi^{5,3}_2 = -2,\ \chi^{5,3}_1 = -4.$$

The complete Foulkes table of S_5 is:

$F_5 =$

$k\backslash j$	5	4	3	2	1
0	1	−1	1	−1	1
1	26	−10	2	2	−4
2	66	0	−6	0	6
3	26	10	2	−2	−4
4	1	1	1	1	1

Further Foulkes tables can be found in the appendix. We note that 8.5.4 allows a recursive evaluation of the entries of F_n, except for the entries in the first column, for which we can use any recursion for the Eulerian numbers, see exercise 8.5.1:

8.5.6 Corollary *The Foulkes character values satisfy the recursion*

$$\chi_j^{n,k} = \chi_j^{n-1,k-1} - \chi_j^{n-1,k}, \text{ if } j < n, k > 0.$$

For $j = n$ we have

$$\chi_n^{n,k} = E(n,k),$$

while, for $k = 0$, we have

$$\chi_j^{n,0} = (-1)^{n-j}.$$

The starting value for the recursion is, of course,

$$\chi_1^{1,0} = 1.$$

Moreover we remark that from 8.4.4 it follows that

8.5.7 $$[\chi^{n,k} \mid \zeta^\alpha] > 0 \Longrightarrow \alpha_0 \le k+1, \alpha_0' \le n-k.$$

Hence the particular hook constituent $[k+1, 1^{n-k-1}]$ is the only hook that occurs in $\chi^{n,k}$, and it occurs with multiplicity $\binom{n-1}{k}$. This implies

8.5.8 Theorem *The Foulkes characters $\chi^{n,k}$ are linearly independent over $\mathbb{Q}$, and each character $\chi\colon S_n \to \mathbb{Q}$, the value of which on π does only depend on $c(\pi)$, is a unique $\mathbb{Q}$–linear combination*

$$\chi = \sum_{k \in n} m_k \cdot \chi^{n,k}$$

of the Foulkes characters. Moreover the coefficients m_k satisfy the condition

$$m_k = \left[\chi \,\middle|\, \zeta^{(k+1,1^{n-k-1})}\right] / \binom{n-1}{k} \in \mathbb{N}.$$

An example is provided by 8.5.5. It is important to notice that this example shows that also

$$\frac{1}{2}\chi^{5,3}$$

is in fact a character the values of which do only depend on the numbers $c(\pi)$. Hence we *cannot* hope that characters with this property are $\mathbb{Z}$–linear combinations of Foulkes characters, and hence the preceding result cannot be sharpened in this way.

8.5.9 Application (the character of $_{S_X}(Y^X)$) An important character to which the preceding theorem applies is the character $\chi^{(m)}$ of the natural action of S_n on m^n:

$$\chi^{(m)}(\pi) := m^{c(\pi)}.$$

If we assume the following theorem from representation theory of symmetric groups (cf. 5.2.20 in the book [72] by James and Kerber):

8.5.10 $$[\chi^{(m)} \mid \zeta^\alpha] = \frac{f^\alpha}{n!} \prod_{(i,j)\in[\alpha]} (m-i+j),$$

then we obtain (since $\binom{n-1}{k}$ is the dimension of $[k+1, 1^{n-k-1}]$):

8.5.11 Corollary *The character $\chi^{(m)}$ of the natural action of S_n on m^n has the following decompositions into irreducibles and Foulkes characters:*

$$\chi^{(m)} = \sum_{\alpha \vdash n} \Big(\prod_{(i,j)\in[\alpha]} \frac{m-i+j}{h^{\alpha}_{ij}} \Big) \zeta^{\alpha} = \sum_{k=0}^{n-1} \binom{m+k}{n} \chi^{n,k}.$$

Hence in particular we have, for each $m, n \in \mathbb{N}^*$:

$$m^n = \sum_{k=0}^{n-1} \binom{m+k}{n} E(n,k).$$

◇

Exercises

Exercise 8.5.1 Derive the following recursion for the Eulerian numbers:

$$E(n,k) = (n-k+1)E(n-1,k-1) + (k+1)E(n-1,k), \text{ if } k > 0,$$

the starting value is $E(n,0) = 1$.

Exercise 8.5.2 Prove 8.5.6.

8.6 Schubert Polynomials

Now we are going to associate with each permutation π a Schubert polynomial X_π. These polynomials have very interesting properties. For example they form a $\mathbb{Z}$–basis of the union of the polynomial rings $\mathbb{Z}[x_0, \ldots, x_{n-1}]$. Moreover, X_π is a monomial if $L(\pi)^+$ is a weakly decreasing sequence, and it is a Schur polynomial, if $L(\pi)$ is a weakly increasing sequence. Hence the Schubert polynomials generalize the Schur polynomials. Since they form a basis of the union of the $\mathbb{Z}[x_0, \ldots, x_{n-1}]$, they can also serve for a different approach to the Littelwood–Richardson Rule, but this will not be described in detail here. But it should be mentioned that this approach (introduced by Lascoux and Schützenberger) is very well suited for computer calculations, and therefore it was chosen for the corresponding procedure in the program system SYMMETRICA.

Consider the polynomial ring $\mathbb{Z}[x_0, \ldots, x_{n-1}]$ and the natural action

$$S_n \times \mathbb{Z}[x_0, \ldots, x_{n-1}] \to \mathbb{Z}[x_0, \ldots, x_{n-1}] : (\pi, f) \mapsto f(x_{\pi 0}, \ldots, x_{\pi(n-1)}).$$

Hence, for each $f \in \mathbb{Z}[x_0, \ldots, x_{n-1}]$, and every elementary transposition σ_i in Σ_n, we have a well defined $\sigma_i f \in \mathbb{Z}[x_0, \ldots, x_{n-1}]$, and so we can introduce, for each $i < n-1$, the linear operator ∂_i on $\mathbb{Z}[x_0, \ldots, x_{n-1}]$ by putting

$$\partial_i f := \frac{f - \sigma_i f}{x_i - x_{i+1}}.$$

The resulting $\partial_i f$ is a polynomial which is symmetric in x_i and x_{i+1} (check this). Moreover, if f is already symmetric in x_i and x_{i+1}, then clearly $\partial_i f = 0$. In the case when f is homogeneous, then $\partial_i f$ is homogeneous, too, if it is nonzero, then its degree is the degree of f minus 1. The polynomial $\partial_i f$ is called *divided difference.* For example

$$\partial_1 x_1^3 x_2^2 x_3^1 = \frac{x_1^3 x_2^2 x_3^1 - x_1^2 x_2^3 x_3^1}{x_1 - x_2} = x_1^2 x_2^2 x_3.$$

A straightforward check shows that these operators ∂_i satisfy the relations

8.6.1 $$\partial_i \partial_j = \begin{cases} 0, & \text{if } i = j, \\ \partial_j \partial_i, & \text{if } |i-j| > 1 \end{cases}, \text{ and } \partial_i \partial_{i+1} \partial_i = \partial_{i+1} \partial_i \partial_{i+1}.$$

They are crucial for

8.6.2 Theorem *For any finite sequence $(i) := (i_0, \ldots, i_{l-1})$, $i_\nu \in n-1$, the following holds:*

- *If both $(i_0, \ldots, i_{l-1})$ and $(j_0, \ldots, j_{l-1})$ are contained in $RS(\pi^{-1})$, the set of reduced sequences (see the appendix) of some $\pi^{-1} \in S_n$, then*

$$\partial_{i_0} \ldots \partial_{i_{l-1}} = \partial_{j_0} \ldots \partial_{j_{l-1}},$$

  and hence to each $\pi \in S_n$ there corresponds a unique operator

$$\partial_\pi := \partial_{(i)} := \partial_{i_0} \ldots \partial_{i_{l-1}}.$$

- *If $(i_0, \ldots, i_{l-1})$ is* not *a reduced sequence, then*

$$\partial_{i_0} \ldots \partial_{i_{l-1}} = 0,$$

  the zero mapping on $\mathbb{Z}[x_0, \ldots, x_{n-1}]$.

Proof: The first item will be proved by induction on the reduced length $l = l(\pi)$. If $l = 0$, then $\pi = \text{id}$, the identity element. The corresponding operator is $\partial_\emptyset$, the identity mapping, and therefore the statement holds in this case.

Let us consider the case when $l > 0$. The induction hypothesis says that, if $i_0 = j_0$ or $i_{l-1} = j_{l-1}$, the corresponding operators $\partial_{(i)}$ and $\partial_{(j)}$ are equal. In the other cases we shall apply the Exchange Lemma 11.3.11. Consider the sequence

$$(\hat{i}) := (j_0, i_0, \ldots, \hat{i}_k, \ldots, i_{l-1}) \in RS(\pi^{-1}),$$

and the corresponding operator $\partial_{(\hat{i})}$.

If $k \neq l-1$, then $\partial_{(j)} = \partial_{(\hat{i})} = \partial_{(i)}$, by induction hypothesis, and the statement holds in this case.

If $k = l-1$, we distinguish two cases:

- If $|j_0 - i_0| > 1$, then $\sigma_{i_0}\sigma_{j_0} = \sigma_{j_0}\sigma_{i_0}$, and so

$$(i') := (i_0, j_0, i_1, \dots, i_{l-2}) \in RS(\pi^{-1}),$$

moreover $\partial_{(i')} = \partial_{(\hat{i})}$, and hence we have

$$\partial_{(i)} = \partial_{(i')} = \partial_{(\hat{i})} = \partial_{(j)},$$

by the induction hypothesis.
- In the case when $|j_0 - i_0| = 1$, we consider $(\hat{i})$, which is now equal to

$$(j_0, i_0, \dots, i_{l-2}),$$

so that, by the Exchange Lemma, at least one of the following three sequences is contained in $RS(\pi^{-1})$, too:

$$(a) := (i_0, i_0, \dots, i_{l-2}),$$

$$(b) := (i_0, j_0, i_1, \dots, i_{l-2}),$$

$$(c) := (i_0, j_0, i_0, \dots, \hat{i}_r, \dots, i_{l-2}),$$

where $r \geq 1$ in the last case. Easy checks show that neither (a) nor (b) correspond to a reduced decomposition of π^{-1}, so that we obtain $(c) \in RS(\pi^{-1})$. Now we use that the assumption $|i_0 - j_0| = 1$ implies $\partial_{i_0}\partial_{j_0}\partial_{i_0} = \partial_{j_0}\partial_{i_0}\partial_{j_0}$, and hence also

$$(d) := (j_0, i_0, j_0, \dots, \hat{i}_r, \dots, i_{l-2}) \in RS(\pi^{-1}).$$

This sequence serves very well for a completion of the proof:

$$\partial_{(j)} = \partial_{(d)} = \partial_{(c)} = \partial_{(i)}.$$

This completes the proof of the first item.

The second item will be proved by induction on the length l of the sequence $(i) := (i_0, \dots, i_{l-1})$, which is now assumed *not to be a reduced sequence*, so that in particular $l \geq 2$.

If $l = 2$, then $i_0 = i_1$, since otherwise $\sigma_{i_0}\sigma_{i_1}$ were a reduced decomposition. Hence $l = 2$ implies $\partial_{(i)} = 0$, by 8.6.1. For the inductive step we can therefore assume that $l \geq 3$ and that $(i_1, \dots, i_{l-1})$ is a reduced sequence, since in all the other cases $\partial_{(i)} = 0$, as it is stated. We put

$$\rho^{-1} := \sigma_{i_1} \cdots \sigma_{i_{l-1}}, \text{ and } \pi^{-1} := \sigma_{i_0} \cdots \sigma_{i_{l-1}} = \sigma_{i_0}\rho^{-1}.$$

Since (i) is not reduced, $l(\pi) = l(\pi^{-1}) \leq l-1$, and so, by 11.3.8, $l(\pi) = l(\rho\sigma_{i_0}) = l-2$, or, equivalently, $l(\rho) = l(\pi\sigma_{i_0}) = l(\pi) + 1$, and

$$\partial_{(i)} = \partial_{i_0}\partial_\rho = (\partial_{i_0})^2\partial_\pi = 0,$$

which completes the proof. □

Now we recall from 11.3.9 that

$$\omega_n = [n-1, \ldots, 0] = (0, n-1)(1, n-2)\ldots = \omega_n^{-1}$$

is the permutation of maximal reduced length in S_n: $l(\omega_n) = \binom{n}{2}$. Using this permutation we can associate with an arbitrary permutation $\pi \in S_n$ the operator $\partial_{\omega_n \pi}$. We apply this operator to the monomial

$$X^E := X^{E_n} := x_0^{n-1} x_1^{n-2} \ldots x_{n-2}^1,$$

obtaining the *Schubert polynomial*

8.6.3 $$X_\pi := \partial_{\omega_n \pi} X^E.$$

For example, if $n := 4$ and $\pi := (132)$, we have $\omega_n \pi = (031) = [3021]$, $L(\omega_n \pi) = 301$, $l(\omega_n \pi) = 4$, and

$$(\omega_n \pi)^{-1} = (23)(01)(12)(23) = \sigma_2 \sigma_0 \sigma_1 \sigma_2$$

is a reduced decomposition. We obtain

$$\begin{aligned} X_{(132)} &= \partial_2 \partial_0 \partial_1 \partial_2 x_0^3 x_1^2 x_2 = \partial_2 \partial_0 \partial_1 \frac{x_0^3 x_1^2 x_2 - x_0^3 x_1^2 x_3}{x_2 - x_3} \\ &= \partial_2 \partial_0 \partial_1 x_0^3 x_1^2 = \partial_2 \partial_0 \frac{x_0^3 x_1^2 - x_0^3 x_2^2}{x_1 - x_2} \\ &= \partial_2 \partial_0 (x_0^3 x_1 + x_0^3 x_2) = \partial_2 \frac{x_0^3 x_1 + x_0^3 x_2 - x_0 x_1^3 - x_1^3 x_2}{x_0 - x_1} \\ &= \partial_2 (x_0^2 x_1 + x_0 x_1^2 + x_0^2 x_2 + x_0 x_1 x_2 + x_1^2 x_2) = x_0^2 + x_0 x_1 + x_1^2. \end{aligned}$$

8.6.4 Lemma *The Schubert polynomials $X_\pi \in \mathbb{Z}[x_0, \ldots, x_{n-1}]$ have the following properties:*

- *The polynomial X_π is homogeneous with rational integral coefficients.*
- *The degree of X_π is equal to the reduced length of π.*
- *In the case when $\pi i < \pi(i+1)$, the polynomial X_π is symmetric in x_i and x_{i+1}.*
- *The application of ∂_i has the following effect:*

$$\partial_i X_\pi = \begin{cases} X_{\pi \sigma_i}, & \text{if } \pi i > \pi(i+1), \\ 0, & \text{otherwise.} \end{cases}$$

- *In particular we have:*

$$X_{\omega_n} = X^E, \ X_1 = 1.$$

The checks are easy and left as exercise 8.6.2. The main property of Schubert polynomials is that they generalize Schur polynomials and that they form a $\mathbb{Z}$–basis of $\mathbb{Z}[x_0, x_1, x_2, \ldots] := \cup_n \mathbb{Z}[x_0, \ldots, x_{n-1}]$. The last statement will follow from the next two lemmas.

8.6.5 Lemma *For each monomial $X^D := x_0^{d_0} \cdots x_{n-1}^{d_{n-1}}$ that occurs in $\partial_\pi X^E$ with a nonzero coefficient, we have that $D \leq E$, which means that for all ν we have $d_\nu \leq n-1-\nu$.*

Proof: By induction on the reduced length $k := l(\pi)$. The case $k = 0$ is trivial since $\pi = 1$. For the inductive step we assume that there exists a reduced decomposition of the form $\pi^{-1} = \sigma_i \cdots$ and we put $\pi' := \sigma_i \pi$. Since $l(\pi') = l(\pi) - 1$, the induction hypothesis applies to π'. The monomial summand X^D of $\partial_\pi X^E = \partial_i \partial_{\pi'} X^E$ is a summand of some

$$\partial_i X^B = \frac{X^B - \sigma_i X^B}{x_i - x_{i+1}},$$

X^B being a monomial summand of $\partial_{\pi'} X^E$. Hence the induction hypothesis applies to X^B: $B \leq E$. We distinguish three cases:

i) If $b_i = b_{i+1}$, then $\partial_i X^B = 0$, as X^B is symmetric in x_i and x_{i+1}. This cannot happen since X^D was supposed to occur.

ii) If $b_i > b_{i+1}$, then, by long division,

$$\partial_i X^B = \cdots x_i^{b_i - 1} x_{i+1}^{b_{i+1}} \cdots + \cdots x_i^{b_i - 2} x_{i+1}^{b_{i+1}+1} \cdots + \ldots + \cdots x_i^{b_{i+1}} x_{i+1}^{b_i - 1} \cdots.$$

This shows that the exponents d_i and d_{i+1} of a monomial summand X^D in $\partial_i X^B$ are less than or equal to $b_i - 1$ which is, by the induction hypothesis, applied to B, less than or equal to $n-1-(i+1)$, the $(i+1)$-th element of the sequence E, as it is stated.

iii) If $b_i < b_{i+1}$, then

$$\partial_i X^B = - \cdots x_i^{b_{i+1}-1} x_{i+1}^{b_i} \cdots - \cdots x_i^{b_{i+1}-2} x_{i+1}^{b_i+1} \cdots - \ldots - \cdots x_i^{b_i} x_{i+1}^{b_{i+1}-1} \cdots.$$

This shows that both d_i and d_{i+1} are $\leq b_i < b_{i+1}$, and so, by the induction hypothesis, applied to B, they are both $\leq n - 1 - i + 1$, the $(i+1)$-th element of the sequence E, as it is stated. □

8.6.6 Lemma *Assume that the monomial $X^D = x_0^{d_0} \cdots x_{n-1}^{d_{n-1}}$ occurs in $\partial_\pi X^E$ with nonzero coefficient, and suppose that D is the lexicographically smallest sequence of exponents with this property. Then*

– *The sequence D satisfies the equation*

$$D = E - L(\pi) := (n - 1 - l_0(\pi), \ldots, 1 - l_{n-2}(\pi), 0 - l_{n-1}(\pi)),$$

and

– *the coefficient of X^D in $\partial_\pi X_\pi$ is equal to 1.*

Proof: By induction on the reduced length $l(\pi)$ of π. The case $l(\pi) = 0$, which means $\pi = 1$, is trivial, since $X_{\omega_n} = X^E$, as we mentioned already, and $L(1) = \overline{0\ldots 0}$, so that $E - L(\pi) = E$, and the statement holds.

Consider now a permutation with reduced length $l(\pi) = k > 0$. There exists an m such that

$$L(\pi) = \overline{l_0 \ldots l_m 0 \ldots 0}, \text{ and } l_m \neq 0.$$

According to the position of m we distinguish the following two cases:

i) In the case when $m = n-2$, we have that $L(\pi) = \overline{\ldots l_{n-3}10}$ and $\pi(n-2) > \pi(n-1)$. We put

$$\rho := L^{-1}(\overline{\ldots l_{n-3}00}), \text{ i. e. } \rho = \pi\sigma_{n-2}, \text{ by 11.3.4.}$$

Thus $\partial_\pi X^E = \partial_{n-2}\partial_\rho X^E$. The induction hypothesis yields that the monomial X^C in $\partial_\rho X^E$ with lexicographically smallest C satisfies

$$C = E - L(\rho) = (\ldots, 2 - l_{n-3}, 1, 0).$$

Since ∂_{n-2} only affects x_{n-2} and x_{n-1}, we have

$$X^D = \partial_{n-2} X^C = \frac{X^C - \sigma_{n-2} X^C}{x_{n-2} - x_{n-1}} = X^{(\ldots,2-l_{n-3},0,0)} = X^{E-L(\pi)}.$$

ii) In the case when $m < n-2$ we put

$$\rho := L^{-1}(\overline{\ldots, l_{m-1}, 0, l_m - 1, 0 \ldots 0}),$$

so that, by 11.3.4, $\rho = \pi\sigma_m$, and therefore $\partial_\pi X^E = \partial_m \partial_\rho X^E$. The monomial X^C in $\partial_\rho X^E$ with lexicographically smallest C has $C = E - L(\rho)$, which is the sequence

$$(\ldots, n - m - l_{m-1}, n - m - 1, n - m - 1 - l_m, n - m - 3, \ldots, 1, 0).$$

The other monomials X^B occurring in $\partial_\rho X^E$ have lexicographically bigger sequences B of exponents and we distinguish two cases according to the relationship between $(b_0, \ldots, b_{m-1})$ and $(c_0, \ldots, c_{m-1})$.

1. Assume first that $(b_0, \ldots, b_{m-1}) > (c_0, \ldots, c_{m-1})$. As ∂_m only affects x_m and x_{m+1}, all these B are lexicographically bigger than the sequences of the monomials in $\partial_m X^C$, which, by long division, is equal to

$$\cdots x_m^{n-m-2} x_{m+1}^{n-m-1-l_m} \cdots + \cdots x_m^{n-m-3} x_{m+1}^{n-m-l_m} \cdots$$

$$+ \ldots + \cdots x_m^{n-m-1-l_m} x_{m+1}^{n-m-2} \cdots.$$

The last one of these summands is clearly the lexicographically smallest sequence D of exponents, and $D = E - L(\pi)$, moreover, the coefficient is 1, so that the statements hold in this particular case.

2. Finally we consider the case $(b_0, \dots, b_{m-1}) = (c_0, \dots, c_{m-1})$:
   As $c_m = n - m - 1$, b_m cannot be bigger, by 8.6.5, and therefore it suffices to consider sequences B such that

$$(b_0, \dots, b_m) = (c_0, \dots, c_m), \text{ and } b_{m+1} > c_{m+1} = n - m - 1 - l_m.$$

   Consider such a monomial summand X^B of $\partial_\rho X^E$. It contributes $\partial_m X^B$ to $\partial_\pi X^E$. Long division shows that the lexicographically smallest sequence of exponents which thereby occurs is

$$A := (b_0, \dots, b_{m-1}, b_{m+1}, n - m - 2, b_{m+2}, \dots) > C = E - L(\pi),$$

   which completes the proof.

□

Since $L(\omega_n \pi) = E - L(\pi)$, this has the following consequence:

8.6.7 Corollary *The monomial X^D occurring in the Schubert polynomial X_π with lexicographically smallest sequence D of exponents is $X^D = X^{L(\pi)}$, and its coefficient is 1.*

This has very important implications. To begin with we note that, by 8.6.7, the Schubert polynomials $X_\pi, \pi \in S_n$, form a $\mathbb{Z}$–basis of the abelian subgroup generated by the monomials X^B, $B \leq E_n$, in formal terms:

8.6.8 $$\mathbb{Z}_{E_n}[x_0, \dots, x_{n-1}] := \langle X^B \mid B \leq E_n \rangle_{\mathbb{Z}} = \ll X_\pi \mid \pi \in S_n \gg_{\mathbb{Z}} .$$

Now we consider the ring of polynomials

$$\mathbb{Z}\,[x_0, x_1, x_2, \dots] := \bigcup_{n>0} \mathbb{Z}\,[x_0, \dots, x_{n-1}],$$

together with the group

$$\mathcal{S} := \bigcup_{n>0} S_n,$$

using the natural embeddings of $\mathbb{Z}\,[x_0, \dots, x_{n-1}]$ into $\mathbb{Z}[x_0, \dots, x_{m-1}]$ and of S_n into $\mathcal{S}$, for each $n < m$. Let us see what happens with X_π, if we use this natural embedding of S_n into S_{n+1}:

$$S_n \hookrightarrow S_{n+1} : \pi \mapsto \pi',$$

where S_n is mapped onto the set of elements π' that keep the point n fixed. Since

$$X_{\pi'} = \partial_{\omega_{n+1}\pi'} X^{E_{n+1}} = \partial_{\omega_n \pi} \partial_{n-1} \partial_{n-2} \cdots \partial_0 X^{E_{n+1}} = \partial_{\omega_n \pi} X_n^E,$$

we obtain the following helpful result:

8.6.9 $$X_\pi = X_{\pi'},$$

which shows that the Schubert polynomials are invariant under embedding. We can therefore introduce the Schubert polynomial X_π for any $\pi \in \mathcal{S}$. As any monomial X^B satisfies $B \leq E_n$, for n large enough, the following is now obtained from 8.6.8:

8.6.10 Theorem *The Schubert polynomials $X_\pi, \pi \in S_n, n > 0$, form a $\mathbb{Z}$–basis of $\mathbb{Z}[x_0, x_1, x_2, \ldots]$:*

$$\mathbb{Z}[x_0, x_1, x_2, \ldots] = \ll X_\pi \mid \pi \in \mathcal{S} \gg_{\mathbb{Z}} .$$

The coefficients of the $\mathbb{Z}$–linear combination of Schubert polynomials which is equal to $p \in \mathbb{Z}[x_0, x_1, x_2, \ldots]$ can be obtained by successive applications of 8.6.7.

For example,

$$p := x_0 + x_0x_1 + x_0x_1x_2 + x_0x_1x_2x_3$$

has lexicographically smallest monomial summand $x_0 = X_{(01)}$, while the lexicographically smallest monomial which occurs in $p - X_{(01)}$ is $x_0x_1 = X_{(012)}$, and so on. We finally obtain in this way that

$$p = X_{(01)} + X_{(012)} + X_{(0123)} + X_{(01234)}.$$

Certain Schubert polynomials are easily seen to be monomials. In fact, if $D \leq E$ is weakly decreasing, then there exists an i such that $d_i > d_{i+1}$, and so

$$X^D = \partial_i x_0^{d_0} \cdots x_i^{d_i+1} x_{i+1}^{d_i} \cdots ,$$

and therefore, by decreasing induction we obtain

8.6.11 Corollary *For weakly decreasing sequences D of exponents we have that, for $\pi := L^{-1}(\overline{d_0 \ldots d_{n-1}})$, the following is true:*

$$X_\pi = X^D,$$

or, in formal terms,

$$X^D = X_{L^{-1}(D)}.$$

Thus weakly *decreasing* sequences D lead to monomials, and it is surprising to see that weakly *increasing* sequences lead to Schur polynomials. The proof is cumbersome but elementary. Let us consider its main steps.

The determinantal formula for Schur polynomials $\{\alpha, X\}$ over the set of indeterminates X reads as follows (recall 6.3.6):

$$\{\alpha, X\} = \det(\{\alpha_i + j - i, X\}) .$$

(Recall that, by definition, $\{0, X\} := 1$, and $\{m, X\} := 0$, if $m < 0$.)

If we replace the set X of indeterminates of the Schur polynomials in the i-th row by the *subset* $\{x_0, \ldots, x_i\}$, then we obtain the *multi Schur polynomial*. For $\alpha := (\alpha_0, \ldots, \alpha_{h-1}) \vdash n$, where $\alpha_{h-1} > 0$, it is defined to be

$$\{\alpha, \{x_0\}, \ldots, \{x_0, \ldots, x_{h-1}\}\} := \det(\{\alpha_i + j - i, \{x_0, \ldots, x_i\}\}) .$$

These polynomials have very interesting properties. The application of ∂_j amounts to the application of ∂_j to the j-th row of the determinant, since the other rows are symmetric in x_j and x_{j+1}. Moreover, it is not difficult to derive from the definitions of ∂_i and of $\{m\}$, that

$$\partial_j\{m, \{x_0, \ldots, x_j\}\} = \{m-1, \{x_0, \ldots, x_{j+1}\}\}.$$

For example $\{(3,2,1), \{x_0\}, \ldots, \{x_0, x_1, x_2\}\}$ is equal to

$$\det\begin{pmatrix} x_0^3 & x_0^4 & x_0^5 \\ x_0+x_1 & x_0^2+x_0x_1+x_1^2 & x_0^3+x_0^2x_1+x_0x_1^2+x_1^3 \\ 0 & 1 & x_0+x_1+x_2 \end{pmatrix}$$
$$= x_0^3x_1^2x_2^1,$$

and

$$\partial_2\{(3,2,1), \{x_0\}, \ldots, \{x_0, x_1, x_2\}\} = \partial_2 x_0^3x_1^2x_2^1 = x_0^3x_1^2$$
$$= \det\begin{pmatrix} x_0^3 & x_0^4 & x_0^5 \\ x_0+x_1 & x_0^2+x_0x_1+x_1^2 & x_0^3+x_0^2x_1+x_0x_1^2+x_1^3 \\ 0 & 0 & 1 \end{pmatrix}.$$

More generally, the following holds (exercise 8.6.3):

8.6.12 $$\{\alpha, \{x_0\}, \ldots, \{x_0, \ldots, x_{h-1}\}\} = X^\alpha := x_0^{\alpha_0} \cdots x_{h-1}^{\alpha_{h-1}}.$$

From this equation we obtain by induction, since the longest element $\omega_h \in S_h$ and the longest element ω_{h-1} in the subgroup S_{h-1} (the stabilizer of the point $h-1$) satisfy

$$\omega_h = \sigma_0 \cdots \sigma_{h-2}\omega_{h-1},$$

the identity

$$\partial_{\omega_h}\{\alpha, \{x_0\}, \ldots, \{x_0, \ldots, x_{h-1}\}\} = \{\alpha_0-(h-1), \ldots, \alpha_{h-2}-1, \alpha_{h-1}, X\},$$

and from this we finally can derive

8.6.13 $$\partial_{\omega_h} X^{E+\alpha} = \{\alpha, X\}.$$

For example,

$$\partial_{\omega_2} X^{E+(3,2)} = \partial_0 x_0^4x_1^2 = x_0^3x_1^2 + x_0^2x_1^3 = \{(3,2), \{x_0, x_1\}\}.$$

We are now in a position to prove that increasing sequences lead to Schur polynomials:

8.6.14 Theorem (Lascoux/Schützenberger) *Let $D = (d_0, \ldots, d_{n-1})$ denote a weakly increasing sequence of natural numbers and α the proper partition obtained by reordering the d_i. If $\pi^{-1} \in \mathcal{S}$ has reduced Lehmer code D:*

$$L(\pi^{-1})^+ = \overline{d_0 \ldots d_{n-1}},$$

then

$$X_\pi = \{\alpha, \{x_0, \ldots, x_{n-1}\}\}.$$

Proof: We put $m := \alpha_0 + n = d_{n-1} + n$, and we define the natural numbers k_i by

$$L((\omega_m \pi)^{-1})^+ = E_m - D = \overline{k_0, \ldots, k_{n-1}, m-n-1, \ldots, 1}.$$

Thus $k_j = m - (j+1) - d_j > m - (j+2) - d_{j+1} = k_{j+1}$, which means that the k_i form a strictly decreasing sequence of natural numbers. We can therefore apply 11.3.4, obtaining

$$(\omega_m \pi)^{-1} = L^{-1}(\overline{k_{n-1}, \ldots, k_0 - (n-1), m-n-1, \ldots, 1})\omega_n.$$

We now define a permutation ρ by

$$(\omega_m \rho)^{-1} := L^{-1}(\overline{k_{n-1}, \ldots, k_0 - (n-1), m-n-1, \ldots, 1}).$$

The above equation shows in particular, that a reduced decomposition of $(\omega_m \pi)^{-1}$ can be obtained by multiplying a reduced decomposition of $(\omega_m \rho)^{-1}$ by a reduced decomposition of ω_n, and therefore $\partial_{\omega_m \pi} = \partial_{\omega_n} \partial_{\omega_m \rho}$. This gives

$$X_\pi = \partial_{\omega_n} \partial_{\omega_m \rho} X^{E_m} = \partial_{\omega_n} X_\rho.$$

Hence, by 8.6.11, it suffices to prove that

$$X_\rho = X^{E_n + \alpha}.$$

This identity follows, by an application of 8.6.13, from

$$L(\rho)^+ = \overline{m-1-k_{n-1}, \ldots, m-n-(k_0-(n-1)), 0, \ldots, 0}$$
$$= \overline{n-1+d_{n-1}, \ldots, n-n+d_{n-n}},$$

which is a decreasing sequence. □

Hence the Schubert polynomials generalize the Schur polynomials. Since they form a basis of the union of the $\mathbb{Z}[x_0, \ldots, x_{n-1}]$, they can also serve for a different approach to the Littelwood–Richardson Rule. This will not be described in detail here since lack of space, but it should be mentioned that this approach (introduced by Lascoux and Schützenberger) is very well suited for computer calculations, and therefore it was chosen for the corresponding procedure in the program system SYMMETRICA. For this reason I add a section with the 120 first Schubert polynomials to the appendix of tables.

Exercises

Exercise 8.6.1 Show that, for each $f, g \in \mathbb{Z}[x_0, x_1, x_2, \ldots]$ and $0 \le i \le n-2$,

$$\partial_i(fg) = (\partial_i f)\sigma_i g + f(\partial_i g).$$

Exercise 8.6.2 Prove 8.6.4.

Exercise 8.6.3 Show that 8.6.12 is true.

9. Construction and Generation

We have counted the orbits of finite groups on finite sets and successively refined our methods by introducing enumeration by weight as well as enumeration by stabilizer class. Moreover we discussed actions on structured sets like posets and semigroups. Later on the permutation group representations were refined by introducing linear representations, which led to applications in both directions. It remains to discuss the most difficult problem, the *construction of a transversal of the orbits.* We shall briefly discuss the general case of this problem, in order to introduce the concept of Sims chains, for cases when the acting group is given by generators and relations. We shall then use it in a detailed description of a direct evaluation of a transversal of the orbits of G on Y^X with prescribed content λ. After that we describe a recursive method, using recursion on $|Y|$, and combining this recursion with the orderly generation method that was introduced by R. C. Read.

These methods can be used for the evaluation of catalogs of discrete structures that can be defined as orbits of finite groups on finite sets, and in particular of discrete structures which can be considered as symmetry classes of mappings. For example, a catalog of all the graphs on $v \leq 12$ vertices was obtained in this way, as well as a catalog of 0-1-matrices under the action of the direct product of the symmetric groups on the rows and columns (the *contexts* that are of interest for the concept analysis, for example, or the isomorphism classes of *incidence structures*).

It is clear that for higher v it is nearly impossible to get such a catalog of graphs, its cardinality is much too big. But nevertheless there are cases where one wants to try a hypothesis on graphs on 15 or 20 vertices, say. In these cases we can apply a recent and very important method of *generating orbit representatives uniformly at random,* which will also be described. It can be used, for example, in order to test graph invariants, and to do all kinds of examinations of structures that can be defined as orbits of finite actions by inspection of big sets of examples. It helps, say, easily to get nonisomorphic labeled graphs with same edge degree sequence and same characteristic polynomial, if we want to demonstrate that these two invariants are not complete, even if we put them together.

Finally we shall describe the corresponding problem in linear representation theory which is the evaluation of symmetry adapted bases. Such bases serve very well whenever there are symmetries.

9.1 Orbit Evaluation

We consider a finite action ${}_GX$ in order to evaluate particular orbits, the whole set of orbits, a transversal of the orbits, stabilizers, and so on. It is clear that for all these calculations the orbit evaluation is basic, hence let us discuss this first.

In the case when both $|X|$ and $|G|$ or $|\bar{G}|$ are very small and G or $\bar{G}$ is given as a *set* together with the operation of each of its elements, then we may just apply this set to x in order to get the desired orbit $G(x)$. But quite often it is so that G is given by a set of generators: $G = \langle g_0, \ldots, g_{r-1}\rangle$, together with the actions of the g_i on X. Then, for $x \in X$, we can put

$$\Omega_0 := \{x\},\ \Omega_1 := \bigcup_{j \in r} g_j\Omega_0 = \bigcup_{j \in r}\{g_jx\},$$

and

$$\forall\, i \geq 2\colon\ \Omega_i := \Omega_{i-1} \bigcup \left(\bigcup_{j \in r} g_j\,(\Omega_{i-1}\backslash\Omega_{i-2}) \right).$$

It is obvious that the smallest i such that $\Omega_i = \Omega_{i-1}$ satisfies

9.1.1 $$G(x) = \Omega_{i-1}.$$

The concrete implementation of this way of evaluating $G(x)$ of course may heavily depend on our knowledge of X, G and ${}_GX$. For example, if $|X| = 1$, then $G(x) = \Omega_0$, while $G(x) = X$, if $|G \backslash\!\backslash X| = 1$, so that the Cauchy-Frobenius lemma can serve as a *stopping rule*. The knowledge of $|G \backslash\!\backslash X|$ is helpful in particular if we are after the whole set of orbits $G \backslash\!\backslash X$, in which case we proceed with the remaining subset $X\backslash G(x)$ correspondingly. The implementation of this method is obvious.

It is clear that a careful implementation of this procedure also yields products of the generators which lead from x to any other element of its orbit, and which therefore form a transversal of G/G_x. Thus we can also obtain generators of the stabilizer G_x by an application of the following fact:

9.1.2 Lemma (Schreier) *If U is a subgroup of $G = \langle g_0, \ldots, g_{r-1}\rangle$ which is finite and which decomposes as follows into left cosets of U:*

$$G = \bigcup_{i \in s} h_iU,\ \text{where } h_0 = 1,$$

and if the mapping ϕ is defined by

$$\phi\colon G \to \{h_0, \ldots, h_{s-1}\}\colon g \mapsto h_i,\ \text{if } g \in h_iU,$$

then U is generated by the elements $\phi(g_ih_k)^{-1}g_ih_k$:

$$U = \langle \phi(g_ih_k)^{-1}g_ih_k \mid i \in r, k \in s\rangle.$$

Proof: Assume that $u = a_0 \dots a_{t-1} \in U$, where the a_i are elements of the generating set $\{g_0, \dots, g_{r-1}\}$. We put

$$u_i := a_i \dots a_{t-1}, \text{ and } x_i := \phi(u_i),$$

so that in particular $x_0 = \phi(u) = 1$. Putting $x_t := 1$, we can rewrite u in the form

$$u = (x_0^{-1} a_0 x_1)(x_1^{-1} a_1 x_2) \dots (x_{t-1}^{-1} a_{t-1} x_t).$$

Now $x_i = \phi(u_i) = \phi(a_i u_{i+1}) = \phi(a_i x_{i+1})$, for $x_{i+1} = \phi(u_{i+1})$ means that $u_{i+1} = x_{i+1} u'$, for a suitable $u' \in U$, and hence $a_i u_{i+1} = a_i x_{i+1} u'$. Thus

$$u = \phi(a_0 x_1)^{-1} a_0 x_1 \phi(a_1 x_2)^{-1} a_1 x_2 \dots,$$

which completes the proof. □

A direct evaluation of the stabilizer G_x will be discussed later. In the case when X is too big to be stored, we can do the following in order to evaluate the set of orbits. We number the elements of X, being now faced with an operation of G on n, say, so that there is a *canonic* transversal of $G \backslash\backslash n$, consisting of the smallest orbit elements. So we take $0 \in n$ as the starting element of the desired transversal. Then we evaluate the minimal $i \in n$ which is not contained in $G(0)$, take this point i as the next element of the transversal and evaluate the minimal $j \in n$ that is not in $G(i)$ and bigger than i. There are now two cases. Either there is a $g \in G$ such that $gj < j$, in which case $j \in G(0)$, or there is no such g, in which case $j \notin G(0) \cup G(i)$, and therefore j has to be added to the transversal, and so on.

In order to evaluate this canonic transversal of $G \backslash\backslash n$ in an economic way we can use a problem oriented description of the elements of $\bar{G}$ so that not too many checks are necessary. For this purpose we use the *pointwise* stabilizer of the subset $k \subseteq n$,

$$C_{\bar{G}}(k) := \{\pi \in \bar{G} \mid \forall\, i \in k\colon\ \pi i = i\}.$$

We called this group the *centralizer* of the set k in order to distinguish it from the *setwise* stabilizer

$$N_{\bar{G}}(k) := \{\pi \in \bar{G} \mid \forall\, i \in k\colon\ \pi i \in k\},$$

which we called the *normalizer* of the set k. These centralizers form the chain of subgroups

$$\{1\} = C_{\bar{G}}(n) \leq C_{\bar{G}}(n-1) \leq \dots \leq C_{\bar{G}}(0) = \bar{G}.$$

Hence there exists a smallest b such that

9.1.3 $$\{1\} = C_{\bar{G}}(b) \leq \dots \leq C_{\bar{G}}(0) = \bar{G}.$$

We call the set b the *base* of the action of G on n, its cardinality b will be called its *length*. The chain 9.1.3 is called the *Sims chain* of this action. Now we consider left coset decompositions

9.1.4 $$C_{\bar{G}}(i-1) = \bigcup_{j \in \underline{r(i)}} \pi_j^{(i)} C_{\bar{G}}(i),\ \pi_1^{(i)} = 1.$$

We note that each $\pi \in \bar{G}$ can *uniquely* be written in terms of these coset representatives as follows:

9.1.5 $$\pi = \pi_{j_1}^{(1)} \cdots \pi_{j_b}^{(b)}.$$

Hence in particular the following holds:

9.1.6 $$|\bar{G}| = \prod_{i \in \underline{b}} r(i).$$

This shows that storing the $\pi_j^{(i)}, i \in \underline{b}, j \in \underline{r(i)}$, allows an economic way to deal with the elements of $\bar{G}$, we have to run through the elements of the following tree:

$$\begin{array}{ccccccc} & \pi_1^{(1)} & \ldots & \ldots & \ldots & \pi_{r(1)}^{(1)} & \\ & \swarrow \ldots \searrow & & \ldots & & \swarrow \ldots \searrow & \\ \pi_1^{(1)}\pi_1^{(2)} & \ldots & \pi_1^{(1)}\pi_{r(2)}^{(2)} & \ldots & \pi_{r(1)}^{(1)}\pi_1^{(2)} & \ldots & \pi_{r(1)}^{(1)}\pi_{r(2)}^{(2)} \\ \swarrow \ldots \searrow & \ldots & \swarrow \ldots \searrow & \ldots & \swarrow \ldots \searrow & \ldots & \swarrow \ldots \searrow \end{array}$$

The goal is to cut this tree as much as possible. A first remark concerning this shows, that in each orbit evaluation we can cut this tree below a certain level, depending on the point the orbit of which is to be evaluated: As $\pi_j^{(i)} k = k$, if $i > k$, the following is true for the orbit of $k \in n$ under G:

9.1.7 $$G(k) = \{\pi_{j_1}^{(1)} \ldots \pi_{j_k}^{(k)} k \mid j_1 \in \underline{r(1)}, \ldots, j_k \in \underline{r(k)}\}.$$

Thus the knowledge of the Sims chain considerably reduces the number of necessary checks for an orbit evaluation. The calculation of the base is therefore very important.

9.1.8 Example Let us consider for example the action of S_p on the set of 2–element subsets of p. In order to evaluate the base of $\bar{G} = S_p^{[2]}$ we first of all embed the set of 2–element subsets into n, where $n := \binom{p}{2}$, by ordering the pairs lexicographically:

$$\{0,1\} < \ldots < \{0, p-1\} < \{1,2\} < \ldots < \{1, p-1\} < \ldots < \{p-2, p-1\},$$

and by replacing the pairs correspondingly by the natural numbers $0, \ldots, \binom{p}{2} - 1$. An easy check shows that, according to this numbering, the following chain of canonic Young subgroups

$$S_p \geq S_{(2,p-2)} \geq S_{(1,1,1,p-3)} \geq \ldots \geq S_{(1,\ldots,1,2)} \geq \{1\}$$

yields, via embedding, the Sims chain

$$\bar{G} = S_p^{[2]} \geq S_{(2,p-2)}^{[2]} \geq S_{(1,1,1,p-3)}^{[2]} \geq \ldots \geq S_{(1,\ldots,1,2)}^{[2]} \geq \{1\}.$$

Thus $p-2$ is the base for the canonic action of $S_p^{[2]}$ on the set of 2–element subsets and its lexicographic ordering. Hence in particular for $p := 5$ we obtain the Sims chain (with respect to the above numbering of the pairs of points)

$$S_5^{[2]} \geq S_{(2,3)}^{[2]} \geq S_{(1,1,1,2)}^{[2]} \geq \{1\}.$$

◇

Another helpful property of the base is described in

9.1.9 Lemma *The elements $\pi \in \bar{G}$ are uniquely determined by their action on the base b.*

Proof: If $\pi i = \rho i$, for each $i \in b$, then $\pi^{-1}\rho \in C_{\bar{G}}(b) = 1$, so that $\pi = \rho$. □

Once the base b is at hand we can evaluate the orbits $\bar{G}(k)$, $k \in b$, as it is described above by 9.1.7. These orbits can serve very well for a check if $\pi \in S_n$ is contained in $\bar{G}$ or not: Consider $\pi 0$. If $\pi 0 \notin \bar{G}(0)$, then obviously $\pi \notin \bar{G}$, and we can stop. If otherwise $\pi 0$ is contained in the orbit $\bar{G}(0) = C_{\bar{G}}(0)(0)$, then there exists a unique left coset representative $\pi_{j_1}^{(1)}$ such that $\pi_{j_1}^{(1)} 0 = \pi 0 \in C_{\bar{G}}(0)(0)$, and therefore $(\pi_{j_1}^{(1)})^{-1}\pi \in C_{\bar{G}}(1)$. In the next step we check if $(\pi_{j_1}^{(1)})^{-1}\pi 1$ lies in the orbit $C_{\bar{G}}(1)(1)$ of 1 under the action of $C_{\bar{G}}(1)$. If this is not the case, then $\pi \notin \bar{G}$. In the other case there is exactly one orbit representative $\pi_{j_2}^{(2)}$ such that $(\pi_{j_1}^{(1)})^{-1}\pi 1 = \pi_{j_2}^{(2)} 1$. Hence, by induction, we obtain

9.1.10 Corollary *Either there exist $\pi_{j_i}^{(i)} \in C_{\bar{G}}(i)$, $i \in \underline{b}$, for which*

$$(\pi_{j_i}^{(i)})^{-1} \cdots (\pi_{j_1}^{(1)})^{-1}\pi i \in C_{\bar{G}}(i),$$

and

$$(\pi_{j_i}^{(i)})^{-1} \cdots (\pi_{j_1}^{(1)})^{-1}\pi i = \pi_{j_{i+1}}^{(i+1)} i,$$

so that

$$\pi = \pi_{j_1}^{(1)} \cdots \pi_{j_b}^{(b)} \in \bar{G},$$

or π is not *an element of $\bar{G}$.*

Exercises

Exercise 9.1.1 Apply Schreier's Lemma in order to derive from

$$S_n = \langle (0, 1), (0, \ldots, n-1) \rangle$$

a two generator system for the alternating group A_n.

9.2 Transversals of Symmetry Classes

In this section we shall restrict attention to a redundancy free construction of a transversal of the G–classes on $Y^X := m^n$. Moreover, in order to decrease the complexity we restrict attention to G–classes of fixed content $\lambda = (\lambda_0, \ldots, \lambda_{m-1}) \models n$, starting off from the *canonic mapping* f with this content:

$$f_\lambda := (f_\lambda(0), \dots, f_\lambda(n-1)) := (\underbrace{0, \dots, 0}_{\lambda_0}, \underbrace{1, \dots, 1}_{\lambda_1}, \dots, \underbrace{m-1, \dots, m-1}_{\lambda_{m-1}}).$$

The set of all the mappings of this content will be indicated as follows:

$$m^n_\lambda := \{\pi f_\lambda = f_\lambda \circ \pi^{-1} \mid \pi \in S_n\}.$$

Since a permutation of the arguments does not change the content, this set m^n_λ is a union of orbits of G on m^n:

$$G \backslash\!\backslash m^n_\lambda \subseteq G \backslash\!\backslash m^n.$$

We would like to construct a transversal of m^n_λ. This reduces the complexity since the desired transversal of $G \backslash\!\backslash m^n$ is the union of the transversals of the sets $G \backslash\!\backslash m^n_\lambda$, taken over all the $\lambda \models n$. The crucial point is now that in each orbit of G on m^n_λ there is a unique representative which is the *lexicographically smallest* element of its orbit. Therefore we may call these lexicographically smallest elements a *canonic transversal* of $G \backslash\!\backslash m^n_\lambda$. A mapping $f \in m^n_\lambda$ can be displayed by its λ*–tabloid*

$$\begin{array}{l} \overline{\underline{i_0 \dots i_{\lambda_0 - 1}}} \\ \underline{j_0 \dots j_{\lambda_1 - 1}} \\ \dots \end{array},$$

where we put the elements of the inverse image $f^{-1}(k)$ of $k \in m$ into the k-th row, in increasing order, so that e. g.

$$f(i_0) = \dots = f(i_{\lambda_0 - 1}) = 0, \text{ and } i_0 < \dots < i_{\lambda_0 - 1}.$$

Let us consider as an example the graphs on 4 vertices which contain exactly 2 edges, so that $n = \binom{4}{2} = 6$, $m = 2$, and $\lambda = (4, 2)$. Before we write down all the 15 (4, 2)–tabloids in full detail, it is practical to note that in each tabloid the first (or any other) row is uniquely determined by the remaining rows which form the *truncated* tabloid. Hence in the present case we need only to display the second rows, here they are:

$$\overline{\underline{01}},\ \overline{\underline{02}},\ \overline{\underline{03}},\ \overline{\underline{04}},\ \overline{\underline{05}},\ \overline{\underline{12}},\ \overline{\underline{13}},\ \overline{\underline{14}},\ \overline{\underline{15}},\ \overline{\underline{23}},\ \overline{\underline{24}},\ \overline{\underline{25}},\ \overline{\underline{34}},\ \overline{\underline{35}},\ \overline{\underline{45}}.$$

According to the above arguments we have to find the orbits of $\bar{S}_4$ on this set. Recall that the 6 entries $0, \dots, 5 \in 6$ (of the tabloids) on which S_6 acts, stand for the $6 = \binom{4}{2}$ pairs $\{0, 1\}, \dots, \{2, 3\}$ of vertices. The subgroup

$$\bar{S}_4 := S_4^{[2]} \hookrightarrow S_6$$

is the permutation group induced by $S_4 = \langle (01), (0123) \rangle$ on this set of pairs of vertices. Thus

$$\bar{S}_4 = \langle (13)(24), (0352)(14) \rangle,$$

and therefore one of the two orbits of $\bar{S_4}$ on the set of (4, 2)–tabloids is

$$\omega_1 := \{\overline{\underline{05}}, \overline{\underline{14}}, \overline{\underline{23}}\},$$

the other orbit ω_0 consists of the remaining 12 truncated tabloids. (Note that in a preprocessing calculation we can obtain, by an application of the enumeration theory described above, that the number of graphs with 4 vertices and 2 edges is *two*, a result which is very helpful as *stopping rule*, as the present example already shows!)

The lexicographically smallest elements of these orbits ω_0, ω_1 are the tabloids corresponding to $\overline{\underline{45}} \in \omega_0$ and $\overline{\underline{23}} \in \omega_1$. These tabloids are

$$\begin{array}{c}\overline{\underline{0123}}\\ \underline{45}\end{array} \quad \text{and} \quad \begin{array}{c}\overline{\underline{0145}}\\ \underline{23}\end{array}.$$

The corresponding mappings $f, f' \in 2^6$ are

$$f = (0, 0, 0, 0, 1, 1), \text{ and } f' = (0, 0, 1, 1, 0, 0).$$

Hence the graphs shown in figure 9.1 form a complete set of graphs on 4 vertices containing 2 edges! This method is cumbersome, it can be used for cataloging the

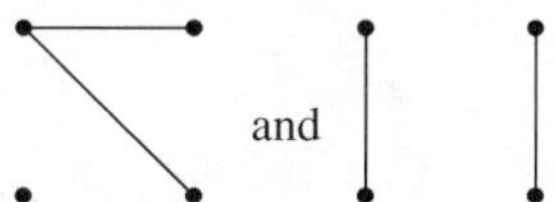

Fig. 9.1. The graphs with 4 vertices and 2 edges

graphs on ≤ 7 vertices, say. We want therefore to describe further refinements which allow to catalog the graphs with up to 11 or maybe even 12 vertices (there are exactly 165 091 172 592 unlabeled graphs on 12 vertices).

Having displayed an example in full, let us go into detail in order to make the procedure of finding the minimal representatives more efficient. Recall that we have to check if a given f is the lexicographically smallest element in its orbit, i. e. if the following is true:

9.2.1 $$\forall\, \sigma \in \bar{G}\colon\ f \leq f \circ \sigma, \quad \text{for short: } f \leq \bar{G}(f).$$

The verification of this condition is one of the crucial parts of the whole procedure. Let us call this check the *minimality test* for f. In order to carry it out in a reasonable way we recall what has been said about Sims chains. Assume that, as above, $b \subseteq n$ is the base for the action of $\bar{G}$ on n (still $f \in m^n$), so that

$$\{1\} = C_{\bar{G}}(b) \leq C_{\bar{G}}(b-1) \leq \ldots \leq C_{\bar{G}}(0) = \bar{G},$$

and we assume that left coset representatives $\pi_j^{(i)}$ ar at hand such that

$$C_{\bar{G}}(i-1) = \bigcup_{j=1}^{r(i)} \pi_j^{(i)} C_{\bar{G}}(i), \text{ where } \pi_1^{(i)} := 1 \in C_{\bar{G}}(i).$$

Thus, in order to apply the minimality test 9.2.1, we have to run through the elements of $\bar{G}$, each of which can uniquely be written in the form

$$\sigma = \pi_{j_1}^{(1)} \cdots \pi_{j_b}^{(b)}.$$

This means that in order to apply *each* element of $\bar{G}$ to f, say, we have to run through the following tree that was mentioned above already:

9.2.2
$$\begin{array}{ccccccccc} & \pi_1^{(1)} & & \cdots & \cdots & \cdots & & \pi_{r(1)}^{(1)} & \\ & \swarrow \cdots \searrow & & & \cdots & & & \swarrow \cdots \searrow & \\ \pi_1^{(1)}\pi_1^{(2)} & & \cdots & \pi_1^{(1)}\pi_{r(2)}^{(2)} & \cdots & \pi_{r(1)}^{(1)}\pi_1^{(2)} & \cdots & & \pi_{r(1)}^{(1)}\pi_{r(2)}^{(2)} \\ \swarrow \cdots \searrow & & \cdots & \swarrow \cdots \searrow & \cdots & \swarrow \cdots \searrow & \cdots & & \swarrow \cdots \searrow \end{array}$$

It is very important to cut this tree down as much as possible. This can be done, for example, by an application of

9.2.3 Lemma *If $f < f \circ \pi_j^{(i)}$ and $f(i) < f \circ \pi_j^{(i)}(i)$, then we have, for the orbit of f under the action of $C_{\bar{G}}(i) \leq \bar{G}$:*

$$f < C_{\bar{G}}(i)(f \circ \pi_j^{(i)}).$$

Proof: For $\tau \in C_{\bar{G}}(i)$ and $k \in i-1$ we have

$$f \circ \pi_j^{(i)} \circ \tau^{-1}(k) = f(k),$$

while

$$f \circ \pi_j^{(i)} \circ \tau^{-1}(i) = f \circ \pi_j^{(i)}(i) > f(i).$$

□

This result allows us to *cut off the branch starting with the element* $\cdots \pi_j^{(i)}$ from the tree 9.2.2 formed by the elements of $\bar{G}$.

Besides this we can choose a suitable *way in which we work through the tree* 9.2.2 in order to find the smallest element in the orbit $G(f)$. A remark that will lead us to a good such choice uses the fact that $f \in m^n$ will not in general be injective. Assuming $\tau, \sigma \in S_n \setminus \{1\}$ such that $f \circ \sigma = f \circ \tau$ we have, for each $i \in n$, the equation

9.2.4
$$\sigma^{-1} C_{\bar{G}}(i)(f) = \tau^{-1} C_{\bar{G}}(i)(f).$$

We can therefore subdivide $\bar{G}$ into set theoretic differences of centralizers according to the Sims chain and the base b:

9.2.5
$$\bar{G} = \bigcup_{i=1}^{b}(C_{\bar{G}}(b-i) - C_{\bar{G}}(b-i+1)) \bigcup \{1\}.$$

Correspondingly we shall run through $\bar{G}$ (in any case, whether f is injective or not) by working through the following differences of centralizers, one row after the other and from top to bottom:

1. $C_{\bar{G}}(b-1) - C_{\bar{G}}(b)$,
2. $C_{\bar{G}}(b-2) - C_{\bar{G}}(b-1)$,

$\vdots$

b. $C_{\bar{G}}(\emptyset) - C_{\bar{G}}(1)$.

The identity element can be left out since $f \circ 1 = f$. We note that in step number $b - i + 1$ we are running through the difference set

$$C_{\bar{G}}(i-1) - C_{\bar{G}}(i) = \bigcup_{j=2}^{r(i)} \pi_j^{(i)} C_{\bar{G}}(i),$$

if we had $f \leq C_{\bar{G}}(i)(f)$ before, assuming that $\pi_1^{(i)} = 1$, for each i, so that we can always start with $j = 2$. This way to proceed allows to use the following remark: In the case when there exists *any* $\sigma \in C_{\bar{G}}(i)$ *for which* $f \circ \pi_j^{(i)} \circ \sigma = f$, then $\pi_j^{(i)} C_{\bar{G}}(i)(f) \geq f$, and so we can jump over the whole coset $\pi_j^{(i)} C_{\bar{G}}(i)$, by 9.2.4. This idea can also be used in order to derive *the Sims chain of the stabilizer of* f. Let us indicate this stabilizer as follows:

$$A := \{\rho \in \bar{G} \mid f \circ \rho = f\} \leq \bar{G},$$

and note that, for each $i \in b$:

9.2.6
$$C_A(i) \leq C_{\bar{G}}(i).$$

A transversal of the set of left cosets $C_A(i-1)/C_A(i)$ can be obtained from the transversal $\{\pi_j^{(i)} \mid j \in \underline{r(i)}\}$ of $C_{\bar{G}}(i-1)/C_{\bar{G}}(i)$ in the following way:

9.2.7 Lemma *If we denote by* $J(i)$ *the set*

$$J(i) := \{j \in \underline{r(i)} \mid \exists\, \sigma_j \in C_{\bar{G}}(i)\colon \pi_j^{(i)} \sigma_j \in A\},$$

and if we fix, for each $j \in J(i)$, *such an element* σ_j, *then*

$$\{\tau_j := \pi_j^{(i)} \sigma_j \mid j \in J(i)\}$$

is a transversal of $C_A(i-1)/C_A(i)$.

Proof: If $\sigma \in C_A(i-1)$, then, by 9.2.6, there exists a left coset representative $\pi_j^{(i)}$ and a suitable $\sigma'_j \in C_{\bar{G}}(i)$ such that $\sigma = \pi_j^{(i)} \sigma'_j$. We have to show that

$$\pi_j^{(i)}\sigma_j' C_A(i) = \pi_j^{(i)}\sigma_j C_A(i).$$

But this follows from the fact that both $\pi_j^{(i)}\sigma_j$ and $\pi_j^{(i)}\sigma_j'$ are in $A \cap C_{\bar{G}}(i) = C_A(i)$. □

9.2.8 Corollary *The set*

$$\{\tau_j = \pi_j^{(i)}\sigma_j \mid j \in J(i)\}$$

yields the Sims chain of the stabilizer A of f.

Another byproduct is the following test:

9.2.9 Corollary *If, for each i, j, $f'(i) \neq f \circ \pi_j^{(i)}(i)$, then*

$$f' \notin G(f).$$

Let us now summarize what has been said so far about the minimality test which we have to carry out in order to find the lexicographically smallest element of the orbit $G(f)$. We discussed the Sims chain for the action of G on the set n, it leads to the tree 9.2.2, and we saw in 9.2.3 that certain cuts and jumps can be made while running through this tree. But still this does *not suffice* to carry out this test in an efficient way, say if we want to catalog the graphs on 12 vertices. We shall therefore continue by considering the evaluation of the orbits of the centralizers $C_{\bar{G}}(i)$.

9.3 Orbits of Centralizers

We say that $\bar{G} \leq S_n$ is *compatible* (with the natural order on n), if and only if the following holds, for the centralizers of the subsets $i \subseteq n$,

9.3.1 $$\forall\, i \subseteq n\colon\ C_{\bar{G}}(i) \backslash\backslash n \text{ consists of intervals of } n.$$

In particular Young subgroups are compatible.

9.3.2 Lemma *For each subgroup $P \leq S_n$ there exist $\pi \in S_n$ such that $\pi P \pi^{-1}$ is compatible. For short: each subgroup P of S_n is compatible up to conjugation.*

Proof: We shall inductively construct a suitable π that does the appropriate conjugation. Before we start doing so we note that the following is true:

9.3.3 $$\pi C_P(i)\pi^{-1} = C_{\pi P \pi^{-1}}(\pi i).$$

The basis of the induction is trivial: the symmetric group S_n is n-fold transitive and hence there is a $\pi_0 \in S_n$ such that $Q_0 := \pi_0 P \pi_0^{-1} = C_{Q_0}(0)$ has a set

$$C_{Q_0}(0) \backslash\backslash n$$

of intervals as orbits.

In order to carry out the inductive step we assume that we have managed to find permutations π_j, $j \in i$, such that π_j fixes each element of $j \subseteq n$ while, for

$$Q_j := \pi_j \dots \pi_0 P \pi_0^{-1} \dots \pi_j^{-1},$$

$C_{Q_j}(i) \backslash\!\backslash n$ consists of intervals, for $0 \leq i \leq j$.

Consider now the subgroup $C_{Q_j}(j+1)$ of $C_{Q_j}(j)$. Being a subgroup, its orbits are subsets of the orbits of $C_{Q_j}(j)$. We can find an element

$$\pi_{j+1} \in \bigoplus_{\omega \in C_{Q_j}(j) \backslash\!\backslash n} S_\omega$$

fixing the point j (and hence each element of the set $j+1$) and such that the centralizer of the set $j+1$ in $Q_{j+1} := \pi_{j+1} Q_j \pi_{j+1}^{-1}$ has intervals as orbits. As π_{j+1} centralizes $j+1$, Q_{j+1} also satisfies

$$C_{Q_{j+1}}(i) \backslash\!\backslash n \text{ consists of intervals, for } i \subseteq j+1.$$

□

We can therefore *assume without restriction that* $\bar{G}$ *is compatible*, and hence there is also a natural way of numbering the orbits $C_{\bar{G}}(i)$ as follows:

$$\omega_0^{(i)} = \{0\}, \dots, \omega_{i-1}^{(i)} = \{i-1\}, \dots, \omega_{t(i)-1}^{(i)}.$$

Let us consider *contents of restrictions* of mappings $f, g \in m_\lambda^n$ to these orbits. We call $\omega_k^{(i)}$, $k < t(i) - 1$, *content critical* for f and g, if and only if

$$\forall\, s < k\colon\ c(f \downarrow \omega_s^{(i)}) = c(g \downarrow \omega_s^{(i)}),$$

while

$$c(f \downarrow \omega_k^{(i)}) \neq c(g \downarrow \omega_k^{(i)}).$$

9.3.4 Lemma *If* $\omega_k^{(i)}$ *is content critical for* f *and* f'*, then we have, for each* $\sigma \in C_{\bar{G}}(i)$*, that*

$$f \downarrow \omega_0^{(i)} \cup \dots \cup \omega_k^{(i)} \neq f' \circ \sigma \downarrow \omega_0^{(i)} \cup \dots \cup \omega_k^{(i)}.$$

Proof: σ fixes each $\omega_s^{(i)}$ setwise, and hence the contents of the restrictions $f \downarrow \omega_s^{(i)}$ and of $f' \circ \sigma \downarrow \omega_s^{(i)}$ are the same, this also holds for $\omega_k^{(i)}$. Since $\omega_k^{(i)}$ is content critical for f and f', it is also content critical for f and $f' \circ \sigma$. □

Note that $\omega_0^{(i)}, \dots, \omega_{i-1}^{(i)}$ are one element orbits, so $f = f \circ \pi$ and $f \circ \sigma = f \circ \pi \circ \sigma$ are *identical* there, so the preceding lemma means first of all that in order to carry out the minimality test, if $f < f \circ \sigma$, $\sigma \in C_{\bar{G}}(i)$, say, we can start from checking $f(i)$ instead of starting from the very beginning $f(0)$. Moreover the decision if $f \leq f \circ \sigma$ will be made at last by checking if $f(m) < f(\sigma m)$, where $m = \omega_0^{(i)} \cup \dots \cup \omega_k^{(i)}$.

The next step will therefore be a discussion of content critical orbits. Let us say that $f \in m^n$ *lies below* $g \in m^n$ *in* $X \subseteq n$ if and only if

$$\forall\, x \in X\colon\ f(x) \leq g(x),\ \text{ for short: } f \leq_X g,$$

and we say that f lies *strictly* below g in X if $f \leq_X g$ and

$$\exists\, x_0 \in X\colon f(x_0) < g(x_0).$$

This will be abbreviated by

$$f <_X g.$$

9.3.5 Lemma *Let $\pi, \rho \in \bar{G}$ and $\omega_k^{(i)}$ be content critical for $f := f_\lambda \circ \pi$ and $f' := f_\lambda \circ \rho$, while we put $m := \omega_0^{(i)} \cup \ldots \cup \omega_{k-1}^{(i)}$. Now, if*

$$f \downarrow m = f' \downarrow m,\ \textit{and}\ f <_{\omega_k^{(i)}} f',$$

then the following is true:

$$f \leq C_{\bar{G}}(i)(f) \Rightarrow f' \leq C_{\bar{G}}(i)(f').$$

Proof: Take $\sigma \in C_{\bar{G}}(i)$. The assumed equality of the restrictions to m yields, first of all, that also

$$f \circ \sigma \downarrow m = f' \circ \sigma \downarrow m.$$

Moreover we assume that $f \downarrow \omega_k^{(i)} < f' \downarrow \omega_k^{(i)}$ so that there exists an $x_0 \in \omega_k^{(i)}$ for which $f(x_0) < f'(x_0)$. Since $y := \sigma^{-1}(x_0) \in \omega_k^{(i)}$ we have $f(\sigma(y)) < f'(\sigma(y))$, and hence

$$f \circ \sigma \downarrow \omega_k^{(i)} < f' \circ \sigma \downarrow \omega_k^{(i)},$$

which proves the statement. □

Note that 9.3.5 turns out to be a generalization of 9.2.3 if we identify the point i with the orbit $\{i\}$. Finally it should be mentioned that we can also *learn from a negative result in a minimality test*, which is absolutely necessary, since otherwise we would not have the slightest chance to overcome the complexity:

9.3.6 Lemma *If $f \circ \sigma < f$, so that there exists a $j < n$ such that*

$$\forall x < j\colon f(\sigma x) = f(x),\ \textit{while}\ f(\sigma j) < f(j),$$

we put

$$y := \max\{\sigma x \mid x < j \wedge f(\sigma x) < m\},\ \textit{and}\ z := \max\{j, \sigma j, y\}.$$

Then, for each $f' \in m_\lambda^n$ with $f' \downarrow z = f \downarrow z$ the following is true:

$$f' \circ \sigma < f'.$$

Proof: Since j and σj are contained in z, we have

$$f' \circ \sigma(j) = f'(\sigma j) = f(\sigma j) < f(j) = f'(j).$$

Hence $f' \circ \sigma \geq f'$ would imply the existence of an $x < j$ such that $f' \circ \sigma(x) > f'(x)$. This leads to a contradiction:

- If, for this $x < j$, we had $f \circ \sigma(x) < m$, then, as x and $\sigma x \in z$:

  $$f' \circ \sigma(x) = f \circ \sigma(x) = f \circ \sigma(x) = f(x) = f'(x),$$

  a contradiction.
- On the other hand, if $f \circ \sigma(x) = m$, then $x \in z$ gives

  $$m = f'(x) = f(x) = f \circ \sigma(x),$$

  which is a contradiction, too.

□

9.3.7 Corollary *Assume π, σ and z as in 9.3.6, and suppose that the restrictions of f and f' satisfy $f' \downarrow z = f \downarrow z$. Then also f' is not a canonical representative of its orbit. The lexicographically next candidate f' satisfies $f'(z) > f(z)$.*

9.4 The Homomorphism Principle

Besides direct methods of evaluating a transversal of the orbits, there exists also a recursive method using *homomorphic* actions. Recall that we write

$${}_GX \sim {}_HY$$

if and only if there exist a mapping $\theta\colon X \to Y$ and a homomorphism $\eta\colon G \to H$ such that $\theta(gx) = \eta(g)\theta(x)$. Here is a sketch of this situation:

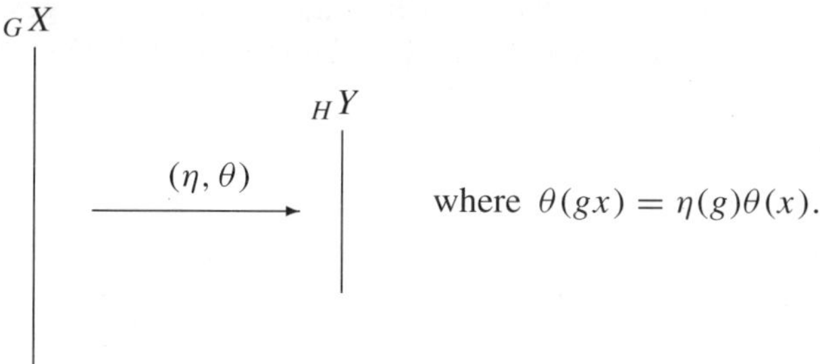

The question arises if we can evaluate, from a transversal T_H of $H \backslash\backslash Y$, a transversal T_G of $G \backslash\backslash X$. An answer is given by the following theorem:

9.4.1 The Homomorphism Principle for group actions *We assume two* epimorphic *actions* ${}_GX$ *und* ${}_HY$*, i. e. we are given a surjective mapping* $\theta\colon X \to Y$ *together with an epimorphism* $\eta\colon G \to H$ *which satify* $\theta(gx) = \eta(g)\theta(x)$*, for each* $x \in X$*,* $g \in G$*. Moreover, we assume that we have a transversal* T_H *of* $H \backslash\backslash Y$ *at hand. Then the following is true:*

- *Each orbit* $\omega \in G \backslash\backslash X$ *is represented in exactly one of the inverse images* $\theta^{-1}(y)$ *of the elements* y *of the transversal* T_H*,*

$$\forall\, \omega \in G \backslash\backslash X\ \exists_1\ y \in T_H\colon\ \omega \cap \theta^{-1}(y) \neq \emptyset.$$

- *The action of* G *on the inverse image of* y *can be replaced by the action of the inverse image of the stabilizer of* y*,*

$$G \backslash\backslash \theta^{-1}(y) = \eta^{-1}(H_y) \backslash\backslash \theta^{-1}(y).$$

Hence we can obtain a transversal T_G *of* $G \backslash\backslash X$ *by simply taking the (disjoint) union of the transversals of these inverse images under the :*

$$T_G := \bigcup_{y \in T_H} T(y),\ \textit{where}\ T(y) \in \mathcal{T}(\eta^{-1}(H_y) \backslash\backslash \theta^{-1}(y)).$$

Proof: To begin with we recall the assumption that θ is surjective. It implies that each orbit $\omega \in G \backslash\backslash X$ has a nonempty intersection with an inverse image $\theta^{-1}(y)$, for a suitable $y \in Y$, say

$$x \in \omega \cap \theta^{-1}(y).$$

Now, if the orbit of this y is represented by $y' = hy \in T_H$, then we can find, since η is surjective, an element g of G such that $\eta(g) = h$, obtaining $\theta(gx) = \eta(g)\theta(x) = hy \in H(y) = H(y')$, so that

$$gx \in \theta^{-1}(y').$$

Thus each orbit is represented in the inverse image of an element in the transversal T_H. Moreover, there is clearly only one such inverse image, and hence there is *exactly one* element in the transversal, the inverse image of which meets ω.

This shows that we need only to examine the inverse images $\theta^{-1}(y)$ of the elements in T_H and get transversals of the orbit sets $G \backslash\backslash \theta^{-1}(y)$ on these sets.

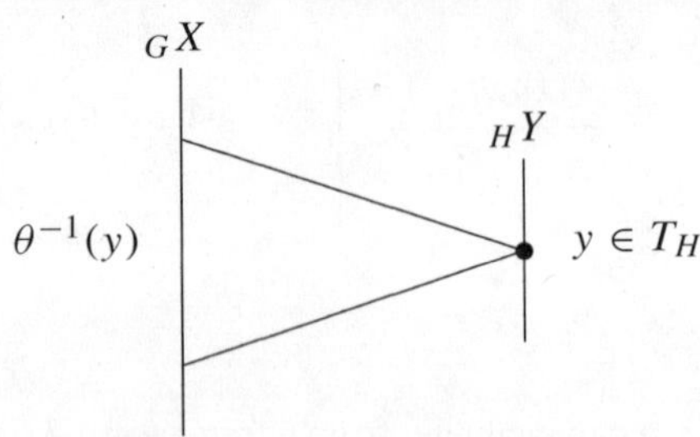

In fact, as it is stated, we can reduce the group G to the inverse imagee of the stabilizer H_y, as we are going to show next. Assume $x, x' \in \theta^{-1}(y)$, where $x' = gx$. Then

$$y = \theta(x) = \theta(gx) = \eta(g)\theta(x) = \eta(g)y,$$

which implies that $\eta(g) \in H_y$, i. e. $g \in \eta^{-1}(H_y)$. This completes the proof. □

In applications we mostly have $G = H$ and $\eta = \text{id}$. An immediate consequence of the Homomorphism Principle is the following systematic method of recursively evaluating transversals:

9.4.2 Surjective Resolution *Assume that G acts on the sets X_i, $i \in m$, and that we are given surjective mappings*

$$\theta_{i+1} : X_{i+1} \to X_i, \textit{ for each } i \in m-2,$$

that commute with the action:

$$\theta_{i+1}(gx_{i+1}) = g\theta_{i+1}(x_{i+1}), \textit{ for each } i \in m-2, g \in G, x_{i+1} \in X_{i+1}.$$

Then we can obtain a transversal T of $G \backslash\backslash X_{m-1}$ by successively working backwards, starting by evaluating a transversal of $G \backslash\backslash X_0$ and the stabilizers of its elements, and then evaluating the inverse images $\theta_1^{-1}(x)$ of the elements x in this transversal, as well as the transversals $G_x \backslash\backslash \theta_1^{-1}(x)$, and so on, in accordance with 9.4.1.

We would like now to apply this to symmetry classes in order to obtain a recursive evaluation of a transversal, using recursion on $|Y|$.

9.4.3 Application (representatives of symmetry classes, recursion on $|Y|$) For sake of simplicity we assume that

$$Y^X = m^n,$$

and that G acts on n and so on $Y^X = m^n$ in the canonic way.

The Surjective Resolution method 9.4.2 can be applied since the following mapping commutes with the action of G on the G–sets $X_K := k^n$,

$$\theta_k : k^n \to (k-1)^n : f \mapsto f',$$

where f' is defined by

$$f'(i) := \begin{cases} f(i), & \text{if } f(i) \in k-1, \\ k-2, & \text{otherwise.} \end{cases}$$

We can therefore start from $X_1 := 1^n$, which consists of the single and constant mapping $i \mapsto 0$, $i \in n$. The stabilizer is G, and the inverse image is the whole set $X_2 := 2^n$. We therefore have to evaluate in the next step a transversal T_2 of $G \backslash\backslash 2^n$. Assume that this is done, and let f_2 denote an element of T_2, G_{f_2} its stabilizer. It remains to evaluate, for each such f_2 and its stabilizer a transversal T_3 of

$$G_{f_2}\backslash\!\backslash\theta_3^{-1}(f_2).$$

This inductive step can be carried out as follows. Two elements f_3 and f_3' of $\theta_3^{-1}(f_2)$ can differ only on the inverse image $f_2^{-1}(1)$ of the point 1, and the values there can only be 1 or 2 We therefore need only to evaluate a transversal T of

$$G_{f_2}\backslash\!\backslash\{1,2\}^{f_2^{-1}(1)}.$$

The desired transversal T_3 then consists of the mappings $f_3'' \in 3^n$ such that

$$f_3'' \downarrow f_2^{-1}(1) \in T, \text{ and } f_3'' \downarrow n\backslash f_2^{-1}(1) = f_2 \downarrow n\backslash f_2^{-1}(1).$$

This can be used, for example, in order to evaluate catalogs of k-graphs, i. e. of multigraphs with multiplicity of edges restricted by k ([11]). Recall 2.3.4.

A more exciting application is Th. Grüner's graph generator that allows to generate connected multigraphs with given vertex degree sequence ([66]). ◇

Another very important application is the successive evaluation of transversals of symmetry classes of given weight:

9.4.4 Application (transversals of symmetry classes by weight) The application which we are going to describe now is due to B. Schmalz ([137]). He introduced it when he was evaluating the $t-(v,k,\lambda)$-designs with a prescribed group of automorphisms and called it the *ladder game*. It uses a *ladder* or *folder* or *leporello,* as we may call it, of Young subgroups in the symmetric group on X in order to evaluate transversals of the orbits of G on Y^X successively and by weight. The crucial fact which allows this is the bijection to double cosets, obtained by Breaking of Symmetry (see 1.2.11) or by Ruch's Lemma 7.1.10: The mapping

$$G\backslash\!\backslash_f Y^X \to \bar{G}\backslash S_X/(S_X)_f\colon\ G(f\circ\pi^{-1}) \mapsto \bar{G}\pi(S_X)_f$$

is a bijection between the set $G\backslash\!\backslash_f Y^X$ of orbits of G on Y^X consisting of mappings with the same content $c(f,-)$ as f (recall 3.1.5) and the set of $\bar{G}\backslash S_X/(S_X)_f$ of $(\bar{G},(S_X)_f)$-double cosets in the symmetric group S_X, where $(S_X)_f$ is the stabilizer of f in S_X.

Hence it remains to show how we can successively evaluate transversals of the sets of double cosets $\bar{G}\backslash S_X/(S_X)_f$, i. e. of double cosets

$$\bar{G}\backslash S_X/S_\lambda,$$

where S_λ denotes a Young subgroup corresponding to the (proper) partition $\lambda \vdash |X|$.

Let us illustrate this with an example that was already mentioned above: The evaluation of the 22 isomers of dioxin. The chemical formula is $C_{12}O_2Cl_4H_4$. And dioxin means that there is a substructure of the following form:

```
            0                     1
            |                     |
  7 \       C      O              C         / 2
     \ C  /   \\ C  /  \  C  //     \  C  /
       ||        |        |            ||
       C         C        C            C
  6 /    \  C  //   \  O  /  \\  C  /    \ 3
            |                    |
            5                    4
```

We have seen that its isomers correspond to the set of orbits of the Kleinian four group V_4 (since we assume this to be the symmetry group of the skeleton) which are of content $(4, 4)$, i. e. they contain exactly four hydrogen and four chlorine atoms.

Thus we are faced with the problem of evaluating a transversal of

$$V_4 \backslash\!\backslash_{(4,4)} \{H, Cl\}^8,$$

which is, according to Ruch's Lemma, bijectively related to the set of double cosets

$$V_4 \backslash S_8 / S_4 \oplus S_4.$$

We can solve that problem successively using the following ladder of subgroups in the symmetric group, where the interesting Young subgroups are emphasized by underlining:

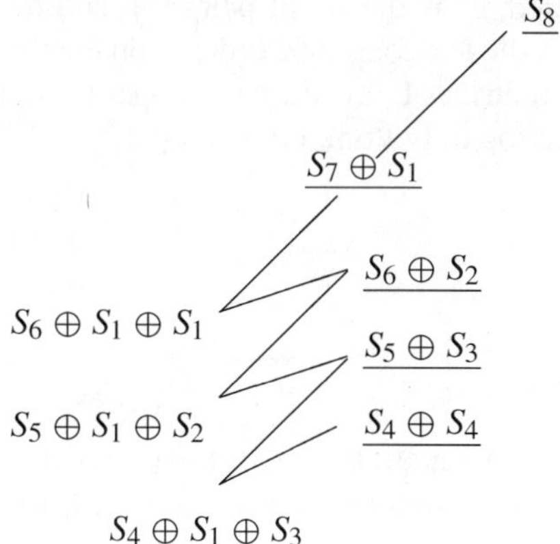

The example shows that such a folder *cannot* be replaced by a chain of Young subgroups, and so we have in fact to go up and down. The upward step from S_α, say, to $S_\beta \geq S_\alpha$, is easy, we simply have to check which left cosets πS_α, in a given transversal of $V_4 \backslash\!\backslash S_8 / S_\alpha$, yield left cosets πS_β that are in the same orbit of V_4 on S_X / S_β.

The downward step is a step from a Young subgroup S_α to a Young subgroup $S_\beta \leq S_\alpha$, and so it is a step from a smaller set S_X / S_α on which G acts to a bigger

G-set S_X/S_β. Hence the Homomorphism Principle applies, we have the following G-epimorphism at hand:

$$\theta\colon S_X/S_\beta \longrightarrow S_X/S_\alpha\colon \pi S_\beta \mapsto \pi S_\alpha.$$

◇

9.5 Orderly Generation

The method of Surjective Resolution as well as the ladder game can be combined with the method of *orderly generation* as described by R. C. Read. This way of generation is based on the fact that total orders on X and Y induce a canonic total order on Y^X, so that a *canonic transversal*

$$T_>(G \backslash\backslash Y^X),$$

consisting of the biggest elements of the orbits exists. Moreover, we can break this problem into pieces, if weights are at hand.

9.5.1 Example Recall that the graphs on v vertices correspond to the set of orbits

$$S_v \backslash\backslash 2^{\binom{v}{2}}.$$

As v is totally ordered, so is the set of pairs $\binom{v}{2}$, and hence also the set of labeled graphs. This is in fact the lexicographic order $\leq$ on the set of 0-1-sequences obtained from the adjacency matrices by reading the upper triangular part row by row from left to right and from top to bottom. For example

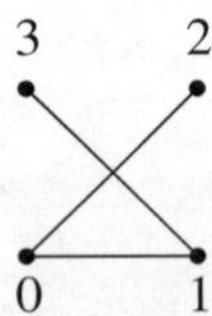

is mapped onto the sequence $(1, 1, 0, 0, 1, 0)$, and hence this labeled graph is a member of the canonic transversal, but $(0, 0, 1, 1, 1, 0)$, which corresponds to the isomorphic labeled graph

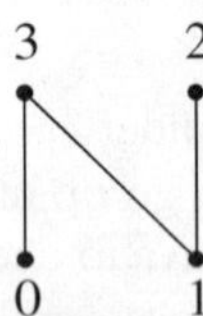

is *not* a canonic representative.

The problem of orderly generation now means the following. We would like to find an economic way of generating 0-1-sequences of length 6 in such a way that not too many sequences are generated, which do not correspond to the desired canonic representatives. Moreover, we should like to do this by increasing the number of ones successively by 1, which means by successively increasing the number of edges.

One way of doing this is to start off from the sequence (0, 0, 0, 0, 0, 0), which forms the canonic transversal of the labeled graphs of total weight 0:

$$T_{>}^{(0)} = \{(0,0,0,0,0,0)\}.$$

Now we will evaluate $T_{>}^{(1)}$, then $T_{>}^{(2)}$, and so on, in the following way. Assume that $T_{>}^{(i)}$ is at hand. Then, in order to obtain $T_{>}^{(i+1)}$, we shall apply to each member of $T_{>}^{(i)}$ the following *augmentation*. We change in each element of $T_{>}^{(i)}$ in all the possible ways just one entry zero that is *to the right of the rightmost entry 1*. The resulting set of labeled graphs is either empty, if no such entry exists, or it consists of labeled graphs with one more edge. These sets have to be checked carefully, the canonic sequences have to be kept, the others have to be thrown away before going to the next step. The remaining set of labeled graphs is in fact the desired transversal $T_{>}^{(i+1)}$. In our example this process runs as follows:

1. In the first step we get the six sequences containing exactly one entry 1. The canonic representative is, of course, (1, 0, 0, 0, 0, 0), and so, clearly
$$T_{>}^{(1)} = \{(1,0,0,0,0,0)\}.$$
2. In the second step we start with (1, 0, 0, 0, 0, 0) and obtain, after canceling the sequences which are not canonic, that
$$T_{>}^{(2)} = \{(1,1,0,0,0,0), (1,0,0,0,0,1)\}.$$
3. The third step gives, from the first canonic representative of the second step, the sequences (1, 1, 1, 0, 0, 0), (1, 1, 0, 1, 0, 0) and (1, 1, 0, 0, 1, 0), while the other representative of $T_{>}^{(2)}$ needs not to be considered, as there is no zero entry to the right of the rightmost entry 1. Thus
$$T_{>}^{(3)} = \{(1,1,1,0,0,0), (1,1,0,1,0,0), (1,1,0,0,1,0)\}.$$
4. In the fourth step we get (1, 1, 1, 1, 0, 0), from the first element of $T_{>}^{(3)}$, nothing new from the second, but (1, 1, 0, 0, 1, 1) from the third element, and hence
$$T_{>}^{(4)} = \{(1,1,1,1,0,0), (1,1,0,0,1,1)\}.$$
5. The next step gives
$$T_{>}^{(5)} = \{(1,1,1,1,1,0)\},$$
from which we obtain in

6. the final step that
$$T_>^{(6)} = \{(1,1,1,1,1,1)\}.$$

And so, summing up, we have obtained the desired canonic transversal of the isomorphism classes of labeled graphs on 4 vertices:

$$T_> = \{(0,0,0,0,0,0), (1,0,0,0,0,0), (1,1,0,0,0,0),$$
$$(1,0,0,0,0,1), (1,1,1,0,0,0), (1,1,0,1,0,0), (1,1,0,0,1,0),$$
$$(1,1,1,1,0,0), (1,1,0,0,1,1), (1,1,1,1,1,0), (1,1,1,1,1,1)\}.$$

◇

It is important to note that we had to check 24 sequences for canonicity, out of $2^6 = 64$ labeled graphs on 4 vertices, which is not bad. Moreover, we note that each of the canonic representatives arises *just once* from an element of smaller total weight. The corresponding generalization reads as follows:

9.5.2 Theorem (Read) *Assume a finite action ${}_GX$ where X is totally ordered by $\leq$ and a disjoint union*
$$X = \bigcup_{i \in n} X_i,$$
of invariant and nonempty subsets X_i. Let A denote an algorithm that produces for each $x \in X$ either the empty set or a set $A(x) \subseteq X$ in descending order, such that the following conditions hold for the canonic transversals $T_>^{(i)}$ of the orbit sets $G \backslash\backslash X_i$, for each $i \in n-1$:

- *$T_>^{(i+1)} \subseteq \bigcup A(x)$, union over the $x \in T_>^{(i)}$,*
- *for each $x \in T_>^{(i+1)}$ there exists a unique element $x' \in T_>^{(i)}$ such that $x \in A(x')$, and finally,*
- *for any $x_1, x_2 \in T_>^{(i+1)}$, $x_1 < x_2$, we have the implication*
$$x_1 \in A(x_1'), x_2 \in A(x_2') \text{ imply } x_1' < x_2'.$$

Then, computing $T_>^{(1)}$ and recursively running through $T_>^{(i)}$ with x, producing the augmentation $A(x)$, and eliminating representatives which are not canonic from $A(x)$, for $i \in n-1$, gives the desired canonic transversal $T_>$ of $G \backslash\backslash X$ as the following union:
$$T_> = \bigcup_{i \in n} T_>^{(i)}.$$

□

This orderly generation obviously will work as soon as we have found an augmentation procedure A that has the properties listed in the items of the above theorem of Read. We already saw such an augmentation that can be used for the graphs and which amounts to adding an edge in a certain set of places. The necessary canonicity test can be done quite efficiently using any suitable invariant that allows to put the canonic representatives in what is called an AVL–tree in computer science. For example, in the molecular graphs case, we can use the *Morgan table*, which is a numbering well known to chemists, and used in the coding procedure for chemical substances in the *Chemical Abstracts*. It reduces the set of checks to a reasonably small number.

Summing up we should like to mention that an efficient way of evaluating a transversal of the symmetry of G on Y^X can be implemented by using the following methods:

- A recursion procedure, using recursion on $|Y|$.
- Read's orderly generation, by applying lexicographic or any other ordering of Y^X.
- A canonic form of the elements $f \in Y^X$ which allows to put the canonic representatives (canonic with respect to the ordering mentioned in the second item) together into an AVL-tree and which reduces the necessary number of checks to a reasonably small one.

9.6 Generating Orbit Representatives

The evaluation of an orbit transversal is of limited use since these sets are usually very big (for example there are approximately 10^9 graphs on $v = 11$ vertices). Thus we are looking for a method which allows to examine further cases, e. g. graphs of more than ten vertices, say. In order to cover such cases with some success, we have to accept redundancy and hence we shall ran into the isomorphism disease. Invariants, isomorphism tests and all that will have to be discussed. To begin with we need to check that we can generate orbit representatives *uniformly at random*.

9.6.1 The Dixon/Wilf Algorithm *If ${}_GX$ denotes a finite action, then we can generate orbit representatives uniformly at random in the following way:*

- *Choose a conjugacy class C of G with probability*

$$p(C) := \frac{|C||X_g|}{|G||G \backslash\backslash X|}, \quad \text{where } g \in C.$$

- *Pick any $g \in C$ and generate a fixed point x of g, uniformly at random.*

Then the probability that x is an element of the orbit $\omega \in G \backslash\backslash X$ is $1/|G \backslash\backslash X|$, i. e. x is uniformly distributed over the orbits of G on X.

Proof: Let $C_0, \ldots, C_{r-1}$ denote the conjugacy classes of G, with representatives $g_i \in C_i$. Then, by the Cauchy-Frobenius lemma, we have

$$\sum_{i \in r} p(C_i) = \frac{\sum_i |C_i||X_{g_i}|}{\sum_g |X_g|} = 1,$$

so that $p(-)$ defines in fact a probability distribution. If $\omega \in G \backslash\backslash X$, then

$$p(x \in \omega) = \sum_i p(C_i) p(x \in X_{g_i} \cap \omega)$$

$$= \sum_i p(C_i) \frac{|X_{g_i} \cap \omega|}{|X_{g_i}|} = \sum_i \frac{|C_i||X_{g_i}|}{|G||G \backslash\!\backslash X|} \frac{|X_{g_i} \cap \omega|}{|X_{g_i}|}$$

$$= \frac{1}{|G||G \backslash\!\backslash X|} \sum_i |C_i||X_{g_i} \cap \omega| = \frac{1}{|G||G \backslash\!\backslash X|} \sum_g |X_g \cap \omega|,$$

where the last identity follows from exercise 9.6.1. Now

$$\sum_g |X_g \cap \omega| = \sum_{g \in G} \sum_{x \in X_g \cap \omega} 1 = \sum_{x \in \omega} \sum_{g \in G_x} 1 = \sum_{x \in \omega} |G_x|,$$

which is equal to $|G_y||\omega| = |G|$, for any $y \in \omega$, and we are done. □

The application of this method to the generation of representatives of G–classes, H–classes, $H \times G$–classes and $H \wr_X G$–classes on Y^X is easy, since we know the fixed points very well. Let us consider the G–classes, for example.

9.6.2 Corollary *For finite ${}_GX$ and Y the following procedure yields elements $f \in Y^X$ that are distributed over the G–classes on Y^X uniformly at random:*

- *Choose a conjugacy class C of G with the probability*

$$p(C) := \frac{|C||Y|^{c(\bar{g}')}}{\sum_g |Y|^{c(\bar{g})}}, \; g' \in C.$$

- *Pick any $g \in C$, evaluate its cycle decomposition and construct an $f \in Y^X$ that takes values $y \in Y$ on these cycles which are distributed uniformly at random over Y.*

9.6.3 Application (graphs with given number of vertices) Consider our standard example for this situation: We would like to generate graphs on 4 vertices, numbered from 0 to 3, uniformly at random. The conjugacy classes are parametrized by the proper partitions (4), $(3,1)$, (2^2), $(2,1^2)$ and (1^4) of 4. Their orders are 6,8,3,6 and 1. The corresponding numbers of cyclic factors of the permutations induced on the set $\binom{4}{2}$ of pairs of vertices are 2,2,4,4 and 6, so that the numbers of fixed points on the set $2^{\binom{4}{2}}$ of labeled graphs amount to 4,4,16,16 and 64. This yields for the probabilities $p(C)$ the values

$$3/33, 4/33, 6/33, 12/33, 8/33.$$

These numbers may be multiplied by the common denominator 33 in order to get the natural numbers 3,4,6,12,8, which we accumulate obtaining

$$3, 7, 13, 25, 33.$$

A generator that yields natural numbers between 1 and 33 uniformly at random is now used in order to choose C. As soon as it generates one of the numbers 1,2 or 3 this means that the conjugacy classe C_0 is chosen. If it generates 4,5,6 or 7, C_1, and so on. Assume that it generates 12, then the conjugacy class C_2 is picked, an element of which is the permutation $(01)(23)$. It induces on the set of pairs of vertices

$$\{a = \{0, 1\}, b = \{0, 2\}, c = \{0, 3\}, d = \{1, 2\}, e = \{1, 3\}, f = \{2, 3\}\}$$

the permutation

$$\overline{(01)(23)} = (a)(be)(cd)(f).$$

A random generator of zeros and ones is now used in order to associate values 0 or 1 with the cyclic factors of $(a)(be)(cd)(f)$. If it generates the sequence 1,0,0,1,say, we obtain the labeled graph that has edges joining the elements of the pairs a and f, while all the other pairs of vertices remain disjoint. Its orbit is represented by the graph shown in figure 9.2. ◇

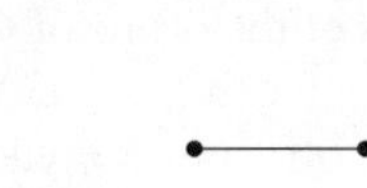

Fig. 9.2. The generated graph

This method can easily be refined in order to generate graphs with prescribed number of edges or, more generally, representatives of orbits with a given weight, uniformly at random. If ${}_GX$ is a finite action, $w\colon X \to R$ a weight function, then G acts on the union of orbits with a prescribed weight $r \in R$, i. e. on the inverse image $w^{-1}[\{r\}]$ of r, and we need only to replace in 9.6.1 the corresponding factors in the numerator and in the denominator obtaining

9.6.4 Corollary *If we choose the conjugacy class $C \subseteq G$ with the probability*

$$p(C) := \frac{|C||w^{-1}(r)_g|}{|G||G \backslash\backslash w^{-1}(r)|},$$

pick any element $g \in C$ and generate a fixed point of weight r of g uniformly at random, then we obtain an element of X that is uniformly distributed over the orbits of weight r.

In the case when the acting group is a symmetric group, the choice of a conjugacy class amounts to the choice of a proper partition. We therefore should not forget that the number of proper partitions of $n \in \mathbb{N}$ is rapidly increasing with n. Here are a few of these numbers:

n	no. of proper partitions
10	42
20	627
40	37338
60	$\approx 10^6$
100	$\approx 2 \cdot 10^8$

This shows that it is worthwile to try to avoid that this amount of probabilities is stored throughout the whole generating process. For practical purposes one can start the generation and evaluate probabilities only if required. This means that we need to evaluate $p(C_i)$ only if the generated random number exceeds $\sum_{j=0}^{i-1} p(C_j)$. The efficiency of this revised method heavily depends on the numbering of the conjugacy classes. The numbering should clearly be chosen in such a way that $p(C_i) \geq p(C_{i+1})$. Asymptotic considerations show (cf. Oberschelp, Dixon/Wilf) that in case of graph generation it is not bad to number the conjugacy classes of the symmetric group according to the lexicographic order of partitions.

In order to generate representatives of the orbits of G on X with prescribed type $\widetilde{U}$, we can do the following. As each such orbit $\omega \in G \backslash\backslash X$ does contain elements x which have U as stabilizer, we can cut off from X_U each element that has a bigger stabilizer. It clearly suffices to remove from X_U the elements belonging to the X_V, where U is maximal in V. The remaining subset of X_U will be indicated as follows:

$$\hat{X}_U := X_U \backslash \bigcup_{U max. V} X_V.$$

The intersection of $\omega \in G \backslash\backslash_{\widetilde{U}} X$ with $\hat{X}_U$ is the orbit of an $x \in \omega$, where $G_x = U$, under the normalizer subgroup $N_G(U)$, this is easy to check. As $G_x = U$ we can even consider the action of the factor group $N_G(U)/U$ on this intersection $\omega \cap \hat{X}_U$. This proves

9.6.5 Lemma *Each transversal of $(N_G(U)/U) \backslash\backslash \hat{X}_U$ is a transversal of the orbits of type $\widetilde{U}$ of G on X.*

Moreover, if we want to apply probabilistic methods, we can use that the orbits of $N_G(U)/U$ on $\hat{X}_U$ are all of the same order $|N_G(U)/U|$. Hence we obtain

9.6.6 Theorem (Laue) *By picking elements from $\hat{X}_U$ uniformly at random, we obtain elements that are uniformly distributed over the orbits of type $\widetilde{U}$ of G on X.*

For example, if $U := \{1\}$, we have that $\hat{X}_U = \hat{X}_1$ is equal to X minus the union of the fixed point sets X_V, taken over all the minimal subgroups $V \leq G$. If in particular $G := C_n$, then the minimal subgroups V are just the cyclic subgroups $C_p \leq C_n$, where p is prime. the same holds for $G := S_n$, but in this case there may be, of course, several such C_p. As $N_G(1) = G$, we can, according to 9.6.6, easily generate representatives of the orbits of maximal length $|\bar{G}|$, uniformly at random. In order to display a concrete example, we recall the enumeration of irreducible polynomials of degree n over $GF(q)$ with leading coefficient 1 (recall example 4.2.2). They correspond to the orbits of maximal length n of the Galois group C_n of $GF(q)^n$ over $GF(q)$. The above reasoning shows that an orbit of $G = N_G(1)$ on $\hat{X}_1$ is also a transversal of these orbits of maximal length n of G on X. But $\hat{X}_1$ is equal to $GF(q)^n$ minus the set of fixed points $GF(q)^n_{C_p}$, where p runs through the prime divisors of n. These subsets which have to be subtracted, can easily be spotted. For example, if $n := 6, q := 2, C_6 := \langle(0 \dots 5)\rangle$, we obtain

$$\begin{aligned} GF(2)^6_{C_2} &= \{000000, 100100, 010010, \\ &\quad 001001, 110110, 101101, 011011, 111111\}, \\ GF(2)^6_{C_3} &= \{000000, 101010, 010101, 111111\}. \end{aligned}$$

Subtracting the union of these sets from $GF(2)^6$, we get $\hat{X}_1$, consisting of the remaining $2^6 - 10$ binary sequences over $GF(2)$. A transversal of the orbits of C_6 on $\hat{X}_1$ is, for example. the set

$$100000, 110000, 101000, 10010, 111000, 110100, 110010, 111100, 111110.$$

Assuming a normal basis given, we can translate these sequences into the 9 corresponding polynomials that are listed in many books.

Exercises

Exercise 9.6.1 Prove that, for conjugate $g, g' \in G$, a finite action ${}_GX$ and an orbit $\omega \in G \backslash\backslash X$, we have

$$|X_g \cap \omega| = |X_{g'} \cap \omega|.$$

9.7 Symmetry Adapted Bases

We discussed finite actions ${}_GX$ and their decomposition into orbits. The corresponding problem in linear representation theory is the decomposition of a representation space V, of $D\colon G \to GL(V)$, over $\mathbb{C}$ say, into its irreducible constituents. If $D = \sum_{i \in r} n_i D_i$ is the decomposition into irreducible constituents, then

$$V = V^0 \oplus \dots \oplus V^{r-1}, V^i = V^i_0 \oplus \dots \oplus V^i_{n_i - 1}, i \in r,$$

where V^i_j affords the irreducible representation D^i of G. This means that we have to find an *adapted basis* of V, i. e. a basis consisting of elements $b^{i,j}_k$ such that

$$V^i_j = \ll b^{i,j}_k \mid k \in f^i \gg, i \in r, j \in n_i.$$

In the case when we are given a complete set $\mathbf{D}^0, \dots, \mathbf{D}^{s-1}$ of pairwise inequivalent and irreducible *matrix representations* of G, which correspond to the irreducible constituents D^i of D, say

$$\mathbf{D}^i(g) = (d^i_{kl}(g)),$$

then we can do even better. We can in fact find basis elements $b^{i,j}_k$ with the property

9.7.1 $$D(g)b^{i,j}_l = \sum_{k \in f^i} d^i_{kl}(g) b^{i,j}_k .$$

Lower down we shall see how that can be done. A basis of V which has this property, is called a *symmetry adapted basis* of V with respect to the given system $\mathbf{D}^0, \dots, \mathbf{D}^{s-1}$ of irreducible matrix representations. Such a symmetry adapted basis has great advantages, since we know in advance, how each of the vectors $v \in V$ transforms under the action of $g \in G$ with respect to this particular basis:

$$v := \sum_{i,j,l} z^{i,j}_l b^{i,j}_l \Longrightarrow D(g)v = \sum_{i,j,l} z^{i,j}_l \sum_k d^i_{kl}(g) b^{i,j}_k .$$

Note that we assume the coefficients $d^i_{kl}(g)$ to be tabulated already! (The problem is, of course, that a complete system of ordinary irreducible matrix representations is available only in very few cases. Such a rare case is the symmetric group, as we have already seen. There are other cases, too, for example cyclic and dihedral groups, alternating groups, wreath products of such groups.)

Before we go into details about the evaluation of symmetry adapted bases, let us briefly describe the kind of problem we would like to apply it to.

We again assume V to be the representation space of $D\colon G \to GL(V)$, completely reducible into subspaces V^i_j, as above. We now consider a linear operator M on V, i. e. a linear mapping on V. We say that M *has the symmetry* D if and only if it commutes with this representation:

$$\forall\, g \in G\colon\ M \cdot D(g) = D(g) \cdot M.$$

It is our aim to use this fact in order *directly to decompose* V *into subspaces that are invariant under* M, so that a symmetry adapted basis of V with respect to D yields a matrix decomposition of M that is a direct sum of matrices and therefore contains many zeros! The argument used is in fact more or less the same as that applied in order to introduce the symmetrization of representations.

Assume that we are given a basis of elements $b^{i,j}_k$ in V for which 9.7.1 holds. For later purposes we write its elements as follows in a scheme that Stiefel/Fäßler ([43]) call a *block scheme:*

9.7.2 $$B_i :\quad \begin{array}{ccccc} V^i_0 & : & b^{i,0}_0 & \dots & b^{i,0}_{f^i-1} \\ \vdots & & & & \vdots \\ V^i_{n_i-1} & : & b^{i,n_i-1}_0 & \dots & b^{i,n_i-1}_{f^i-1} \end{array}$$

9.7.3 Theorem *The columns* $\{b^{i,j}_k \mid j \in n_i\}$ *of the block scheme 9.7.2 generate subspaces*

$$W_k^i := \ll b_k^{i,j} \mid j \in n_i \gg, k \in f^i,$$

of the vector space V which are invariant unter the linear operator M :

$$MW_k^i = W_k^i,$$

for each i and k.

Proof: We consider the following numbering of the elements of the basis $\mathcal{B}$ of V :

$$\mathcal{B} = \{\ldots, b_0^{i,0}, b_1^{i,0}, \ldots, b_{f^i-1}^{i,0}, b_0^{i,1}, b_1^{i,1}, \ldots, b_{f^i-1}^{i,1}, \ldots, b_0^{i,n_i-1}, \ldots, b_{f^i-1}^{i,n_i-1}, \ldots\}.$$

which means reading the basic elements *by the line* in the blocks from top to bottom, and one block scheme after the other. According to this numbering, the matrix corresponding to $D(g)$ is of the form

$$\mathbf{D}(g) = \begin{pmatrix} \mathbf{D}_0(g) & & & & & 0 \\ & \ddots & & & & \\ & & \mathbf{D}_0(g) & & & \\ & & & \ddots & & \\ & & & & \mathbf{D}_{r-1}(g) & \\ & & & & & \ddots \\ 0 & & & & & \mathbf{D}_{r-1}(g) \end{pmatrix}.$$

This same basis yields a matrix corresponding to the linear operator M which is of the following form:

$$\mathbf{M} = \begin{pmatrix} \mathbf{M}_{00}^{00} & \ldots & \mathbf{M}_{0,n_0-1}^{00} & \mathbf{M}_{00}^{01} & \ldots & \mathbf{M}_{0,n_1-1}^{01} & \ldots \\ \vdots & & \vdots & \vdots & & \vdots & \\ \mathbf{M}_{n_0-1,0}^{00} & \ldots & \mathbf{M}_{n_0-1,n_0-1}^{00} & \mathbf{M}_{n_0-1,0}^{01} & \ldots & \mathbf{M}_{n_0-1,n_1-1}^{01} & \ldots \\ \vdots & & \vdots & \vdots & & \vdots & \end{pmatrix}$$

The symmetry of M gives

$$\mathbf{M}_{kl}^{ij} \cdot \mathbf{D}_j(g) = \mathbf{D}_i(g) \cdot \mathbf{M}_{kl}^{ij},$$

for each $g \in G$, and hence we obtain from Schur's Lemma that

$$\mathbf{M}_{kl}^{ij} = \begin{cases} 0\text{–matrix}, & \text{if } i \neq j, \\ m_{kl}^i \cdot \mathbf{I}, & \text{otherwise}, \end{cases}$$

where $\mathbf{I}$ denotes the identity matrix and $m_{kl}^i \in \mathbb{C}$ is a numerical factor. Using these numbers, we define

$$\mathbf{M}_i := (m_{kl}^i), k, l \in n_i, i \in r.$$

Hence there exists a permutation matrix P such that, for each $g \in G$, $P\mathbf{M}P^{-1}$ is equal to

$$\begin{pmatrix} \mathbf{M}_0 & & & & & & \\ & \ddots & & & 0 & & \\ & & \mathbf{M}_0 & & & & \\ & & & \ddots & & & \\ & & & & \mathbf{M}_{r-1} & & \\ & 0 & & & & \ddots & \\ & & & & & & \mathbf{M}_{r-1} \end{pmatrix} = \sum_{i \in r} f^i \mathbf{M}_i .$$

The construction clearly shows that in order to obtain this form of the matrix we need only take the elements of the block schemes *by the column,* as it is stated. □

Now we are going to describe how a symmetry adapted basis can be evaluated for each representation space of G. *The assumption is that we are given a complete set of pairwise inequivalent ordinary irreducible matrix representations* $\mathbf{D}_i$, $i \in s$, *of* G. And we would like to emphasize that this means that in fact *for each* $g \in G$ *and for each* $i \in \underline{s}$ *we can look up each coefficient of the matrix* $\mathbf{D}_i(g)$. (It will turn out that in fact we need only to have access to the first rows of all these matrices.) We consider the representation space of the ordinary representation $D: G \to GL(V)$, say, a representation that is assumed to be reducible. We denote by n_i the multiplicity of the irreducible representation D_i in D, and f^i indicates the dimension of D_i. Note that both these numbers can be read off from D and the tabulated $\mathbf{D}_i(g)$. It is our aim to construct a basis

$$\mathcal{B} := \{b_k^{i,j} \mid i \in s,\ j \in n_i,\ k \in f^i\}$$

such that 9.7.1 is true,

$$D(g) b_l^{i,j} = \sum_{k \in f^i} d_{kl}^i(g) b_k^{i,j} .$$

First of all we note that such bases do in fact exist, and that this is obvious from linear algebra. In order to construct such a basis from given tabulated matrix representations $\mathbf{D}_i$, i. e. from given numbers $d_{kl}^i(g)$, we introduce the following linear operators on V:

$$P_{kj}^i := \frac{f^i}{|G|} \sum_{g \in G} d_{jk}^i(g^{-1}) D(g).$$

Assume, to begin with, that the vectors $b_k^{i,j}$ form a symmetry adapted basis. Then

$$P_{kj}^i b_n^{l,m} = \frac{f^i}{|G|} \sum_g d_{jk}^i(g^{-1}) D(g) b_n^{l,m}$$

$$= \sum_{p \in f^i} \underbrace{\left(\frac{f^i}{|G|} \sum_g d^i_{jk}(g^{-1}) d^l_{pn}(g)\right)}_{=\delta_{il}\delta_{jn}\delta_{kp},\ by\ 11.4.5} b^{l,m}_p = \begin{cases} b^{i,m}_k, & \text{if } i = l,\ j = n, \\ 0, & \text{otherwise.} \end{cases}$$

This shows that the following is true for the product of two such operators:

9.7.4 $$P^i_{kj} P^u_{wv} = \delta_{iu} \delta_{jw} P^i_{kv}.$$

9.7.5 Corollary *The linear operators P^i_{kj} have in particular the following properties:*

- *The elements of a block scheme B_j are mapped onto zero if $i \neq j$.*
- *The j-th column of B_i is mapped onto the k-th column of B_i, the other columns are mapped onto zero.*
- *The linear operator P^i_{kk} is a projection of V onto the subspace*

$$W^i_k := \ll b^{i,j}_k \mid 1 \leq k \leq n_i \gg$$

generated by the elements in the k-th column of B_i.

With the aid of these operators we shall now compute a symmetry adapted basis of the representation space V of D. In order to evaluate a symmetry adapted basis for the subspace

$$V^i = V^i_0 \oplus \ldots \oplus V^i_{n_i - 1},$$

we use the last item of 9.7.5, according to which the dimension of the image W^i_0 of P^i_{00} is n_i. Hence we can find a basis

$$\{v^{i,0}_0, \ldots, v^{i,n_i-1}_0\},$$

say, of this space by taking *any* basis of V, evaluating the corresponding matrix for P^i_{00} and choosing n_i linearly independent columns of that matrix. Each of these vectors can be written in terms of the (fictive) symmetry adapted basis used above: $v^{i,j}_0 = \sum_k z^{i,j}_k b^{i,k}_0$, and they yield further vectors

9.7.6 $$v^{i,j}_l := P^i_{l0} v^{i,j}_0 = \sum_k z^{i,j}_k b^{i,k}_l.$$

Now we note that, since the $v^{i,j}_0$ are linearly independent, the vectors $v^{i,j}_l$, $l \in n_i$, are linearly independent, too (they have the same coefficients but on different places). Moreover the total set of vectors

$$v^{i,j}_l,\ i \in r,\ j \in n_i,\ l \in f^i,$$

is linearly independent. A very easy check shows that $D(g)v^{i,j}_l = \sum_k d^i_{kl}(g) v^{ij}_k$, and so we have obtained

9.7.7 Corollary *The vectors $v_k^{i,j}$ form a symmetry adapted basis of V. Thus we can use the following algorithm for the evaluation of a symmetry adapted basis:*

- *Choose an arbitrary basis of V and evaluate the matrices $\mathbf{P}^i_{00}$ corresponding to the linear mappings P^i_{00}, $i \in s$.*
- *Evaluate a basis $v_0^{i,j}$, $j \in n_i$, of the column space of $\mathbf{P}^i_{00}$.*
- *Evaluate the vectors $v_k^{i,j} := P^i_{k0} v_0^{i,j}$, $k \in f^i$.*

The resulting vectors form a basis of V which is symmetry adapted to D with respect to the tabulated irreducible matrix representations $\mathbf{D}_i$ of G. This means that the following is true for each $g \in G$ and any triple of indices i, j, k :

$$D(g)v_k^{ij} = \sum_l d^i_{lk} v_l^{ij}.$$

Proof: By definition of the vector v_k^{ij} we have

$$D(g)v_k^{ij} = D(g)P^i_{k0}v_0^{ij} = \frac{f^i}{|G|}\sum_{h\in G} \underbrace{d^i_{0k}(h^{-1})}_{=\sum_l d^i_{0l}(h^{-1}g^{-1})d^i_{lk}(g)} D(gh)v_0^{ij}$$

$$\frac{f^i}{|G|}\sum_l d^i_{lk}(g)\sum_h d^i_{0l}(h^{-1}g^{-1})D(gh)v_0^{ij} = \sum_l d^i_{lk}(g)P^i_{l0}v_0^{ij}$$

$$= \sum_l d^i_{lk}(g)v_l^{ij},$$

as it is stated. □

Exercises

Exercise 9.7.1 Prove that we obtain an orthonormal symmetry adapted basis, if the tabulated ordinary irreducible matrix representations are unitary and if we start from orthonormal columns of the matrix corresponding to P^i_{00}.

9.8 Applications of Symmetry Adapted Bases

Some of the many interesting applications combine symmetry adapted bases with eigenvalue considerations. We therefore note the following:

9.8.1 Lemma *If M has symmetry D, then each eigenspace of M is an invariant subspace of each $D(g)$, $g \in G$. Conversely, each eigenspace of $D(g)$ is invariant under M.*

Proof: Let $E_\lambda(M)$ be the eigenspace of M corresponding to the eigenvalue λ of M. For each of its elements v and for any $g \in G$ we have that

$$MD(g)v = D(g)Mv = D(g)\lambda v = \lambda D(g)v,$$

and so $D(g)v$ is in $E_\lambda(M)$, too. The corresponding result for $v \in E_\lambda(D(g))$ follows analogously. □

For example, the natural representation of the cyclic group $C_n = \langle(0, \ldots, n-1)\rangle$ is generated by the matrix

$$\mathbf{D}((0, \ldots, n-1)) := \begin{pmatrix} 0 & 0 & \ldots & 0 & 1 \\ 1 & 0 & \ldots & 0 & 0 \\ \vdots & \vdots & \ddots & \vdots & \vdots \\ 0 & 0 & \ldots & 1 & 0 \end{pmatrix}.$$

Its characteristic polynomial

$$\det(x\mathbf{I} - \mathbf{D}((0, \ldots, n-1))) = x^n - 1$$

shows that the *eigenvalues* are exactly the powers of a primitive n-th root of unity $\epsilon := e^{2\pi i/n}$:

$$\lambda_j := \epsilon^j = e^{2\pi ij/n}.$$

In particular they are pairwise different, and so a basis of eigenvectors exists. An eigenvector for λ_j is

$$v_j := \begin{pmatrix} 1 \\ (\lambda_j)^{n-1} \\ (\lambda_j)^{n-2} \\ \vdots \\ (\lambda_j)^1 \end{pmatrix}.$$

(This representation is in fact the regular representation of the cyclic group.) We obtain

9.8.2 Corollary *Each matrix M that commutes with $\mathbf{D}((0, \ldots, n-1))$ becomes diagonal if we change the basis into a symmetry adapted one.*

A nice application is the following symmetry approach to algebraic equations (see [43]):

9.8.3 Application (Cardano's formulae) Assuming, for the moment, that we were not aware of the solution of a quadratic equation, we consider the matrix

$$\mathbf{M} := \begin{pmatrix} a & b \\ b & a \end{pmatrix}.$$

It commutes with the transposition $\mathbf{D}((01))$,

$$\mathbf{D}((01)) \cdot \mathbf{M} = \mathbf{M} \cdot \mathbf{D}((01)) = \begin{pmatrix} b & a \\ a & b \end{pmatrix}.$$

The characteristic polynomial of $\mathbf{M}$ is

$$x^2 - 2ax + (a^2 - b^2).$$

From the above discussion we know that $\mathbf{D}((01))$ and therefore also $\mathbf{M}$ has the eigenvectors

$$\begin{pmatrix} 1 \\ 1 \end{pmatrix}, \begin{pmatrix} 1 \\ -1 \end{pmatrix}.$$

Hence $a \pm b$ are the eigenvalues of $\mathbf{M}$. Let us apply that to the quadratic equation

$$x^2 + px + q = 0,$$

which we would like to solve. Comparing it with the characteristic polynomial we obtain $p = -2a$ and $q = a^2 - b^2$. This yields $a = -\frac{p}{2}$ and so $b = \sqrt{\frac{p^2}{4} - q}$. Hence the solutions are

$$a \pm b = -\frac{p}{2} \pm \sqrt{\frac{p^2}{4} - q}.$$

A similar argument works also in the case 3: The following matrix clearly commutes with the matrix $\mathbf{D}((012))$:

$$M := \begin{pmatrix} 0 & a & b \\ b & 0 & a \\ a & b & 0 \end{pmatrix}.$$

The characteristic polynomial of this matrix is

$$x^3 - 3abx - a^3 - b^3.$$

In order to derive its roots we can use 9.8.2. Putting $\omega := e^{2\pi i/3}$, we obtain from 9.8.2 the following eigenvectors of M:

$$v_0 := \begin{pmatrix} 1 \\ 1 \\ 1 \end{pmatrix}, \quad v_1 := \begin{pmatrix} 1 \\ \omega^2 \\ \omega \end{pmatrix}, \quad v_2 := \begin{pmatrix} 1 \\ \omega \\ \omega^2 \end{pmatrix}.$$

Hence the eigenvalues of the matrix M turn out to be

$$\lambda_0 = a + b, \lambda_1 = a\omega^2 + b\omega, \lambda_2 = a\omega + b\omega^2.$$

We can apply this now to a cubic equation of the form

$$x^3 + px + q = 0,$$

obtaining $p = -3ab, q = -(a^3 + b^3)$, from which we derive *Cardano's formulae:* The equations for p and q give the following identity, where σ stands for b^3 :

$$q = \frac{p^3 - 27\sigma^2}{27\sigma},$$

a quadratic equation in σ, the solutions of which are

$$\sigma_{0,1} = -\frac{q}{2} \pm \sqrt{\left(\frac{q}{2}\right)^2 + \left(\frac{p}{3}\right)^3},$$

which yields via resubstitution that

$$b = \sqrt[3]{-\frac{q}{2} - \sqrt{\left(\frac{q}{2}\right)^2 + \left(\frac{p}{3}\right)^3}}.$$

The corresponding equation for a is obtained analogously:

$$a = \sqrt[3]{-\frac{q}{2} + \sqrt{\left(\frac{q}{2}\right)^2 + \left(\frac{p}{3}\right)^3}}.$$

The above equations for the eigenvalues $\lambda_{0,1,2}$ of the matrix M therefore show how the roots of the equation $x^3+px+q = 0$ can be expressed in terms of the coefficients p and q. ◇

9.8.4 Application (eigenvalues of graphs) Another field of applications is that of evaluation of spectra of graphs by using their symmetry groups. For example, the complete graph has the adjacency matrix

$$A = \begin{pmatrix} 0 & 1 & 1 & \dots & 1 \\ 1 & 0 & 1 & \dots & 1 \\ \vdots & \vdots & \vdots & \vdots & \vdots \\ 1 & 1 & 1 & \dots & 0 \end{pmatrix}.$$

The considerations above show that, since A again has the symmetry of the regular representation of the cyclic group, the following is true:

9.8.5 Corollary *The eigenvalues of the adjacency matrix of the complete graph on n vertices are $n - 1$ (with multiplicity 1) and -1 (with multiplicity $n - 1$).*

9.8.6 Application (polynomial representations of general linear groups) Another important application of symmetry adapted bases is the evaluation of the irreducible polynomial representations $\langle\alpha\rangle$ of the general linear group $GL_m(\mathbb{C})$, which we are going to describe next.

We take the set m^n as basis of a complex vector space, obtaining

$$V := \mathbb{C}^{(m^n)} = \ll f \mid f \in m^n \gg$$

This space is isomorphic to the n−fold tensor power of $\mathbb{C}^m$ via the isomorphism

$$\mathbb{C}^{(m^n)} \simeq \otimes^n \mathbb{C}^m = \ll e_f \mid f \in m^n \gg: f \mapsto e_f,$$

where

$$e_f := e_{f(0)} \otimes \ldots \otimes e_{f(n-1)},$$

for the standard basis $\{e_0, \ldots, e_{m-1}\}$ of $\mathbb{C}^n$. Both the symmetric group S_n and the general linear group $GL_m(\mathbb{C})$ act on V in a canonic way: The action of S_n is obvious: $\pi f = f \circ \pi^{-1}$. The action of $GL_m(\mathbb{C})$ is also clear, it is the *diagonal action* on the factors of the tensor power: If $(x_{ik}) \in GL_m(\mathbb{C})$, then

$$(x_{ik})f := \sum_{g \in m^n} \underbrace{\left(\prod_i x_{g(i),f(i)}\right)}_{=:X_{g,f}} = \sum_g X_{g,f} \cdot g.$$

These actions clearly commute:

$$(x_{ik})\pi f = \pi(x_{ik})f,$$

and so we can apply the method of symmetry adapted bases in order to decompose the linear operator (x_{ik}).

9.8.7 Example Take $m := n := 2$. The basis of the space $\mathbb{C}^{(2^2)}$ consists of the mappings $f_i = (f_i(0), f_i(1)) \in 2^2, i = 0, 1, 2, 3$, which we number lexicographically:

$$f_0 := (0, 0),\ f_1 := (0, 1),\ f_2 := (1, 0),\ f_3 := (1, 1).$$

Hence, according to this numbering, the matrix representation $\mathbf{D}$ of S_2 on the tensor space looks as follows:

$$\mathbf{D}(1) = \begin{pmatrix} 1 & & & 0 \\ & 1 & & \\ & & 1 & \\ 0 & & & 1 \end{pmatrix}, \mathbf{D}((01)) = \begin{pmatrix} 1 & & & \\ & 0 & 1 & \\ & 1 & 0 & \\ & & & 1 \end{pmatrix}$$

The ordinary irreducible matrix representation of $S_n = S_2$ are the identity representation $\pi \mapsto (1)$ and the alternating representation $\pi \mapsto (sgn(\pi))$, and hence the decomposition of D into irreducible representations is $D = 3 \cdot [2] + [1^2]$, so that $n_0 = 3, n_1 = 1$. The matrices representing the projection operators in that basis are

$$\mathbf{P}^0_{00} = \frac{1}{2}\begin{pmatrix} 2 & & & \\ & 1 & 1 & \\ & 1 & 1 & \\ & & & 2 \end{pmatrix}, \ \mathbf{P}^1_{00} = \frac{1}{2}\begin{pmatrix} 0 & & & \\ & 1 & -1 & \\ & -1 & 1 & \\ & & & 0 \end{pmatrix}.$$

According to 9.7.3 we now pick n_i linearly independent columns in the column space of $\mathbf{P}^i_{00}$, for example the vectors

$$v^{0,0}_0 = \begin{pmatrix} 1 \\ 0 \\ 0 \\ 0 \end{pmatrix}, v^{0,1}_0 = \begin{pmatrix} 0 \\ 1/2 \\ 1/2 \\ 0 \end{pmatrix}, v^{0,2}_0 = \begin{pmatrix} 0 \\ 0 \\ 0 \\ 1 \end{pmatrix}, v^{1,0}_0 = \begin{pmatrix} 0 \\ 1/2 \\ -1/2 \\ 0 \end{pmatrix}$$

According to 9.7.3, since $f^2 = 1, i = 1, 2$, these vectors form a symmetry adapter basis, the elements of which we take as columns of the desired transformation matrix T, obtaining

$$\begin{pmatrix} 1 & 0 & 0 & 0 \\ 0 & 1/2 & 0 & 1/2 \\ 0 & 1/2 & 0 & -1/2 \\ 0 & 0 & 1 & 0 \end{pmatrix}, T^{-1} = \begin{pmatrix} 1 & 0 & 0 & 0 \\ 0 & 1 & 1 & 0 \\ 0 & 0 & 0 & 1 \\ 0 & 1 & -1 & 0 \end{pmatrix}.$$

These matrixes decompose any linear operator that commutes with the action of the symmetric group. A particular such operator it the theses square of a generic element of $GL_2(\mathbb{C})$:

$$\begin{pmatrix} x_{00} & x_{01} \\ x_{10} & x_{11} \end{pmatrix} \otimes \begin{pmatrix} x_{00} & x_{01} \\ x_{10} & x_{11} \end{pmatrix} =$$

$$\left(\begin{array}{c|c|c|c} x_{00}^2 & x_{00}x_{01} & x_{01}x_{00} & x_{01}^2 \\ \hline x_{00}x_{10} & x_{00}x_{11} & x_{01}x_{10} & x_{01}x_{11} \\ \hline x_{10}x_{00} & x_{10}x_{01} & x_{11}x_{00} & x_{11}x_{01} \\ \hline x_{11}^2 & x_{10}x_{11} & x_{11}x_{10} & x_{11}^2 \end{array}\right)$$

The transformal matrix $T^{-1}AT$ is

$$\left(\begin{array}{ccc|c} x_{00}^2 & x_{00}x_{01} & x_{01}^2 & 0 \\ 2x_{00}x_{10} & x_{00}x_{11} + x_{01}x_{10} & 2x_{01}x_{11} & 0 \\ x_{10}^2 & x_{10}x_{11} & x_{11}^2 & 0 \\ \hline 0 & 0 & 0 & x_{00}x_{11} - x_{01}x_{10} \end{array}\right)$$

and so we have obtained the following two matrix representations of $GL_2(\mathbb{C})$:

$$\langle 2\rangle : GL_2(\mathbb{C}) \to GL_3(\mathbb{C}),$$

defined by

$$\begin{pmatrix} x_{00} & x_{01} \\ x_{10} & x_{11} \end{pmatrix} \mapsto \begin{pmatrix} x_{00}^2 & x_{00}x_{01} & x_{01}^2 \\ 2x_{00}x_{10} & x_{00}x_{11} + x_{01}x_{10} & 2x_{01}x_{11} \\ x_{10}^2 & x_{10}x_{11} & x_{11}^2 \end{pmatrix},$$

and

$$\langle 1^2 \rangle : GL_2(\mathbb{C}) \to \mathbb{C} : \begin{pmatrix} x_{00}x_{01} \\ x_{10}x_{11} \end{pmatrix} \mapsto x_{00}x_{11} - x_{01}x_{10} = \det \begin{pmatrix} x_{01}x_{01} \\ x_{10}x_{11} \end{pmatrix}.$$

◇

The elements of the representing matrices an polynomials in the elements x_{ik} of the represented matrix (x_{jl}), and theefore these representations are called *polynomial representations* of $GL_2(\mathbb{C})$. One of the most famous application of the representation theory of symmetric groups is concerned with these:

9.8.8 Schur's Theorem *If*

$$\mathbf{P}_m^n = \sum_{\alpha \vdash n} m_\alpha [\alpha]$$

is the decompositin of the representation $\mathbf{P}_m^n$ *of* S_n *in* $\otimes^n \mathbb{C}^m \simeq \mathbb{C}^{(m^n)}$ *into irreducibles, then the corresponding representation* $\otimes^n \, id_{GL_m(\mathbb{C})}$ *of* $GL_m(\mathbb{C})$ *of the general linear group on this space decomposes (with respect to any symmetry adapted basis) as follows into polynomial representations* $\langle \alpha \rangle$*:*

$$\otimes^n \, id_{GL_m(\mathbb{C})} = \sum_{\alpha \vdash n :\, \alpha_0' \leq m} f^\alpha \, \langle \alpha \rangle.$$

These polynomial representation $\langle \alpha \rangle$ *are irreducible and each irreducible polynomial representation of* $GL_m(\mathbb{C})$ *is obtained n this way.*

◇

10. Tables

This chapter contains tables of marks and Burnside matrices of the smallest cyclic, dihedral, symmetric and alternating groups. Moreover, the interested reader will find characters (reducible and irreducible) of symmetric groups, in order to support further applications by providing some numerical information. The irreducible characters are given, together with the Young characters and their decompositions. Also the scalar products of Young characters and products of Young characters and the alternating character are tabulated, since these numbers are numbers of 0-1–matrices with prescribed row and column sums. In addition the smallest Foulkes tables are shown since the characters occuring in enumeration theory of symmetry classes of mappings are linear combinations of Foulkes characters. Then the first 44 character polynomials are listed, each of them allows to evaluate an infinite series of ordinary irreducible characters of symmetric groups. The final section contains the 120 first Schubert polynomials.

The tables were obtained using different program systems like CAYLEY for the tables of marks and MAPLE for inverting the tables of marks in order to obtain the Burnside matrices. SYMMETRICA (which is due in particular to A. Kohnert) was used for the irreducible characters of the symmetric groups, the Young characters and their decompositions, the Foulkes tables and the Schubert polynomials. The character polynomials were taken from the diploma thesis of P. Zeiss. The reader who would like to do more can obtain SYMMETRICA (which is still in progress, and which can do many other things, too, like handling Schur and Schubert polynomials, cycle indices and group reduction functions) from its home page

http://www.mathe2.uni-bayreuth.de/axel/symneu_engl.html

A computer algebra package designed (mainly by A. Betten, R. Laue and A. Wassermann) especially for the constructive theory of discrete structures, in particular for designs, can be obtained from

http://www.mathe2.uni-bayreuth.de/betten/DISCRETA

10.1 Tables of Marks and Burnside Matrices

This part of the appendix contains tables of marks and Burnside matrices of cyclic, dihedral, symmetric and alternating groups.

10.1.1 Cyclic Groups

This section contains the system of subgroups, the table of marks and the Burnside matrix of the cyclic groups C_3, up to C_{20}. The groups C_p, where p is a prime number are trivial since they contain trivial subgroups only, and so we have for the table of marks and its inverse, the Burnside matrix of this particular groups:

$$M(C_p) = \begin{pmatrix} p & 1 \\ & 1 \end{pmatrix}, \quad B(C_p) = \begin{pmatrix} 1/p & -1/p \\ & 1 \end{pmatrix}.$$

There is also an explicit expression for the table of marks in the general case since the subgroup lattice of C_n, $n > 0$, is the lattice of divisors of n, but in order to make life easier for the interested reader we show these matrices and their inverses, which are less trivial.

The group C_4: The subgroups are

$$U_0 = \langle 1 \rangle, U_1 = \langle (02)(13) \rangle, U_2 = \langle (0123) \rangle = C_4.$$

The table of marks and the Burnside matrix:

$$\begin{pmatrix} 4 & 2 & 1 \\ & 2 & 1 \\ & & 1 \end{pmatrix}, \begin{pmatrix} 1/4 & -1/4 & . \\ & 1/2 & -1/2 \\ & & 1 \end{pmatrix}$$

The group C_6: The subgroups are

$$U_0 = \langle 1 \rangle, U_1 = \langle (03)(14)(25) \rangle, U_2 = \langle (024)(135) \rangle, U_3 = C_6.$$

The table of marks and the Burnside matrix:

$$\begin{pmatrix} 6 & 3 & 2 & 1 \\ & 3 & . & 1 \\ & & 2 & 1 \\ & & & 1 \end{pmatrix}, \begin{pmatrix} 1/6 & -1/6 & -1/6 & 1/6 \\ & 1/3 & . & -1/3 \\ & & 1/2 & -1/2 \\ & & & 1 \end{pmatrix}.$$

The group C_8: The subgroups are

$$U_0 = \langle 1 \rangle, U_1 = \langle (04)(15)(26)(37) \rangle, U_2 = \langle (0246)(1367) \rangle, U_3 = C_8.$$

The table of marks and the Burnside matrix:

$$\begin{pmatrix} 8 & 4 & 2 & 1 \\ & 4 & 2 & 1 \\ & & 2 & 1 \\ & & & 1 \end{pmatrix}, \begin{pmatrix} 1/8 & -1/8 & . & . \\ & 1/4 & -1/4 & . \\ & & 1/2 & -1/2 \\ & & & 1 \end{pmatrix}.$$

The group C_9: The subgroups are

$$U_0 = \langle 1 \rangle, U_1 = \langle (036)(147)(258) \rangle, U_2 = C_9.$$

The table of marks and the Burnside matrix:

$$\begin{pmatrix} 9 & 3 & 1 \\ & 3 & 1 \\ & & 1 \end{pmatrix}, \begin{pmatrix} 1/9 & -1/9 & . \\ & 1/3 & -1/3 \\ & & 1 \end{pmatrix}.$$

The group C_{10}: The subgroups are

$$U_0 = \langle 1 \rangle, U_1 = \langle (05)(16)(27)(38)(49) \rangle, U_2 = \langle (02468)(13579) \rangle, U_3 = C_{10}.$$

The table of marks and the Burnside matrix:

$$\begin{pmatrix} 10 & 5 & 2 & 1 \\ & 5 & . & 1 \\ & & 2 & 1 \\ & & & 1 \end{pmatrix}, \begin{pmatrix} 1/10 & -1/10 & -1/10 & 1/10 \\ & 1/5 & . & -1/5 \\ & & 1/2 & -1/2 \\ & & & 1 \end{pmatrix}.$$

The group C_{12}: The subgroups are

$$U_0 = \langle 1 \rangle, U_1 = \langle (0,6)(1,7)(2,8)(3,9)(4,10)(5,11) \rangle,$$

$$U_2 = \langle (0,4,8)(1,5,9)(2,6,10)(3,7,11) \rangle,$$

$$U_3 = \langle (0,3,6,9)(1,4,7,10)(2,5,8,11) \rangle,$$

$$U_4 = \langle (0,4,8)(1,5,9)(2,6,10)(3,7,11), (0,6)(1,7)(2,8)(3,9)(4,10)(5,11) \rangle,$$

$$U_5 = C_{12}.$$

The table of marks is

$$\begin{pmatrix} 12 & 6 & 4 & 3 & 2 & 1 \\ & 6 & . & 3 & 2 & 1 \\ & & 4 & . & 2 & 1 \\ & & & 3 & . & 1 \\ & & & & 2 & 1 \\ & & & & & 1 \end{pmatrix}.$$

The Burnsidematrix looks as follows

$$\begin{pmatrix} 1/12 & -1/12 & -1/12 & . & 1/12 & . \\ & 1/6 & . & -1/6 & -1/6 & 1/6 \\ & & 1/4 & . & -1/4 & . \\ & & & 1/3 & . & -1/3 \\ & & & & 1/2 & -1/2 \\ & & & & & 1 \end{pmatrix}.$$

The group C_{14}: The subgroups are

$$U_0 = \langle 1\rangle, U_1 = \langle (0,7)(1,8)(2,9)(3,10)(4,11)(5,12)(6,13)\rangle,$$

$$U_2 = \langle (0,2,4,6,8,10,12)(1,3,5,7,9,11,13)\rangle, U_3 = C_{14}.$$

The table of marks and the Burnside matrix:

$$\begin{pmatrix} 14 & 7 & 2 & 1 \\ & 7 & . & 1 \\ & & 2 & 1 \\ & & & 1 \end{pmatrix}, \begin{pmatrix} 1/14 & -1/14 & -1/14 & 1/14 \\ & 1/7 & . & -1/7 \\ & & 1/2 & -1/2 \\ & & & 1 \end{pmatrix}.$$

The group C_{15}: The subgroups are

$$U_0 = \langle 1\rangle, U_1 = \langle (0,5,10)(1,6,11)(2,7,12)(3,8,13)(4,9,14)\rangle,$$

$$U_2 = \langle (0,3,6,9,12)(1,4,7,10,13)(2,5,8,11,14)\rangle, U_3 = C_{15}.$$

The table of marks and the Burnside matrix:

$$\begin{pmatrix} 15 & 5 & 3 & 1 \\ & 5 & . & 1 \\ & & 3 & 1 \\ & & & 1 \end{pmatrix}, \begin{pmatrix} 1/15 & -1/15 & -1/15 & 1/15 \\ & 1/5 & . & -1/5 \\ & & 1/3 & -1/3 \\ & & & 1 \end{pmatrix}.$$

The group C_{16}: The subgroups are

$$U_0 = \langle 1\rangle, U_1 = \langle (0,8)(1,9)(2,10)(3,11)(4,12)(5,13)(6,14)(7,15)\rangle,$$

$$U_2 = \langle (0,4,8,12)(1,5,9,13)(2,6,10,14)(3,7,11,15)\rangle,$$

$$U_3 = \langle (0,2,4,6,8,10,12,14)(1,3,5,7,9,11,13,15)\rangle, U_4 = C_{16}.$$

The table of marks is

$$\begin{pmatrix} 16 & 8 & 4 & 2 & 1 \\ & 8 & 4 & 2 & 1 \\ & & 4 & 2 & 1 \\ & & & 2 & 1 \\ & & & & 1 \end{pmatrix}.$$

The Burnsidematrix looks as follows

$$\begin{pmatrix} 1/16 & -1/16 & . & . & . \\ & 1/8 & -1/8 & . & . \\ & & 1/4 & -1/4 & . \\ & & & 1/2 & -1/2 \\ & & & & 1 \end{pmatrix}.$$

The group C_{18}: The subgroups are

$$\begin{aligned}
U_0 &= \langle 1\rangle, U_1 = \langle (0,9)(1,10)(2,11)(3,12)(4,13)(5,14)(6,15)(7,16)(8,17)\rangle,\\
U_2 &= \langle (0,6,12)(1,7,13)(2,8,14)(3,9,15)(4,10,16)(5,11,17)\rangle,\\
U_3 &= \langle (0,6,12)(1,7,13)(2,8,14)(3,9,15)(4,10,16)(5,11,17),\\
&\qquad (0,9)(1,10)(2,11)(3,12)(4,13)(5,14)(6,15)(7,16)(8,17)\rangle,\\
U_4 &= \langle (0,2,4,6,8,10,12,14,16)(1,3,5,7,9,11,13,15,17)\rangle, U_5 = C_{18}.
\end{aligned}$$

The table of marks is

$$\begin{pmatrix}
18 & 9 & 6 & 3 & 2 & 1\\
 & 9 & . & 3 & . & 1\\
 & & 6 & 3 & 2 & 1\\
 & & & 3 & . & 1\\
 & & & & 2 & 1\\
 & & & & & 1
\end{pmatrix}.$$

The Burnsidematrix looks as follows

$$\begin{pmatrix}
1/18 & -1/18 & -1/18 & 1/18 & . & .\\
 & 1/9 & . & -1/9 & . & .\\
 & & 1/6 & -1/6 & -1/6 & 1/6\\
 & & & 1/3 & . & -1/3\\
 & & & & 1/2 & -1/2\\
 & & & & & 1
\end{pmatrix}.$$

The group C_{20}: The subgroups are

$$\begin{aligned}
U_0 &= \langle 1\rangle, U_1 = \langle (0,10)(1,11)(2,12)\cdots(9,19)\rangle,\\
U_2 &= \langle (0,4,8,12,16)(1,5,9,13,17)(2,6,10,14,18)(3,7,11,15,19)\rangle,\\
U_3 &= \langle (0,5,10,15)(1,6,11,16)(2,7,12,17)(3,8,13,18)(4,9,14,19)\rangle,\\
U_4 &= \langle (0,4,8,12,16)(1,5,9,13,17)(2,6,10,14,18)(3,7,11,15,19),\\
&(0,10)(1,11)(2,12)(3,13)(4,14)(5,15)(6,16)(7,17)(8,18)(9,19)\rangle, U_5 = C_{20}.
\end{aligned}$$

The table of marks is

$$\begin{pmatrix}
20 & 10 & 4 & 5 & 2 & 1\\
 & 10 & . & 5 & 2 & 1\\
 & & 4 & . & 2 & 1\\
 & & & 5 & . & 1\\
 & & & & 2 & 1\\
 & & & & & 1
\end{pmatrix}.$$

The Burnsidematrix looks as follows

$$\begin{pmatrix}
1/20 & -1/20 & -1/20 & . & 1/20 & .\\
 & 1/10 & . & -1/10 & -1/10 & 1/10\\
 & & 1/4 & . & -1/4 & .\\
 & & & 1/5 & . & -1/5\\
 & & & & 1/2 & -1/2\\
 & & & & & 1
\end{pmatrix}.$$

10.1.2 Dihedral Groups

This section contains a system of representatives of the conjugacy classes of subgroups, the table of marks and the Burnside matrix of the dihedral groups D_3 up to D_{11}.

The group D_3: A transversal of the conjugacy classes of subgroups is

$$U_0 = \langle 1\rangle, U_1 = \langle(01)\rangle, U_2 = \langle(012)\rangle, U_3 = \langle(012), (01)\rangle = D_3.$$

The table of marks is

$$\begin{pmatrix} 6 & 3 & 2 & 1 \\ & 1 & . & 1 \\ & & 2 & 1 \\ & & & 1 \end{pmatrix}.$$

The Burnside matrix looks as follows

$$\begin{pmatrix} 1/6 & -1/2 & -1/6 & 1/2 \\ & 1 & . & -1 \\ & & 1/2 & -1/2 \\ & & & 1 \end{pmatrix}.$$

The group D_4: A transversal of the conjugacy classes of subgroups is

$$U_0 = \langle 1\rangle, U_1 = \langle(02)(13)\rangle, U_2 = \langle(03)(12)\rangle, U_3 = \langle(02)\rangle,$$

$$U_4 = \langle(03)(12), (01)(23)\rangle, U_5 = \langle(02), (13)\rangle, U_6 = \langle(0123)\rangle, U_7 = D_4.$$

The table of marks is

$$\begin{pmatrix} 8 & 4 & 4 & 4 & 2 & 2 & 2 & 1 \\ & 4 & . & . & 2 & 2 & 2 & 1 \\ & & 2 & . & 2 & . & . & 1 \\ & & & 2 & . & 2 & . & 1 \\ & & & & 2 & . & . & 1 \\ & & & & & 2 & . & 1 \\ & & & & & & 2 & 1 \\ & & & & & & & 1 \end{pmatrix}.$$

The Burnside matrix looks as follows

$$\begin{pmatrix} 1/8 & -1/8 & -1/4 & -1/4 & 1/4 & 1/4 & . & . \\ & 1/4 & . & . & -1/4 & -1/4 & -1/4 & 1/2 \\ & & 1/2 & . & -1/2 & . & . & . \\ & & & 1/2 & . & -1/2 & . & . \\ & & & & 1/2 & . & . & -1/2 \\ & & & & & 1/2 & . & -1/2 \\ & & & & & & 1/2 & -1/2 \\ & & & & & & & 1 \end{pmatrix}.$$

The group D_5: A transversal of the conjugacy classes of subgroups is

$$U_0 = \langle 1 \rangle, U_1 = \langle (01)(24) \rangle, U_2 = \langle (01234) \rangle, U_3 = D_5.$$

The table of marks is

$$\begin{pmatrix} 10 & 5 & 2 & 1 \\ & 1 & . & 1 \\ & & 2 & 1 \\ & & & 1 \end{pmatrix}.$$

The Burnside matrix looks as follows

$$\begin{pmatrix} 1/10 & -1/2 & -1/10 & 1/2 \\ & 1 & . & -1 \\ & & 1/2 & -1/2 \\ & & & 1 \end{pmatrix}.$$

The group D_6: A transversal of the conjugacy classes of subgroups is

$$U_0 = \langle 1 \rangle, U_1 = \langle (03)(14)(25) \rangle, U_2 = \langle (05)(14)(23) \rangle,$$
$$U_3 = \langle (04)(13) \rangle, U_4 = \langle (024)(135) \rangle, U_5 = \langle (05)(14)(23), (02)(35) \rangle,$$
$$U_6 = \langle (024)(135), (03)(14)(25) \rangle, U_7 = \langle (024)(135), (05)(14)(23) \rangle,$$
$$U_8 = \langle (024)(135), (04)(13) \rangle, U_9 = D_6.$$

The table of marks is

$$\begin{pmatrix} 12 & 6 & 6 & 6 & 4 & 3 & 2 & 2 & 2 & 1 \\ & 6 & . & . & . & 3 & 2 & . & . & 1 \\ & & 2 & . & . & 1 & . & 2 & . & 1 \\ & & & 2 & . & 1 & . & . & 2 & 1 \\ & & & & 4 & . & 2 & 2 & 2 & 1 \\ & & & & & 1 & . & . & . & 1 \\ & & & & & & 2 & . & . & 1 \\ & & & & & & & 2 & . & 1 \\ & & & & & & & & 2 & 1 \\ & & & & & & & & & 1 \end{pmatrix}.$$

The Burnside matrix is

$$\begin{pmatrix} 1/12 & -1/12 & -1/4 & -1/4 & -1/12 & 1/2 & 1/12 & 1/4 & 1/4 & -1/2 \\ & 1/6 & . & . & . & -1/2 & -1/6 & . & . & 1/2 \\ & & 1/2 & . & . & -1/2 & . & -1/2 & . & 1/2 \\ & & & 1/2 & . & -1/2 & . & . & -1/2 & 1/2 \\ & & & & 1/4 & . & -1/4 & -1/4 & -1/4 & 1/2 \\ & & & & & 1 & . & . & . & -1 \\ & & & & & & 1/2 & . & . & -1/2 \\ & & & & & & & 1/2 & . & -1/2 \\ & & & & & & & & 1/2 & -1/2 \\ & & & & & & & & & 1 \end{pmatrix}$$

The group D_7: A transversal of the conjugacy classes of subgroups is

$$U_0 = \langle 1 \rangle, U_1 = \langle (01)(26)(35) \rangle, U_2 = \langle (0123456) \rangle, U_3 = D_7.$$

The table of marks is

$$\begin{pmatrix} 14 & 7 & 2 & 1 \\ & 1 & . & 1 \\ & & 2 & 1 \\ & & & 1 \end{pmatrix}.$$

The Burnside matrix looks as follows

$$\begin{pmatrix} 1/14 & -1/2 & -1/14 & 1/2 \\ & 1 & . & -1 \\ & & 1/2 & -1/2 \\ & & & 1 \end{pmatrix}.$$

The group D_8: A transversal of the conjugacy classes of subgroups is

$$U_0 = \langle 1 \rangle, U_1 = \langle (04)(15)(26)(37) \rangle, U_2 = \langle (07)(16)(25)(34) \rangle,$$

$$U_3 = \langle (04)(13)(57) \rangle, U_4 = \langle (07)(16)(25)(34), (03)(12)(47)(56) \rangle,$$

$$U_5 = \langle (04)(13)(57), (17)(26)(35) \rangle, U_6 = \langle (0246)(1357) \rangle,$$

$$U_7 = \langle (0246)(1357), (07)(16)(25)(34) \rangle, U_8 = \langle (0246)(1357), (04)(13)(57) \rangle,$$

$$U_9 = \langle (01234567) \rangle, U_{10} = D_8.$$

The table of marks is

$$\begin{pmatrix} 16 & 8 & 8 & 8 & 4 & 4 & 4 & 2 & 2 & 2 & 1 \\ & 8 & . & . & 4 & 4 & 4 & 2 & 2 & 2 & 1 \\ & & 2 & . & 2 & . & . & 2 & . & . & 1 \\ & & & 2 & . & 2 & . & . & 2 & . & 1 \\ & & & & 2 & . & . & 2 & . & . & 1 \\ & & & & & 2 & . & . & 2 & . & 1 \\ & & & & & & 4 & 2 & 2 & 2 & 1 \\ & & & & & & & 2 & . & . & 1 \\ & & & & & & & & 2 & . & 1 \\ & & & & & & & & & 2 & 1 \\ & & & & & & & & & & 1 \end{pmatrix}.$$

The Burnside matrix looks as follows

$$\begin{pmatrix}
1/16 & -1/16 & -1/4 & -1/4 & 1/4 & 1/4 & . & . & . & . & . \\
 & 1/8 & . & . & -1/4 & -1/4 & -1/8 & 1/4 & 1/4 & . & . \\
 & & 1/2 & . & -1/2 & . & . & . & . & . & . \\
 & & & 1/2 & . & -1/2 & . & . & . & . & . \\
 & & & & 1/2 & . & . & -1/2 & . & . & . \\
 & & & & & 1/2 & . & . & -1/2 & . & . \\
 & & & & & & 1/4 & -1/4 & -1/4 & -1/4 & 1/2 \\
 & & & & & & & 1/2 & . & . & -1/2 \\
 & & & & & & & & 1/2 & . & -1/2 \\
 & & & & & & & & & 1/2 & -1/2 \\
 & & & & & & & & & & 1
\end{pmatrix}.$$

The group D_9: A transversal of the conjugacy classes of subgroups is

$$U_0 = \langle 1 \rangle,\, U_1 = \langle (01)(28)(37)(46) \rangle,\, U_2 = \langle (036)(147)(258) \rangle,$$

$$U_3 = \langle (036)(147)(258),\, (01)(28)(37)(46) \rangle,\, U_4 = \langle (012345678) \rangle,\, U_5 = D_9.$$

The table of marks is

$$\begin{pmatrix}
18 & 9 & 6 & 3 & 2 & 1 \\
 & 1 & . & 1 & . & 1 \\
 & & 6 & 3 & 2 & 1 \\
 & & & 1 & . & 1 \\
 & & & & 2 & 1 \\
 & & & & & 1
\end{pmatrix}.$$

The Burnside matrix looks as follows

$$\begin{pmatrix}
1/18 & -1/2 & -1/18 & 1/2 & . & . \\
 & 1 & . & -1 & . & . \\
 & & 1/6 & -1/2 & -1/6 & 1/2 \\
 & & & 1 & . & -1 \\
 & & & & 1/2 & -1/2 \\
 & & & & & 1
\end{pmatrix}.$$

The group D_{10}: A transversal of the conjugacy classes of subgroups is

$$U_0 = \langle 1 \rangle,\, U_1 = \langle (05)(16)(27)(38)(49) \rangle,\, U_2 = \langle (09)(18)(27)(36)(45) \rangle,$$

$$U_3 = \langle (06)(15)(24)(79) \rangle,\, U_4 = \langle (02468)(13579) \rangle,$$

$$U_5 = \langle (09)(18)(27)(36)(45),\, (04)(13)(59)(68) \rangle,$$

$$U_6 = \langle (02468)(13579),\, (05)(16)(27)(38)(49) \rangle,$$

$$U_7 = \langle (02468)(13579),\, (09)(18)(27)(36)(45) \rangle,$$

$$U_8 = \langle (02468)(13579),\, (06)(15)(24)(79) \rangle,\, U_9 = D_{10}.$$

The table of marks is

$$\begin{pmatrix}
20 & 10 & 10 & 10 & 4 & 5 & 2 & 2 & 2 & 1 \\
 & 10 & . & . & . & 5 & 2 & . & . & 1 \\
 & & 2 & . & . & 1 & . & 2 & . & 1 \\
 & & & 2 & . & 1 & . & . & 2 & 1 \\
 & & & & 4 & . & 2 & 2 & 2 & 1 \\
 & & & & & 1 & . & . & . & 1 \\
 & & & & & & 2 & . & . & 1 \\
 & & & & & & & 2 & . & 1 \\
 & & & & & & & & 2 & 1 \\
 & & & & & & & & & 1
\end{pmatrix}.$$

The Burnside matrix looks as follows

$$\begin{pmatrix}
1/20 & -1/20 & -1/4 & -1/4 & -1/20 & 1/2 & 1/20 & 1/4 & 1/4 & -1/2 \\
 & 1/10 & . & . & . & -1/2 & -1/10 & . & . & 1/2 \\
 & & 1/2 & . & . & -1/2 & . & -1/2 & . & 1/2 \\
 & & & 1/2 & . & -1/2 & . & . & -1/2 & 1/2 \\
 & & & & 1/4 & . & -1/4 & -1/4 & -1/4 & 1/2 \\
 & & & & & 1 & . & . & . & -1 \\
 & & & & & & 1/2 & . & . & -1/2 \\
 & & & & & & & 1/2 & . & -1/2 \\
 & & & & & & & & 1/2 & -1/2 \\
 & & & & & & & & & 1
\end{pmatrix}.$$

The group D_{11}: A transversal of the conjugacy classes of subgroups is

$$U_0 = \langle 1 \rangle,\ U_1 = \langle (0,1)(2,10)(3,9)(4,8)(5,7) \rangle,$$

$$U_2 = \langle (0,1,2,3,4,5,6,7,8,9,10) \rangle,\ U_3 = D_{11}.$$

The table of marks is

$$\begin{pmatrix}
22 & 11 & 2 & 1 \\
 & 1 & . & 1 \\
 & & 2 & 1 \\
 & & & 1
\end{pmatrix}.$$

The Burnside matrix looks as follows

$$\begin{pmatrix}
1/22 & -1/2 & -1/22 & 1/2 \\
 & 1 & . & -1 \\
 & & 1/2 & -1/2 \\
 & & & 1
\end{pmatrix}.$$

10.1.3 Alternating Groups

This section contains a system of representatives of the conjugacy classes of subgroups, the table of marks and the Burnside matrix of A_3, A_4 and A_5. From A_6 only the table of marks can be shown in a reasonable way.

The group A_3**:** A transversal of the conjugacy classes of subgroups is

$$U_0 = \langle 1 \rangle, U_1 = \langle (012) \rangle.$$

The table of marks is

$$\begin{pmatrix} 3 & 1 \\ & 1 \end{pmatrix}.$$

The Burnside matrix looks as follows

$$\begin{pmatrix} 1/3 & -1/3 \\ & 1 \end{pmatrix}.$$

The group A_4**:** A transversal of the conjugacy classes of subgroups is

$$U_0 = \langle 1 \rangle, U_1 = \langle (02)(13) \rangle, U_2 = \langle (012) \rangle,$$

$$U_3 = \langle (02)(13), (03)(12) \rangle, U_4 = \langle (012), (031) \rangle = A_4.$$

The table of marks is

$$\begin{pmatrix} 12 & 6 & 4 & 3 & 1 \\ & 2 & . & 3 & 1 \\ & & 1 & . & 1 \\ & & & 3 & 1 \\ & & & & 1 \end{pmatrix}.$$

The Burnside matrix looks as follows

$$\begin{pmatrix} 1/12 & -1/4 & -1/3 & 1/6 & 1/3 \\ & 1/2 & . & -1/2 & . \\ & & 1 & . & -1 \\ & & & 1/3 & -1/3 \\ & & & & 1 \end{pmatrix}.$$

The group A_5**:** A transversal of the conjugacy classes of subgroups is

$$U_0 = \langle 1 \rangle, U_1 = \langle (04)(23) \rangle, U_2 = \langle (123) \rangle, U_3 = \langle (02314) \rangle,$$

$$U_4 = \langle (04)(23), (02)(34) \rangle, U_5 = \langle (123), (04)(23) \rangle, U_6 = \langle (02314), (02)(34) \rangle,$$

$$U_7 = \langle (023), (042) \rangle, U_8 = \langle (02314), (04213) \rangle = A_5.$$

The table of marks is

$$\begin{pmatrix} 60 & 30 & 20 & 12 & 15 & 10 & 6 & 5 & 1 \\ & 2 & . & . & 3 & 2 & 2 & 1 & 1 \\ & & 2 & . & . & 1 & . & 2 & 1 \\ & & & 2 & . & . & 1 & . & 1 \\ & & & & 3 & . & . & 1 & 1 \\ & & & & & 1 & . & . & 1 \\ & & & & & & 1 & . & 1 \\ & & & & & & & 1 & 1 \\ & & & & & & & & 1 \end{pmatrix}.$$

The Burnside matrix looks as follows

$$\begin{pmatrix} 1/60 & -1/4 & -1/6 & -1/10 & 1/6 & 1/2 & 1/2 & 1/3 & -1 \\ & 1/2 & . & . & -1/2 & -1 & -1 & . & 2 \\ & & 1/2 & . & . & -1/2 & . & -1 & 1 \\ & & & 1/2 & . & . & -1/2 & . & . \\ & & & & 1/3 & . & . & -1/3 & . \\ & & & & & 1 & . & . & -1 \\ & & & & & & 1 & . & -1 \\ & & & & & & & 1 & -1 \\ & & & & & & & & 1 \end{pmatrix}.$$

The group A_6: A transversal of the conjugacy classes of subgroups is

$$U_0 = \langle 1 \rangle, U_1 = \langle (12)(45) \rangle, U_2 = \langle (054)(132) \rangle, U_3 = \langle (235) \rangle, U_4 = \langle (04235) \rangle,$$

$$U_5 = \langle (03)(1524) \rangle, U_6 = \langle (12)(45), (03)(45) \rangle, U_7 = \langle (12)(45), (14)(25) \rangle,$$

$$U_8 = \langle (054)(132), (12)(45) \rangle, U_9 = \langle (235), (01)(23) \rangle,$$

$$U_{10} = \langle (054)(132), (054)(123) \rangle, U_{11} = \langle (04235), (05)(34) \rangle,$$

$$U_{12} = \langle (03)(1524), (03)(45) \rangle, U_{13} = \langle (025)(143), (015)(243) \rangle,$$

$$U_{14} = \langle (254), (154) \rangle, U_{15} = \langle (054)(132), (054)(123), (12)(45) \rangle,$$

$$U_{16} = \langle (03)(1524), (0534)(12) \rangle, U_{17} = \langle (03)(1524), (03)(1245) \rangle,$$

$$U_{18} = \langle (03)(1524), (01)(2534) \rangle, U_{19} = \langle (04235), (02453) \rangle,$$

$$U_{20} = \langle (03514), (02135) \rangle, U_{21} = A_6.$$

The table of marks is

360	180	120	120	72	90	90	90	60	60	40	36	45	30	30	20	15	15	10	6	6	1
	4	.	.	.	2	6	6	4	4	.	4	5	2	2	4	3	3	2	2	2	1
		6	.	.	.	.	.	3	.	4	.	.	6	.	2	3	.	1	.	3	1
			6	.	.	.	.	.	3	4	.	.	.	6	2	.	3	1	3	.	1
				2	.	.	.	.	.	.	1	.	.	.	.	.	.	.	1	1	1
					2	.	.	.	.	.	.	1	.	.	.	1	1	2	.	.	1
						6	.	.	.	.	.	3	2	.	.	1	3	.	.	2	1
							6	.	.	.	.	3	.	2	.	3	1	.	2	.	1
								1	.	.	.	.	.	.	2	1	.	1	.	1	1
									1	.	.	.	.	.	2	.	1	1	1	.	1
										4	.	.	.	.	2	.	.	1	.	.	1
											1	.	.	.	.	.	.	.	1	1	1
												1	.	.	.	1	1	.	.	.	1
													2	.	.	1	.	.	.	2	1
														2	.	.	1	.	2	.	1
															2	.	.	1	.	.	1
																1	.	.	.	.	1
																	1	.	.	.	1
																		1	.	.	1
																			1	.	1
																				1	1
																					1

The Burnsidematrix is 1/360 times the matrix

1	−45	−20	−20	−36	.	30	30	180	180	30	180	.	60	60	−270	−180	−180	.	−360	−360	720
	90	.	.	.	−90	−90	−90	−360	−360	.	−360	180	.	.	540	360	360	180	720	720	−1800
		60	.	.	.	.	.	−180	.	−60	.	.	−180	.	180	180	.	.	.	360	−360
			60	.	.	.	.	.	−180	−60	.	.	.	−180	180	.	180	.	360	.	−360
				180	.	.	.	.	.	.	−180	.	.	.	.	.	.	.	.	.	.
					180	.	.	.	.	.	.	−180	.	.	.	.	.	−360	.	.	360
						60	.	.	.	.	.	−180	−60	.	.	180	.	.	.	.	.
							60	.	.	.	.	−180	.	−60	.	.	180	.	.	.	.
								360	.	.	.	.	.	.	−360	−360	.	.	.	−360	720
									360	.	.	.	.	.	−360	.	−360	.	−360	.	720
										90	.	.	.	.	−90	.	.	.	.	.	.
											360	.	.	.	.	.	.	.	−360	−360	360
												360	.	.	.	−360	−360	.	.	.	360
													180	.	.	−180	.	.	.	−360	360
														180	.	.	−180	.	−360	.	360
															180	.	.	−180	.	.	.
																360	.	.	.	.	−360
																	360	.	.	.	−360
																		360	.	.	−360
																			360	.	−360
																				360	−360
																					360

10.1.4 Symmetric Groups

This section contains a system of representatives of the conjugacy classes of subgroups, the table of marks and the Burnside matrix of S_3, S_4 and S_5, as well as a system of representatives of the conjugacy classes of their subgroups.

The group S_3: A transversal of the conjugacy classes of subgroups is

$$U_0 = \langle 1\rangle, U_1 = \langle (0,1)\rangle, U_2 = \langle (0,1,2)\rangle, U_3 = \langle (0,1,2),(0,1)\rangle = S_3.$$

The table of marks is

$$\begin{pmatrix} 6 & 3 & 2 & 1 \\ & 1 & \cdot & 1 \\ & & 2 & 1 \\ & & & 1 \end{pmatrix}.$$

The Burnside matrix looks as follows

$$\begin{pmatrix} 1/6 & -1/2 & -1/6 & 1/2 \\ & 1 & \cdot & -1 \\ & & 1/2 & -1/2 \\ & & & 1 \end{pmatrix}.$$

The group S_4: A transversal of the conjugacy classes of subgroups is

$$U_0 = \langle 1\rangle, U_1 = \langle (1,3)\rangle, U_2 = \langle (0,2)(1,3)\rangle, U_3 = \langle (0,2,1)\rangle,$$
$$U_4 = \langle (0,2),(1,3)\rangle, U_5 = \langle (0,1,2,3)\rangle,$$
$$U_6 = \langle (0,1)(2,3),(0,3)(1,2)\rangle, U_7 = \langle (0,2,1),(0,2)\rangle,$$
$$U_8 = \langle (0,1,2,3),(1,3)\rangle, U_9 = \langle (0,2,1),(0,3,1)\rangle, U_{10} = S_4.$$

The table of marks is

$$\begin{pmatrix} 24 & 12 & 12 & 8 & 6 & 6 & 6 & 4 & 3 & 2 & 1 \\ & 2 & \cdot & \cdot & 2 & \cdot & \cdot & 2 & 1 & \cdot & 1 \\ & & 4 & \cdot & 2 & 2 & 6 & \cdot & 3 & 2 & 1 \\ & & & 2 & \cdot & \cdot & \cdot & 1 & \cdot & 2 & 1 \\ & & & & 2 & \cdot & \cdot & \cdot & 1 & \cdot & 1 \\ & & & & & 2 & \cdot & \cdot & 1 & \cdot & 1 \\ & & & & & & 6 & \cdot & 3 & 2 & 1 \\ & & & & & & & 1 & \cdot & \cdot & 1 \\ & & & & & & & & 1 & \cdot & 1 \\ & & & & & & & & & 2 & 1 \\ & & & & & & & & & & 1 \end{pmatrix}.$$

The Burnside matrix looks as follows

$$\begin{pmatrix}
1/24 & -1/4 & -1/8 & -1/6 & 1/4 & . & 1/12 & 1/2 & . & 1/6 & -1/2 \\
 & 1/2 & . & . & -1/2 & . & . & -1 & . & . & 1 \\
 & & 1/4 & . & -1/4 & -1/4 & -1/4 & . & 1/2 & . & . \\
 & & & 1/2 & . & . & . & -1/2 & . & -1/2 & 1/2 \\
 & & & & 1/2 & . & . & . & -1/2 & . & . \\
 & & & & & 1/2 & . & . & -1/2 & . & . \\
 & & & & & & 1/6 & . & -1/2 & -1/6 & 1/2 \\
 & & & & & & & 1 & . & . & -1 \\
 & & & & & & & & 1 & . & -1 \\
 & & & & & & & & & 1/2 & -1/2 \\
 & & & & & & & & & & 1
\end{pmatrix}$$

The group S_5: A transversal of the conjugacy classes of subgroups is

$$U_0 = \langle 1 \rangle, U_1 = \langle (3,4) \rangle, U_2 = \langle (0,1)(3,4) \rangle, U_4 = \langle (0,2,1) \rangle,$$
$$U_4 = \langle (0,4,1,2,3) \rangle, U_5 = \langle (3,4), (0,2) \rangle, U_6 = \langle (0,4,1,3) \rangle,$$
$$U_7 = \langle (0,1)(3,4), (0,4)(1,3) \rangle, U_8 = \langle (0,2,1), (3,4) \rangle,$$
$$U_9 = \langle (0,2,1), (0,1) \rangle, U_{10} = \langle (0,2,1), (0,1)(3,4) \rangle,$$
$$U_{11} = \langle (0,4,1,2,3), (1,2)(3,4) \rangle, U_{12} = \langle (0,4,1,3), (3,4) \rangle,$$
$$U_{13} = \langle (0,2,1), (3,4), (0,2) \rangle, U_{14} = \langle (0,4,1), (0,4,3) \rangle,$$
$$U_{15} = \langle (1,4,2,3), (0,3,1,2) \rangle, U_{16} = \langle (0,4,1,3), (0,1,4,3) \rangle,$$
$$U_{17} = \langle (0,4,1,2,3), (0,1,2,4,3) \rangle, U_{18} = S_5.$$

The table of marks is

$$\left(\begin{array}{rrrrrrrrrrrrrrrrrrr}
120 & 60 & 60 & 40 & 24 & 30 & 30 & 30 & 20 & 20 & 20 & 12 & 15 & 10 & 10 & 6 & 5 & 2 & 1 \\
 & 6 & . & . & . & 6 & . & . & 2 & 6 & . & . & 3 & 4 & . & . & 3 & . & 1 \\
 & & 4 & . & . & 2 & 2 & 6 & . & . & 4 & 4 & 3 & 2 & 2 & 2 & 1 & 2 & 1 \\
 & & & 4 & . & . & . & . & 2 & 2 & 2 & . & . & 1 & 4 & . & 2 & 2 & 1 \\
 & & & & 4 & . & . & . & . & . & . & . & 2 & . & . & . & 1 & . & 2 & 1 \\
 & & & & & 2 & . & . & . & . & . & . & 1 & 2 & . & . & 1 & . & 1 \\
 & & & & & & 2 & . & . & . & . & . & 1 & . & . & 2 & 1 & . & 1 \\
 & & & & & & & 6 & . & . & . & . & 3 & . & 2 & . & 1 & 2 & 1 \\
 & & & & & & & & 2 & . & . & . & . & 1 & . & . & . & . & 1 \\
 & & & & & & & & & 2 & . & . & . & 1 & . & . & 2 & . & 1 \\
 & & & & & & & & & & 2 & . & . & 1 & . & . & . & 2 & 1 \\
 & & & & & & & & & & & 2 & . & . & . & 1 & . & 2 & 1 \\
 & & & & & & & & & & & & 1 & . & . & . & 1 & . & 1 \\
 & & & & & & & & & & & & & 1 & . & . & . & . & 1 \\
 & & & & & & & & & & & & & & 2 & . & 1 & 2 & 1 \\
 & & & & & & & & & & & & & & & 1 & . & . & 1 \\
 & & & & & & & & & & & & & & & & 1 & . & 1 \\
 & & & & & & & & & & & & & & & & & 2 & 1 \\
 & & & & & & & & & & & & & & & & & & 1
\end{array}\right)$$

The Burnsidematrix is 1/120 times the matrix

$$\left(\begin{array}{rrrrrrrrrrrrrrrrrrr}
1 & -10 & -15 & -10 & -6 & 30 & . & 10 & 10 & 30 & 30 & 30 & . & -60 & 20 & . & -60 & -60 & 60 \\
 & 20 & . & . & . & -60 & . & . & -20 & -60 & . & . & . & 120 & . & . & 120 & . & -120 \\
 & & 30 & . & . & -30 & -30 & -30 & . & . & -60 & -60 & 60 & 60 & . & 60 & . & 120 & -120 \\
 & & & 30 & . & . & . & . & -30 & -30 & -30 & . & . & 60 & -60 & . & 60 & 60 & -60 \\
 & & & & 30 & . & . & . & . & . & . & -30 & . & . & . & . & . & . & . \\
 & & & & & 60 & . & . & . & . & . & . & -60 & -120 & . & . & . & . & 120 \\
 & & & & & & 60 & . & . & . & . & . & -60 & . & . & -120 & . & . & 120 \\
 & & & & & & & 20 & . & . & . & . & -60 & . & -20 & . & 60 & . & . \\
 & & & & & & & & 60 & . & . & . & . & -60 & . & . & . & . & . \\
 & & & & & & & & & 60 & . & . & . & -60 & . & . & -120 & . & 120 \\
 & & & & & & & & & & 60 & . & . & -60 & . & . & . & -60 & 60 \\
 & & & & & & & & & & & 60 & . & . & . & -60 & . & -60 & 60 \\
 & & & & & & & & & & & & 120 & . & . & . & -120 & . & . \\
 & & & & & & & & & & & & & 120 & . & . & . & . & -120 \\
 & & & & & & & & & & & & & & 60 & . & -60 & -60 & 60 \\
 & & & & & & & & & & & & & & & 120 & . & . & -120 \\
 & & & & & & & & & & & & & & & & 120 & . & -120 \\
 & & & & & & & & & & & & & & & & & 60 & -60 \\
 & & & & & & & & & & & & & & & & & & 120
\end{array}\right)$$

10.2 Characters of Symmetric Groups

10.2.1 Irreducible Characters and Young Characters

This section contains the irreducible characters ζ^α of the symmetric groups S_n, where $n \leq 7$, together with the Young characters ξ^α and the scalar products or multiplicities

$$[\xi^\alpha \mid \zeta^\beta] = \kappa_{\alpha\beta} = st^\beta(\alpha).$$

Moreover the reader will find tables containing the scalar products

$$[\xi^\alpha \mid \xi^\beta] = m_{\alpha\beta}, \text{ and } [\epsilon\xi^\alpha \mid \xi^\beta] = m'_{\alpha\beta},$$

where ϵ denotes the alternating or sign character. Hence these tables contain the numbers of matrices over $\mathbb{N}$ and over $\{0, 1\}$, resepectively, with prescribed row and column sum vectors α and β.

Each subsection starts with a table that shows the partition α corresponding to the irreducible character ζ^i, to the Young character ξ^i, and the conjugacy class C^i. (We now use *numbers as indices* of the irreducible characters, of the Young characters and of the conjugacy classes. The index i in ζ^i, ξ^i and C^i should not be mixed up with the index α used in the notation ζ^α, ξ^α and C^α, since α means a proper partition as parameter, while i means a number!) The last but one column of this table contains the orders of the conjugacy class while the last column contains the order of the centralizer of an element in this class.

The second table is the character table, while the third one contains the multiplicities $[\xi^i \mid \zeta^j]$ of the irreducible characters ζ^j in the Young characters ξ^i. The values of the Young characters are shown in the fourth table.

The fifth and sixth table of each subsection gives the scalar products $[\xi^i \mid \xi^j]$ and the scalar products $[\epsilon\xi^i \mid \xi^j]$, respectively, which can be interpreted as numbers of matrices with prescribed row and column sums, as it was mentioned above already.

The group S_2:

ζ^0	ξ^0	2	C^1	1	2
ζ^1	ξ^2	11	C^0	1	2

	C^0	C^1
ζ^0	1	1
ζ^1	1	-1

	ζ^0	ζ^1
ξ^0	1	0
ξ^1	1	1

	C^0	C^1
ξ^0	1	1
ξ^1	2	0

	ξ^0	ξ^1
ξ^0	1	1
ξ^1	1	2

	ξ^0	ξ^1
$\epsilon\xi^0$	0	1
$\epsilon\xi^1$	1	2

The group S_3:

ζ^0	ξ^0	3	C^2	2	3
ζ^1	ξ^2	21	C^1	3	2
ζ^2	ξ^2	111	C^0	1	6

	C^0	C^1	C^2
ζ^0	1	1	1
ζ^1	2	0	-1
ζ^2	1	-1	1

	ζ^0	ζ^1	ζ^2
ξ^0	1	0	0
ξ^1	1	1	0
ξ^2	1	2	1

	C^0	C^1	C^2
ξ^0	1	1	1
ξ^1	3	1	0
ξ^2	6	0	0

	ξ^0	ξ^1	ξ^2
ξ^0	1	1	1
ξ^1	1	2	3
ξ^2	1	3	6

	ξ^0	ξ^1	ξ^2
$\epsilon\xi^0$	0	0	1
$\epsilon\xi^1$	0	1	3
$\epsilon\xi^2$	1	3	6

The group S_4:

ζ^0	ξ^0	4	C^4	6	4
ζ^1	ξ^2	31	C^3	8	3
ζ^2	ξ^2	22	C^2	3	8
ζ^3	ξ^4	211	C^1	6	4
ζ^4	ξ^4	1111	C^0	1	24

	C^0	C^1	C^2	C^3	C^4
ζ^0	1	1	1	1	1
ζ^1	3	1	-1	0	-1
ζ^2	2	0	2	-1	0
ζ^3	3	-1	-1	0	1
ζ^4	1	-1	1	1	-1

	ζ^0	ζ^1	ζ^2	ζ^3	ζ^4
ξ^0	1	0	0	0	0
ξ^1	1	1	0	0	0
ξ^2	1	1	1	0	0
ξ^3	1	2	1	1	0
ξ^4	1	3	2	3	1

	C^0	C^1	C^2	C^3	C^4
ξ^0	1	1	1	1	1
ξ^1	4	2	0	1	0
ξ^2	6	2	2	0	0
ξ^3	12	2	0	0	0
ξ^4	24	0	0	0	0

	ξ^0	ξ^1	ξ^2	ξ^3	ξ^4
ξ^0	1	1	1	1	1
ξ^1	1	2	2	3	4
ξ^2	1	2	3	4	6
ξ^3	1	3	4	7	12
ξ^4	1	4	6	12	24

	ξ^0	ξ^1	ξ^2	ξ^3	ξ^4
$\epsilon\xi^0$	0	0	0	0	1
$\epsilon\xi^1$	0	0	0	1	4
$\epsilon\xi^2$	0	0	1	2	6
$\epsilon\xi^3$	0	1	2	5	12
$\epsilon\xi^4$	1	4	6	12	24

The group S_5:

ζ^0	ξ^0	5	C^6	24	5
ζ^1	ξ^2	41	C^5	30	4
ζ^2	ξ^2	32	C^4	20	6
ζ^3	ξ^4	311	C^3	20	6
ζ^4	ξ^4	221	C^2	15	8
ζ^5	ξ^6	2111	C^1	10	12
ζ^6	ξ^6	11111	C^0	1	120

	C^0	C^1	C^2	C^3	C^4	C^5	C^6
ζ^0	1	1	1	1	1	1	1
ζ^1	4	2	0	1	−1	0	−1
ζ^2	5	1	1	−1	1	−1	0
ζ^3	6	0	−2	0	0	0	1
ζ^4	5	−1	1	−1	−1	1	0
ζ^5	4	−2	0	1	1	0	−1
ζ^6	1	−1	1	1	−1	−1	1

	ζ^0	ζ^1	ζ^2	ζ^3	ζ^4	ζ^5	ζ^6
ξ^0	1	0	0	0	0	0	0
ξ^1	1	1	0	0	0	0	0
ξ^2	1	1	1	0	0	0	0
ξ^3	1	2	1	1	0	0	0
ξ^4	1	2	2	1	1	0	0
ξ^5	1	3	3	3	2	1	0
ξ^6	1	4	5	6	5	4	1

	C^0	C^1	C^2	C^3	C^4	C^5	C^6
ξ^0	1	1	1	1	1	1	1
ξ^1	5	3	1	2	0	1	0
ξ^2	10	4	2	1	1	0	0
ξ^3	20	6	0	2	0	0	0
ξ^4	30	6	2	0	0	0	0
ξ^5	60	6	0	0	0	0	0
ξ^6	120	0	0	0	0	0	0

	ξ^0	ξ^1	ξ^2	ξ^3	ξ^4	ξ^5	ξ^6
ξ^0	1	1	1	1	1	1	1
ξ^1	1	2	2	3	3	4	5
ξ^2	1	2	3	4	5	7	10
ξ^3	1	3	4	7	8	13	20
ξ^4	1	3	5	8	11	18	30
ξ^5	1	4	7	13	18	33	60
ξ^6	1	5	10	20	30	60	120

	ξ^0	ξ^1	ξ^2	ξ^3	ξ^4	ξ^5	ξ^6
$\epsilon\xi^0$	0	0	0	0	0	0	1
$\epsilon\xi^1$	0	0	0	0	0	1	5
$\epsilon\xi^2$	0	0	0	0	1	3	10
$\epsilon\xi^3$	0	0	0	1	2	7	20
$\epsilon\xi^4$	0	0	1	2	5	12	30
$\epsilon\xi^5$	0	1	3	7	12	27	60
$\epsilon\xi^6$	1	5	10	20	30	60	120

The group S_6:

ζ^0	ξ^0	6	C^{10}	120	6
ζ^1	ξ^2	51	C^9	144	5
ζ^2	ξ^2	42	C^8	90	8
ζ^3	ξ^4	411	C^7	90	8
ζ^4	ξ^4	33	C^6	40	18
ζ^5	ξ^6	321	C^5	120	6
ζ^6	ξ^6	3111	C^4	40	18
ζ^7	ξ^8	222	C^3	15	48
ζ^8	ξ^8	2211	C^2	45	16
ζ^9	ξ^{10}	21111	C^1	15	48
ζ^{10}	ξ^{10}	111111	C^0	1	720

	C^0	C^1	C^2	C^3	C^4	C^5	C^6	C^7	C^8	C^9	C^{10}
ζ^0	1	1	1	1	1	1	1	1	1	1	1
ζ^1	5	3	1	−1	2	0	−1	1	−1	0	−1
ζ^2	9	3	1	3	0	0	0	−1	1	−1	0
ζ^3	10	2	−2	−2	1	−1	1	0	0	0	1
ζ^4	5	1	1	−3	−1	1	2	−1	−1	0	0
ζ^5	16	0	0	0	−2	0	−2	0	0	1	0
ζ^6	10	−2	−2	2	1	1	1	0	0	0	−1
ζ^7	5	−1	1	3	−1	−1	2	1	−1	0	0
ζ^8	9	−3	1	−3	0	0	0	1	1	−1	0
ζ^9	5	−3	1	1	2	0	−1	−1	−1	0	1
ζ^{10}	1	−1	1	−1	1	−1	1	−1	1	1	−1

	ζ^0	ζ^1	ζ^2	ζ^3	ζ^4	ζ^5	ζ^6	ζ^7	ζ^8	ζ^9	ζ^{10}
ξ^0	1	0	0	0	0	0	0	0	0	0	0
ξ^1	1	1	0	0	0	0	0	0	0	0	0
ξ^2	1	1	1	0	0	0	0	0	0	0	0
ξ^3	1	2	1	1	0	0	0	0	0	0	0
ξ^4	1	1	1	0	1	0	0	0	0	0	0
ξ^5	1	2	2	1	1	1	0	0	0	0	0
ξ^6	1	3	3	3	1	2	1	0	0	0	0
ξ^7	1	2	3	1	1	2	0	1	0	0	0
ξ^8	1	3	4	3	2	4	1	1	1	0	0
ξ^9	1	4	6	6	3	8	4	2	3	1	0
ξ^{10}	1	5	9	10	5	16	10	5	9	5	1

	C^0	C^1	C^2	C^3	C^4	C^5	C^6	C^7	C^8	C^9	C^{10}
ξ^0	1	1	1	1	1	1	1	1	1	1	1
ξ^1	6	4	2	0	3	1	0	2	0	1	0
ξ^2	15	7	3	3	3	1	0	1	1	0	0
ξ^3	30	12	2	0	6	0	0	2	0	0	0
ξ^4	20	8	4	0	2	2	2	0	0	0	0
ξ^5	60	16	4	0	3	1	0	0	0	0	0
ξ^6	120	24	0	0	6	0	0	0	0	0	0
ξ^7	90	18	6	6	0	0	0	0	0	0	0
ξ^8	180	24	4	0	0	0	0	0	0	0	0
ξ^9	360	24	0	0	0	0	0	0	0	0	0
ξ^{10}	720	0	0	0	0	0	0	0	0	0	0

	ξ^0	ξ^1	ξ^2	ξ^3	ξ^4	ξ^5	ξ^6	ξ^7	ξ^8	ξ^9	ξ^{10}
ξ^0	1	1	1	1	1	1	1	1	1	1	1
ξ^1	1	2	2	3	2	3	4	3	4	5	6
ξ^2	1	2	3	4	3	5	7	6	8	11	15
ξ^3	1	3	4	7	4	8	13	9	14	21	30
ξ^4	1	2	3	4	4	6	8	7	10	14	20
ξ^5	1	3	5	8	6	12	19	15	24	38	60
ξ^6	1	4	7	13	8	19	34	24	42	72	120
ξ^7	1	3	6	9	7	15	24	21	33	54	90
ξ^8	1	4	8	14	10	24	42	33	58	102	180
ξ^9	1	5	11	21	14	38	72	54	102	192	360
ξ^{10}	1	6	15	30	20	60	120	90	180	360	720

	ξ^0	ξ^1	ξ^2	ξ^3	ξ^4	ξ^5	ξ^6	ξ^7	ξ^8	ξ^9	ξ^{10}
$\epsilon\xi^0$	0	0	0	0	0	0	0	0	0	0	1
$\epsilon\xi^1$	0	0	0	0	0	0	0	0	0	1	6
$\epsilon\xi^2$	0	0	0	0	0	0	0	0	1	4	15
$\epsilon\xi^3$	0	0	0	0	0	0	1	0	2	9	30
$\epsilon\xi^4$	0	0	0	0	0	0	0	1	2	6	20
$\epsilon\xi^5$	0	0	0	0	0	1	3	3	8	22	60
$\epsilon\xi^6$	0	0	0	1	0	3	10	6	18	48	120
$\epsilon\xi^7$	0	0	0	0	1	3	6	6	15	36	90
$\epsilon\xi^8$	0	0	1	2	2	8	18	15	34	78	180
$\epsilon\xi^9$	0	1	4	9	6	22	48	36	78	168	360
$\epsilon\xi^{10}$	1	6	15	30	20	60	120	90	180	360	720

The group S_7:

ζ^0	ξ^0	7	C^{14}	720	7
ζ^1	ξ^2	61	C^{13}	840	6
ζ^2	ξ^2	52	C^{12}	504	10
ζ^3	ξ^4	511	C^{11}	504	10
ζ^4	ξ^4	43	C^{10}	420	12
ζ^5	ξ^6	421	C^9	630	8
ζ^6	ξ^6	4111	C^8	210	24
ζ^7	ξ^8	331	C^7	280	18
ζ^8	ξ^8	322	C^6	210	24
ζ^9	ξ^{10}	3211	C^5	420	12
ζ^{10}	ξ^{10}	31111	C^4	70	72
ζ^{11}	ξ^{12}	2221	C^3	105	48
ζ^{12}	ξ^{12}	22111	C^2	105	48
ζ^{13}	ξ^{14}	211111	C^1	21	240
ζ^{14}	ξ^{14}	1111111	C^0	1	5040

	C^0	C^1	C^2	C^3	C^4	C^5	C^6	C^7	C^8	C^9	C^{10}	C^{11}	C^{12}	C^{13}	C^{14}
ζ^0	1	1	1	1	1	1	1	1	1	1	1	1	1	1	1
ζ^1	6	4	2	0	3	1	−1	0	2	0	−1	1	−1	0	−1
ζ^2	14	6	2	2	2	0	2	−1	0	0	0	−1	1	−1	0
ζ^3	15	5	−1	−3	3	−1	−1	0	1	−1	1	0	0	0	1
ζ^4	14	4	2	0	−1	1	−1	2	−2	0	1	−1	−1	0	0
ζ^5	35	5	−1	1	−1	−1	−1	−1	−1	1	−1	0	0	1	0
ζ^6	20	0	−4	0	2	0	2	2	0	0	0	0	0	0	−1
ζ^7	21	1	1	−3	−3	1	1	0	−1	−1	−1	1	1	0	0
ζ^8	21	−1	1	3	−3	−1	1	0	1	−1	1	1	−1	0	0
ζ^9	35	−5	−1	−1	−1	1	−1	−1	1	1	1	0	0	−1	0
ζ^{10}	15	−5	−1	3	3	1	−1	0	−1	−1	−1	0	0	0	1
ζ^{11}	14	−4	2	0	−1	−1	−1	2	2	0	−1	−1	1	0	0
ζ^{12}	14	−6	2	−2	2	0	2	−1	0	0	0	−1	−1	1	0
ζ^{13}	6	−4	2	0	3	−1	−1	0	−2	0	1	1	1	0	−1
ζ^{14}	1	−1	1	−1	1	−1	1	1	−1	1	−1	1	−1	−1	1

	ζ^0	ζ^1	ζ^2	ζ^3	ζ^4	ζ^5	ζ^6	ζ^7	ζ^8	ζ^9	ζ^{10}	ζ^{11}	ζ^{12}	ζ^{13}	ζ^{14}
ξ^0	1	0	0	0	0	0	0	0	0	0	0	0	0	0	0
ξ^1	1	1	0	0	0	0	0	0	0	0	0	0	0	0	0
ξ^2	1	1	1	0	0	0	0	0	0	0	0	0	0	0	0
ξ^3	1	2	1	1	0	0	0	0	0	0	0	0	0	0	0
ξ^4	1	1	1	0	1	0	0	0	0	0	0	0	0	0	0
ξ^5	1	2	2	1	1	1	0	0	0	0	0	0	0	0	0
ξ^6	1	3	3	3	1	2	1	0	0	0	0	0	0	0	0
ξ^7	1	2	2	1	2	1	0	1	0	0	0	0	0	0	0
ξ^8	1	2	3	1	2	2	0	1	1	0	0	0	0	0	0
ξ^9	1	3	4	3	3	4	1	2	1	1	0	0	0	0	0
ξ^{10}	1	4	6	6	4	8	4	3	2	3	1	0	0	0	0
ξ^{11}	1	3	5	3	4	6	1	3	3	2	0	1	0	0	0
ξ^{12}	1	4	7	6	6	11	4	6	5	6	1	2	1	0	0
ξ^{13}	1	5	10	10	9	20	10	11	10	15	5	5	4	1	0
ξ^{14}	1	6	14	15	14	35	20	21	21	35	15	14	14	6	1

	C^0	C^1	C^2	C^3	C^4	C^5	C^6	C^7	C^8	C^9	C^{10}	C^{11}	C^{12}	C^{13}	C^{14}
ξ^0	1	1	1	1	1	1	1	1	1	1	1	1	1	1	1
ξ^1	7	5	3	1	4	2	0	1	3	1	0	2	0	1	0
ξ^2	21	11	5	3	6	2	2	0	3	1	0	1	1	0	0
ξ^3	42	20	6	0	12	2	0	0	6	0	0	2	0	0	0
ξ^4	35	15	7	3	5	3	1	2	1	1	1	0	0	0	0
ξ^5	105	35	9	3	12	2	0	0	3	1	0	0	0	0	0
ξ^6	210	60	6	0	24	0	0	0	6	0	0	0	0	0	0
ξ^7	140	40	12	0	8	4	0	2	0	0	0	0	0	0	0
ξ^8	210	50	14	6	6	2	2	0	0	0	0	0	0	0	0
ξ^9	420	80	12	0	12	2	0	0	0	0	0	0	0	0	0
ξ^{10}	840	120	0	0	24	0	0	0	0	0	0	0	0	0	0
ξ^{11}	630	90	18	6	0	0	0	0	0	0	0	0	0	0	0
ξ^{12}	1260	120	12	0	0	0	0	0	0	0	0	0	0	0	0
ξ^{13}	2520	120	0	0	0	0	0	0	0	0	0	0	0	0	0
ξ^{14}	5040	0	0	0	0	0	0	0	0	0	0	0	0	0	0

	ξ^0	ξ^1	ξ^2	ξ^3	ξ^4	ξ^5	ξ^6	ξ^7	ξ^8	ξ^9	ξ^{10}	ξ^{11}	ξ^{12}	ξ^{13}	ξ^{14}
ξ^0	1	1	1	1	1	1	1	1	1	1	1	1	1	1	1
ξ^1	1	2	2	3	2	3	4	3	3	4	5	4	5	6	7
ξ^2	1	2	3	4	3	5	7	5	6	8	11	9	12	16	21
ξ^3	1	3	4	7	4	8	13	8	9	14	21	15	22	31	42
ξ^4	1	2	3	4	4	6	8	7	8	11	15	13	18	25	35
ξ^5	1	3	5	8	6	12	19	13	16	25	39	30	46	70	105
ξ^6	1	4	7	13	8	19	34	20	25	43	73	51	84	135	210
ξ^7	1	3	5	8	7	13	20	16	19	30	46	37	58	90	140
ξ^8	1	3	6	9	8	16	25	19	25	39	62	51	81	130	210
ξ^9	1	4	8	14	11	25	43	30	39	67	114	87	148	250	420
ξ^{10}	1	5	11	21	15	39	73	46	62	114	208	150	270	480	840
ξ^{11}	1	4	9	15	13	30	51	37	51	87	150	120	207	360	630
ξ^{12}	1	5	12	22	18	46	84	58	81	148	270	207	378	690	1260
ξ^{13}	1	6	16	31	25	70	135	90	130	250	480	360	690	1320	2520
ξ^{14}	1	7	21	42	35	105	210	140	210	420	840	630	1260	2520	5040

	ξ^0	ξ^1	ξ^2	ξ^3	ξ^4	ξ^5	ξ^6	ξ^7	ξ^8	ξ^9	ξ^{10}	ξ^{11}	ξ^{12}	ξ^{13}	ξ^{14}
$\epsilon\xi^0$	0	0	0	0	0	0	0	0	0	0	0	0	0	0	1
$\epsilon\xi^1$	0	0	0	0	0	0	0	0	0	0	0	0	0	1	7
$\epsilon\xi^2$	0	0	0	0	0	0	0	0	0	0	0	0	1	5	21
$\epsilon\xi^3$	0	0	0	0	0	0	0	0	0	0	1	0	2	11	42
$\epsilon\xi^4$	0	0	0	0	0	0	0	0	0	0	0	1	3	10	35
$\epsilon\xi^5$	0	0	0	0	0	0	0	0	0	1	4	3	11	35	105
$\epsilon\xi^6$	0	0	0	0	0	0	1	0	0	3	13	6	24	75	210
$\epsilon\xi^7$	0	0	0	0	0	0	0	0	1	2	6	7	18	50	140
$\epsilon\xi^8$	0	0	0	0	0	0	0	1	2	5	12	12	31	80	210
$\epsilon\xi^9$	0	0	0	0	0	1	3	2	5	13	34	27	68	170	420
$\epsilon\xi^{10}$	0	0	0	1	0	4	13	6	12	34	88	60	150	360	840
$\epsilon\xi^{11}$	0	0	0	0	1	3	6	7	12	27	60	51	117	270	630
$\epsilon\xi^{12}$	0	0	1	2	3	11	24	18	31	68	150	117	258	570	1260
$\epsilon\xi^{13}$	0	1	5	11	10	35	75	50	80	170	360	270	570	1200	2520
$\epsilon\xi^{14}$	1	7	21	42	35	105	210	140	210	420	840	630	1260	2520	5040

10.2.2 Foulkes Tables

This section contains the Foulkes F_n tables of the symmetric groups S_n, for $n \leq 10$. We recall that the j-th column of the Foulkes table F_n contains in its i-th row and j-th column the value of the Foulkes character $\chi^{n,i}$ of S_n on the classes of elements which consist of j cyclic factors.

$F_1 =$

$i\backslash j$	1
0	1

, $F_2 =$

$i\backslash j$	2	1
0	1	−1
1	1	1

, $F_3 =$

$i\backslash j$	3	2	1
0	1	−1	1
1	4	0	−2
2	1	1	1

,

$F_4 =$

$i\backslash j$	4	3	2	1
0	1	−1	1	−1
1	11	−3	−1	3
2	11	3	−1	−3
3	1	1	1	1

, $F_5 =$

$i\backslash j$	5	4	3	2	1
0	1	−1	1	−1	1
1	26	−10	2	2	−4
2	66	0	−6	0	6
3	26	10	2	−2	−4
4	1	1	1	1	1

,

$F_6 =$

$i\backslash j$	6	5	4	3	2	1
0	1	−1	1	−1	1	−1
1	57	−25	9	−1	−3	5
2	302	−40	−10	8	2	−10
3	302	40	−10	−8	2	10
4	57	25	9	1	−3	−5
5	1	1	1	1	1	1

,

$F_7 =$

$i\backslash j$	7	6	5	4	3	2	1
0	1	−1	1	−1	1	−1	1
1	120	−56	24	−8	0	4	−6
2	1191	−245	15	19	−9	−5	15
3	2416	0	−80	0	16	0	−20
4	1191	245	15	−19	−9	5	15
5	120	56	24	8	0	−4	−6
6	1	1	1	1	1	1	1

,

$F_8 =$

$i\backslash j$	8	7	6	5	4	3	2	1
0	1	−1	1	−1	1	−1	1	−1
1	247	−119	55	−23	7	1	−5	7
2	4293	−1071	189	9	−27	9	9	−21
3	15619	−1225	−245	95	19	−25	−5	35
4	15619	1225	−245	−95	19	25	−5	−35
5	4293	1071	189	−9	−27	−9	9	21
6	247	119	55	23	7	−1	−5	−7
7	1	1	1	1	1	1	1	1

,

$F_9 =$

$i\backslash j$	9	8	7	6	5	4	3	2	1
0	1	−1	1	−1	1	−1	1	−1	1
1	502	−246	118	−54	22	−6	−2	6	−8
2	14608	−4046	952	−134	−32	34	−8	−14	28
3	88234	−11326	154	434	−86	−46	34	14	−56
4	156190	0	−2450	0	190	0	−50	0	70
5	88234	11326	154	−434	−86	46	34	−14	−56
6	14608	4046	952	134	−32	−34	−8	14	28
7	502	246	118	54	22	6	−2	−6	−8
8	1	1	1	1	1	1	1	1	1

,

$F_{10} =$

$i\backslash j$	10	9	8	7	6	5	4	3	2	1
0	1	−1	1	−1	1	−1	1	−1	1	−1
1	1013	−501	245	−117	53	−21	5	3	−7	9
2	47840	−14106	3800	−834	80	54	−40	6	20	−36
3	455192	−73626	7280	798	−568	54	80	−42	−28	84
4	1310354	−67956	−11326	2604	434	−276	−46	84	14	−126
5	1310354	67956	−11326	−2604	434	276	−46	−84	14	126
6	455192	73626	7280	−798	−568	−54	80	42	−28	−84
7	47840	14106	3800	834	80	−54	−40	−6	20	36
8	1013	501	245	117	53	21	5	−3	−7	−9
9	1	1	1	1	1	1	1	1	1	1

.

10.2.3 Character Polynomials

This subsection contains character polynomials for the symmetric group. Recall that these are polynomial functions in variables a_i, corresponding to *truncated partitions* $\hat{\alpha}$. This means the following: If you want to obtain the value of the ordinary irreducible character ζ^{α}, where $\alpha = (\alpha_0, \alpha_1, \ldots)$, on the conjugacy class C^{γ}, you can proceed as follows: put $\hat{\alpha} := (\alpha_1, \alpha_2, \ldots)$, and denote by $a = (a_1, \ldots)$ the cycle type corresponding to the cycle partition γ. Then ζ_γ^α is the the value of the character polynomial $\Xi^{\hat{\alpha}}$ at $a = (a_1, \ldots)$:

$$\zeta_\gamma^\alpha = \Xi^{\hat{\alpha}}(a_1, \ldots).$$

On this and the following pages you will find the first character polynomials listed, they cover the character tables of S_n, for $n \leq 10$, and each of these polynomials allows to evaluate an *infinite series* of ordinary irreducible character of symmetric groups! I did not try to simplify the expressions of these polynomials as sums of products of binomial expressions.

The character polynomial $\Xi^{\hat{\alpha}}$, for $\hat{\alpha} := (\alpha_1, \alpha_2, \ldots) := (0)$:

$$1$$

The character polynomial $\Xi^{\hat{\alpha}}$, for $\hat{\alpha} := (\alpha_1, \alpha_2, \ldots) := (1)$:

$$\binom{a_1}{1} - 1$$

The character polynomial $\Xi^{\hat{\alpha}}$, for $\hat{\alpha} := (\alpha_1, \alpha_2, \ldots) := (2)$:

$$\binom{a_2}{1} + \binom{a_1}{2} - \binom{a_1}{1}$$

The character polynomial $\Xi^{\hat{\alpha}}$, for $\hat{\alpha} := (\alpha_1, \alpha_2, \ldots) := (1^2)$:

$$\binom{a_1}{2} - \binom{a_2}{1} - \binom{a_1}{1} + 1$$

The character polynomial $\Xi^{\hat{\alpha}}$, for $\hat{\alpha} := (\alpha_1, \alpha_2, \ldots) := (3)$:

$$\binom{a_3}{1} + \binom{a_1}{1}\binom{a_2}{1} + \binom{a_1}{3} - \binom{a_2}{1} - \binom{a_1}{2}$$

The character polynomial $\Xi^{\hat{\alpha}}$, for $\hat{\alpha} := (\alpha_1, \alpha_2, \ldots) := (2, 1)$:

$$2\binom{a_1}{3} - \binom{a_3}{1} - 2\binom{a_1}{2} + \binom{a_1}{1}$$

The character polynomial $\Xi^{\hat{\alpha}}$, for $\hat{\alpha} := (\alpha_1, \alpha_2, \ldots) := (1^3)$:

$$\binom{a_1}{3} - \binom{a_1}{1}\binom{a_2}{1} + \binom{a_3}{1} - \binom{a_1}{2} + \binom{a_2}{1} + \binom{a_1}{1} - 1$$

The character polynomial $\Xi^{\hat{\alpha}}$, for $\hat{\alpha} := (\alpha_1, \alpha_2, \ldots) := (4)$:

$$\binom{a_4}{1} + \binom{a_1}{1}\binom{a_3}{1} + \binom{a_2}{2} + \binom{a_1}{2}\binom{a_2}{1} + \binom{a_1}{4} - \binom{a_3}{1} - \binom{a_1}{1}\binom{a_2}{1} - \binom{a_1}{3}$$

The character polynomial $\Xi^{\hat{\alpha}}$, for $\hat{\alpha} := (\alpha_1, \alpha_2, \ldots) := (3, 1)$:

$$\binom{a_1}{2}\binom{a_2}{1} + 3\binom{a_1}{4} - \binom{a_4}{1} - \binom{a_2}{2} - \binom{a_1}{1}\binom{a_2}{1} - 3\binom{a_1}{3} + \binom{a_2}{1} + \binom{a_1}{2}$$

The character polynomial $\Xi^{\hat{\alpha}}$, for $\hat{\alpha} := (\alpha_1, \alpha_2, \ldots) := (2^2)$:

$$2\binom{a_2}{2} + 2\binom{a_1}{4} - \binom{a_1}{1}\binom{a_3}{1} - 2\binom{a_1}{3} + \binom{a_1}{2} + \binom{a_3}{1} - \binom{a_2}{1}$$

The character polynomial $\Xi^{\hat{\alpha}}$, for $\hat{\alpha} := (\alpha_1, \alpha_2, \ldots) := (2, 1^2)$:

$$-\binom{a_1}{2}\binom{a_2}{1} + 3\binom{a_1}{4} - \binom{a_2}{2} + \binom{a_4}{1} - 3\binom{a_1}{3} + \binom{a_1}{1}\binom{a_2}{1} + 2\binom{a_1}{2} - \binom{a_1}{1}$$

The character polynomial $\Xi^{\hat{\alpha}}$, for $\hat{\alpha} := (\alpha_1, \alpha_2, \ldots) := (1^4)$:

$$\binom{a_1}{4} - \binom{a_1}{2}\binom{a_2}{1} + \binom{a_1}{1}\binom{a_3}{1} + \binom{a_2}{2} - \binom{a_4}{1}$$
$$-\binom{a_1}{3} + \binom{a_1}{1}\binom{a_2}{1} - \binom{a_3}{1} + \binom{a_1}{2} - \binom{a_2}{1} - \binom{a_1}{1} + 1$$

The character polynomial $\Xi^{\hat{\alpha}}$, for $\hat{\alpha} := (\alpha_1, \alpha_2, \ldots) := (5)$:

$$\binom{a_5}{1} + \binom{a_1}{1}\binom{a_4}{1} + \binom{a_2}{1}\binom{a_3}{1} + \binom{a_1}{2}\binom{a_3}{1} + \binom{a_1}{1}\binom{a_2}{2}$$
$$+\binom{a_1}{3}\binom{a_2}{1} + \binom{a_1}{5} - \binom{a_4}{1} - \binom{a_1}{1}\binom{a_3}{1} - \binom{a_2}{2} - \binom{a_1}{2}\binom{a_2}{1} - \binom{a_1}{4}$$

The character polynomial $\Xi^{\hat{\alpha}}$, for $\hat{\alpha} := (\alpha_1, \alpha_2, \ldots) := (4, 1)$:

$$\binom{a_1}{2}\binom{a_3}{1} + 2\binom{a_1}{3}\binom{a_2}{1} + 4\binom{a_1}{5} - \binom{a_5}{1} - \binom{a_2}{1}\binom{a_3}{1}$$
$$-\binom{a_1}{1}\binom{a_3}{1} - 2\binom{a_1}{2}\binom{a_2}{1} - 4\binom{a_1}{4} + \binom{a_3}{1} + \binom{a_1}{1}\binom{a_2}{1} + \binom{a_1}{3}$$

The character polynomial $\Xi^{\hat{\alpha}}$, for $\hat{\alpha} := (\alpha_1, \alpha_2, \ldots) := (3, 2)$:

$$\binom{a_2}{1}\binom{a_3}{1} - \binom{a_1}{2}\binom{a_3}{1} + \binom{a_1}{1}\binom{a_2}{2} + \binom{a_1}{3}\binom{a_2}{1} + 5\binom{a_1}{5} - \binom{a_1}{1}\binom{a_4}{1}$$
$$-\binom{a_2}{2} - \binom{a_1}{2}\binom{a_2}{1} - 5\binom{a_1}{4} + 2\binom{a_1}{3} + \binom{a_4}{1} + \binom{a_1}{1}\binom{a_3}{1} - \binom{a_3}{1}$$

The character polynomial $\Xi^{\hat{\alpha}}$, for $\hat{\alpha} := (\alpha_1, \alpha_2, \ldots) := (3, 1^2)$:

$$6\binom{a_1}{5} - 2\binom{a_1}{1}\binom{a_2}{2} + \binom{a_5}{1} - 6\binom{a_1}{4} + 2\binom{a_2}{2} + \binom{a_1}{1}\binom{a_2}{1}$$
$$+3\binom{a_1}{3} - \binom{a_2}{1} - \binom{a_1}{2}$$

The character polynomial $\Xi^{\hat{\alpha}}$, for $\hat{\alpha} := (\alpha_1, \alpha_2, \ldots) := (2^2, 1)$:

$$\binom{a_1}{1}\binom{a_2}{2} - \binom{a_1}{3}\binom{a_2}{1} + 5\binom{a_1}{5} - \binom{a_2}{1}\binom{a_3}{1}$$
$$-\binom{a_1}{2}\binom{a_3}{1} + \binom{a_1}{1}\binom{a_4}{1} + \binom{a_1}{2}\binom{a_2}{1} - 5\binom{a_1}{4} + \binom{a_1}{1}\binom{a_3}{1}$$
$$+3\binom{a_1}{3} - \binom{a_1}{2} - \binom{a_4}{1} - \binom{a_2}{2} - \binom{a_1}{1}\binom{a_2}{1} + \binom{a_2}{1}$$

The character polynomial $\Xi^{\hat{\alpha}}$, for $\hat{\alpha} := (\alpha_1, \alpha_2, \ldots) := (2, 1^3)$:

$$-2\binom{a_1}{3}\binom{a_2}{1} + 4\binom{a_1}{5} + \binom{a_2}{1}\binom{a_3}{1} + \binom{a_1}{2}\binom{a_3}{1} - \binom{a_5}{1} - 4\binom{a_1}{4}$$
$$+2\binom{a_1}{2}\binom{a_2}{1} - \binom{a_1}{1}\binom{a_3}{1} + 3\binom{a_1}{3} - \binom{a_1}{1}\binom{a_2}{1} - 2\binom{a_1}{2} + \binom{a_1}{1}$$

The character polynomial $\Xi^{\hat{\alpha}}$, for $\hat{\alpha} := (\alpha_1, \alpha_2, \ldots) := (1^5)$:

$$\binom{a_1}{5} - \binom{a_1}{3}\binom{a_2}{1} + \binom{a_1}{2}\binom{a_3}{1} + \binom{a_1}{1}\binom{a_2}{2} - \binom{a_1}{1}\binom{a_4}{1}$$
$$-\binom{a_2}{1}\binom{a_3}{1} + \binom{a_5}{1} - \binom{a_1}{4} + \binom{a_1}{2}\binom{a_2}{1} - \binom{a_1}{1}\binom{a_3}{1} - \binom{a_2}{2} + \binom{a_4}{1}$$
$$+\binom{a_1}{3} - \binom{a_1}{1}\binom{a_2}{1} + \binom{a_3}{1} - \binom{a_1}{2} + \binom{a_2}{1} + \binom{a_1}{1} - 1$$

The character polynomial $\Xi^{\hat{\alpha}}$, for $\hat{\alpha} := (\alpha_1, \alpha_2, \ldots) := (6)$:

$$\binom{a_6}{1} + \binom{a_1}{1}\binom{a_5}{1} + \binom{a_2}{1}\binom{a_4}{1} + \binom{a_1}{2}\binom{a_4}{1} + \binom{a_3}{2}$$
$$+\binom{a_1}{1}\binom{a_2}{1}\binom{a_3}{1} + \binom{a_1}{3}\binom{a_3}{1} + \binom{a_2}{3} + \binom{a_1}{2}\binom{a_2}{2} + \binom{a_1}{4}\binom{a_2}{1}$$
$$+\binom{a_1}{6} - \binom{a_5}{1} - \binom{a_1}{1}\binom{a_4}{1} - \binom{a_2}{1}\binom{a_3}{1} - \binom{a_1}{2}\binom{a_3}{1}$$
$$-\binom{a_1}{1}\binom{a_2}{2} - \binom{a_1}{3}\binom{a_2}{1} - \binom{a_1}{5}$$

The character polynomial $\Xi^{\hat{\alpha}}$, for $\hat{\alpha} := (\alpha_1, \alpha_2, \ldots) := (5, 1)$:

$$\binom{a_1}{2}\binom{a_4}{1} + 2\binom{a_1}{3}\binom{a_3}{1} + \binom{a_1}{2}\binom{a_2}{2} + 3\binom{a_1}{4}\binom{a_2}{1} + 5\binom{a_1}{6} - \binom{a_6}{1}$$
$$-\binom{a_2}{1}\binom{a_4}{1} - \binom{a_3}{2} - \binom{a_2}{3} - \binom{a_1}{1}\binom{a_4}{1} - 2\binom{a_1}{2}\binom{a_3}{1} - \binom{a_1}{1}\binom{a_2}{2}$$
$$-3\binom{a_1}{3}\binom{a_2}{1} - 5\binom{a_1}{5} + \binom{a_4}{1} + \binom{a_1}{1}\binom{a_3}{1}$$
$$+\binom{a_2}{2} + \binom{a_1}{2}\binom{a_2}{1} + \binom{a_1}{4}$$

The character polynomial $\Xi^{\hat{\alpha}}$, for $\hat{\alpha} := (\alpha_1, \alpha_2, \ldots) := (4, 2)$:

$$\binom{a_2}{1}\binom{a_4}{1} - \binom{a_1}{2}\binom{a_4}{1} + 3\binom{a_2}{3} + \binom{a_1}{2}\binom{a_2}{2} + 3\binom{a_1}{4}\binom{a_2}{1} + 9\binom{a_1}{6}$$
$$-\binom{a_1}{1}\binom{a_5}{1} - \binom{a_1}{1}\binom{a_2}{2} - 3\binom{a_1}{3}\binom{a_2}{1} - 9\binom{a_1}{5} + \binom{a_1}{2}\binom{a_2}{1}$$
$$+3\binom{a_1}{4} + \binom{a_5}{1} + \binom{a_1}{1}\binom{a_4}{1} - \binom{a_4}{1} - \binom{a_2}{2}$$

The character polynomial $\Xi^{\hat{\alpha}}$, for $\hat{\alpha} := (\alpha_1, \alpha_2, \ldots) := (4, 1^2)$:

$$\binom{a_1}{3}\binom{a_3}{1} - 2\binom{a_1}{2}\binom{a_2}{2} + 2\binom{a_1}{4}\binom{a_2}{1} + 10\binom{a_1}{6} - \binom{a_1}{1}\binom{a_2}{1}\binom{a_3}{1}$$
$$-2\binom{a_2}{3} + \binom{a_6}{1} + \binom{a_3}{2} - \binom{a_1}{2}\binom{a_3}{1} - 2\binom{a_1}{3}\binom{a_2}{1} - 10\binom{a_1}{5}$$
$$+\binom{a_2}{1}\binom{a_3}{1} + 2\binom{a_1}{1}\binom{a_2}{2} + \binom{a_1}{1}\binom{a_3}{1} + 2\binom{a_1}{2}\binom{a_2}{1}$$
$$+4\binom{a_1}{4} - \binom{a_3}{1} - \binom{a_1}{1}\binom{a_2}{1} - \binom{a_1}{3}$$

The character polynomial $\Xi^{\hat{\alpha}}$, for $\hat{\alpha} := (\alpha_1, \alpha_2, \ldots) := (3^2)$:

$$2\binom{a_3}{2} + \binom{a_1}{1}\binom{a_2}{1}\binom{a_3}{1} - \binom{a_1}{3}\binom{a_3}{1} + \binom{a_1}{2}\binom{a_2}{2} + \binom{a_1}{4}\binom{a_2}{1}$$
$$+5\binom{a_1}{6} - \binom{a_2}{1}\binom{a_4}{1} - \binom{a_1}{2}\binom{a_4}{1} - 3\binom{a_2}{3} - \binom{a_2}{1}\binom{a_3}{1} + \binom{a_1}{2}\binom{a_3}{1}$$
$$-\binom{a_1}{1}\binom{a_2}{2} - \binom{a_1}{3}\binom{a_2}{1} - 5\binom{a_1}{5} + 2\binom{a_2}{2} + 2\binom{a_1}{4}$$

$$+\binom{a_1}{1}\binom{a_4}{1}-\binom{a_1}{1}\binom{a_3}{1}$$

The character polynomial $\Xi^{\hat{\alpha}}$, for $\hat{\alpha} := (\alpha_1, \alpha_2, \ldots) := (3, 2, 1)$:

$$-2\binom{a_1}{3}\binom{a_3}{1}+16\binom{a_1}{6}-2\binom{a_3}{2}+\binom{a_1}{1}\binom{a_5}{1}-16\binom{a_1}{5}+2\binom{a_1}{2}\binom{a_3}{1}$$
$$+8\binom{a_1}{4}-2\binom{a_1}{3}-\binom{a_5}{1}-\binom{a_1}{1}\binom{a_3}{1}+\binom{a_3}{1}$$

The character polynomial $\Xi^{\hat{\alpha}}$, for $\hat{\alpha} := (\alpha_1, \alpha_2, \ldots) := (3, 1^3)$:

$$\binom{a_1}{3}\binom{a_3}{1}-2\binom{a_1}{4}\binom{a_2}{1}+10\binom{a_1}{6}+\binom{a_1}{1}\binom{a_2}{1}\binom{a_3}{1}-2\binom{a_1}{2}\binom{a_2}{2}$$
$$+\binom{a_3}{2}+2\binom{a_2}{3}-\binom{a_6}{1}+2\binom{a_1}{3}\binom{a_2}{1}-10\binom{a_1}{5}+2\binom{a_1}{1}\binom{a_2}{2}$$
$$-\binom{a_2}{1}\binom{a_3}{1}-\binom{a_1}{2}\binom{a_3}{1}+6\binom{a_1}{4}-2\binom{a_2}{2}-\binom{a_1}{1}\binom{a_2}{1}$$
$$-3\binom{a_1}{3}+\binom{a_2}{1}+\binom{a_1}{2}$$

The character polynomial $\Xi^{\hat{\alpha}}$, for $\hat{\alpha} := (\alpha_1, \alpha_2, \ldots) := (2^3)$:

$$3\binom{a_2}{3}+\binom{a_1}{2}\binom{a_2}{2}-\binom{a_1}{4}\binom{a_2}{1}+5\binom{a_1}{6}-\binom{a_1}{1}\binom{a_2}{1}\binom{a_3}{1}$$
$$-\binom{a_1}{3}\binom{a_3}{1}+\binom{a_1}{2}\binom{a_4}{1}+2\binom{a_3}{2}-\binom{a_2}{1}\binom{a_4}{1}-\binom{a_1}{1}\binom{a_2}{2}$$
$$+\binom{a_1}{3}\binom{a_2}{1}-5\binom{a_1}{5}+\binom{a_1}{2}\binom{a_3}{1}-\binom{a_1}{2}\binom{a_2}{1}+3\binom{a_1}{4}-\binom{a_1}{3}$$
$$+\binom{a_1}{1}\binom{a_2}{1}+\binom{a_2}{1}\binom{a_3}{1}-\binom{a_1}{1}\binom{a_4}{1}-\binom{a_2}{2}+\binom{a_4}{1}-\binom{a_3}{1}$$

The character polynomial $\Xi^{\hat{\alpha}}$, for $\hat{\alpha} := (\alpha_1, \alpha_2, \ldots) := (2^2, 1^2)$:

$$\binom{a_1}{2}\binom{a_2}{2}-3\binom{a_1}{4}\binom{a_2}{1}+9\binom{a_1}{6}-3\binom{a_2}{3}+\binom{a_2}{1}\binom{a_4}{1}$$
$$+\binom{a_1}{2}\binom{a_4}{1}-\binom{a_1}{1}\binom{a_5}{1}+3\binom{a_1}{3}\binom{a_2}{1}-9\binom{a_1}{5}-\binom{a_1}{1}\binom{a_2}{2}$$
$$-\binom{a_1}{1}\binom{a_4}{1}+6\binom{a_1}{4}-2\binom{a_1}{2}\binom{a_2}{1}-3\binom{a_1}{3}+\binom{a_1}{2}+\binom{a_5}{1}$$

$$+2\binom{a_2}{2}+\binom{a_1}{1}\binom{a_2}{1}-\binom{a_2}{1}$$

The character polynomial $\Xi^{\hat{\alpha}}$, for $\hat{\alpha} := (\alpha_1, \alpha_2, \ldots) := (2, 1^4)$:

$$-3\binom{a_1}{4}\binom{a_2}{1}+5\binom{a_1}{6}+\binom{a_1}{2}\binom{a_2}{2}+2\binom{a_1}{3}\binom{a_3}{1}+\binom{a_2}{3}$$
$$-\binom{a_2}{1}\binom{a_4}{1}-\binom{a_1}{2}\binom{a_4}{1}-\binom{a_3}{2}+\binom{a_6}{1}-5\binom{a_1}{5}+3\binom{a_1}{3}\binom{a_2}{1}$$
$$-2\binom{a_1}{2}\binom{a_3}{1}-\binom{a_1}{1}\binom{a_2}{2}+\binom{a_1}{1}\binom{a_4}{1}+4\binom{a_1}{4}-2\binom{a_1}{2}\binom{a_2}{1}$$
$$+\binom{a_1}{1}\binom{a_3}{1}-3\binom{a_1}{3}+\binom{a_1}{1}\binom{a_2}{1}+2\binom{a_1}{2}-\binom{a_1}{1}$$

The character polynomial $\Xi^{\hat{\alpha}}$, for $\hat{\alpha} := (\alpha_1, \alpha_2, \ldots) := (1^6)$:

$$\binom{a_1}{6}-\binom{a_1}{4}\binom{a_2}{1}+\binom{a_1}{3}\binom{a_3}{1}+\binom{a_1}{2}\binom{a_2}{2}-\binom{a_1}{2}\binom{a_4}{1}$$
$$-\binom{a_1}{1}\binom{a_2}{1}\binom{a_3}{1}+\binom{a_1}{1}\binom{a_5}{1}-\binom{a_2}{3}+\binom{a_2}{1}\binom{a_4}{1}+\binom{a_3}{2}-\binom{a_6}{1}$$
$$-\binom{a_1}{5}+\binom{a_1}{3}\binom{a_2}{1}-\binom{a_1}{2}\binom{a_3}{1}-\binom{a_1}{1}\binom{a_2}{2}+\binom{a_1}{1}\binom{a_4}{1}$$
$$+\binom{a_2}{1}\binom{a_3}{1}-\binom{a_5}{1}+\binom{a_1}{4}-\binom{a_1}{2}\binom{a_2}{1}+\binom{a_1}{1}\binom{a_3}{1}+\binom{a_2}{2}-\binom{a_4}{1}$$
$$-\binom{a_1}{3}+\binom{a_1}{1}\binom{a_2}{1}-\binom{a_3}{1}+\binom{a_1}{2}-\binom{a_2}{1}-\binom{a_1}{1}+1$$

The character polynomial $\Xi^{\hat{\alpha}}$, for $\hat{\alpha} := (\alpha_1, \alpha_2, \ldots) := (7)$:

$$\binom{a_7}{1}+\binom{a_1}{1}\binom{a_6}{1}+\binom{a_2}{1}\binom{a_5}{1}+\binom{a_1}{2}\binom{a_5}{1}+\binom{a_3}{1}\binom{a_4}{1}$$
$$+\binom{a_1}{1}\binom{a_2}{1}\binom{a_4}{1}+\binom{a_1}{3}\binom{a_4}{1}+\binom{a_1}{1}\binom{a_3}{2}+\binom{a_2}{2}\binom{a_3}{1}$$
$$+\binom{a_1}{2}\binom{a_2}{1}\binom{a_3}{1}+\binom{a_1}{4}\binom{a_3}{1}+\binom{a_1}{1}\binom{a_2}{3}+\binom{a_1}{3}\binom{a_2}{2}$$
$$+\binom{a_1}{5}\binom{a_2}{1}+\binom{a_1}{7}-\binom{a_6}{1}-\binom{a_1}{1}\binom{a_5}{1}-\binom{a_2}{1}\binom{a_4}{1}$$
$$-\binom{a_1}{2}\binom{a_4}{1}-\binom{a_3}{2}-\binom{a_1}{1}\binom{a_2}{1}\binom{a_3}{1}-\binom{a_1}{3}\binom{a_3}{1}$$

$$-\binom{a_2}{3}-\binom{a_1}{2}\binom{a_2}{2}-\binom{a_1}{4}\binom{a_2}{1}-\binom{a_1}{6}$$

The character polynomial $\Xi^{\hat{\alpha}}$, for $\hat{\alpha} := (\alpha_1, \alpha_2, \ldots) := (6, 1)$:

$$\binom{a_1}{2}\binom{a_5}{1}+2\binom{a_1}{3}\binom{a_4}{1}+\binom{a_1}{2}\binom{a_2}{1}\binom{a_3}{1}+3\binom{a_1}{4}\binom{a_3}{1}$$
$$+2\binom{a_1}{3}\binom{a_2}{2}+4\binom{a_1}{5}\binom{a_2}{1}+6\binom{a_1}{7}-\binom{a_7}{1}-\binom{a_2}{1}\binom{a_5}{1}$$
$$-\binom{a_3}{1}\binom{a_4}{1}-\binom{a_2}{2}\binom{a_3}{1}-\binom{a_1}{1}\binom{a_5}{1}-2\binom{a_1}{2}\binom{a_4}{1}$$
$$-\binom{a_1}{1}\binom{a_2}{1}\binom{a_3}{1}-3\binom{a_1}{3}\binom{a_3}{1}-2\binom{a_1}{2}\binom{a_2}{2}-4\binom{a_1}{4}\binom{a_2}{1}$$
$$-6\binom{a_1}{6}+\binom{a_5}{1}+\binom{a_1}{1}\binom{a_4}{1}+\binom{a_2}{1}\binom{a_3}{1}+\binom{a_1}{2}\binom{a_3}{1}$$
$$+\binom{a_1}{1}\binom{a_2}{2}+\binom{a_1}{3}\binom{a_2}{1}+\binom{a_1}{5}$$

The character polynomial $\Xi^{\hat{\alpha}}$, for $\hat{\alpha} := (\alpha_1, \alpha_2, \ldots) := (5, 2)$:

$$\binom{a_2}{1}\binom{a_5}{1}-\binom{a_1}{2}\binom{a_5}{1}+2\binom{a_2}{2}\binom{a_3}{1}+2\binom{a_1}{4}\binom{a_3}{1}+2\binom{a_1}{1}\binom{a_2}{3}$$
$$+2\binom{a_1}{3}\binom{a_2}{2}+6\binom{a_1}{5}\binom{a_2}{1}+14\binom{a_1}{7}-\binom{a_1}{1}\binom{a_6}{1}-\binom{a_1}{1}\binom{a_3}{2}$$
$$-2\binom{a_1}{3}\binom{a_3}{1}-2\binom{a_2}{3}-2\binom{a_1}{2}\binom{a_2}{2}-6\binom{a_1}{4}\binom{a_2}{1}-14\binom{a_1}{6}$$
$$+\binom{a_1}{2}\binom{a_3}{1}+2\binom{a_1}{3}\binom{a_2}{1}+4\binom{a_1}{5}+\binom{a_6}{1}$$
$$+\binom{a_1}{1}\binom{a_5}{1}+\binom{a_3}{2}-\binom{a_5}{1}-\binom{a_2}{1}\binom{a_3}{1}$$

The character polynomial $\Xi^{\hat{\alpha}}$, for $\hat{\alpha} := (\alpha_1, \alpha_2, \ldots) := (5, 1^2)$:

$$\binom{a_1}{3}\binom{a_4}{1}-\binom{a_1}{2}\binom{a_2}{1}\binom{a_3}{1}+3\binom{a_1}{4}\binom{a_3}{1}-\binom{a_1}{3}\binom{a_2}{2}$$
$$+5\binom{a_1}{5}\binom{a_2}{1}+15\binom{a_1}{7}-\binom{a_1}{1}\binom{a_2}{1}\binom{a_4}{1}-\binom{a_2}{2}\binom{a_3}{1}$$
$$-3\binom{a_1}{1}\binom{a_2}{3}+\binom{a_7}{1}+\binom{a_3}{1}\binom{a_4}{1}-\binom{a_1}{2}\binom{a_4}{1}-3\binom{a_1}{3}\binom{a_3}{1}$$

$$+\binom{a_1}{2}\binom{a_2}{2}-5\binom{a_1}{4}\binom{a_2}{1}-15\binom{a_1}{6}+\binom{a_2}{1}\binom{a_4}{1}$$

$$+\binom{a_1}{1}\binom{a_2}{1}\binom{a_3}{1}+3\binom{a_2}{3}+\binom{a_1}{1}\binom{a_4}{1}+2\binom{a_1}{2}\binom{a_3}{1}$$

$$+\binom{a_1}{1}\binom{a_2}{2}+3\binom{a_1}{3}\binom{a_2}{1}+5\binom{a_1}{5}-\binom{a_4}{1}$$

$$-\binom{a_1}{1}\binom{a_3}{1}-\binom{a_2}{2}-\binom{a_1}{2}\binom{a_2}{1}-\binom{a_1}{4}$$

The character polynomial $\Xi^{\hat{\alpha}}$, for $\hat{\alpha} := (\alpha_1, \alpha_2, \ldots) := (4, 3)$:

$$\binom{a_3}{1}\binom{a_4}{1}-2\binom{a_1}{3}\binom{a_4}{1}+2\binom{a_1}{1}\binom{a_3}{2}+\binom{a_1}{2}\binom{a_2}{1}\binom{a_3}{1}$$

$$-\binom{a_1}{4}\binom{a_3}{1}-\binom{a_2}{2}\binom{a_3}{1}+2\binom{a_1}{3}\binom{a_2}{2}+4\binom{a_1}{5}\binom{a_2}{1}$$

$$+14\binom{a_1}{7}-\binom{a_2}{1}\binom{a_5}{1}-\binom{a_1}{2}\binom{a_5}{1}-2\binom{a_3}{2}-\binom{a_1}{1}\binom{a_2}{1}\binom{a_3}{1}$$

$$+\binom{a_1}{3}\binom{a_3}{1}-2\binom{a_1}{2}\binom{a_2}{2}-4\binom{a_1}{4}\binom{a_2}{1}-14\binom{a_1}{6}+\binom{a_2}{1}\binom{a_3}{1}$$

$$-\binom{a_1}{2}\binom{a_3}{1}+\binom{a_1}{1}\binom{a_2}{2}+\binom{a_1}{3}\binom{a_2}{1}+5\binom{a_1}{5}+\binom{a_1}{1}\binom{a_5}{1}$$

$$+2\binom{a_1}{2}\binom{a_4}{1}-\binom{a_1}{1}\binom{a_4}{1}$$

The character polynomial $\Xi^{\hat{\alpha}}$, for $\hat{\alpha} := (\alpha_1, \alpha_2, \ldots) := (4, 2, 1)$:

$$\binom{a_1}{1}\binom{a_2}{1}\binom{a_4}{1}-\binom{a_1}{3}\binom{a_4}{1}-\binom{a_1}{2}\binom{a_2}{1}\binom{a_3}{1}-\binom{a_1}{4}\binom{a_3}{1}$$

$$+\binom{a_1}{1}\binom{a_2}{3}-\binom{a_1}{3}\binom{a_2}{2}+5\binom{a_1}{5}\binom{a_2}{1}+35\binom{a_1}{7}-\binom{a_3}{1}\binom{a_4}{1}$$

$$-\binom{a_1}{1}\binom{a_3}{2}-\binom{a_2}{2}\binom{a_3}{1}+\binom{a_1}{1}\binom{a_6}{1}+\binom{a_1}{1}\binom{a_2}{1}\binom{a_3}{1}$$

$$+\binom{a_1}{3}\binom{a_3}{1}+\binom{a_1}{2}\binom{a_2}{2}-5\binom{a_1}{4}\binom{a_2}{1}-35\binom{a_1}{6}+\binom{a_3}{2}$$

$$+3\binom{a_1}{3}\binom{a_2}{1}+15\binom{a_1}{5}-\binom{a_1}{2}\binom{a_2}{1}$$

$$-3\binom{a_1}{4}+\binom{a_1}{2}\binom{a_4}{1}-\binom{a_6}{1}-\binom{a_2}{1}\binom{a_4}{1}-\binom{a_2}{3}-\binom{a_1}{1}\binom{a_4}{1}$$

$$-\binom{a_1}{1}\binom{a_2}{2}+\binom{a_4}{1}+\binom{a_2}{2}$$

The character polynomial $\Xi^{\hat{\alpha}}$, for $\hat{\alpha} := (\alpha_1, \alpha_2, \ldots) := (4, 1^3)$:

$$2\binom{a_1}{4}\binom{a_3}{1}-4\binom{a_1}{3}\binom{a_2}{2}+20\binom{a_1}{7}+2\binom{a_1}{1}\binom{a_3}{2}+2\binom{a_2}{2}\binom{a_3}{1}$$
$$-\binom{a_7}{1}-2\binom{a_1}{3}\binom{a_3}{1}-20\binom{a_1}{6}+4\binom{a_1}{2}\binom{a_2}{2}-2\binom{a_3}{2}+\binom{a_1}{2}\binom{a_3}{1}$$
$$+2\binom{a_1}{3}\binom{a_2}{1}+10\binom{a_1}{5}-\binom{a_2}{1}\binom{a_3}{1}-2\binom{a_1}{1}\binom{a_2}{2}-\binom{a_1}{1}\binom{a_3}{1}$$
$$-2\binom{a_1}{2}\binom{a_2}{1}-4\binom{a_1}{4}+\binom{a_3}{1}+\binom{a_1}{1}\binom{a_2}{1}+\binom{a_1}{3}$$

The character polynomial $\Xi^{\hat{\alpha}}$, for $\hat{\alpha} := (\alpha_1, \alpha_2, \ldots) := (3^2, 1)$:

$$\binom{a_1}{2}\binom{a_2}{1}\binom{a_3}{1}-3\binom{a_1}{4}\binom{a_3}{1}+\binom{a_1}{3}\binom{a_2}{2}+\binom{a_1}{5}\binom{a_2}{1}+21\binom{a_1}{7}$$
$$-\binom{a_3}{1}\binom{a_4}{1}-\binom{a_1}{1}\binom{a_2}{1}\binom{a_4}{1}-\binom{a_1}{3}\binom{a_4}{1}+\binom{a_2}{2}\binom{a_3}{1}$$
$$-3\binom{a_1}{1}\binom{a_2}{3}+\binom{a_2}{1}\binom{a_5}{1}+\binom{a_1}{2}\binom{a_5}{1}-\binom{a_1}{1}\binom{a_2}{1}\binom{a_3}{1}$$
$$+3\binom{a_1}{3}\binom{a_3}{1}-\binom{a_1}{2}\binom{a_2}{2}-\binom{a_1}{4}\binom{a_2}{1}-21\binom{a_1}{6}+\binom{a_2}{1}\binom{a_4}{1}$$
$$+\binom{a_1}{2}\binom{a_4}{1}+3\binom{a_2}{3}+2\binom{a_1}{1}\binom{a_2}{2}+10\binom{a_1}{5}-2\binom{a_2}{2}-2\binom{a_1}{4}$$
$$-\binom{a_1}{1}\binom{a_5}{1}-2\binom{a_1}{2}\binom{a_3}{1}+\binom{a_1}{1}\binom{a_3}{1}$$

The character polynomial $\Xi^{\hat{\alpha}}$, for $\hat{\alpha} := (\alpha_1, \alpha_2, \ldots) := (3, 2^2)$:

$$\binom{a_2}{2}\binom{a_3}{1}-\binom{a_1}{2}\binom{a_2}{1}\binom{a_3}{1}-3\binom{a_1}{4}\binom{a_3}{1}+3\binom{a_1}{1}\binom{a_2}{3}$$
$$+\binom{a_1}{3}\binom{a_2}{2}-\binom{a_1}{5}\binom{a_2}{1}+21\binom{a_1}{7}-\binom{a_1}{1}\binom{a_2}{1}\binom{a_4}{1}+\binom{a_1}{3}\binom{a_4}{1}$$
$$+\binom{a_1}{2}\binom{a_5}{1}+\binom{a_3}{1}\binom{a_4}{1}-\binom{a_2}{1}\binom{a_5}{1}-3\binom{a_2}{3}-\binom{a_1}{2}\binom{a_2}{2}$$
$$+\binom{a_1}{4}\binom{a_2}{1}-21\binom{a_1}{6}+\binom{a_1}{1}\binom{a_2}{1}\binom{a_3}{1}+3\binom{a_1}{3}\binom{a_3}{1}-\binom{a_1}{1}\binom{a_2}{2}$$

$$-\binom{a_1}{3}\binom{a_2}{1}+11\binom{a_1}{5}+\binom{a_1}{2}\binom{a_2}{1}-3\binom{a_1}{4}-\binom{a_2}{1}\binom{a_3}{1}$$
$$-\binom{a_1}{2}\binom{a_3}{1}+\binom{a_2}{2}+\binom{a_2}{1}\binom{a_4}{1}-\binom{a_1}{2}\binom{a_4}{1}-\binom{a_1}{1}\binom{a_5}{1}+\binom{a_5}{1}$$
$$+\binom{a_1}{1}\binom{a_4}{1}-\binom{a_4}{1}$$

The character polynomial $\Xi^{\hat{\alpha}}$, for $\hat{\alpha} := (\alpha_1, \alpha_2, \ldots) := (3, 2, 1^2)$:

$$\binom{a_1}{2}\binom{a_2}{1}\binom{a_3}{1}-\binom{a_1}{4}\binom{a_3}{1}-\binom{a_1}{3}\binom{a_2}{2}-5\binom{a_1}{5}\binom{a_2}{1}+35\binom{a_1}{7}$$
$$-\binom{a_2}{2}\binom{a_3}{1}-\binom{a_1}{1}\binom{a_2}{3}-\binom{a_1}{1}\binom{a_3}{2}+\binom{a_3}{1}\binom{a_4}{1}$$
$$+\binom{a_1}{1}\binom{a_2}{1}\binom{a_4}{1}+\binom{a_1}{3}\binom{a_4}{1}-\binom{a_1}{1}\binom{a_6}{1}+\binom{a_1}{2}\binom{a_2}{2}$$
$$+5\binom{a_1}{4}\binom{a_2}{1}-35\binom{a_1}{6}+\binom{a_2}{3}-\binom{a_1}{1}\binom{a_2}{1}\binom{a_3}{1}+\binom{a_1}{3}\binom{a_3}{1}$$
$$-\binom{a_2}{1}\binom{a_4}{1}-\binom{a_1}{2}\binom{a_4}{1}-2\binom{a_1}{3}\binom{a_2}{1}+20\binom{a_1}{5}-8\binom{a_1}{4}+2\binom{a_1}{3}$$
$$+\binom{a_6}{1}+\binom{a_3}{2}-\binom{a_1}{2}\binom{a_3}{1}+\binom{a_2}{1}\binom{a_3}{1}+\binom{a_1}{1}\binom{a_3}{1}-\binom{a_3}{1}$$

The character polynomial $\Xi^{\hat{\alpha}}$, for $\hat{\alpha} := (\alpha_1, \alpha_2, \ldots) := (3, 1^4)$:

$$3\binom{a_1}{4}\binom{a_3}{1}-5\binom{a_1}{5}\binom{a_2}{1}+15\binom{a_1}{7}+\binom{a_1}{2}\binom{a_2}{1}\binom{a_3}{1}-\binom{a_1}{3}\binom{a_2}{2}$$
$$-\binom{a_2}{2}\binom{a_3}{1}+3\binom{a_1}{1}\binom{a_2}{3}-\binom{a_3}{1}\binom{a_4}{1}-\binom{a_1}{1}\binom{a_2}{1}\binom{a_4}{1}$$
$$-\binom{a_1}{3}\binom{a_4}{1}+\binom{a_7}{1}+5\binom{a_1}{4}\binom{a_2}{1}-15\binom{a_1}{6}+\binom{a_1}{2}\binom{a_2}{2}$$
$$-\binom{a_1}{1}\binom{a_2}{1}\binom{a_3}{1}-3\binom{a_1}{3}\binom{a_3}{1}-3\binom{a_2}{3}+\binom{a_2}{1}\binom{a_4}{1}$$
$$+\binom{a_1}{2}\binom{a_4}{1}-2\binom{a_1}{3}\binom{a_2}{1}+10\binom{a_1}{5}-2\binom{a_1}{1}\binom{a_2}{2}+\binom{a_2}{1}\binom{a_3}{1}$$
$$+\binom{a_1}{2}\binom{a_3}{1}-6\binom{a_1}{4}+2\binom{a_2}{2}+\binom{a_1}{1}\binom{a_2}{1}+3\binom{a_1}{3}-\binom{a_2}{1}-\binom{a_1}{2}$$

The character polynomial $\Xi^{\hat{\alpha}}$, for $\hat{\alpha} := (\alpha_1, \alpha_2, \ldots) := (2^3, 1)$:

$$2\binom{a_1}{3}\binom{a_2}{2}-4\binom{a_1}{5}\binom{a_2}{1}+14\binom{a_1}{7}-\binom{a_2}{2}\binom{a_3}{1}-\binom{a_1}{2}\binom{a_2}{1}\binom{a_3}{1}$$
$$-\binom{a_1}{4}\binom{a_3}{1}+2\binom{a_1}{3}\binom{a_4}{1}+2\binom{a_1}{1}\binom{a_3}{2}-\binom{a_1}{2}\binom{a_5}{1}-\binom{a_3}{1}\binom{a_4}{1}$$
$$+\binom{a_2}{1}\binom{a_5}{1}-2\binom{a_1}{2}\binom{a_2}{2}+4\binom{a_1}{4}\binom{a_2}{1}-14\binom{a_1}{6}$$
$$+\binom{a_1}{1}\binom{a_2}{1}\binom{a_3}{1}+\binom{a_1}{3}\binom{a_3}{1}-2\binom{a_1}{2}\binom{a_4}{1}-3\binom{a_1}{3}\binom{a_2}{1}$$
$$+9\binom{a_1}{5}-4\binom{a_1}{4}+\binom{a_1}{3}+\binom{a_1}{1}\binom{a_4}{1}+\binom{a_1}{1}\binom{a_2}{2}+2\binom{a_1}{2}\binom{a_2}{1}$$
$$-\binom{a_1}{1}\binom{a_2}{1}-2\binom{a_3}{2}+\binom{a_1}{1}\binom{a_5}{1}-\binom{a_5}{1}-\binom{a_1}{1}\binom{a_3}{1}+\binom{a_3}{1}$$

The character polynomial $\Xi^{\hat{\alpha}}$, for $\hat{\alpha} := (\alpha_1, \alpha_2, \ldots) := (2^2, 1^3)$:

$$2\binom{a_1}{3}\binom{a_2}{2}-6\binom{a_1}{5}\binom{a_2}{1}+14\binom{a_1}{7}-2\binom{a_1}{1}\binom{a_2}{3}+2\binom{a_2}{2}\binom{a_3}{1}$$
$$+2\binom{a_1}{4}\binom{a_3}{1}-\binom{a_2}{1}\binom{a_5}{1}-\binom{a_1}{2}\binom{a_5}{1}-\binom{a_1}{1}\binom{a_3}{2}+\binom{a_1}{1}\binom{a_6}{1}$$
$$+6\binom{a_1}{4}\binom{a_2}{1}-14\binom{a_1}{6}-2\binom{a_1}{2}\binom{a_2}{2}-2\binom{a_1}{3}\binom{a_3}{1}+\binom{a_1}{1}\binom{a_5}{1}$$
$$+10\binom{a_1}{5}-4\binom{a_1}{3}\binom{a_2}{1}+\binom{a_1}{2}\binom{a_3}{1}-6\binom{a_1}{4}+2\binom{a_1}{2}\binom{a_2}{1}+3\binom{a_1}{3}$$
$$-\binom{a_1}{2}+\binom{a_3}{2}+2\binom{a_2}{3}-\binom{a_6}{1}+2\binom{a_1}{1}\binom{a_2}{2}-\binom{a_2}{1}\binom{a_3}{1}$$
$$-2\binom{a_2}{2}-\binom{a_1}{1}\binom{a_2}{1}+\binom{a_2}{1}$$

The character polynomial $\Xi^{\hat{\alpha}}$, for $\hat{\alpha} := (\alpha_1, \alpha_2, \ldots) := (2, 1^5)$:

$$-4\binom{a_1}{5}\binom{a_2}{1}+6\binom{a_1}{7}+2\binom{a_1}{3}\binom{a_2}{2}-\binom{a_1}{2}\binom{a_2}{1}\binom{a_3}{1}$$
$$+3\binom{a_1}{4}\binom{a_3}{1}-2\binom{a_1}{3}\binom{a_4}{1}-\binom{a_2}{2}\binom{a_3}{1}+\binom{a_2}{1}\binom{a_5}{1}$$
$$+\binom{a_1}{2}\binom{a_5}{1}+\binom{a_3}{1}\binom{a_4}{1}-\binom{a_7}{1}-6\binom{a_1}{6}+4\binom{a_1}{4}\binom{a_2}{1}$$
$$-3\binom{a_1}{3}\binom{a_3}{1}-2\binom{a_1}{2}\binom{a_2}{2}+2\binom{a_1}{2}\binom{a_4}{1}+\binom{a_1}{1}\binom{a_2}{1}\binom{a_3}{1}$$

$$-\binom{a_1}{1}\binom{a_5}{1}+5\binom{a_1}{5}-3\binom{a_1}{3}\binom{a_2}{1}+2\binom{a_1}{2}\binom{a_3}{1}+\binom{a_1}{1}\binom{a_2}{2}$$
$$-\binom{a_1}{1}\binom{a_4}{1}-4\binom{a_1}{4}+2\binom{a_1}{2}\binom{a_2}{1}-\binom{a_1}{1}\binom{a_3}{1}+3\binom{a_1}{3}$$
$$-\binom{a_1}{1}\binom{a_2}{1}-2\binom{a_1}{2}+\binom{a_1}{1}$$

The character polynomial $\Xi^{\hat{\alpha}}$, for $\hat{\alpha} := (\alpha_1, \alpha_2, \ldots) := (1^7)$:

$$\binom{a_1}{7}-\binom{a_1}{5}\binom{a_2}{1}+\binom{a_1}{4}\binom{a_3}{1}+\binom{a_1}{3}\binom{a_2}{2}-\binom{a_1}{3}\binom{a_4}{1}$$
$$-\binom{a_1}{2}\binom{a_2}{1}\binom{a_3}{1}+\binom{a_1}{2}\binom{a_5}{1}-\binom{a_1}{1}\binom{a_2}{3}+\binom{a_1}{1}\binom{a_2}{1}\binom{a_4}{1}$$
$$+\binom{a_1}{1}\binom{a_3}{2}-\binom{a_1}{1}\binom{a_6}{1}+\binom{a_2}{2}\binom{a_3}{1}-\binom{a_2}{1}\binom{a_5}{1}-\binom{a_3}{1}\binom{a_4}{1}$$
$$+\binom{a_7}{1}-\binom{a_1}{6}+\binom{a_1}{4}\binom{a_2}{1}-\binom{a_1}{3}\binom{a_3}{1}-\binom{a_1}{2}\binom{a_2}{2}+\binom{a_1}{2}\binom{a_4}{1}$$
$$+\binom{a_1}{1}\binom{a_2}{1}\binom{a_3}{1}-\binom{a_1}{1}\binom{a_5}{1}+\binom{a_2}{3}-\binom{a_2}{1}\binom{a_4}{1}-\binom{a_3}{2}+\binom{a_6}{1}$$
$$+\binom{a_1}{5}-\binom{a_1}{3}\binom{a_2}{1}+\binom{a_1}{2}\binom{a_3}{1}+\binom{a_1}{1}\binom{a_2}{2}-\binom{a_1}{1}\binom{a_4}{1}$$
$$-\binom{a_2}{1}\binom{a_3}{1}+\binom{a_5}{1}-\binom{a_1}{4}+\binom{a_1}{2}\binom{a_2}{1}-\binom{a_1}{1}\binom{a_3}{1}-\binom{a_2}{2}+\binom{a_4}{1}$$
$$+\binom{a_1}{3}-\binom{a_1}{1}\binom{a_2}{1}+\binom{a_3}{1}-\binom{a_1}{2}+\binom{a_2}{1}+\binom{a_1}{1}-1$$

10.3 Schubert Polynomials

This section contains the 120 Schubert polynomials that correspond to the elements of the symmetric group S_5. The permutations are given first by the Lehmer code and by their list, then the Schubert polynomial follows.

$\overline{00000} \approx [01234]$: 1
$\overline{00010} \approx [01243]$: $x_3 + x_2 + x_1 + x_0$
$\overline{00100} \approx [01324]$: $x_2 + x_1 + x_0$
$\overline{00110} \approx [01342]$: $x_2x_3 + x_1x_3 + x_1x_2 + x_0x_3 + x_0x_2 + x_0x_1$
$\overline{00200} \approx [01423]$: $x_2^2 + x_1x_2 + x_1^2 + x_0x_2 + x_0x_1 + x_0^2$

$\overline{00210} \approx [01432]$: $x_2^2x_3 + x_1x_2x_3 + x_1x_2^2 + x_1^2x_3 + x_1^2x_2 + x_0x_2x_3 + x_0x_2^2 + x_0x_1x_3 + 2\,x_0x_1x_2 + x_0x_1^2 + x_0^2x_3 + x_0^2x_2 + x_0^2x_1$

$\overline{01000} \approx [02134]$: $x_1 + x_0$

$\overline{01010} \approx [02143]$: $x_1x_3 + x_1x_2 + x_1^2 + x_0x_3 + x_0x_2 + 2\,x_0x_1 + x_0^2$

$\overline{01100} \approx [02314]$: $x_1x_2 + x_0x_2 + x_0x_1$

$\overline{01110} \approx [02341]$: $x_1x_2x_3 + x_0x_2x_3 + x_0x_1x_3 + x_0x_1x_2$

$\overline{01200} \approx [02413]$: $x_1x_2^2 + x_1^2x_2 + x_0x_2^2 + 2x_0x_1x_2 + x_0x_1^2 + x_0^2x_2 + x_0^2x_1$

$\overline{01210} \approx [02431]$: $x_1x_2^2x_3 + x_1^2x_2x_3 + x_0x_2^2x_3 + 2\,x_0x_1x_2x_3 + x_0x_1x_2^2 + x_0x_1^2x_3 + x_0x_1^2x_2 + x_0^2x_2x_3 + x_0^2x_1x_3 + x_0^2x_1x_2$

$\overline{02000} \approx [03124]$: $x_1^2 + x_0x_1 + x_0^2$

$\overline{02010} \approx [03142]$: $x_1^2x_3 + x_1^2x_2 + x_0x_1x_3 + x_0x_1x_2 + x_0x_1^2 + x_0^2x_3 + x_0^2x_2 + x_0^2x_1$

$\overline{02100} \approx [03214]$: $x_1^2x_2 + x_0x_1x_2 + x_0x_1^2 + x_0^2x_2 + x_0^2x_1$

$\overline{02110} \approx [03241]$: $x_1^2x_2x_3 + x_0x_1x_2x_3 + x_0x_1^2x_3 + x_0x_1^2x_2 + x_0^2x_2x_3 + x_0^2x_1x_3 + x_0^2x_1x_2$

$\overline{02200} \approx [03412]$: $x_1^2x_2^2 + x_0x_1x_2^2 + x_0x_1^2x_2 + x_0^2x_2^2 + x_0^2x_1x_2 + x_0^2x_1^2$

$\overline{02210} \approx [03421]$: $x_1^2x_2^2x_3 + x_0x_1x_2^2x_3 + x_0x_1^2x_2x_3 + x_0x_1^2x_2^2 + x_0^2x_2^2x_3 + x_0^2x_1x_2x_3 + x_0^2x_1x_2^2 + x_0^2x_1^2x_3 + x_0^2x_1^2x_2$

$\overline{03000} \approx [04123]$: $x_1^3 + x_0x_1^2 + x_0^2x_1 + x_0^3$

$\overline{03010} \approx [04132]$: $x_1^3x_3 + x_1^3x_2 + x_0x_1^2x_3 + x_0x_1^2x_2 + x_0x_1^3 + x_0^2x_1x_3 + x_0^2x_1x_2 + x_0^2x_1^2 + x_0^3x_3 + x_0^3x_2 + x_0^3x_1$

$\overline{03100} \approx [04213]$: $x_1^3x_2 + x_0x_1^2x_2 + x_0x_1^3 + x_0^2x_1x_2 + x_0^2x_1^2 + x_0^3x_2 + x_0^3x_1$

$\overline{03110} \approx [04231]$: $x_1^3x_2x_3 + x_0x_1^2x_2x_3 + x_0x_1^3x_3 + x_0x_1^3x_2 + x_0^2x_1x_2x_3 + x_0^2x_1^2x_3 + x_0^2x_1^2x_2 + x_0^3x_2x_3 + x_0^3x_1x_3 + x_0^3x_1x_2$

$\overline{03200} \approx [04312]$: $x_1^3x_2^2 + x_0x_1^2x_2^2 + x_0x_1^3x_2 + x_0^2x_1x_2^2 + x_0^2x_1^2x_2 + x_0^2x_1^3 + x_0^3x_2^2 + x_0^3x_1x_2 + x_0^3x_1^2$

$\overline{03210} \approx [04321]$: $x_1^3x_2^2x_3 + x_0x_1^2x_2^2x_3 + x_0x_1^3x_2x_3 + x_0x_1^3x_2^2 + x_0^2x_1x_2^2x_3 + x_0^2x_1^2x_2x_3 + x_0^2x_1^2x_2^2 + x_0^2x_1^3x_3 + x_0^2x_1^3x_2 + x_0^3x_2^2x_3 + x_0^3x_1x_2x_3 + x_0^3x_1x_2^2 + x_0^3x_1^2x_3 + x_0^3x_1^2x_2$

$\overline{10000} \approx [10234]$: x_0

$\overline{10010} \approx [10243]$: $x_0x_3 + x_0x_2 + x_0x_1 + x_0^2$

$\overline{10100} \approx [10324]$: $x_0x_2 + x_0x_1 + x_0^2$

$\overline{10110} \approx [10342]$: $x_0x_2x_3 + x_0x_1x_3 + x_0x_1x_2 + x_0^2x_3 + x_0^2x_2 + x_0^2x_1$

$\overline{10200} \approx [10423]$: $x_0x_2^2 + x_0x_1x_2 + x_0x_1^2 + x_0^2x_2 + x_0^2x_1 + x_0^3$

$\overline{10210} \approx [10432]$: $x_0x_2^2x_3 + x_0x_1x_2x_3 + x_0x_1x_2^2 + x_0x_1^2x_3 + x_0x_1^2x_2 + x_0^2x_2x_3 + x_0^2x_2^2 + x_0^2x_1x_3 + 2x_0^2x_1x_2 + x_0^2x_1^2 + x_0^3x_3 + x_0^3x_2 + x_0^3x_1$

$\overline{11000} \approx [12034]$: x_0x_1

$\overline{11010} \approx [12043]$: $x_0x_1x_3 + x_0x_1x_2 + x_0x_1^2 + x_0^2x_1$

$\overline{11100} \approx [12304]$: $x_0x_1x_2$

$\overline{11110} \approx [12340]$: $x_0x_1x_2x_3$
$\overline{11200} \approx [12403]$: $x_0x_1x_2^2 + x_0x_1^2x_2 + x_0^2x_1x_2$
$\overline{11210} \approx [12430]$: $x_0x_1x_2^2x_3 + x_0x_1^2x_2x_3 + x_0^2x_1x_2x_3$
$\overline{12000} \approx [13024]$: $x_0x_1^2 + x_0^2x_1$
$\overline{12010} \approx [13042]$: $x_0x_1^2x_3 + x_0x_1^2x_2 + x_0^2x_1x_3 + x_0^2x_1x_2 + x_0^2x_1^2$
$\overline{12100} \approx [13204]$: $x_0x_1^2x_2 + x_0^2x_1x_2$
$\overline{12110} \approx [13240]$: $x_0x_1^2x_2x_3 + x_0^2x_1x_2x_3$
$\overline{12200} \approx [13402]$: $x_0x_1^2x_2^2 + x_0^2x_1x_2^2 + x_0^2x_1^2x_2$
$\overline{12210} \approx [13420]$: $x_0x_1^2x_2^2x_3 + x_0^2x_1x_2^2x_3 + x_0^2x_1^2x_2x_3$
$\overline{13000} \approx [14023]$: $x_0x_1^3 + x_0^2x_1^2 + x_0^3x_1$
$\overline{13010} \approx [14032]$: $x_0x_1^3x_3 + x_0x_1^3x_2 + x_0^2x_1^2x_3 + x_0^2x_1^2x_2 + x_0^2x_1^3 + x_0^3x_1x_3$
$+ x_0^3x_1x_2 + x_0^3x_1^2$
$\overline{13100} \approx [14203]$: $x_0x_1^3x_2 + x_0^2x_1^2x_2 + x_0^3x_1x_2$
$\overline{13110} \approx [14230]$: $x_0x_1^3x_2x_3 + x_0^2x_1^2x_2x_3 + x_0^3x_1x_2x_3$
$\overline{13200} \approx [14302]$: $x_0x_1^3x_2^2 + x_0^2x_1^2x_2^2 + x_0^2x_1^3x_2 + x_0^3x_1x_2^2 + x_0^3x_1^2x_2$
$\overline{13210} \approx [14320]$: $x_0x_1^3x_2^2x_3 + x_0^2x_1^2x_2^2x_3 + x_0^2x_1^3x_2x_3 + x_0^3x_1x_2^2x_3$
$+ x_0^3x_1^2x_2x_3$
$\overline{20000} \approx [20134]$: x_0^2
$\overline{20010} \approx [20143]$: $x_0^2x_3 + x_0^2x_2 + x_0^2x_1 + x_0^3$
$\overline{20100} \approx [20314]$: $x_0^2x_2 + x_0^2x_1$
$\overline{20110} \approx [20341]$: $x_0^2x_2x_3 + x_0^2x_1x_3 + x_0^2x_1x_2$
$\overline{20200} \approx [20413]$: $x_0^2x_2^2 + x_0^2x_1x_2 + x_0^2x_1^2 + x_0^3x_2 + x_0^3x_1$
$\overline{20210} \approx [20431]$: $x_0^2x_2^2x_3 + x_0^2x_1x_2x_3 + x_0^2x_1x_2^2 + x_0^2x_1^2x_3 + x_0^2x_1^2x_2$
$+ x_0^3x_2x_3 + x_0^3x_1x_3 + x_0^3x_1x_2$
$\overline{21000} \approx [21034]$: $x_0^2x_1$
$\overline{21010} \approx [21043]$: $x_0^2x_1x_3 + x_0^2x_1x_2 + x_0^2x_1^2 + x_0^3x_1$
$\overline{21100} \approx [21304]$: $x_0^2x_1x_2$
$\overline{21110} \approx [21340]$: $x_0^2x_1x_2x_3$
$\overline{21200} \approx [21403]$: $x_0^2x_1x_2^2 + x_0^2x_1^2x_2 + x_0^3x_1x_2$
$\overline{21210} \approx [21430]$: $x_0^2x_1x_2^2x_3 + x_0^2x_1^2x_2x_3 + x_0^3x_1x_2x_3$
$\overline{22000} \approx [23014]$: $x_0^2x_1^2$
$\overline{22010} \approx [23041]$: $x_0^2x_1^2x_3 + x_0^2x_1^2x_2$
$\overline{22100} \approx [23104]$: $x_0^2x_1^2x_2$
$\overline{22110} \approx [23140]$: $x_0^2x_1^2x_2x_3$
$\overline{22200} \approx [23401]$: $x_0^2x_1^2x_2^2$
$\overline{22210} \approx [23410]$: $x_0^2x_1^2x_2^2x_3$
$\overline{23000} \approx [24013]$: $x_0^2x_1^3 + x_0^3x_1^2$
$\overline{23010} \approx [24031]$: $x_0^2x_1^3x_3 + x_0^2x_1^3x_2 + x_0^3x_1^2x_3 + x_0^3x_1^2x_2$
$\overline{23100} \approx [24103]$: $x_0^2x_1^3x_2 + x_0^3x_1^2x_2$
$\overline{23110} \approx [24130]$: $x_0^2x_1^3x_2x_3 + x_0^3x_1^2x_2x_3$

$\overline{23200} \approx [24301]$: $x_0^2x_1^3x_2^2 + x_0^3x_1^2x_2^2$
$\overline{23210} \approx [24310]$: $x_0^2x_1^3x_2^2x_3 + x_0^3x_1^2x_2^2x_3$
$\overline{30000} \approx [30124]$: x_0^3
$\overline{30010} \approx [30142]$: $x_0^3x_3 + x_0^3x_2 + x_0^3x_1$
$\overline{30100} \approx [30214]$: $x_0^3x_2 + x_0^3x_1$
$\overline{30110} \approx [30241]$: $x_0^3x_2x_3 + x_0^3x_1x_3 + x_0^3x_1x_2$
$\overline{30200} \approx [30412]$: $x_0^3x_2^2 + x_0^3x_1x_2 + x_0^3x_1^2$
$\overline{30210} \approx [30421]$: $x_0^3x_2^2x_3 + x_0^3x_1x_2x_3 + x_0^3x_1x_2^2 + x_0^3x_1^2x_3 + x_0^3x_1^2x_2$
$\overline{31000} \approx [31024]$: $x_0^3x_1$
$\overline{31010} \approx [31042]$: $x_0^3x_1x_3 + x_0^3x_1x_2 + x_0^3x_1^2$
$\overline{31100} \approx [31204]$: $x_0^3x_1x_2$
$\overline{31110} \approx [31240]$: $x_0^3x_1x_2x_3$
$\overline{31200} \approx [31402]$: $x_0^3x_1x_2^2 + x_0^3x_1^2x_2$
$\overline{31210} \approx [31420]$: $x_0^3x_1x_2^2x_3 + x_0^3x_1^2x_2x_3$
$\overline{32000} \approx [32014]$: $x_0^3x_1^2$
$\overline{32010} \approx [32041]$: $x_0^3x_1^2x_3 + x_0^3x_1^2x_2$
$\overline{32100} \approx [32104]$: $x_0^3x_1^2x_2$
$\overline{32110} \approx [32140]$: $x_0^3x_1^2x_2x_3$
$\overline{32200} \approx [32401]$: $x_0^3x_1^2x_2^2$
$\overline{32210} \approx [32410]$: $x_0^3x_1^2x_2^2x_3$
$\overline{33000} \approx [34012]$: $x_0^3x_1^3$
$\overline{33010} \approx [34021]$: $x_0^3x_1^3x_3 + x_0^3x_1^3x_2$
$\overline{33100} \approx [34102]$: $x_0^3x_1^3x_2$
$\overline{33110} \approx [34120]$: $x_0^3x_1^3x_2x_3$
$\overline{33200} \approx [34201]$: $x_0^3x_1^3x_2^2$
$\overline{33210} \approx [34210]$: $x_0^3x_1^3x_2^2x_3$
$\overline{40000} \approx [40123]$: x_0^4
$\overline{40010} \approx [40132]$: $x_0^4x_3 + x_0^4x_2 + x_0^4x_1$
$\overline{40100} \approx [40213]$: $x_0^4x_2 + x_0^4x_1$
$\overline{40110} \approx [40231]$: $x_0^4x_2x_3 + x_0^4x_1x_3 + x_0^4x_1x_2$
$\overline{40200} \approx [40312]$: $x_0^4x_2^2 + x_0^4x_1x_2 + x_0^4x_1^2$
$\overline{40210} \approx [40321]$: $x_0^4x_2^2x_3 + x_0^4x_1x_2x_3 + x_0^4x_1x_2^2 + x_0^4x_1^2x_3 + x_0^4x_1^2x_2$
$\overline{41000} \approx [41023]$: $x_0^4x_1$
$\overline{41010} \approx [41032]$: $x_0^4x_1x_3 + x_0^4x_1x_2 + x_0^4x_1^2$
$\overline{41100} \approx [41203]$: $x_0^4x_1x_2$
$\overline{41110} \approx [41230]$: $x_0^4x_1x_2x_3$
$\overline{41200} \approx [41302]$: $x_0^4x_1x_2^2 + x_0^4x_1^2x_2$
$\overline{41210} \approx [41320]$: $x_0^4x_1x_2^2x_3 + x_0^4x_1^2x_2x_3$
$\overline{42000} \approx [42013]$: $x_0^4x_1^2$
$\overline{42010} \approx [42031]$: $x_0^4x_1^2x_3 + x_0^4x_1^2x_2$

$\overline{42100} \approx [42103]$: $x_0^4 x_1^2 x_2$
$\overline{42110} \approx [42130]$: $x_0^4 x_1^2 x_2 x_3$
$\overline{42200} \approx [42301]$: $x_0^4 x_1^2 x_2^2$
$\overline{42210} \approx [42310]$: $x_0^4 x_1^2 x_2^2 x_3$
$\overline{43000} \approx [43012]$: $x_0^4 x_1^3$
$\overline{43010} \approx [43021]$: $x_0^4 x_1^3 x_3 + x_0^4 x_1^3 x_2$
$\overline{43100} \approx [43102]$: $x_0^4 x_1^3 x_2$
$\overline{43110} \approx [43120]$: $x_0^4 x_1^3 x_2 x_3$
$\overline{43200} \approx [43201]$: $x_0^4 x_1^3 x_2^2$
$\overline{43210} \approx [43210]$: $x_0^4 x_1^3 x_2^2 x_3$

11. Appendix

In this appendix a few very basic or technical things are recalled in order to make the book selfcontained in a reasonable sense.

11.1 Groups

A *group* is a set, say G, together with a law of composition for which we shall mostly use the common symbol "$\cdot$" for multiplication:

$$\cdot : G \times G \to G : (g, g') \mapsto g \cdot g' \quad (\text{or } gg' \text{ for sake of simplicity}).$$

If we use this notation, we call the group G, or, better say, the pair $(G, \cdot)$, consisting of the set and the composition, a *multiplicative* group. At most in the commutative case, i. e. when $gg' = g'g$ for each $g, g' \in G$, we use the additive notation "+" and speak of an *additive* group G or $(G, +)$ in this case. Groups with a commutative composition are called *abelian.* The composition is assumed to be *associative:*

$$g(g'g'') = (gg')g'',$$

for each $g, g', g'' \in G$. Moreover, we assume that there exists an element (and so G is not empty!) which is a *left unit:* $eg = g$, for every $g \in G$. Besides this, each group element g must possess a *left inverse* g' with respect to $e : g'g = e$. It is easy to check that the left unit is also a right unit, and that it is uniquely determined. It will therefore be denoted by 1 or, more explicitly, by 1_G in the multiplicative case, by 0 or 0_G in the additive notation. A left inverse is also a right inverse, and it is uniquely determined, but it, of course, depends on g. We shall indicate it by g^{-1} in the multiplicative case, by $-g$ in the additive notation.

It is not difficult to see that the same sets turn out to be groups if we assume a *right* unit together with *right inverses* instead.

In a multiplicative (additive) group we can multiply (add) also *subsets:* For $M, N \subseteq G$ we simply put

$$M \cdot N := \{m \cdot n \mid m \in M, n \in N\},$$

or, in the additive case,

$$M + N := \{m + n \mid m \in M, n \in N\}.$$

The result is called the *complex product* (*complex sum*) of these subsets.

A mapping $\varphi: G \to H$ between two groups G and H is called a *homomorphism* if the following is true:

$$\varphi(g \cdot g') = \varphi(g) \cdot \varphi(g').$$

Note that the dot on the left hand side means the multiplication in G while that on the right hand side indicates the multiplication in the group H. A homomorphism δ from G into the *symmetric group*

$$S_X := \{\pi \mid \pi: X \to X, \text{ bijectively}\}$$

(the multiplication is the composition of mappings) is called a *permutation representation of* G *on* X. A homomorphism into a general linear group of a vector space V is called a *linear representation of* G *on* V.

A subset H of a group $(G, \cdot)$ is called a *subgroup,* for short: $H \leq G$, if and only if $(H, \cdot)$ is a group. This is equivalent to

11.1.1 $$H \neq \emptyset, \text{ and } [h, h' \in H \Rightarrow h \cdot h'^{-1} \in H].$$

The sets

$$gH := \{gh \mid h \in H\}$$

are called *left cosets,* and the sets

$$Hg := \{hg \mid h \in H\}$$

are called *right cosets* of the subgroup H. Subgroups $N \subseteq G$ for which left and right cosets are the same, $gN = Ng$, for each g, are of particular importance, they are called *normal* subgroups. We abbreviate this by writing $N \trianglelefteq G$. Each homomorphism $\varphi: G \to H$ defines (and is defined by) a particular normal subgroup, the *kernel* of φ :

$$\ker(\varphi) := \{g \in G \mid \varphi(g) = 1_H\}.$$

A normal subgroup $N \trianglelefteq G$ gives rise to a group, the *factor group* G/N, that consists of the cosets of the normal subgroup:

$$G/N := \{gN \mid g \in G\},$$

together with complex multiplication:

$$(gN)(g'N) = gg'N.$$

Bijective homomorphisms are called *isomorphisms,* and we indicate the existence of isomorphisms between two groups G and H by writing

$$G \simeq H.$$

One of the most important facts in group theory is that a homomorphism $\varphi: G \to H$ gives an isomorphism $\widetilde{\varphi}$ between the factor group $G/\ker(\varphi)$ and the image $\mathrm{im}(\varphi)$:

11.1.2 $$\widetilde{\varphi}\colon G/\ker(\varphi) \simeq \mathrm{im}(\varphi)\colon g\ker(\varphi) \mapsto \varphi(g).$$

This result is called the *homomorphism theorem.*

Exercises

Exercise 11.1.1 Check the criterion 11.1.1 and show that, for finite groups G, it suffices, for a nonempty subset H of G, to verify for each $h, h' \in H$,

$$hh' \in H.$$

Exercise 11.1.2 Prove that a complex product $MN \subseteq G$ is a subgroup if $M \leq G$ and $N \trianglelefteq G$. Give an example which shows that the assumption of normality of one of the factors is necessary.

Exercise 11.1.3 Prove the homomorphism theorem 11.1.2.

11.2 Finite Symmetric Groups

An action ${}_GX$ of a group G on a set X is essentially the same as a homomorphism from G into the symmetric group S_X (see 1.1.2). We therefore need to consider symmetric groups on sets to some detail, in particular symmetric groups S_X of finite sets X. A first remark shows that it is only the order of X which really matters:

11.2.1 Lemma *For any two finite and nonempty sets X and Y, the natural actions of S_X on X and S_Y on Y are isomorphic if and only if $|X| = |Y|$.*

This is very easy to check and therefore left as exercise 11.2.1. We call $n := |X|$ the *degree* of S_X, of any subgroup $P \leq S_X$ and of any $\pi \in S_X$. In order to examine permutations of degree n it therefore suffices to consider a particular set of order n and its symmetric group. For technical reasons we introduce two sets of order n:

$$n := \{0, \ldots, n-1\} \text{ and } \underline{n} := \{1, \ldots, n\},$$

hoping that it will be always clear from the context if this *set* n is meant or its *cardinality* n. It is an old tradition to prefer the set $\underline{n}$ and its symmetric group which we should denote by $S_{\underline{n}}$ in order to be consistent, but there is a modern tendency — strongly influenced by computer science —, to be more consequent and start numbering from 0 onwards. Hence let us fix the notation for the elements of S_n, the corresponding notation for the elements of $S_{\underline{n}}$ is then obvious.

A permutation $\pi \in S_n$ is written down in full detail by putting the images πi in a row under the points $i \in n$, say

$$\pi = \begin{pmatrix} 0 & \ldots & n-1 \\ \pi 0 & \ldots & \pi(n-1) \end{pmatrix}.$$

This will be abbreviated by

$$\pi = \binom{i}{\pi i}.$$

Hence, for example, S_3 consists of the following elements:

$$\binom{012}{012}, \binom{012}{102}, \binom{012}{210}, \binom{012}{021}, \binom{012}{120}, \binom{012}{201}.$$

The points which form the first row need not be written in their natural order, e. g.

$$\binom{012}{120} = \binom{102}{210}.$$

Keeping this in mind, we call a permutation $\pi \in S_n$ a *cyclic* permutation or a *cycle* if and only if it can be written in the form

$$\begin{pmatrix} i_0 & i_1 & \dots & i_{r-2} & i_{r-1} & i_r & \dots & i_{n-1} \\ i_1 & i_2 & \dots & i_{r-1} & i_0 & i_r & \dots & i_{n-1} \end{pmatrix},$$

where $r > 0$. In order to emphasize r we also call it an *r-cycle*. We note that in this case the orbits of the subgroup generated by this permutation are the following subsets of n: $\{i_0, \dots, i_{r-1}\}, \{i_r\}, \dots, \{i_{n-1}\}$. We therefore abbreviate this cycle by $(i_0, \dots, i_{r-1})(i_r) \dots (i_{n-1})$, where the points which are *cyclically permuted* are put together in round brackets. For example $\binom{0\,1\,2}{0\,2\,1} = (1, 2)(0)$. Commas which seperate the points may be omitted if no confusion can arise (e. g. if $n \leq 10$), and 1-cycles can be left out if it is clear which n is meant. Hence we can write $\pi = (i_0 \dots i_{r-1})$ for the r–cycle introduced above. This cycle π can also be expressed in terms of i_0 alone: $\pi = (i_0\, \pi i_0 \dots \pi^{r-1} i_0)$. Using all these abbreviations and denoting by $1 := (0) \dots (n-1)$ the identity element, we obtain for example

$$S_3 = \{1, (01), (02), (12), (012), (021)\}.$$

The notation for a cyclic permutation is not uniquely determined, since

11.2.2 $$(i_0 \dots i_{r-1}) = (i_1 \dots i_{r-1} i_0) = \dots = (i_{r-1} i_0 \dots i_{r-2}).$$

2-cycles are called *transpositions*. The *order* of a cycle $(i_0 \dots i_{r-1})$, i. e. the order of the cyclic group $\langle (i_0 \dots i_{r-1}) \rangle$ generated by this cycle, is equal to its *length*:

11.2.3 $$|\langle (i_0 \dots i_{r-1}) \rangle| = r.$$

Two cycles π and ρ are called *disjoint*, if the two sets of points which are not fixed by π and ρ are disjoint sets. Note that, for example, $1 = (0)(1)(2)$ *and* (012) are disjoint cycles. Disjoint cycles π and ρ commute, i. e. $\pi\rho = \rho\pi$. (We read compositions of mappings from right to left, so that $(\pi\rho)x = \pi(\rho x)$.) *Each permutation of a finite set can be written as a product of pairwise different disjoint cycles, e. g.*

$$\binom{01234567}{64152307} = (06)(142)(53)(7).$$

The disjoint cyclic factors $\neq 1$ of $\pi \in S_n$ are uniquely determined by π and therefore we call these factors together with the fixed point cycles of π *the cyclic factors* of π. Let $c(\pi)$ denote the number of these cyclic factors of π (including 1-cycles), let l_ν be their lengths, $\nu \in c(\pi) = \{0, \dots, c(\pi) - 1\}$, and choose, for each ν an element j_ν of the ν-th cyclic factor. Then

11.2.4 $$\pi = \prod_{\nu \in c(\pi)} (j_\nu \, \pi j_\nu \dots \pi^{l_\nu - 1} j_\nu).$$

This notation becomes uniquely determined if we choose the j_ν so that

11.2.5 $$\forall\, m \in \mathbb{N}\colon\ j_\nu \le \pi^m j_\nu, \text{ and } \forall\, \nu < c(\pi) - 1\colon\ j_\nu < j_{\nu+1}.$$

If this holds, then 11.2.4 is called the *(standard) cycle notation* for π. We note in passing that the sets $\{j_\nu, \pi j_\nu, \dots, \pi^{l_\nu - 1} j_\nu\}$ of points which are cyclically permuted by π are just the orbits of the cyclic group $\langle \pi \rangle$ generated by π :

$$\langle \pi \rangle \backslash\!\backslash\, n = \{\{j_\nu \, \pi j_\nu \dots \pi^{l_\nu - 1} j_\nu\} \mid \nu \in c(\pi)\}.$$

Having described the elements of S_n, we show which of them are in the same conjugacy class, i. e. in the same orbit of the group S_n on the set S_n under the conjugation action (cf. 1.1.11). In order to do this, we first note how $\rho\pi\rho^{-1}$ is obtained from the permutation π:

$$\rho\pi\rho^{-1} = \begin{pmatrix} i \\ \rho i \end{pmatrix} \begin{pmatrix} i \\ \pi i \end{pmatrix} \begin{pmatrix} \rho i \\ i \end{pmatrix} = \begin{pmatrix} \rho i \\ \rho(\pi i) \end{pmatrix}.$$

Thus, in terms of cyclic factors of π, $\rho\pi\rho^{-1}$ arises from

$$\pi \ = \ \dots (\dots i \, \pi i \dots) \dots$$

by simply applying the mapping ρ to the points in the cycles of π:

11.2.6 $$\rho\pi\rho^{-1} = \dots (\dots \rho i \; \rho(\pi i) \dots) \dots .$$

(For any mapping $\varphi\colon S \to S$ we mean by $\varphi(s) = t$ that s has to be replaced by t and *not* that t is replaced by s, as it is sometimes understood!) This equation shows that the lengths of the cyclic factors of π are the same as those of $\rho\pi\rho^{-1}$. It is easy to see that, conversely, for any two elements $\pi, \sigma \in S_n$ with the same lengths l_ν of cyclic factors there exists a $\rho \in S_n$ such that $\rho\pi\rho^{-1} = \sigma$. Hence the lengths of the cyclic factors of π characterize its conjugacy class. To make this more explicit, we introduce the notion of a *(proper) partition* of $n \in \mathbb{N}$, by which we mean any sequence $\alpha = (\alpha_0, \alpha_1, \dots)$ of natural numbers α_i which satisfy

$$\forall\, i\colon\ \alpha_i \ge \alpha_{i+1}, \text{ and } \sum_i \alpha_i = n.$$

The α_i are called the *parts* of α. The fact that α is a partition of n is abbreviated by

$$\alpha \vdash n.$$

If $\alpha \vdash n$ then there exists an h such that $\alpha_i = 0$ for all $i > h - 1$. We may therefore write

$$\alpha = (\alpha_0, \ldots, \alpha_{h-1}),$$

for any such h. The minimal h with this property will be denoted by $l(\alpha)$ and called the *length* of α. The following abbreviation is useful in the case when several nonzero parts of α are equal, say a_i parts are equal to $i, i \in \underline{n}$:

$$\alpha = (n^{a_n}, (n-1)^{a_{n-1}}, \ldots, 1^{a_1}).$$

If $a_i = 0$, then i^{a_i} is usually omitted, e. g. $(3, 1^2) = (3, 1, 1, 0, \ldots)$. For $\pi \in S_n$ the ordered lengths $\alpha_i(\pi), i \in c(\pi)$, of the cyclic factors of π in cycle notation form a uniquely determined proper partition

$$\alpha(\pi) = (\alpha_0(\pi), \alpha_1(\pi), \ldots, \alpha_{c(\pi)-1}(\pi)) \vdash n,$$

which we call the *cycle partition* of π. We note in passing that $l(\alpha(\pi)) = c(\pi)$. The corresponding n-tuple

$$a(\pi) := (a_1(\pi), \ldots, a_n(\pi))$$

consisting of the multiplicities $a_i(\pi)$ of the parts of length i in $\alpha(\pi)$ is called the *cycle type* of π. Correspondingly we call an n-tuple $a := (a_1, \ldots, a_n)$ a cycle type of n if and only if each $a_i \in \mathbb{N}$, and $\sum i \cdot a_i = n$. This will be abbreviated by

$$a \vdash\!\dashv n.$$

The conjugacy class of $\pi \in S_n$ will be denoted by $C^S(\pi)$, the centralizer by $C_S(\pi)$, so that we obtain the following descriptions and properties of conjugacy classes and centralizers of elements of S_n (or $S_{\underline{n}}$, if the reader prefers):

11.2.7 Corollary *Let π and σ denote elements of S_n. Then*

- $C^S(\pi) = C^S(\sigma) \iff \alpha(\pi) = \alpha(\sigma) \iff a(\pi) = a(\sigma)$.
- $C^S(\pi) = C^S(\pi^{-1})$, *i. e. S_n is* ambivalent, *which means that each element is a conjugate of its inverse.*
- $|C_S(\pi)| = \prod_i i^{a_i(\pi)} a_i(\pi)!$, *and* $|C^S(\pi)| = n!/\prod_i i^{a_i(\pi)} a_i(\pi)!$.
- $|\langle\pi\rangle| = \operatorname{lcm}\{\alpha_i(\pi) \mid i \in c(\pi)\} = \operatorname{lcm}\{i \mid a_i(\pi) > 0\}$.
- *Each proper partition $\alpha \vdash n$ occurs as the cycle partition of some permutation π in S_n.*

(The first, second, fourth and fifth item is clear from the foregoing, while the third follows from the fact that there are exactly $i^{a_i} a_i!$ mappings which map a set of a_i i-tuples onto this same set up to cyclic permutations inside each i-tuple.) For the sake of simplicity we can therefore parametrize the conjugacy classes of elements in S_n (and correspondingly in $S_{\underline{n}}$) by partitions or cycle types putting

$$C^\alpha := C^a := C^S(\pi), \text{ when } \alpha(\pi) = \alpha, \text{ and } a(\pi) = a.$$

Since we can decompose a cycle into 2-cycles,

11.2.8 $$(i_0 \dots i_{r-1}) = (i_0 i_1)(i_1 i_2) \dots (i_{r-2} i_{r-1})$$

each cycle, and hence every element of S_n, can be written as a product of transpositions. Thus S_n as well as $S_{\underline{n}}$ is generated by its subset of transpositions (if this is empty, then $n \leq 1$, and both $S_1 = S_0 = S_{\underline{1}} = S_{\underline{0}} = \{1\}$ are generated by the empty set $\emptyset$). But, except for the case when $n = 2$, we do not need every transposition in order to generate the symmetric group, since, for $0 \leq j < k < n - 1$, we derive from 11.2.6 that

$$(j, k+1) = (k, k+1)(j, k)(k, k+1).$$

Thus the transposition $(j, k+1)$ can be obtained from (j, k) by conjugation with the transposition $(k, k+1)$ of adjacent points. Therefore the subset

$$\Sigma_n := \{\sigma_i := (i, i+1) \mid 0 \leq i < n - 1\} = \{(01), (12), \dots, (n-2, n-1)\},$$

consisting of the *elementary* transpositions σ_i, generates S_n, and

$$\Sigma_{\underline{n}} := \{\sigma_i := (i, i+1) \mid 1 \leq i < n\} = \{(12), (23), \dots, (n-1, n)\}$$

generates $S_{\underline{n}}$. A further system of generators of S_n is obtained from

11.2.9 $$(0 \dots n-1)^i (01)(0 \dots n-1)^{-i} = (i, i+1), 0 \leq i \leq n-2,$$

so that we have proved

11.2.10 $$S_n = \langle (01), (12), \dots, (n-2, n-1) \rangle = \langle (01), (0 \dots n-1) \rangle,$$

and, correspondingly,

11.2.11 $$S_{\underline{n}} = \langle (12), (23), \dots, (n-1, n) \rangle = \langle (12), (1 \dots n) \rangle.$$

Another useful result describes the cycle structure of a power of a cycle:

11.2.12 Lemma *For each natural number m the power $(i_0 \dots i_{r-1})^m$ consists of exactly* $\gcd(r, m)$ *disjoint cyclic factors, they all are of length $r/\gcd(r, m)$.*

Proof: Let l denote the length of the cyclic factor of $(i_0 \dots i_{r-1})^m$ containing i_0. This cycle is

$$(i_0\, i_{\overline{m}}\, i_{\overline{2m}} \dots i_{\overline{(l-1)m}}),$$

where $\overline{k}$ denotes the residue class of k modulo r. Correspondingly the cyclic factor containing i_1 (if $r > 1$) must be $(i_1 \dots i_{\overline{(l-1)m+1}})$, so that all the cyclic factors of $(i_0, \dots, i_{r-1})^m$ have the same length l. Thus l is the order of $(i_0 \dots i_{r-1})^m$, an element of the group $\langle (i_0 \dots i_{r-1}) \rangle$ which is of order r. Hence r divides $l \cdot m$ and $r/\gcd(r, m)$ divides $l \cdot (m/\gcd(r, m))$ and therefore also l. But

$$(i_0 \dots i_{r-1})^{m \cdot (r/\gcd(r,m))} = (i_0 \dots i_{r-1})^{r \cdot (m/\gcd(r,m))} = 1,$$

so that also l must divide $r/\gcd(r, m)$, which proves that in fact $l = r/\gcd(r, m)$. □

An easy application of 11.2.12 gives

11.2.13 Corollary *The elements of order d in the group generated by the cycle $(0\dots n-1)$ are the powers $(0\dots n-1)^i$, where i is of the form $\frac{n\cdot j}{d}$, and $0 \le j < d-1$ is relatively prime to d.*

A direct consequence of this is

11.2.14 Corollary *The group $C_m := \langle(0\dots m-1)\rangle$ contains, for each divisor d of m exactly one subgroup U of order d. Furthermore, this subgroup contains $\phi(d)$ elements consisting of d–cycles only, if $\phi(-)$ denotes the* Euler function,

$$\phi(d) := |\{i \mid 0 \le i < d, \gcd(d,i) = 1\}|.$$

These $\phi(d)$ elements form the set of generators of U.

As finite cyclic groups of the same order are isomorphic, they have the same properties:

11.2.15 Corollary *A finite cyclic group G has, for each divisor d of its order $|G|$, exactly one subgroup U of order d. Furthermore, this subgroup contains $\phi(d)$ generators, and so, G has exactly $\phi(d)$ elements of this particular order. Moreover*

$$\sum_{d|n} \phi(d) = n.$$

Another important fact exhibits normal subgroups A_n of S_n. In order to show this we introduce the *sign* function ϵ as follows:

$$\epsilon(\pi) := \prod_{0 \le i < j \le n-1} \frac{\pi j - \pi i}{j - i} \in \mathbb{Z}, \text{ if } n \ge 2, \text{ while } \epsilon(1_{S_0}) := \epsilon(1_{S_1}) := 1_{\mathbb{Z}}.$$

As $i \neq j$ implies $\pi i \neq \pi j$, we have $\epsilon(\pi) \neq 0$. Moreover, the following sets of pairs are equal:

$$\{\{i,j\} \mid 0 \le i < j \le n-1\} = \{\{\pi i, \pi j\} \mid 0 \le i < j \le n-1\},$$

and so we have $\epsilon(\pi) = \pm 1_{\mathbb{Z}}$. Furthermore ϵ is a homomorphism of $S_{\underline{n}}$ into $\{1,-1\}$:

$$\epsilon(\pi\rho) = \prod_{i<j} \frac{\pi\rho j - \pi\rho i}{j-i} = \prod_{i<j} \frac{\pi\rho j - \pi\rho i}{\rho j - \rho i} \prod_{i<j} \frac{\rho j - \rho i}{j - i} = \epsilon(\pi)\epsilon(\rho).$$

This proves

11.2.16 Corollary *The sign map $\epsilon \colon S_n \to \{1,-1\} \colon \pi \mapsto \epsilon(\pi)$ is a homomorphism which is surjective for each $n \ge 2$. Hence its kernel*

$$A_n := \ker \epsilon = \{\pi \in S_n \mid \epsilon(\pi) = 1\}$$

is a normal subgroup of S_n:

$$A_n \trianglelefteq S_n, \ |A_n| = |S_n|/2 = n!/2, \ \text{if } n \ge 2.$$

The elements of A_n are called *even* permutations, while the elements of $S_n \setminus A_n$ are called *odd* permutations. Correspondingly, an r–cycle is even if and only if r is odd. The sign of a cyclic permutation is easily obtained from the equation 11.2.8 and the homomorphism property of the sign,

$$(i_0 \dots i_{r-1}) \in A_n \iff r \text{ is odd.}$$

But we need not check the lengths of the cyclic factors of π since an easy calculation shows (exercise 11.2.3) that we need only the number $c(\pi)$ of cyclic factors of π,

11.2.17 $$\epsilon(\pi) = (-1)^{n-c(\pi)}, \text{ if } \pi \in S_n.$$

Another way of describing permutations $\pi \in S_n$ is the *list notation,* where we put the list of their values in square brackets:

$$\pi = [\pi 0, \dots, \pi(n-1)], \text{ or } \pi = [\pi 0 \dots \pi(n-1)], \text{ if unambiguous.}$$

The i-th position, $i \leq n-1$, of this list is called an *up* if $\pi i < \pi(i+1)$, otherwise it is called a *down*. Replacing an up by "+", a down by "−", we obtain the *up–down sequence* $U\pi$ of π, for example

$$U[02137654] = (+ - + + - - -).$$

Permutations with exactly one down are called *grassmannian,* while permutations with exactly one up are called *co-grassmannian.* Here are the elements of S_4 in list-notation, together with their up-down sequences:

[0123]	1	$+++$
[0132]	(23)	$++-$
[0213]	(12)	$+-+$
[0231]	(123)	$++-$
[0312]	(132)	$+-+$
[0321]	(13)	$+--$
[1023]	(01)	$-++$
[1032]	(01)(23)	$-+-$
[1203]	(012)	$+-+$
[1230]	(0123)	$++-$
[1302]	(0132)	$+-+$
[1320]	(013)	$+--$
[2013]	(021)	$-++$
[2031]	(0231)	$-+-$
[2103]	(02)	$--+$
[2130]	(023)	$-+-$
[2301]	(02)(13)	$+-+$

[2310]	(0213)	$+--$
[3012]	(0321)	$-++$
[3021]	(031)	$-+-$
[3102]	(032)	$--+$
[3120]	(03)	$-+-$
[3201]	(0312)	$--+$
[3210]	(03)(12)	$---$

Permutations π where both π and π^{-1} are grassmannian permutations are called *bi-grassmannian.* Here are the bi-grassmannians of S_4 gathered into pairs $\{\pi, \pi^{-1}\}$:

$$\{[0132]\}, \{[0213]\}, \{[0231], [0312]\}, \{[1023]\},$$

$$\{[1203], [2013]\}, \{[1230], [3012]\}, \{[2301]\}.$$

Correspondingly, there are the *bi-co-grassmannian:*

$$\{[0321]\}, \{[1032]\}, \{[1320], [3021]\}, \{[2103]\},$$

$$\{[2130], [3102]\}, \{[2310], [3201]\}, \{[3120]\}.$$

These permutations are related to Schubert polynomials.

Exercises

Exercise 11.2.1 Prove lemma 11.2.1.

Exercise 11.2.2 Show that, for each $m, n \in \mathbb{N}^*$ and $\pi \in S_n$, the permutations π and π^m are conjugate, if and only if m and each length of a cyclic factor of π are relatively prime.

Exercise 11.2.3 Check 11.2.17.

11.3 Rothe Diagram and Lehmer Code

In order to visualize permutations we can use a permutation matrix or, equivalently, a board consisting of boxes in the second quadrant. We put a black button • into the box with the pair of coordinates (i, k) if and only $i = \pi k$, obtaining the *Rothe diagram* $D(\pi)$ of π. This diagram was introduced by H. A. Rothe as early as in 1800. For example $D([43021])$ is equal to

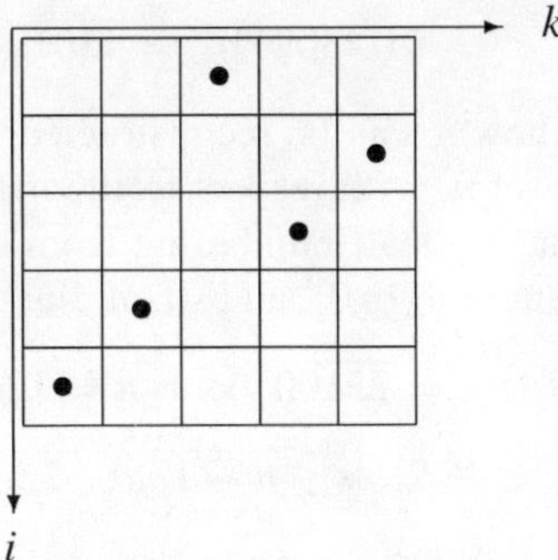

The list notation gives rise to the *Lehmer code* which we are going to describe next. In order to do this we consider the *set* $I(\pi)$ of inversions of π:

$$I(\pi) := \{(i, j) \in n^2 \mid i < j,\ \pi i > \pi j\}.$$

An easy check (exercise 11.3.1) shows that the following holds:

11.3.1 Lemma *If we represent the inversion (i, j) of π by a white button $\circ$ at $(\pi j, i)$, then we fill exactly the boxes of the Rothe diagram of π where the black button of the row is in a column to the right and where the black button of the column is lower down.*

The resulting diagram $\Lambda(\pi)$ consisting of the white buttons is called the *Lehmer diagram* of π. For example, the superposition of the Rothe diagram and the Lehmer diagram of $\pi := [43021]$ is

$$D([43021]) + \Lambda([43021]) \quad =$$

○	○	●		
○	○		○	●
○	○		●	
○	●			
●				

The *Lehmer code* is the sequence of numbers of white buttons in the columns of $\Lambda(\pi)$:

$$L(\pi) := \overline{l_0(\pi)\ldots l_{n-1}(\pi)},\ \text{ where } l_i(\pi) := |\{j \mid i < j, \pi i > \pi j\}|.$$

For example

$$L([43021]) = \overline{43010}.$$

If it is clear which n is meant, then we may replace $L(\pi)$ by the shorter sequence

$$L(\pi)^+$$

arising from $L(\pi)$ by canceling the zeros at the end of $L(\pi)$, for example

$$L([43021])^+ = \overline{4301}.$$

It is not difficult to see how π can be reconstructed from $L(\pi)$. Clearly $\pi 0$ is the $l_0(\pi)$-th element of the set n, with respect to the natural order of n, which is the element $l_0(\pi)$ (recall that we start numbering from 0 onwards). Correspondingly, $\pi 1$ is the $l_1(\pi)$-th element of $n \backslash \{\pi 0\}$, and so on. For example

$$L(\pi) := \overline{23100} \text{ yields } \pi = [24103].$$

Furthermore, each $L(\pi)$, $\pi \in S_n$, is $\leq \overline{n-1, n-2, \ldots, 1, 0}$, i. e.

$$\forall\, i \in n\colon\ l_i(\pi) \leq n - i - 1.$$

Since there are $n!$ such sequences in $\mathbb{N}^n$, we have obtained:

11.3.2 Corollary *The Lehmer code establishes a bijection between S_n and the set of sequences $\overline{l_0(\pi)\ldots l_{n-1}(\pi)}$ in $\mathbb{N}^n$, $l_i(\pi) \leq n - i - 1$, for each i.*

This shows that in fact the Lehmer code is a code for permutations. But there is much more to say about Lehmer codes. Applying the equivalences

$$l_i(\pi) \leq l_{i+1}(\pi) \iff \pi i < \pi(i+1),\ l_i(\pi) > l_{i+1}(\pi) \iff \pi i > \pi(i+1),$$

we can easily derive (exercise 11.3.2) that right multiplication of π by the elementary transposition $\sigma_i = (i, i+1)$ has the following effect on π:

11.3.3 $\quad \pi \circ \sigma_i = [\pi 0 \ldots \pi(i-1)\, \pi(i+1)\, \pi(i)\, \pi(i+2) \ldots \pi(n-1)]\,.$

Hence right multiplication of π by an elementary transpostion *interchanges* the adjacent entries in the places i and $i+1$ of the list representing π. This yields

11.3.4 Corollary *For each $\pi \in S_n$ and each $\sigma_i \in \Sigma_n$ we have*

$$L(\pi\sigma_i) = \overline{l_0(\pi), \ldots, l_{i-1}(\pi), l_{i+1}(\pi), l_i(\pi) - 1, l_{i+2}(\pi), \ldots},$$

if $l_i(\pi) > l_{i+1}(\pi)$, while

$$L(\pi\sigma_i) = \overline{l_0(\pi), \ldots, l_{i-1}(\pi), l_{i+1}(\pi) + 1, l_i(\pi), l_{i+2}(\pi), \ldots},$$

if $l_i(\pi) \leq l_{i+1}(\pi)$.

The main point is that the shift from π to $\pi\sigma_i$ amounts (cf. 11.3.4) to an increase or a decrease in

$$l(\pi) := \sum_i l_i(\pi)$$

by 1. Hence the following holds:

11.3.5 Corollary *The sum $l(\pi)$ of the entries of the Lehmer code $L(\pi)$ of $\pi \in S_n$ is equal to the minimal number of elementary transpositions σ_i which are needed in order to express π as a product of elementary transpositions. Moreover, this yields a canonic way of expressing π as a product of $l(\pi)$ elementary transpositions by decreasing the rightmost nonzero entry of the Lehmer code by 1 and iterating this process until each entry is equal to zero.*

We therefore call $l(\pi)$ the *reduced length* of π and we call such an expression of π as a product of $l(\pi)$ elementary transpositions, a *reduced decomposition.* A reduced decomposition of [43021], for example, is obtained from the following sequence of manipulations of Lehmer codes, according to 11.3.4:

$$\overline{43010} \overset{\cdot\sigma_3}{\longmapsto} \overline{43000} \overset{\cdot\sigma_1}{\longmapsto} \overline{40200} \overset{\cdot\sigma_2}{\longmapsto} \overline{40010} \overset{\cdot\sigma_3}{\longmapsto} \overline{40000}$$

$$\overset{\cdot\sigma_0}{\longmapsto} \overline{03000} \overset{\cdot\sigma_1}{\longmapsto} \overline{00200} \overset{\cdot\sigma_2}{\longmapsto} \overline{00010} \overset{\cdot\sigma_3}{\longmapsto} \overline{00000}.$$

This shows that

$$[43021] = \sigma_3\sigma_2\sigma_1\sigma_0\sigma_3\sigma_2\sigma_1\sigma_3$$

is a reduced decomposition. This particular reduced composition can directly be read off from the superposition of the Rothe diagram and the Lehmer diagram:

<table>
<tr><td>σ_0</td><td>σ_1</td><td>•</td><td></td><td></td></tr>
<tr><td>σ_1</td><td>σ_2</td><td></td><td>σ_3</td><td>•</td></tr>
<tr><td>σ_2</td><td>σ_3</td><td></td><td>•</td><td></td></tr>
<tr><td>σ_3</td><td>•</td><td></td><td></td><td></td></tr>
<tr><td>•</td><td></td><td></td><td></td><td></td></tr>
</table>

(Read the entries columnwise upwards from left to right.) The corollary 11.3.5 also shows how we can obtain all the reduced decompositions of a given permutation:

11.3.6 Corollary *The complete set of reduced decompositions of π is obtained by carrying out all the successive multiplications that reduce the Lehmer sequence by 1, according to 11.3.5.*

Reduced decompositions play a role also in the theory of other groups. The symmetric group S_n is in fact a *Coxeter group,* which means that there exists a set of generators s_i such that the relations are of the particular form $(s_i s_j)^{m_{ij}} = 1$, where

$$m_{ij} \in \mathbb{N},\ m_{ii} = 1,\ m_{ij} = m_{ji} > 1,\ \text{if } i \neq j.$$

In the case of the symmetric group S_n, the Coxeter system of generators is the subset Σ_n of elementary transpositions $\sigma_i = (i, i+1)$, as we have seen already. Furthermore, this leads us to introduce the *weak Bruhat order* $\preceq$ on S_n, which is the transitive closure of

$$\pi \prec\!\cdot\, \rho \;:\Longleftrightarrow\; l(\rho) = l(\pi) + 1, \text{ and } \exists\, \sigma_i : \rho = \pi\sigma_i.$$

The Bruhat order of the symmetric group S_4 is shown in figure 11.1. We note that the number of reduced decompositions (and also the *set* of reduced decompositions) can be obtained by going from the identity upwards in all the possible ways until the permutation in question is reached.

We are now approaching a famous result on reduced sequences the proof of which requires a better knowledge of the set of inversions. In order to describe how

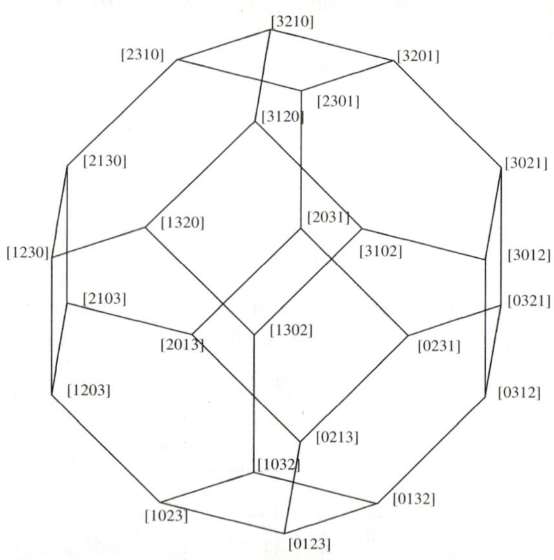

Fig. 11.1. The Bruhat order of S_4

$I(\pi\sigma_i)$, $\sigma_i \in \Sigma_n$, arises from $I(\pi)$, we note that S_n acts on the set n^2 in a canonic way: $\rho(i, j) = (\rho i, \rho j)$. Keeping this in mind, we easily obtain from 11.3.3:

11.3.7 $$I(\pi\sigma_i) = \begin{cases} \sigma_i I(\pi) \cup \{(i, i+1)\}, & \text{if } \pi i < \pi(i+1), \\ \sigma_i I(\pi) \backslash \{(i, i+1)\}, & \text{if } \pi i > \pi(i+1). \end{cases}$$

This yields for the corresponding reduced lengths:

11.3.8 $$l(\pi\sigma_i) = \begin{cases} l(\pi) + 1, & \text{if } \pi i < \pi(i+1), \\ l(\pi) - 1, & \text{if } \pi i > \pi(i+1), \end{cases}$$

in accordance with 11.3.4. If ω_n denotes the permutation $[n-1, \ldots, 0]$ of maximal length, then

11.3.9 $$l(\omega_n) = \binom{n}{2}, \; l(\omega_n \rho) = \binom{n}{2} - l(\rho) = l(\rho\omega_n).$$

Proof: Clearly $I(\omega_n) = \{(i, j) \in n^2 \mid i < j\}$, and hence $l(\omega_n) = \binom{n}{2}$. Moreover

$$\rho\omega_n = [\rho(n-1), \ldots, \rho 0],$$

so $I(\rho\omega_n) = I(\omega_n) \backslash I(\rho)$, and $l(\rho\omega_n) = \binom{n}{2} - l(\rho)$. Finally we note that

$$l(\omega_n \rho) = l((\omega_n \rho)^{-1}) = l(\rho^{-1}\omega_n) = \binom{n}{2} - l(\rho^{-1}) = \binom{n}{2} - l(\rho),$$

which completes the proof. □

An expression

$$\pi = \sigma_{i_0} \dots \sigma_{i_{l-1}}, \sigma_i \in \Sigma_n,$$

of π in terms of elementary transpositions and *minimal* $l = l(\pi)$ was called a reduced decomposition of π. The set of corresponding *sequences* of indices is indicated as follows:

$$RS(\pi) := \{(i_0, \dots, i_{l-1}) \mid \pi = \sigma_{i_0} \dots \sigma_{i_{l-1}}, \sigma_i \in \Sigma_n, l \text{ minimal}\}.$$

These sequences are called *reduced sequences* of π.

11.3.10 Lemma *If* $(i_0, \dots, i_{l-1}) \in RS(\pi)$*, then*

$$I(\pi) = \{\sigma_{i_{l-1}} \dots \sigma_{i_{r+1}}(i_r, i_r + 1) \mid r \in l = l(\pi)\}.$$

Proof: By induction on $l = l(\pi)$. The case $l = 0$ yields the empty set which is in fact the set of inversions of $\pi = 1$. If $l \geq 1$, then we can consider $\pi' := \pi\sigma_{i_{l-1}} = \sigma_{i_0} \dots \sigma_{i_{l-2}}$, which is of reduced length $l(\pi) - 1$. The induction hypothesis gives

$$I(\pi') = \{\sigma_{i_{l-2}} \dots \sigma_{i_{r+1}}(i_r, i_r + 1) \mid r \in l - 1\},$$

and from 11.3.7 we know how $I(\pi)$ can be obtained from $I(\pi')$, since $l(\pi) = l(\pi') + 1$ shows which of the two cases holds:

$$\begin{aligned} I(\pi = \pi'\sigma_{i_{l-1}}) &= \sigma_{i_{l-1}} I(\pi') \cup \{(i_{l-1}, i_{l-1} + 1)\} \\ &= \{\sigma_{i_{l-1}} \dots \sigma_{i_{r+1}}(i_r, i_r + 1) \mid r \in l - 1\} \cup \{(i_{l-1}, i_{l-1} + 1)\} \\ &= \{\sigma_{i_{l-1}} \dots \sigma_{i_{r+1}}(i_r, i_r + 1) \mid r \in l\}, \end{aligned}$$

as it is stated. □

We are now in a position to prove the following important result:

11.3.11 The Exchange Lemma *If* $(i_0, \dots, i_{l-1}), (j_0, \dots, j_{l-1})$ *are elements of* $RS(\pi)$*, then there exist* $k \in l = l(\pi)$ *such that*

$$(j_0, i_0, \dots, \hat{i}_k, \dots, i_{l-1}) \in RS(\pi),$$

where $\hat{i}_k$ *means that* i_k *is left out.*

Proof: We know that $I(\pi^{-1}) = \{\sigma_{i_0} \dots \sigma_{i_{k-1}}(i_k, i_k + 1) \mid k \in l\}$, and hence there exists a k such that

$$(j_0, j_0 + 1) = \sigma_{i_0} \dots \sigma_{i_{k-1}}(i_k, i_k + 1) = (\sigma_{i_0} \dots \sigma_{i_{k-1}}(i_k), \sigma_{i_0} \dots \sigma_{i_{k-1}}(i_k + 1)).$$

But this implies (since σ_{j_0} *transposes* j_0 and $j_0 + 1$):

$$\sigma_{j_0} = \sigma_{i_0} \dots \sigma_{i_{k-1}} \sigma_{i_k} (\sigma_{i_0} \dots \sigma_{i_{k-1}})^{-1},$$

and so $\sigma_{j_0}\sigma_{i_0} \dots \sigma_{i_{k-1}} = \sigma_{i_0} \dots \sigma_{i_{k-1}}\sigma_{i_k}$, which proves the statement. □

This result is crucial for the introduction of Schubert polynomials. They correspond to the permutations and form an important basis of the polynomial ring $\bigcup_{n>0}\mathbb{Z}\,[x_0, \ldots, x_{n-1}]$.

Exercises

Exercise 11.3.1 Prove 11.3.1.

Exercise 11.3.2 Check 11.3.4.

11.4 Linear Representations

Recall that a finite action ${}_GX$ is essentially the same as a certain permutation representation of G on X, namely the homomorphism

$$\delta: G \to S_X: g \mapsto (x \mapsto gx)$$

from G into S_X, where $g \in G$ is mapped onto $\bar{g}: x \mapsto gx$, an element of S_X.

Now we consider, for an arbitrary field $\mathbb{F}$, the free vector space over $\mathbb{F}$ with basis X, i. e. we consider $\mathbb{F}^X = \{f: X \to \mathbb{F}\}$. The linear mapping

$$D(g): \mathbb{F}^X \to \mathbb{F}^X : f \mapsto f \circ \bar{g}^{-1}$$

generalizes $\delta(g)$ since $D(g)$ permutes the basis X in the same way as $\delta(g) = \bar{g}$ does with the set X. After having numbered the basis elements $x_0, \ldots, x_{|X|-1}$, we obtain a matrix which represents this linear mapping:

$$\mathbf{D}(g) := (d_{ik}(g)), \text{ where } d_{ik}(g) := \begin{cases} 1, & \text{if } gx_k = x_i, \\ 0, & \text{otherwise.} \end{cases}$$

(Such matrices are called *permutation matrices* since they describe permutations of the basis.) This shows that the notions of finite group action and that of permutation representation are particular cases of the general concept of linear representation of finite groups. A *linear representation* D of G over a field $\mathbb{F}$ is defined to be a homomorphism D from G into the group $GL(V)$ of invertible linear mappings of a finite dimensional vector space V over $\mathbb{F}$ onto itself. The vector space V is called the *representation space* and its dimension d is also called the *dimension* of D, $\mathbb{F}$ is said to be the *ground field* of D.

Two representations $D: G \to GL(V)$ and $D': G \to GL(V')$ of G over $\mathbb{F}$ are considered to be essentially the same and are called *equivalent* if and only if there exists an invertible linear $T: V \to V'$ such that

$$\forall\, g \in G : TD(g) = D'(g)T.$$

Each choice of a basis $B = \{b_0, \ldots, b_{d-1}\}$ yields invertible matrices $\mathbf{D}(g)$ which describe the $D(g)$ with respect to B. Therefore a *matrix representation* $\mathbf{D}$ of G over

$\mathbb{F}$ is defined to be a homomorphism from G into a *general linear group* $GL(d, \mathbb{F})$, the group consisting of all the invertible matrices over $\mathbb{F}$ with d rows and columns. It is clear that, conversely, each matrix representation $\mathbf{D}: G \to GL(d, \mathbb{F})$ yields a representation $D: G \to GL(V)$, V being the d–dimensional vector space $\mathbb{F}^d$ over $\mathbb{F}$. Equivalence of matrix representations is defined correspondingly, so that we are free to consider either representations or matrix representations. Which concept we choose will depend on the situation in question.

We already have a wealth of examples at hand since each finite action ${}_GX$ yields a representation as described above. The trivial representation of G arising from the *trivial action* ${}_G\{x\}$ of G on a one element set $\{x\}$, where $gx := x$, is called the *identity representation* and it is indicated as follows:

$$IG: G \to GL(\mathbb{F}): g \mapsto id_{\mathbb{F}},$$

where $\mathbb{F}$ is the 1–dimensional vector space over $\mathbb{F}$.

Furthermore there are several ways to get new representations from old ones. For example $D: G \to GL(V)$ and $D': G \to GL(V')$ yield the *sum* $D + D'$ which we can define by giving a corresponding matrix representation $\mathbf{D} + \mathbf{D}'$, arising from $\mathbf{D}$ and $\mathbf{D}'$ as follows:

$$(\mathbf{D} + \mathbf{D}')(g) := \begin{pmatrix} \mathbf{D}(g) & 0 \\ 0 & \mathbf{D}'(g) \end{pmatrix}.$$

We can also form the *inner tensor product* $D \otimes D'$, corresponding to $\mathbf{D} \otimes \mathbf{D}'$ which is defined by

$$(\mathbf{D} \otimes \mathbf{D}')(g) := \begin{pmatrix} d_{00}(g)\mathbf{D}'(g) & d_{01}(g)\mathbf{D}'(g) & \dots \\ d_{10}(g)\mathbf{D}'(g) & d_{11}(g)\mathbf{D}'(g) & \dots \\ \dots & \dots & \dots \end{pmatrix}.$$

This matrix is called a *Kronecker product* of the matrices $\mathbf{D}(g)$ and $\mathbf{D}'(g)$ and it is denoted by $\mathbf{D}(g) \otimes \mathbf{D}'(g)$. This product of matrices can also be used to define the *outer tensor product* $D \# D'$ of D of G and D' of G', which is a representation of the direct product $G \times G'$ and which has as one of its corresponding matrix representations the matrix representation $\mathbf{D} \# \mathbf{D}'$, the representing matrices $(\mathbf{D} \# \mathbf{D}')(g, g')$ of which are defined to be

$$\mathbf{D}(g) \otimes \mathbf{D}'(g') := \begin{pmatrix} d_{00}(g)\mathbf{D}'(g') & d_{01}(g)\mathbf{D}'(g') & \dots \\ d_{10}(g)\mathbf{D}'(g') & d_{11}(g)\mathbf{D}'(g') & \dots \\ \dots & \dots & \dots \end{pmatrix}.$$

Besides these constructions we can get a representation $D \downarrow H$ of $H \leq G$ by restricting a representation D of G to the subgroup H:

$$(D \downarrow H(h)) := D(h)$$

is called the *restriction* of D to H. Conversely, we can *induce* from a representation D of a subgroup $H \leq G$ obtaining the *induced representation* $D \uparrow G$ of

G as follows: Take a corresponding matrix representation $\mathbf{D}$ of H and consider a decomposition of G into left cosets of H, say

$$G = \bigcup_{i=0}^{|G/H|-1} g_i H.$$

Then put

$$(\mathbf{D} \uparrow G)(g) := \begin{pmatrix} & \vdots & \\ \dots & \dot{\mathbf{D}}(g_i^{-1} g g_k) & \dots \\ & \vdots & \end{pmatrix},$$

where

$$\dot{\mathbf{D}}(g_i^{-1} g g_k) := \begin{cases} \mathbf{D}(g_i^{-1} g g_k), & \text{if } g_i^{-1} g g_k \in H, \\ \text{zero matrix}, & \text{otherwise.} \end{cases}$$

For our purposes here it is very important to notice that *the particular representation $IH \uparrow G$ of G, induced by the identity representation IH of the subgroup $H \leq G$, is equivalent to the representation D of G which corresponds to the action ${}_G(G/H)$*. This follows from the fact that $gg_kH = g_iH$ is equivalent to $g_i^{-1} g g_k \in H$. Further constructions will be discussed later.

The representation $D \colon G \to GL(V)$ and also V are called *reducible* if and only if there exists a subspace W such that $\{0_V\} \neq W \neq V$ and

$$\forall\, g \in G : \; D(g)(W) = W,$$

i. e. if and only if there exists a nontrivial *G–invariant subspace* W of V. Otherwise D and V are called *irreducible* . The restriction of D to such an invariant subspace W of V is denoted by

$$D|_W.$$

It is a representation of G and called a *subrepresentation* of D. We can choose a basis $B = \{b_0, \dots, b_{d-1}\}$ of V which is *adapted* to W, i. e. the subset $\{b_0, \dots, b_{e-1}\} \subseteq B$ is a basis of W. The corresponding matrix representation $\mathbf{D}$ then takes the form

$$\mathbf{D}(g) = \begin{pmatrix} \mathbf{D}_0(g) & \star \\ 0 & \mathbf{D}_1(g) \end{pmatrix},$$

where $\mathbf{D}_0 \colon g \mapsto \mathbf{D}_0(g)$ is an e–dimensional matrix representation of G which is in fact the restriction of $\mathbf{D}$ to W:

$$\mathbf{D}_0 = \mathbf{D}|_W.$$

But also $\mathbf{D}_1$ is a matrix representation of G, namely on the factor space V/W. The representation $\mathbf{D}_1$ is irreducible if and only if W is a maximal invariant subspace of V and hence, since d is finite, we can find a finite chain of subspaces W_i of V such that

$$V = W_0 \supset W_1 \supset \dots \supset W_k = \{0_V\},$$

and where W_i is a maximal invariant subspace of W_{i-1}, $1 \leq i \leq k$. The matrix representation $\mathbf{D}^{(i)}$ of G corresponding to W_{i-1}/W_i is irreducible, and for a suitable choice of a basis B of V, $\mathbf{D}$ takes the form

$$\mathbf{D}(g) = \begin{pmatrix} \mathbf{D}^{(k-1)}(g) & & \star \\ & \ddots & \\ 0 & & \ddots \\ & & & \mathbf{D}^{(0)}(g) \end{pmatrix}.$$

The $D^{(i)}$ are uniquely determined up to equivalence and order of occurrence, they are called the *irreducible constituents* of D. This follows from the *Jordan–Hölder Theorem* for any D and any ground field $\mathbb{F}$. We do not give a proof of this theorem since for the particular case in which we are interested, there is an easier proof.

From now on we restrict our attention to linear representations of finite groups G over the complex field $\mathbb{C}$. Such representations are called *ordinary* representations of G.

Our first observation concerns the decomposition of ordinary representations into irreducible representations. We show that such decompositions are direct decompositions for a suitable choice of basis. Let $D: G \to GL(V)$ denote an ordinary representation of G on V. As we know already from linear algebra, each finite dimensional vector space V over $\mathbb{C}$ carries, for any basis $B = \{b_0, \ldots, b_{d-1}\}$, the hermitian inner product

$$\langle -, - \rangle: V^2 \to \mathbb{C}: (u, v) \mapsto \sum_{i \in d} u_i \overline{v_i}.$$

Using such an inner product on V and the representation D of G on V, we can define

$$(- \mid -): V^2 \to \mathbb{C}: (u, v) \mapsto \sum_{g \in G} \langle D(g)u, D(g)v \rangle.$$

It is easy to check that this is also an hermitian inner product, and that it has the further property

11.4.1 $$\forall\, g \in G\ \forall\, u, v \in V:\ (u \mid v) = (D(g)u \mid D(g)v).$$

This has the remarkable consequence that the orthogonal complement $W^\perp$ (with respect to $(- \mid -)$) of a G–invariant subspace W of V is also G–invariant. Since $v \neq 0$ implies that $(v \mid v) \neq 0$ it follows that the sum $W + W^\perp$ is direct, and so a basis of V which is adapted to this decomposition $V = W \oplus W^\perp$ yields a matrix representation corresponding to D which *splits* :

$$\mathbf{D}(g) = \begin{pmatrix} \mathbf{D}_0(g) & 0 \\ 0 & \mathbf{D}_1(g) \end{pmatrix}.$$

where $\mathbf{D}_0 = \mathbf{D}|_W$, $\mathbf{D}_1 = \mathbf{D}|_{W^\perp}$. This proves

11.4.2 Maschke's Theorem *Each ordinary representation of a finite group splits into a direct sum of irreducible representations.*

Furthermore we can construct a basis of V which is orthonormal with respect to $(- \mid -)$. It is easy to check that a corresponding matrix representation $\mathbf{D}$ consists of unitary matrices $\mathbf{D}(g)$. Thus we have also proved

11.4.3 Theorem *Each ordinary representation of a finite group is equivalent to a unitary matrix representation.*

Our next aim is to prove that a decomposition of an ordinary representation into irreducible ones, which is direct by Maschke's Theorem, is also unique in the sense that the irreducible constituents $D^{(i)}$ are uniquely determined up to equivalence and order of occurrence. We first prove

11.4.4 Schur's Lemma *Let $D: G \to GL(V)$ and $D': G \to GL(V')$ denote ordinary irreducible representations of G and assume that $T: V \to V'$ is an* intertwining operator, *which means a linear mapping satisfying*

$$TD(g) = D'(g)T, \text{ for each } g \in G.$$

Then the following is true:

- *Both the kernel and the image of the intertwining operator T are G-invariant subspaces.*
- *If D is not equivalent to D', then $T = 0$.*
- *If D is equal to D', then T is a multiple of the identity mapping.*
- *If D is equivalent to D', say $D'(g) = SD(g)S^{-1}$, for each g and a suitable $\mathbb{C}$–isomorphism $S: V \simeq V'$, then T is a multiple of S.*

Proof: It is easy to check that both the kernel and the image of T are G–invariant subspaces of V and V', respectively. Hence, if $T \neq 0$, then $T(V) = V'$, since the image $\mathrm{im}T$ is invariant and D' is irreducible. Thus T is surjective. Moreover, as $\ker T$ is invariant and D is irreducible, this kernel must be zero and so T must be injective as well. Therefore T is bijective and hence an isomorphism between V and V', and so D is equivalent to D'. This proves the first two statements.

Let us now assume that $D = D'$, and that $T \neq 0$ (otherwise the statement is trivially true). Thus T is an endomorphism of $V = V'$. As $\mathbb{C}$ is the ground field, there exists an eigenvalue λ, say, of T. Thus $T' := T - \lambda \cdot id$ has a nonzero kernel.

Because of $T'D(g) = D'(g)T'$, for every $g \in G$, and the irreduciblity of V, this kernel, being G–invariant and nonzero, must be V, so that $T' = 0$ and, consequently, $T = \lambda \cdot id$.

Finally if D is equivalent to D', say $D'(g) = SD(g)S^{-1}$, then either $T = 0$, in which case T is, of course, a multiple of S, or we have that $T^{-1}SD(g)S^{-1}T = D(g)$, for each g. In the latter case, by the second statement, $T^{-1}S$ is a multiple of the identity mapping, $T^{-1}S = \mu \cdot id$, say, which completes the proof, since $\mu \neq 0$, as S is invertible. □

We can construct an intertwining operator T from every linear mapping $t: V \to V'$ simply by putting

$$T := \frac{1}{|G|} \sum_{g \in G} D'(g^{-1}) t D(g),$$

since $D'(x^{-1}) T D(x) = T$, for each $x \in G$. In terms of elements of representing matrices we have:

$$T_{il} = \frac{1}{|G|} \sum_{g} \sum_{j,k} d'_{ij}(g^{-1}) t_{jk} d_{kl}(g).$$

Thus Schur's Lemma gives:

11.4.5 Schur's Relations *For any $t_{jk} \in \mathbb{C}$ and ordinary irreducible matrix representations* $\mathbf{D}$ *and* $\mathbf{D}'$ *of G we have, for any i, l and $d := \dim(D)$:*

$$\frac{1}{|G|} \sum_{g} \sum_{j,k} d'_{ij}(g^{-1}) t_{jk} d_{kl}(g) = \begin{cases} 0, & \mathbf{D}, \mathbf{D}' \textit{ inequivalent,} \\ \delta_{il} d^{-1} \sum_j t_{jj}, & \textit{if } \mathbf{D} = \mathbf{D}'. \end{cases}$$

Exercises

Exercise 11.4.1 Find out where in the above proof of Maschke's theorem we made use of the fact that the groundfield is of characteristic 0.

Exercise 11.4.2 Show that the mapping

$$\mathbf{D}: C_p \to GL(2, GF(p)): (0, \ldots, p-1)^i \mapsto \begin{pmatrix} 1 & 1 \\ 0 & 1 \end{pmatrix}^i,$$

p a prime, defines a reducible matrix representation which does not split.

Exercise 11.4.3 Assume that the normal subgroup N of a finite group G is contained in the kernel of the representation D of G.

- Check that $\widetilde{D}(Ng) := D(g)$ defines a representation of the factor group G/N.
- Prove that $\widetilde{D}$ is irreducible if and only if D is.

Exercise 11.4.4 Let G denote a finite group. Its *commutator subgroup* G' is defined to be the subgroup *generated* by the *commutators*, i. e. by the elements of the form $xyx^{-1}y^{-1}$, for $x, y \in G$.

- Prove that G' is the smallest normal subgroup of G such that the corresponding factor group is abelian.
- Show that G' is the kernel intersection of the 1-dimensional representations of G.
- Verify that $|G/G'|$ is the number of 1–dimensional representations of G.

Exercise 11.4.5 Prove that A_n is the commutator subgroup of S_n and moreover that each element of A_n is itself a commutator.

Exercise 11.4.6 How many onedimensional representations of S_n exist?

Exercise 11.4.7 Use the fact that A_n is the only nontrivial proper normal subgroup of S_n, if $n \neq 4$, in order to show that each irreducible representation of S_n, $n \neq 4$, of dimension $\neq 1$ is *faithful*, i. e. it has kernel $\{1\}$.

Exercise 11.4.8 Show that for subgroups U and H of a finite group G such that $U \leq H$ and any representation D of U we have

$$D \uparrow G = (D \uparrow H) \uparrow G$$

(in short: the induction of representations is *transitive*, a notion which must not be confused with the transitivity of actions).

11.5 Ordinary Characters of Finite Groups

We now apply the preceding results to

$$\chi^D\colon G \to \mathbb{C}\colon g \mapsto \operatorname{tr} D(g) = \sum_i d_{ii}(g),$$

which maps $g \in G$ onto the trace of $D(g)$. The map χ^D is called the *character* of D. In fact D is characterized by its character in the sense that we shall be able to derive from it the irreducible constituents of D, which in turn characterize D.

11.5.1 Lemma *For any finite groups G, G'', ordinary representations D, D' of G and D'' of G'' and their characters we have:*

- $\chi^D(1_G) = d = \dim(D)$, $\chi^D(g^{-1}) = \overline{\chi^D(g)}$, *the complex conjugate of* $\chi^D(g)$, *and* $\chi^D(g)$ *is a sum of complex roots of unity.*
- $\chi^D(gg'g^{-1}) = \chi^D(g')$, *for any* $g, g' \in G$, *i. e.* χ^D *is constant on each conjugacy class of elements of* G, *in short:* χ^D *is a* class function.
- $\chi^D = \chi^{D'}$, *if* D *is equivalent to* D'.
- $\chi^{D+D'} = \chi^D + \chi^{D'}$, *i. e.* $\chi^{D+D'}(g) = \chi^D(g) + \chi^{D'}(g)$.
- $\chi^{D \otimes D'} = \chi^D \chi^{D'}$, *i. e.* $\chi^{D\otimes D'}(g) = \chi^D(g)\chi^{D'}(g)$.
- $\chi^{D \# D''} = \chi^D \chi^{D''}$, *i. e.* $\chi^{D \# D''}(g, g'') = \chi^D(g)\chi^{D''}(g'')$.

Proof: These statements are trivial except for the second and third part of the first statement. The third part follows from the fact that the eigenvalues of $D(g^k)$ are the k-th powers of the eigenvalues of $D(g)$ and therefore powers of a primitive $|G|$-th root of unity, as G is supposed to be finite. The second part of the first statement follows from the fact that the eigenvalues of $D(g^{-1})$ are the inverses of the eigenvalues of $D(g)$, and the fact that inverses of roots of unity are the complex conjugates of the roots. □

We are now going to consider a hermitian inner product on the space $\mathbb{C}^G$ consisting of all the complex valued functions on G. The inner product is defined by

$$[\varphi \mid \psi] := \frac{1}{|G|}\sum_g \varphi(g)\overline{\psi(g)}.$$

Applying it to irreducible characters, we get the following remarkable result:

11.5.2 Theorem *The ordinary irreducible characters of G form an orthonormal set with respect to* $[- \mid -]$.

Proof: 11.4.5 yields

$$[\chi^D \mid \chi^{D'}] = \sum_{i,k}\frac{1}{|G|}\sum_g d_{ii}(g)d'_{kk}(g^{-1}) = \begin{cases} 1, & \text{if } D, D' \text{ equivalent,}\\ 0, & \text{otherwise.}\end{cases}$$

□

If χ^D is the character of the ordinary representation D, and if

$$D = D^{(k-1)} + \ldots + D^{(0)}$$

is the decomposition of D into irreducible representations $D^{(j)}$, where exactly a_i of the $D^{(j)}$ are equivalent to an ordinary irreducible representation D_i of G, then, by 11.5.2, this multiplicity is the inner product

$$[\chi^D \mid \chi^{D_i}] = [\sum a_j\chi^{D_j} \mid \chi^{D_i}] = a_i.$$

Thus we have proved

11.5.3 Theorem *If D is an ordinary representation of G and D_i an ordinary irreducible representation, then, for any decomposition of D into irreducible constituents, the number (D, D_i) of constituents equivalent to D_i is equal to the inner product of the corresponding characters:*

$$(D, D_i) = [\chi^D \mid \chi^{D_i}].$$

This multiplicity *of D_i in D is therefore uniquely determined, and so there is essentially one such decomposition only and the multiplicities of the irreducible constituents of D, and hence also the character χ^D, characterize the equivalence class of D.*

We can therefore write

$$D = n_0 D_0 + \ldots + n_{r-1} D_{r-1},$$

where n_i is the multiplicity of the irreducible representation D_i in D.

It remains to determine the D_i and to show how many of them exist. For this purpose we consider the *regular representation* (cf. 1.1.11) which is induced by the identity representation of the identity subgroup, and which corresponds to the action of G on G via left multiplication:

$$R := I\{1\} \uparrow G.$$

It has the character

$$\chi^R(g) = \begin{cases} |G|, & \text{if } g = 1_G, \\ 0, & \text{otherwise.} \end{cases}$$

An application of 11.5.3 to this character yields:

11.5.4 Corollary *The multiplicity of the irreducible representation D_i of G in the regular representation is equal to the dimension d_i of D_i:*

$$(R, D_i) = d_i.$$

We are now in a position to evaluate the number of (equivalence classes of) ordinary irreducible representations of G. Lemma 11.5.1 together with theorem 11.5.2 shows that there are at most as many ordinary irreducible representations as there are classes of conjugate elements in G. We would like to show that this is the exact number. In order to do this we introduce, for each representation $D: G \to GL(V)$ and any $\varphi: G \to \mathbb{C}$ the following linear mapping on V:

$$D^\varphi := \sum_g \overline{\varphi(g)} D(g).$$

11.5.5 Lemma *If D is irreducible and φ a class function, then*

$$D^\varphi = \frac{|G|}{d} [\chi^D \mid \varphi] \cdot \text{id}.$$

Proof: A direct calculation shows that D^φ commutes with every $D(g)$, so that $D^\varphi = \lambda \cdot id$, by Schur's Lemma. But

$$\lambda \cdot d = \text{tr}\,(D^\varphi) = \sum_g \overline{\varphi(g)} \text{tr}\,(D(g)) = |G|[\chi^D \mid \varphi].$$

□

11.5.6 Theorem *The space $CF(G, \mathbb{C})$ of all the complex valued class functions on G has the ordinary irreducible characters of G as an orthogonal basis. Hence the number of conjugacy classes of elements of G, i. e. the dimension of $CF(G, \mathbb{C})$, is equal to the number of ordinary irreducible characters of G.*

Proof: By 11.5.2 it remains to prove that the subspace V of $CF(G, \mathbb{C})$ generated by the ordinary irreducible characters is in fact equal to $CF(G, \mathbb{C})$. Let φ be in the orthogonal complement $V^\perp$ of V. Then, by 11.5.5, $D^\varphi = 0$, for each irreducible representation D. Thus, for the regular representation R of G, we have $R^\varphi = \sum_i d_i D_i^\varphi = 0$, the zero mapping. But, on the other hand, as R is a mapping from G into $GL(\mathbb{C}^G)$, the general linear group of the vector space $\mathbb{C}^G$, and since $\mathbb{C}^G$ has the elements of G as a basis, we obtain:

$$R^\varphi 1_G = \sum_g \overline{\varphi(g)} R(g) 1_G = \sum_g \overline{\varphi(g)} g.$$

Thus $\overline{\varphi(g)} = 0$, for each $g \in G$, and so $V^\perp = \{0\}$ and $V = CF(G, \mathbb{C})$. □

In the sequel we shall denote the ordinary irreducible characters of G by ζ^i and we shall mean by ζ_k^i the value of ζ^i on the k-th conjugacy class of G. Furthermore we shall assume that ζ^0 is the character of the identity representation $D_0 = I$, and that the conjugacy class with number 0 is the one that consists of the identity element alone, so that $\zeta_1^i = d_i$, the dimension. Then the *character table*

$$Z(G) := (\zeta_k^i),\ 0 \le i, k < r := \text{number of conjugacy classes},$$

contains in its highest row only 1's, and in its leftmost column the dimensions d_i of the ordinary irreducible representations:

$$Z(G) = \begin{pmatrix} d_0 = 1 & \dots & 1 \\ \vdots & & \\ d_i & \star & \\ \vdots & & \\ d_{r-1} & & \end{pmatrix}.$$

11.5.7 The second orthogonality relation *For a finite group G, its irreducible characters ζ^i, $i = 0, \dots, r-1$, and elements $g, g' \in G$ we have*

$$\sum_{i=0}^{r-1} \zeta^i(g')\zeta^i(g^{-1}) = \begin{cases} |C_G(g)|, & \textit{if } g \textit{ and } g' \textit{ are conjugate elements,} \\ 0, & \textit{otherwise.} \end{cases}$$

Proof: Consider the function $\chi : G \to \mathbb{C}$ which is characteristic for the conjugacy class $C^G(g)$ of g, i. e. which has value 1 on $C^G(g)$ and zero elsewhere. Then $\chi = \sum_i a_i \zeta^i$, where $a_i = [\chi \mid \zeta^i] = |C^G(g)| \cdot |G|^{-1} \overline{\zeta^i(g)}$. We obtain

$$\chi(g') = \sum_i a_i \zeta^i(g') = \frac{|C^G(g)|}{|G|} \sum_i \zeta^i(g')\zeta^i(g^{-1}),$$

which yields the statement. □

Later we shall need a formula for the evaluation of the induced character. Assume that $H \leq G = \dot{\cup}_i g_i H$ and that χ is a character of H. The definition of induced representation yields

11.5.8 $$(\chi \uparrow G)(g) = \sum_i \dot{\chi}(g_i^{-1} g g_i), \text{ where } \dot{\chi}(g_i^{-1} g g_i) = \operatorname{tr}(\dot{\mathbf{D}}(g_i^{-1} g g_i)).$$

We can express this in terms of conjugacy classes C_i of H and the corresponding values χ_i of χ (exercise 11.5.5):

11.5.9 $$(\chi \uparrow G)(g) = \frac{|G|}{|H|} \sum_{i:\, C_i \subseteq C^G(g)} \frac{|C_i|}{|C^G(g)|} \chi_i .$$

An important particular case is $\chi = \chi^{IH}$, the identity character of H. We get from 11.5.9 that

11.5.10 $$\chi^{IH\uparrow G}(g) := \chi^{IH} \uparrow G(g) = \frac{|G||C^G(g) \cap H|}{|H||C^G(g)|}.$$

Another application of 11.5.9 yields (exercise 11.5.6):

11.5.11 Frobenius' Reciprocity Law *If $H \leq G$ and if χ is a character of H and ψ is a character of G, then we have the following equality of inner products:*

$$[\chi \uparrow G \mid \psi] = [\chi \mid \psi \downarrow H].$$

In many cases we can find the desired irreducible characters of a given finite group by simply considering the powers of a single character and decomposing it. The relevant result (of Burnside) is the following theorem:

11.5.12 Theorem (Burnside) *Assume that the finite group G possesses a faithful character χ which takes exactly k different values in $\mathbb{C}$. Then each irreducible character of G occurs among the irreducible constituents of the powers $\otimes^i \chi$, where $0 \leq i < k$.*

Proof: If this were not the case, we had an irreducible character ζ^j which satifies the following system of equations: If G_i denotes the set of elements in G where χ takes its i-th value χ_i, $1 \leq i \leq k$, and if we put

$$x_i^j := \sum_{g \in G_i} \zeta^j(g^{-1}),$$

then the following holds:

$$[\otimes^\nu \chi \mid \zeta^j] = \sum_{i=1}^k (\chi_i)^\nu x_i^j = 0,\ 0 \leq \nu < k.$$

But this is a system of equations the matrix of coefficients of which is the Vandermonde matrix of the character values:

$$\det((\chi_i)^\nu)_{1\le i\le k, 0\le \nu<k} = \prod_{1\le i<j\le k} (\chi_j - \chi_i) \neq 0.$$

Hence this system of equations has the trivial solution $\chi_i^j = 0$ only, a contradiction to the fact that χ is faithful (which means that, for a suitable i, $\chi_i^j = \zeta^j(1) = d_j \neq 0$). □

11.5.13 Example An easy example is the natural permutation representation of the symmetric group S_n which obviously is faithful. For example, in the case when $n = 3$, its character is

$$\chi(1) = 3, \ \chi((01)) = 1, \ \chi((012)) = 0,$$

as the value of the character of the natural permutation representation is equal to the number of fixed points of the permutation in question. The power $\otimes^0\chi$ of that character is the identity character ζ^0, its values on these representatives $1, (01), (012)$ of the conjugacy classes are $1, 1, 1$. The orthogonality relations show that ζ^0, the identity character, is contained in the natural character exactly once. Therefore, subtracting ζ^0 from the natural character, we obtain the character $\chi - \zeta^0$ with the values $2, 0, -1$ which turns out to be irreducible, and so we indicate it by ζ^1. Thus $\chi = \zeta^0 + \zeta^1$ is the decomposition of $\otimes^1\chi = \chi$ into irreducible characters.

In the next step we form the square $\otimes^2\chi$ of the natural character, its values are $9, 1, 0$, and an application of the orthogonality relations gives that it contains ζ^0 exactly twice and ζ^1 exactly three times. The remaining rest $\otimes^2\chi - 2\cdot\zeta^0 - 3\cdot\zeta^1$ has the values $1, -1, 1$, it is the sign character, and it is irreducible. Hence we have finally obtained the character table of the symmetric group S_3 :

$Z(S_3) =$

	(1^3)	(21)	(3)
ζ^0	1	1	1
ζ^1	2	0	-1
ζ^2	1	-1	1

(The first row gives the cycle partitions which characterize the conjugacy classes.) ◇

Exercises

Exercise 11.5.1 Show that a representation of a finite group G with character χ contains $g \in G$ in its kernel if and only if $\chi(g) = \chi(1_G)$.

Exercise 11.5.2 Rephrase exercise 2.1.1 in terms of characters.

Exercise 11.5.3 Prove that the sum of elements in each row of a character table of a finite group is a nonnegative integer.

Exercise 11.5.4 Consider $G \leq S_n$ and the set

$$\{k_1, \ldots, k_r\} := \{a_1(g) \mid g \in G\backslash\{1\}\}.$$

Show that $|G|$ divides the product $\prod_{i=1}^{r}(n - k_i)$. (Hint: The mapping $\pi \mapsto \prod(a_1(\pi) - k_i)$ is a class function on G, the inner product of which with each irreducible character lies in $\mathbb{Z}$. Such functions are called *generalized characters.*)

Exercise 11.5.5 Prove 11.5.9.

Exercise 11.5.6 Prove 11.5.11.

Exercise 11.5.7 A finite action ${}_GX$ was called k–fold transitive if and only if the corresponding action ${}_GX^k_{inj}$ of G on the injective k–tuples is transitive. Show that the character χ of a 2–fold transitive action ${}_GX$ is of the form $\chi = \zeta^0 + \xi$, where ζ^0 is the identity character and ξ is an irreducible character of G.

Exercise 11.5.8 Use 11.5.10 together with 2.1.6 in order to show that, for each finite group G and its subgroups A and B,

$$|A\backslash G/B| = \left[\chi^{IA\uparrow G} \mid \chi^{IB\uparrow G}\right] = \left[\chi^{IA\uparrow G} \cdot \chi^{IB\uparrow G} \mid \iota\right],$$

i. e. the number of (A, B)-double cosets is the inner product of the characters of G which are induced by the identity representation IA of A and the identity representation IB of B, or, equivalently, the inner product of the product of these two induced characters, and the identity character ι of G.

11.6 The Möbius Inversion

We start by introducing the notion of *incidence algebra*. Let $(P, \leq)$ denote a *poset* (partially ordered set). Its partial order $\leq$ allows us to introduce *intervals*

$$[p, q] := \{r \in P \mid p \leq r \leq q\}.$$

If all these intervals are finite, then $(P, \leq)$ is called a *locally finite* poset. Assuming this and denoting by $\mathbb{F}$ a field, we can turn the set

$$I_{\mathbb{F}}(P) := \{\varphi\colon P^2 \to \mathbb{F} \mid \varphi(p, q) = 0 \text{ unless } p \leq q\}$$

of all the *incidence functions* into an $\mathbb{F}$–algebra. The following addition and scalar multiplication define a vector space structure:

$$(\varphi + \psi)(p, q) := \varphi(p, q) + \psi(p, q)\,,\ (\rho\varphi)(p, q) := \rho \cdot \varphi(p, q),\ \rho \in \mathbb{F},$$

while the local finiteness allows to define the *convolution* product by

$$(\varphi \star \psi)(p,q) := \sum_{r\in[p,q]} \varphi(p,r)\psi(r,q).$$

This turns $I_{\mathbb{F}}(P)$ into a ring (exercise 11.6.1). The identity is Kronecker's δ*–function*

$$\delta(p,q) := \begin{cases} 1, & \text{if } p = q, \\ 0, & \text{otherwise.} \end{cases}$$

As scalar mutiplication and convolution satisfy

$$\rho(\varphi \star \psi) = (\rho\varphi) \star \psi = \varphi \star (\rho\psi),$$

this ring is even an $\mathbb{F}$*–algebra*, the *incidence algebra* over $\mathbb{F}$ of $(P, \leq)$. It is important to characterize its invertible elements, i. e. the φ in $I_{\mathbb{F}}(P)$ for which there exists a $\psi \in I_{\mathbb{F}}(P)$ such that $\psi \star \varphi = \delta$ and also $\varphi \star \psi = \delta$. We obtain these incidence functions by an easy argument from linear algebra that allows to identify the incidence functions with upper triangular matrices as follows. In the case when P is finite, we can embed the partial order into a total order (i. e. we can number the $p \in P$ in a way such that $p_i < p_k$ implies $i < k$). This yields an embedding of $I_{\mathbb{F}}(P)$ into the set of upper triangular matrices over $\mathbb{F}$ which respects addition and scalar multiplication. The convolution product corresponds to the matrix product of the associated matrices

$$\varphi \mapsto \Phi := (\varphi(p_i, p_k)),$$

and so φ is invertible if and only if the values $\varphi(p,p)$ are nonzero. But this is true also in the more general case when we only assume $(P, \leq)$ to be locally finite, since then we can easily show that the incidence function recursively defined in the second item of the following lemma is in fact an inverse with respect to the convolution product:

11.6.1 Lemma *For each locally finite partial order* $(P, \leq)$ *and any* $\varphi \in I_{\mathbb{F}}(P)$ *the following is true:*

- *φ is invertible if and only if, for each $p \in P$, we have*

$$\varphi(p,p) \neq 0.$$

- *If φ is invertible, then φ^{-1} satisfies $\varphi^{-1}(p,p) = \varphi(p,p)^{-1}$, and*

$$\begin{aligned} \varphi^{-1}(p,q) &= -\varphi(q,q)^{-1} \sum_{r\in[p,q)} \varphi^{-1}(p,r)\varphi(r,q) \\ &= -\varphi(p,p)^{-1} \sum_{r\in(p,q]} \varphi(p,r)\varphi^{-1}(r,q), \end{aligned}$$

where as usual $[p,q)$ and $(p,q]$ denote half open intervals.

An important invertible incidence function is the *zeta function* which describes the partial order in question:

$$\zeta(p,q) := \begin{cases} 1, & \text{if } p \leq q, \\ 0, & \text{otherwise.} \end{cases}$$

Its inverse is called the *Möbius function* of $(P, \leq)$:

$$\mu := \zeta^{-1},$$

for which we obtain from the preceding lemma the following recursions:

11.6.2 $$\mu(p,q) = -\sum_{r\in[p,q)} \mu(p,r) = -\sum_{r\in(p,q]} \mu(r,q).$$

This close connection between the zeta and the Möbius function provides a very useful inversion theorem which we introduce next. In a poset $(P, \leq)$ the sets

$$\{q \in P \mid q \leq p\}$$

are called the *principal (order–)ideals* of P, while the sets

$$\{q \in P \mid q \geq p\}$$

are called the *principal filters* . The inversion theorem now reads as follows:

11.6.3 The Möbius Inversion *Let $(P, \leq)$ denote a locally finite poset and let F and G denote mappings from P into the field $\mathbb{F}$. Then*

- *if all the principal ideals of P are finite, we have the following equivalence of systems of equations:*

$$\forall\, p: \; G(p) = \sum_{q\leq p} F(q) \iff \forall\, p: \; F(p) = \sum_{q\leq p} G(q)\mu(q,p).$$

- *If all the principal filters of P are finite, we have the following equivalence of systems of equations:*

$$\forall\, p: \; G(p) = \sum_{q\geq p} F(q) \iff \forall\, p: \; F(p) = \sum_{q\geq p} \mu(p,q)G(q).$$

Proof: The incidence algebra $I_{\mathbb{F}}(P)$ acts linearly on the vector space $\mathbb{F}^P$ from the right:

$$(F\cdot\varphi)(p) := \sum_{q\leq p} F(q)\varphi(q,p).$$

Thus $G(p) = \sum_{q\leq p} F(q)$ means that $G = F\cdot\zeta$, which is equivalent to $F = G\cdot\mu$, i. e. $F(p) = \sum_{q\leq p} G(q)\mu(q,p)$. This proves the first equivalence, the second follows analogously with the aid of the analogous action of the incidence algebra from the left. □

A particular locally finite poset is $(\mathbb{N}^*, |)$, the set of positive natural numbers together with divisibility as its partial order. The corresponding μ is called the *number theoretic* Möbius function. Instead of $\mu(p,q)$ one can write $\mu(q/p)$ in this case since the intervals $[p,q]$ and $[r,s]$ are order isomorphic if $q/p = s/r$, and so $\mu(p,q) = \mu(r,s)$, by the above mentioned recursion. In order to apply the corresponding number theoretic Möbius Inversion, we need to know the values of the number theoretic Möbius function. As ζ is defined by the partial order, the same holds for the Möbius function, and hence the Möbius function is the same for order isomorphic posets. Thus the values of the Möbius function on $(\mathbb{N}^*, |)$ can be calculated by noting that the Möbius function is multiplicative:

11.6.4 Lemma *From locally finite posets P, Q and their Möbius functions μ_P, μ_Q, we obtain the Möbius function $\mu_{P\times Q}$ of $P \times Q$, where*

$$(p_1, q_1) \leq (p_2, q_2) \iff p_1 \leq p_2, \text{ and } q_1 \leq q_2,$$

in the following way (use exercise 11.6.2):

$$\mu_{P\times Q}((p_1, q_1), (p_2, q_2)) = \mu_P(p_1, p_2) \cdot \mu_Q(q_1, q_2).$$

In order to evaluate the Möbius function on $(\mathbb{N}^*, |)$, say, we note that that the interval $[d, n]$, where d divides n, is order isomorphic to the cartesian product

11.6.5 $$[1, p_0^{k_0}] \times \ldots \times [1, p_{r-1}^{k_{r-1}}],$$

if n/d has the prime number decomposition $n/d = p_0^{k_0} \cdot \ldots \cdot p_{r-1}^{k_{r-1}}$, where the p_i are different prime numbers and the $k_i \geq 1$. Thus, for the number theoretic Möbius function we obtain from 11.6.4 and exercise 11.6.3 that

11.6.6 $$\mu(n/d) := \mu(d, n) = \mu(1, n/d) = \prod_{i=0}^{r-1} \mu(1, p_i^{k_i}).$$

But from 11.6.6 we can deduce that $\mu(1) = 1, \mu(p) = -1$, if p is prime, and $\mu(p^r) = 0$, if $r > 1$. This together with 11.6.5 finally yields:

11.6.7 $$\mu(n) = \begin{cases} 1, & \text{if } n = 1, \\ (-1)^r, & \text{if } n \text{ is a product of } r \text{ different primes,} \\ 0, & \text{otherwise.} \end{cases}$$

Exercises

Exercise 11.6.1 Check that addition and convolution in fact define a ring structure on the set $I_{\mathbb{F}}(P)$ of incidence functions.

Exercise 11.6.2 Prove 11.6.4.

Exercise 11.6.3 Let $(L, \wedge, \vee)$ denote a lattice and $(L, \leq)$ the corresponding poset. We call $f \in I_{\mathbb{F}}(L)$ *multiplicative* if and only if, for each x, y in L, an order isomorphism

$$[x \wedge y, x \vee y] \simeq [x \wedge y, x] \times [x \wedge y, y]$$

implies that

$$f(x \wedge y, x \vee y) = f(x \wedge y, x) f(x \wedge y, y).$$

- Prove that the invertible multiplicative $f \in I_{\mathbb{F}}(L)$ form a group.
- Show that the zeta function (and hence also the Möbius function) is multiplicative.
- Verify 11.6.5 and 11.6.7.

Exercise 11.6.4 Show that

$$\phi(n) = \sum_{d|n} d \cdot \mu(n/d), \text{ and } n = \sum_{d|n} \phi(d).$$

Exercise 11.6.5 Let $(M, \leq)$ and $(N, \leq)$ denote partially ordered sets and consider a pair (α, β) of mappings $\alpha: M \to N$ and $\beta: N \to M$. Then (α, β) is called a *Galois connection* if and only if α and β are antitone, and

$$(\beta \circ \alpha)(m) \geq m, \ (\alpha \circ \beta)(n) \geq n.$$

Prove that in this case the following is true:

- $\alpha \circ \beta \circ \alpha = \alpha$, and $\beta \circ \alpha \circ \beta = \beta$.
- $\alpha \circ \beta$ and $\beta \circ \alpha$ are *closure operators*:
  - $m \leq (\beta \circ \alpha)(m)$, and $n \leq (\alpha \circ \beta)(n)$,
  - $m \leq m' \Rightarrow (\beta \circ \alpha)(m) \leq (\beta \circ \alpha)(m')$,
  - $n \leq n' \Rightarrow (\alpha \circ \beta)(n) \leq (\alpha \circ \beta)(n')$,
  - $(\alpha \circ \beta)^2 = \alpha \circ \beta$, and $(\beta \circ \alpha)^2 = \beta \circ \alpha$.
- For the subsets $\bar{M}, \bar{N}$ of *closed elements* (i. e. the m, n with the property $m = (\beta \circ \alpha)(m), n = (\alpha \circ \beta)(n)$) we have

$$\bar{M} = \beta[N], \ \bar{N} = \alpha[M].$$

- $\bar{M}$ and $\bar{N}$ are antiisomorphic, with α and β as inverse mappings.

Exercise 11.6.6 Let $(M, \leq)$ and $(N, \leq)$ denote partially ordered sets. A mapping $\alpha: M \to N$ is called a *Galois function* if there exists a mapping β from N to M such that both these mappings are monotone, while

$$(\beta \circ \alpha)(m) \geq m, \text{ and } (\alpha \circ \beta)(n) \leq n.$$

Prove that in this case the following is true:

- $\beta \circ \alpha$ is a closure operator on M, while $\alpha \circ \beta$is a *co-closure operator* on n, which means:

$$n \geq (\alpha \circ \beta)(n), n \leq n' \Rightarrow (\alpha \circ \beta)(n) \leq (\alpha \circ \beta)(n'), (\alpha \circ \beta)^2 = \alpha \circ \beta.$$

- $\alpha[M]$ is the set of *co-closed elements* (i. e. the set of n such that $n = (\alpha \circ \beta)(n)$).
- The set $\bar{M}$ of co-closed elements of M and the set $\bar{N}$ of co-closed elements of N are isomorphic with α and β as inverse mappings.

12. Comments and References

This chapter begins with remarks on the history of finite group actions. The reader will then find comments that point to certain important articles and books, together with hints for further reading and additional references.

12.1 Historical Remarks, Books and Review Articles

Chapter 0 of the present book is devoted to labeled structures. They are important since unlabeled structures, which are the main subject here, are equivalence classes of labeled ones. In order to give a flavor of a modern treatment of labeled structures, too, the first chapter contains the basics of the theory of species. It should serve as an appetizer for the standard book on this topic, which is the book [7], by Bergeron, Labelle and Leroux. The original paper was by Joyal ([73]). Another important paper on species theory and it applications is the Habilitationsschrift of Strehl ([149]).

The chapter 1 is mainly devoted to the introduction of unlabeled structures as orbits of finite groups on finite sets. The enumeration of such structures, which is described in chapter 3, can be done by an application of the Cauchy–Frobenius Lemma, the history of which is described in articles by Neumann ([110]) and Wright ([168]). This lemma is mostly ascribed to Burnside, who gives it in the first edition of his book on finite groups ([21]), but the lemma is contained in section 118, while the ascription is at the beginning of section 119. In the *second* edition of this book ([20], 1911, reprinted by Dover Publications in 1955) which is mostly quoted, these sections are completely rewritten, and Burnside omits the ascription. This might be the reason for usually attributing this lemma to Burnside. Burnside's reference is to [55], a paper of Frobenius. Frobenius gives credit to Cauchy who proved this lemma for the transitive case in [34]. In fact Burnside proved a much stronger result, which is described in chapter 4 of the present book.

Besides the basic concepts of the theory of finite group actions, the second chapter contains in particular the notion of symmetry classes of mappings and the corresponding enumerative results. The pioneering publication on this topic was the famous paper [117] by Pólya, a masterpiece. It had a predecessor ([125]), a paper by Redfield, which was overlooked for many years. In fact Redfield's paper contains stronger results than Pólya's, but it is very difficult to read since it expresses the results in terms of operations on polynomials that can be understood more or less only in terms of linear representation theory. There was at least one further paper written by Redfield, it had been rejected once but it was published recently ([124]). Another paper, entitled "Enumeration distinguishable arrangements for general frame groups", was found together with an untitled manuscript. They are not published yet. A translation of Pólya's paper into English, together with an article on the fifty years' history of it, can be found in [118]. The paper by Pólya was motivated by the problem of chemical isomerism, which amounts in a certain sense to the enumeration (or better: the construction) of all the connected multigraphs with a given degree sequence, i. e. which correspond to a given chemical formula, the vertices colored by atom names. A good part of the history of this problem and its relationship with graph theory can be found in the book [14] by Biggs, Lloyd and Wilson. But their description — as possibly every other one, too — is incomplete since, for example, they do not mention von Humboldt, who stated the existence of isomerism long before it was verified (see [100],[71]).

This basic problem of constructing the molecular graphs that satisfy a given chemical formula, the *connectivity isomers*, was attacked in a big and successful project called DENDRAL, which was started in 1965 by Lederberg. This project is described in the book [99]. The present author, together with R. Laue and co–workers, also developed a program system that allows to construct molecular graphs. This program system is called MOLGEN, its first version is due to D. Moser (see [78]), a second one was implemented by Grund ([64]), the third version is due to Grüner ([65]). Descriptions of several other generators can be found in volume 27 of MATCH. The relevance of Redfield's work for chemistry is described in [102] by Lloyd.

One of the very first applications of Pólya's methods outside chemistry or graph theory is described by Slepian in [143], he uses it for an enumeration of symmetry types of boolean functions. The nicest short introductions to the enumerative applications of finite group actions to graph theory can be found in the booklet [67], edited by Harary and Beineke, and in the review articles [18] and [17] by de Brujin, which contain enormous collections of interesting examples. Review [17] emphasizes the applications to chemistry. The standard reference for graph theoretical applications is the book [68] by Harary and Palmer. An extensive description of this theory which emphasizes the fact that it is a particular case of finite group action theory was published in three parts by the present author and K.–J. Thürlings ([79], [80], [81]). These papers, books and review articles are also the standard references for the weighted enumeration. There are of course many other review articles on this subject, and it shows up more and more in books on general combinatorics, too.

The enumeration by stabilizer class is due to Burnside, it can be found in the second edition of the book that was already mentioned above: The weighted form of his lemma was proved first in the dissertation [148] by Stockmeyer, see also the papers [160] and [159] by White. The enumeration under finite group operations on posets and lattices is due to Plesken ([116]).

The present introduction to the representation theory of symmetric groups along group actions and a chain of set theoretic bijections was used first in my paper [75]. It is mixed with ideas coming from the paper [41] by Doubilet, Fox and Rota and the dissertation by Clausen ([37]) who was the first to use standard bideterminants in representation theory of symmetric groups (see also his further papers, in particular [35] and [36]). Chapter 7 is devoted to some of the applications of representation theory to combinatorial enumeration and vice versa. The main point is that Schur polynomials are very helpful here and, conversely, enumeration theory can contribute elegant proofs of results on Schur polynomials. These polynomials became important in particular in connection with Schur's famous discovery of the close relationship between the representation theories of symmetric and of general linear groups. But these polynomials were known before, as well as the intertwining matrices between this basis of the space of symmetric polynomials and the other bases mentioned. A reference which shows this clearly is the collection of tables [89], given by Kostka. A complete introduction to the representation theory of symmetric groups along the line of considering Schur polynomials is Littlewood's book [101]. A modern version of this theory and of important extensions of it is Macdonald's book [107], a second edition of which appeared in 1995. The relevance of representation theory and Schur polynomials for combinatorial enumeration was elucidated in particular in the papers [49], [50] by Foulkes and in Read's paper [122]. There are various other books where the representation theory of symmetric groups is described. An introduction which differs from the present one is given in the book [72] by James and the present author.

The Schubert polynomials, which generalize the Schur polynomials, are due to Lascoux and Schützenberger ([91]). Their motivation is the enumerative Schubert calculus, and they are the main tool in the computer algebra system SYMMETRICA for the representation theory, invariant theory and combinatorics of the finite symmetric groups and related classes of groups like wreath products of symmetric groups, alternating groups, general linear groups, and so on ([76]) designed originally by Kohnert ([87]). Various extensions of the system are due to further diploma theses, dissertations and research project results, in particular to Golembiowski ([59]) and the collaboration with Lascoux and his students, see e. g. [33], as well as to the collaboration with McDonough and Morris. The program system covers character decompositions, manipulations with symmetric polynomials, base changes, Schubert polynomials, irreducible matrix representations (ordinary and modular ones), evaluations of symmetry adapted bases, and so on. The program itself can be obtained via Internet from

http://www.mathe2.uni-bayreuth.de/axel/symneu_engl.html

it is free for non commercial purposes. There exists a recent software package ACE that allows to use SYMMETRICA and also to take advantage of the user shell of MAPLE. Another useful software package in MAPLE, for symmetric polynomials, is due to Stembridge (SF).

Further results on the Schubert polynomials can be found in Kohnert's paper [88], where he gives an exciting conjecture about what Schubert polynomials count. It was recently proved by Winkel ([166]). For applications to the Schubert calculus, the interested reader should consult the papers and the book ([56]) of Fulton.

12.2 Further Comments

Having mentioned the history of the Cauchy–Frobenius Lemma and of Burnside's Lemma as well as some basic sources, a few remarks on the other contents may be in order.

A description of Coxeter groups, which are a natural generalization of the symmetric groups (as generated by transpositions), can be found in the part [15] of the Bourbaki series of books. The Lehmer code is ascribed in the article [98] by D. H. Lehmer to his father. The Exchange Lemma is mentioned here since it is basic for the theory of Schubert polynomials.

Concerning the Garsia–Milne bijection it should be mentioned that P. Paule ([115]) pointed to the fact that it is an easy corollary of a lemma due to Ingleton and Piff, which gives a graph theoretic bijection (the aim of Garsia and Milne were bijections between sets of partitions). Besides the papers on enumeration of symmetry classes of mappings which were already mentioned in the section on history, there are hundreds of further ones among which the papers by de Bruijn and Read should be mentioned in particular, as well as the papers of Harary and his school. The interested reader will easily find the precise references in the extremely helpful data bank of the European Mathematical Society

http://www.emis.de/cgi-bin/MATH

based on the *Zentralblatt für Mathematik und ihre Grenzgebiete,* or in the book by Harary and Palmer, which was already mentioned, so that I do not need to list them all, otherwise the next hundred pages would have to be filled.

The weighted enumeration leads to the cycle indicator polynomials. Interesting papers on their enumeration and application are the papers by Oberschelp (in particular[112],[111],[113]). It is important to note that weighted enumeration reduces complexity. It was mentioned that also weighted enumeration leads to congruences, see e. g. [126]. The enumeration by stabilizer class is due to Burnside ([20]). Its disadvantage is that it needs a detailed information on the lattice of subgroups of the acting group, which is usually difficult to obtain. Here the various subgroup lattice programs can be applied that are incorporated in the program systems CAYLEY (which was used for the evaluation of the tables of marks given in the appendix of tables), DISCRETA, GAP as well as MAGMA. It is clear that there is much redundancy in this information, a paper that describes this is [127]. But it should be made clear that the philosophy of the present book is *not* to consider the induced *permutation* group $\bar{G}$ on X, but to consider as long as possible the *abstract* group G instead. Correspondingly, it is important to use the notion of *Burnside ring.* Dress, Kratzer, Morris, Siebeneicher, Solomon, Thévenaz, Wensley and Yoshida have published important papers. The canonic mapping from the Burnside ring $\Omega(C(|G|))$ into $\Omega(G)$ was introduced in [42] by Dress, Siebeneicher and Yoshida. Here the interested reader can find further applications of Burnside ring methods, for which he or she should also consult the paper [157] by B. Wagner.

The following papers discuss the evaluation of *complete catalogs* of graphs, so they are important under the general aspect of the exhaustive construction of discrete structures: [119],[30],[120], [123]. The evaluation of catalogs of graphs is discussed in [8]. An atlas of graphs was recently published by Read ([121]). It should be mentioned that catalogs can also be obtained using different methods, for example the Dixon–Wilf algorithm for generating orbit representatives uniformly at random ([40]). The application of this algorithm to the generation of unlabeled graphs is also described in [77]. The Dixon–Wilf algorithm can be applied to each mathematical structure that is defined as an orbit of a finite group on a finite set. We used it for *contexts*, which are defined to be orbits of 0-1–matrices under the direct product of the symmetric groups on the sets of rows and columns. Another application was the enumeration (due in particular to H. Fripertinger, see [51],[53],[54],[52]) and the generation of representatives (A. Betten) of isometry classes of linear codes (see [8] and the list of references given there). The problem is of course that one needs efficient isomorphism checks in order to build a catalog, but cataloging is in a sense an abuse of the Dixon–Wilf algorithm. This algorithm is in fact much better in just providing big sets of examples, and it solves this very important problem of constructing sets of examples without prejudice in an optimal way.

The most important references for the Robinson–Schensted correspondence, which was generalized by Knuth and embedded into combinatorics by Schützenberger and Lascoux are [83],[136], [138].

The first complete proofs of the Littlewood–Richardson rule I was aware of can be found in Wagner's diploma thesis ([158]) and Thomas' doctoral thesis ([153]). The first published proof (which is based on an idea of Robinson) can be found in Macdonald's book [107], while the proof in the book of James/Kerber is based on Wagner's thesis which uses an idea of Bender and Knuth ([6]).

The results on the root number functions on symmetric groups are due to Scharf ([134],[135]).

12.3 Suggestions for Further Reading

Further reading is of course a matter of personal taste. The following suggestions correspond to *my* personal taste, which is, as I said several times, focussed on *the constructive theory* of finite structures and, as it was also mentioned, we mainly dealt with graphs and their applications, with error correcting linear codes and with combinatorial designs. Hence several aspects were not mentioned at all, for example the knot theory, the enumerative theory of mappings, finite geometry, etc.

If you are primarily interested in foundations of combinatorics, I strongly recommend to read Aigner's book ([1]). If you prefer to concentrate on enumeration, then take Stanley's books ([145],[146]). Besides these books which are theory oriented, I think the reader should consider books on discrete structures, and in particular those that emphasize the *constructive aspects* of this theory. Helpful books which emphasize these aspects are the books by Jungnickel ([74]), Klin, Pöschel, Rosenbaum ([82]). The latter book is of particular importance for people like me who do not speak Russian, it contains a long list of relevant and important references of

Russian groups who were active in this field. [84] by Knuth contains a lot of interesting results and applications of standard tableaux, different aspects are covered by Lüneburg ([105]), Williamson ([165]).

The reader who is interested in group actions in their own right, should consult the booklets by Tamaschke, where the theory of permutation groups is described from a categorical point of view ([150],[151]).

Applications to group theory are described in the classical text by Wielandt which is another masterpiece ([163]), really a diamond. An extended description of applications to groups and geometry is the book by Tsuzuku ([155]). Another booklet with a similar intention is that by Biggs and White ([13]). The algebraic part of graph theory is nicely described in a further book by Biggs ([12]).

For the theory of actions on posets the interested reader is referred to Plesken's paper ([116]) which was already mentioned above.

There are many other methods known that come from group theory, from linear representation theory, character theory, and so on. But there are also many results on the theory of generating functions that can be sharpened to constructions, e. g. the methods described in the book by Goulden and Jackson ([61]). Schur polynomials and related structures like tableaux come up in books on combinatorics, too, e. g. in the books by Krishnamurthy ([90]), Sagan ([133]) and Stanton/White ([90]).

There is also the whole theory of enumeration of permutations with prescribed properties (like rises and falls). The corresponding theory of statistics on the symmetric group is a source of interesting problems and methods. The reader should consult the review article by Foata ([46]). This article refers (among other references) to the book by MacMahon ([108]) which is an important source for all kinds of combinatorial problems and results. But there are many more recent publications by the Lotharingian seminar of combinatorics, the interested reader is strongly recommended to have a look at the corresponding series of proceedings:

*Séminaire Lotharingien de Combinatoire (Bayreuth, Erlangen, Strasbourg), *Publication de l'institut de recherche mathématique avancée, Strasbourg.*

More than fourty volumes have been published already. The address of the home page is

http://cartan.u-strasbg.fr:80/~slc/

Another important source of results and questions is the book by Comtet ([38]). The related theory of words and of statistics on words is considered in two books by Lothaire ([104],[103]).

The Robinson–Schensted Construction and its generalization to matrices over $\mathbb{N}$, given by D. E. Knuth, is a source of interesting results and methods. Besides the review article by Schützenberger (who made this construction an important part of combinatorics) which was already mentioned, I should like to mention papers of Burge ([19]), Gansner ([57]), Remmel. The standard reference for the theory of partitions is Andrews' book ([3]). An interesting thesis, where for each q a bijection on the set of proper partitions of n is derived such that these bijections generate the symmetric group on this set of partitions is [147], by Stockhofe. More generally, replacing combinatorial identities by suitable bijections which give an insight into the relationship between various mathematical structures is of course a very important part of research. Here I strongly recommend to study the papers on the Robinson–Schensted Construction which were mentioned already, and to consider the resulting theory on the enumeration of permutations. The book and the various papers by Andrews are full of interesting identities and bijections between various sets of partitions. The dissertation of Stockhofe invites to rewrite these bijections in terms of his bijections. A more general approach to identities and bijections is beautifully described in various papers of Zeilberger, for example in [169] and [170]. (Recall the above mentioned remarks by P. Paule on a lemma by Ingleton and Piff and the Garsia-Milne bijection!)

The importance of the dominance order for representation theory of symmetric groups, combinatorics and sciences is described in articles by Ruch and Schönhofer ([132]), Ruch ([128]), Ruch and Gutman ([129]) and Aigner ([2]). The theory of unimodality of finite sequences is also a very interesting field of research. A review, by Brenti, of recent developments can be found in [16]. The basic results mentioned here are due to White ([161]), Macdonald, Stanley and the present author.

Anyone who wants to apply the concept of symmetry classes of mappings to a concrete situation, and who wants really to construct representatives, should take the following into account. The method of constructing representatives of the orbits of G on Y^X of content λ, $\lambda \models |X|$, can be interpreted as evaluating a transversal of the $(S_\lambda, \bar{G})$–double cosets in S_X. This is shown in a paper by Ruch, Hässelbarth and Richter ([130]). The more general bilateral classes were introduced in [69] by Hässelbarth, Ruch, Klein and Seligman. A review article on further applications of double cosets in science is [131], by Ruch and Klein. Interesting applications to group theory (constructions of solvable groups) are given by Laue in [95] and [92].

A careful implementation of the applications to symmetry classes of mappings can be found in the thesis by Grund ([64]). Methods for the evaluation of a double cosets transversal in the general case are described in articles by Butler ([22],[23]) and Schmalz ([137]). The question of representatives with given stabilizer is considered in a paper by Laue ([93]). Helpful sources for computer methods using groups are [44] by Felsch, [44], [70] by Hoffman, [141] and [142] by Sims, [43] by Fäßler and Stiefel.

The journals on theoretical physics are full of applications, in particular the Journal of Mathematical Physics. There are many notions that deserve a better mathematical explanation. And there are many open questions the physicists would like to have answered. They want to really put their hands on the structures in question (like on the matrix representations), while the mathematicians quite often do not want to look at them as close as that. But nowadays, in the times of cheap and efficient computers, there is at least no excuse left for being unable to examine them in full detail. For the evaluation of the irreducible polynomial representations of the general linear groups I can refer the interested reader to [62].

If you want to do sports, then you may take the book by Sloane ([144]) that contains an enormous list of integer sequences. Nowadays these sequences and corresponding references can be found under the address

http://www.research.att.com/~njas/sequences/

Pick a sequence, and try to find a corresponding sequence of group actions that have the elements of the sequence as numbers of orbits. Hints, how this can be done, you may find in the paper [28] by Cameron. If you are interested in properties of actions and their applications to general group theory, then further papers of Cameron are a source of interesting results and problems, too (for example [24],[26],[27],[25], [29]). As far as equations and roots in groups are concerned, I recommend the review article [45] by Finkelstein and the papers [135] and [134] by Scharf.

The research on Schubert polynomials was pushed forward since it pays by many applications to classical and modern algebra as well as to combinatorics and sciences, see the papers by Lascoux/Schützenberger and Kohnert which were already listed and, of course, the monograph by I. G. Macdonald ([106]), the Habilitationsschrift by Winkel ([167]), where a proof of a very stimulating conjecture by Kohnert can be found, papers by Fomin and other authors, the book by Fulton ([56]).

Since powerful computers became very cheap, the research on the *construction* of discrete structures in mathematics and sciencies became more and more important. In particular, as these machines are in a sense algebraic machines, mixtures of algebraic and combinatorial approaches, as they are described in this book, seem to allow much more efficient attacks to many problems of enormous complexity. For example to the basic problem of enumeration theory of symmetry classes of mappings, which is the construction of the connectivity isomers corresponding to a given chemical formula. Here I recommend to have a glance at the above mentioned book by Lindsay et al. on the DENDRAL project, in order first of all to get an idea of what is wanted by chemists, and how difficult the problem is. Afterwards the interested reader should have a look at the papers by Read and others on the various methods of cataloging graphs, and then to go on and consider his (or her) pet discrete structure under this aspect, if he or she can construct or even catalog it. If not, then the Dixon/Wilf algorithm should be considered, maybe the structure in question can be defined as an orbit of a finite group on a finite set, and therefore generated uniformly at random. If this is true, then some sort of pattern recognition can be started in order to find interesting properties, hypotheses, invariants, and so on, and finally, if some hypotheses were formulated, he or she should start to prove them. This is part of the very interesting and brand new branch of *experimental mathematics*.

A consequent application of these concepts can be found in the above mentioned book [8] on error correcting linear codes. It is devoted to the enumeration of isometry classes of finite vector spaces, to the construction of transversals of these classes (by evaluating generator matrices), and to the generation of such representatives uniformly at random.

References

1. Aigner, M. *Combinatorial theory. Repr. of the 1979 ed.* Classics in Mathematics. Berlin: Springer–Verlag 1997, x, 484 p. [ISBN 3-540-61787-6].
2. Aigner, M. Uses of the diagram lattice. Mitt. Math. Sem. Univ. Gießen 163, 61-77 (1984).
3. Andrews, G. E. *The theory of partitions.* Encyclopedia of Mathematics, its Applications, Vol. 2. Section: Number Theory. Addison-Wesley Publishing Company. Advanced Book Program (1976). Cambridge University Press (1984), XIV, 255 p. [ISBN 0-521-30222-6].
4. Arnauld, A., Nicole, P. *Die Logik oder die Kunst des Denkens (aus dem Französischen übersetzt und eingeleitet von Christos Axelos).* Wissenschaftliche Buchgesellschaft, Darmstadt, 1972.
5. Arnauld, A., Nicole, P. *La logique ou l'art de penser.* Paris 1662.
6. Bender, E. A., Knuth, D. E. Enumeration of plane partitions. J. Comb. Theory, Ser. A 13, 40-54 (1972).
7. Bergeron, F., Labelle, G., Leroux, P. *Combinatorial species and tree-like structures. Transl. from the French by Margaret Readdy.* Encyclopedia of Mathematics, Its Applications. 67. Cambridge University Press. xx, 457 p. [ISBN 0-521-57323-8].
8. Betten, A., Fripertinger, H., Kerber, A., Wassermann, A., Zimmermann, K.-H. *Codierungstheorie (Konstruktion und Anwendung linearer Codes).* Springer-Verlag 1998 [ISBN 3-540-64502-0].
9. Betten, A., Kerber, A., Laue, R., Wassermann, A. Es gibt 7-Designs mit kleinen Parametern!. Bayreuther Math. Schriften 49, 213 (1995).
10. Betten, A., Kerber, A., Laue, R., Wassermann, A. Simple 8-designs with small parameters. Designs, Codes, Cryptography 15, 5-27 (1998).
11. Biegholt, J. Computerunterstützte Berechnung von Multigraphen mittels Homomorphieprinzip. Diploma thesis, Bayreuth 1995.
12. Biggs, N. *Algebraic graph theory. 2nd ed.* Cambridge Mathematical Library. Cambridge: Univ. Press, 205 p., (1994) [ISBN 0-521-45897-8].
13. Biggs, N.L., White, A.T. *Permutation groups and combinatorial structures.* London Mathematical Society Lecture Note Series. 33. Cambridge University Press. VIII, 140 p. (1979).
14. Biggs, Norman L., Lloyd, E.K. and Wilson, R. J. The history of combinatorics. Graham, R. L. (ed.) et al., Handbook of combinatorics. Elsevier (North-Holland) 1995.
15. Bourbaki, N. *Éléments de mathématique. Fasc. XXXIV. Groupes et algèbres de Lie. Chapitres IV, V et VI: Groupes de Coxeter et systèmes de Tits. Groupes engendrés par des réflexions. Systèmes de racines.* Paris: Hermann & Cie. 288 p. (1968).
16. Brenti, F. Unimodal, log-concave and Pólya frequency sequences in combinatorics. Mem. Am. Math. Soc. 413, 106 p. (1989). [ISSN 0065-9266].
17. Bruijn, N.G. de. Pólya's Abzähl-Theorie: Muster für Graphen und chemische Verbindungen. Selecta math. 3, 1-26 (1971).
18. Bruijn, N.G.de. Pólya's theory of counting. Appl. Comb. Math. 144-184 (1964).

19. Burge, William H. Four correspondences between graphs and generalized Young tableaux. J. Comb. Theory, Ser. A 17, 12-30 (1974).
20. Burnside, W. *The theory of groups of finite order.* Cambridge 1911, reprinted by Dover Publications, 1955.
21. Burnside, W. *Theory of groups of finite order.* Cambridge University Press, 1897.
22. Butler, G. Double cosets and searching small groups. Symbolic, algebraic computation, Proc. ACM Symp., Snowbird/Utah 1981.
23. Butler, G. On computing double coset representatives in permutation groups. Computational group theory, Proc. Symp., Durham/Engl. 1982, 283-290 (1984).
24. Cameron, P. J. A combinatorial toolkit for permutation groups. Relations between combinatorics and other parts of mathematics, Proc. Symp. Pure Math. Am. Math. Soc., Columbus, Ohio 1978, Proc. Symp. Pure Math. 34, 77-96 (1979).
25. Cameron, P. J. Colour schemes. Ann. Discrete Math. 15, 81-95 (1982).
26. Cameron, P. J. Orbits and enumeration. Combinatorial theory, Proc. Conf., Schloß Rauischholzhausen 1982, Lect. Notes Math. 969, 86-99 (1982).
27. Cameron, P. J. Orbits, enumeration and colouring. Combinatorial mathematics IX, Proc. 9th Australian Conf., Brisbane Australia, 1981, Lect. Notes Math. 952, 34-66 (1982).
28. Cameron, P. J. Some sequences of integers. Discrete Math. 75, 89-102 (1989).
29. Cameron, P. J., Thomas, S. Groups acting on unordered sets. Proc. Lond. Math. Soc., III. Ser. 59, 541-557 (1989).
30. Cameron, R.D., Colbourn, C.J., Read, R.C., Wormald, N.C. Cataloguing the graphs on 10 vertices. J. Graph Theory 9, 551-562 (1985).
31. Carell, Th., Wintner, E.A., Bashir-Hashemi, A., Rebek, J. Jr. Neuartiges Verfahren zur Herstellung von Bibliotheken kleiner organischer Moleküle. Angewandte Chemie 106, 2159-2161 (1994).
32. Carell, Th., Wintner, E.A., Sutherland, A.J., Rebek, J. Jr., Dunayevskiy, Y.M., Vouros, P. New promise in combinatorial chemistry: synthesis, characterization, and screening of small-molecule libraries in solution. Chem. & Biol. 2, 171-183 (1995).
33. Carré, C. Plethysm of elementary functions. Bayreuther Math. Schriften 31, 1-18 (1990).
34. Cauchy, A.L. Mémoire sur diverses propriétés remarquables des substitutions régulières ou irrégulières, et des systèmes de substitutions conjuguées (suite). C. R. Acad. Sci. Paris 21, 972-987 (1845).
35. Clausen, M. Letter place algebras and a characteristic-free approach to the representation theory of the general linear and symmetric groups. I. Adv. Math. 33, 161-191 (1979).
36. Clausen, M. Letter place algebras and a characteristic-free approach to the representation theory of the general linear and symmetric groups. II. Adv. Math. 38, 152-177 (1980).
37. Clausen, M. Letter-Place-Algebren und ein charakteristik-freier Zugang zur Darstellungstheorie symmetrischer und voller linearer Gruppen. Bayreuther Math. Schriften 4, 133 S. (1980).
38. Comtet, L. *Advanced combinatorics. The art of finite, infinite expansions. Translated from the French by J. W. Nienhuys. Rev., enlarged ed.* Dordrecht, Holland - Boston, U.S.A.: D. Reidel Publishing Company. X, 343 p. (1974).
39. Desarmenien, J. and Kung, Joseph P.S. and Rota, Gian-Carlo. Invariant theory, Young bitableaux, and combinatorics. Adv. Math. 27, 63-92 (1978).
40. Dixon, J. D., Wilf, H. S. The random selection of unlabeled graphs. J. Algorithms 4, 205-213 (1983).
41. Doubilet, P., Fox, J., Rota, G.-C. The elementary theory of the symmetric group. Combinatorics, representation theory, statistical methods in groups, Young Day Proc., Lect. Notes pure appl. Math., Vol. 57, 31-65 (1980).
42. Dress, A.W.M., Siebeneicher, Ch., Yoshida, T. An application of Burnside rings in elementary finite group theory. Adv. Math. 91, 27-44 (1992).

43. Fäßler, A., Stiefel, E. *Group theoretical methods and their applications. Transl. from the German by Baoswan Dzung Wong.Rev. transl.* Birkhäuser, xii, 296 p. (1992). [ISBN 0-8176-3527-0].
44. Felsch, V. A bibliography on the use of computers in group theory, related topics: algorithms, implementations, and applications. SIGSAM Bull. 12, 23-86.
45. Finkelstein, H. Solving equations in groups: a survey of Frobenius' theorem. Period. Math. Hung. 9, 187-204 (1978).
46. Foata, D. Distributions Eulériennes et Mahoniennes sur le groupe des permutations. Higher Comb. Proc. NATO Adv. Study Inst., Berlin (West) 1976, 27-49 (1977).
47. Foata, D. *La série génératrice exponentielle dans les problèmes d'énumération.* Seminaire de mathématiques supérieures - ete 1971. No.54. Montreal: Les Presses de l'Université de Montreal. 186 p. (1974).
48. Foata, D., Schützenberger, M.-P. *Theorie géométrique des polynômes eulériens.* Lecture Notes in Mathematics. Vol. 138. Berlin-Heidelberg-New York: Springer-Verlag (1970).
49. Foulkes, H.O. On Redfield's group reduction functions. Can. J. Math. 15, 272-284 (1963).
50. Foulkes, H.O. On Redfield's range-correspondences. Can. J. Math. 18, 1060-1071 (1966).
51. Fripertinger, H. Cycle indices of linear, affine, and projective groups. Linear Algebra Appl. 263, 133-156 (1997).
52. Fripertinger, H. Enumeration of isometry-classes of linear (n, k)-codes over $GF(q)$ in SYMMETRICA. Bayreuther Math. Schr. 49, 215-223 (1995).
53. Fripertinger, H. Enumeration of linear codes by applying methods from algebraic combinatorics. Grazer Math. Ber. 328, 31-42 (1996).
54. Fripertinger, H. Zyklenzeiger linearer Gruppen und Abzählung linearer Codes. Semin. Lothar. Comb. 33, 11 p. (1994).
55. Frobenius, G. Über die Congruenz nach einem aus zwei endlichen Gruppen gebildeten Doppelmodul. Crelle's J. 101, 273-299 (1887).
56. Fulton, W. *Young tableaux. With applications to representation theory and geometry.* London Mathematical Society Student Texts. 35. Cambridge: Cambridge University Press. ix, 260 p. (1997). [ISBN 0-521-56724-6/pbk; ISSN 0963-1631].
57. Gansner, E. R. The enumeration of plane partitions via the Burge correspondence. Ill. J. Math. 25, 533-554 (1981).
58. Ganter, Bernhard and Wille, Rudolf. *Formale Begriffsanalyse. Mathematische Grundlagen.* Berlin: Springer. x, 286 p. (1996). [ISBN 3-540-60868-0].
59. Golembiowski, A. Zur Berechnung modular irreduzibler Matrixdarstellungen symmetrischer Gruppen mit Hilfe eines Verfahrens von M. Clausen. Bayreuther Math. Schriften 25, 135-222 (1987).
60. Goodman, F. M., O'Hara, K. M. On the Gaussian polynomials. q-Series and partitions, Proc. Workshop, Minneapolis/MN (USA) 1988, IMA Vol. Math. Appl. 18, 57-66 (1989).
61. Goulden, I.P., Jackson, D.M. *Combinatorial enumeration. With a foreword by Gian-Carlo Rota.* Wiley-Interscience Series in Discrete Mathematics. J. Wiley & Sons. XXIV, 569 p. (1983).
62. Grabmeier, J., Kerber, A. The evaluation of irreducible polynomial representations of the general linear groups and of the unitary groups over fields of characteristic 0. Acta Appl. Math. 8, 271-291 (1987).
63. Grace, J.H. and Young, A. *The Algebra of Invariants.* Cambridge University Press 1903, reprinted by Chelsea Publishing Company 1978.
64. Grund, R. Symmetrieklassen von Abbildungen und die Konstruktion von diskreten Strukturen. Bayreuther Math. Schr. 31, 19-54 (1990).
65. Grüner, T. Strategien zur Konstruktion diskreter Strukturen. Doctoral dissertation, Bayreuth 1998.

66. Grüner, Th., Laue, R., Meringer, M. Algorithms for group actions applied to graph generation. Finkelstein, Larry (ed.) et al., Groups and computation II. Workshop on groups and computation, June 7–10, 1995, New Brunswick, NJ, USA. Providence, RI: American Mathematical Society. DIMACS, Ser. Discrete Math. Theor. Comput. Sci. 28, 113-122 (1997). [ISBN 0-8218-0516-9].
67. Harary, F., Beineke, L. (eds.). *A seminar on graph theory*. Athena Series. Selected Topics in Mathematics. New York-Chicago-San Francisco-Toronto-London: Holt, Rinehart, Winston. VII, 116 p. (1967).
68. Harary, Frank, Palmer, Edgar M. *Graphical enumeration*. New York-London: Academic Press. XIV, 271 p. (1973).
69. Hässelbarth, W., Ruch, E., Klein, D.J., Seligman, T.H. Bilateral classes: A new class concept of group theory. Match 7, 341-348 (1979).
70. Hoffman, C.M. *Group theoretic algorithms and graph isomorphism*. Lecture Notes in Computer Science 136, Springer Verlag, 1982.
71. Humboldt, A. von. *Versuche über die gereizte Muskel- und Nervenfaser, nebst Vermutungen über den chemischen Prozeß des Lebens in der Tier- und Pflanzenwelt*. Rottmann, Leipzig, 1797.
72. James, G. D., Kerber, A. *The representation theory of the symmetric group. Foreword by P. M. Cohn, introduction by G. de B. Robinson*. Encyclopedia of Mathematics, Its Applications, Vol. 16. Reading, Massachusetts, Addison-Wesley Publishing Company, Advanced Book Program. XXVIII, 510 p. (1981).
73. Joyal, A. Une théorie combinatoire des séries formelles. Adv. Math. 42, 1-82 (1981).
74. Jungnickel, Dieter. *Graphen, Netzwerke und Algorithmen. 2nd rev. a. exp. ed.* B.I.-Wissenschaftsverlag 440 p. (1990). [ISBN 3-411-14262-6].
75. Kerber, A. La théorie combinatoire sous–tendant la théorie des représentations linéaires des groupes symétriques finis. M. Lothaire (ed.): Mots, Paris 1990.
76. Kerber, A., Kohnert, A., Lascoux, A. SYMMETRICA, an object oriented computer-algebra system for the symmetric group. J. Symb. Comput. 14, 195-203 (1992).
77. Kerber, A., Laue, R., Moser, D. Cataloging graphs by generating them uniformly at random. J. Graph Theory 14, 559-563 (1990).
78. Kerber, A., Laue, R., Moser, D. Ein Strukturgenerator für molekulare Graphen. Analytica Chimica Acta 235, 221-228 (1990).
79. Kerber, A., Thürlings, K.-J. Symmetrieklassen von Funktionen und ihre Abzählungstheorie. I: Die Grundprobleme. Bayreuther Math. Schriften 12, 236 S. (1983).
80. Kerber, A., Thürlings, K.-J. Symmetrieklassen von Funktionen und ihre Abzählungstheorie. II: Hinzunahme darstellungstheoretischer Begriffsbildungen. Bayreuther Math. Schriften 15, 338 S. (1983).
81. Kerber, A., Thürlings, K.-J. Symmetrieklassen von Funktionen und ihre Abzählungstheorie. III: Der Burnsidering und Verallgemeinerungen, Unimodalitätsfragen. Bayreuther Math. Schriften 21, 156-278 (1986).
82. Klin, M.Ch., Pöschel, R., Rosenbaum, K. *Angewandte Algebra für Mathematiker und Informatiker. Einführung in gruppentheoretisch-kombinatorische Methoden*. Vieweg 208 S. (1988). [ISBN 3-528-08985-7].
83. Knuth, D.E. Permutations, matrices, and generalized Young tableaux. Pac. J. Math. 34, 709-727 (1970).
84. Knuth, D.E. *The Art of Computer Programming, III*. Addison-Wesley, 2nd ed. 1998.
85. Koch, H. von. Die Unimodalität der Gauß'schen Polynome. (Dissertation). Fakultät für Mathematik und Informatik der Technischen Universität München. 52 S. (1982).
86. Koch, H. von. Elementary proof of the unimodality of Gauss polynomials. Riv. Mat. Univ. Parma, IV. Ser. 8, 495-500 (1982).
87. Kohnert, A. Die computerunterstützte Berechnung von Littlewood–Richardson Koeffizienten mit Hilfe von Schubertpolynomen. Diploma thesis, Bayreuth 1987.
88. Kohnert, A. Weintrauben, Polynome, Tableaux. Bayreuther Math. Schriften 38, 1-97 (1991).

89. Kostka, C. Tafeln für symmetrische Funktionen bis zur elften Dimension. (Mit kurzen Erläuterungen). Wissenschaftliche Beilage zum Programm des königlichen Gymnasiums und Realgymnasiums zu Insterburg, Ostern 1908, 10 pp.
90. Krishnamurthy, V. *Combinatorics: theory, applications.* Mathematics, its Applications. Statistics and Operational Research. Chichester: Ellis Horwood Limited; Halsted Press: a division of J. Wiley & Sons. XXXV, 483 p. (Orig. publ. by Affiliated East-West Press Private Ltd., India) (1986).
91. Lascoux, A., Schützenberger, M. P. Tableaux and non commutative Schubert polynomials. Funk. Anal. 23, 63-64 (1989).
92. Laue, R. Computing double coset representatives for the generation of solvable groups. Computer algebra, EUROCAM '82, Conf. Marseille/France 1982, Lect. Notes Comput. Sci. 144, 65-70 (1982).
93. Laue, R. Eine konstruktive Version des Lemmas von Burnside. Bayreuther Math. Schr. 28, 111-125 (1989).
94. Laue, R. Some Simple $9-(28,14,\lambda)$ Designs and Myriads of Other t-Designs. (in preparation).
95. Laue, R. Zur Konstruktion und Klassifikation endlicher auflösbarer Gruppen. Bayreuther Math. Schr. 9, 304 S. (1982).
96. Lehmann, W. Das Abzähltheorem der Exponentialgruppe in gewichteter Form. Mitt. math. Sem. Univ. Gießen 112, 19-33 (1974).
97. Lehmann, W. Ein vereinheitlichender Ansatz für die Redfield-Pólya-de Bruijnsche Abzähltheorie. Doctoral dissertation, Aachen 1976.
98. Lehmer, D.H. Teaching combinatorial tricks to a computer. Proc. Sympos. appl. Math. 10, 179-193 (1960).
99. Lindsay, R. K., Buchanan, B. G., Feigenbaum, E. A., Lederberg, J. *Applications of Artificial Intelligence for Organic Chemistry: The Dendral Project.* McGraw–Hill, 1980.
100. Lippmann, E.O. von. Alexander von Humboldt als Vorläufer der Lehre von der Isomerie. Chemiker-Zeitung 1, 1-2 (1909).
101. Littlewood, D. E. *The theory of group characters and matrix representations of groups.* Oxford University Press, 1940.
102. Lloyd, E. K. Redfield's papers, their relevance to counting isomers and isomerizations. Discrete Appl. Math. 19, 289-304 (1988).
103. Lothaire, M. *Combinatorics on words. Foreword by Roger Lyndon. 2nd ed.* Encyclopedia of Mathematics, Its Applications. 17. Cambridge: Cambridge University Press. xvii, 238 p. (1997). [ISBN 0-521-59924-5].
104. Lothaire, M. *Mots. Melanges offerts a M.-P. Schützenberger.* Langue, Raisonnement, Calcul. Paris: Editions Hermes. 401 p. (1990). [ISBN 2-86601-206-2].
105. Lüneburg, Heinz. *Tools and fundamental constructions of combinatorial mathematics.* B.I.-Wissenschaftsverlag 525 S. (1989). [ISBN 3-411-03194-8].
106. Macdonald, I.G. Schubert polynomials. Surveys in combinatorics, Proc. 13th Br. Comb. Conf., Guildford/UK 1991, Lond. Math. Soc. Lect. Note Ser. 166, 73-99 (1991).
107. Macdonald, I.G. *Symmetric functions and Hall polynomials.* Oxford Mathematical Monographs. Oxford: Clarendon Press. VIII, 180 p. (1979).
108. MacMahon, P. A. *Combinatory analysis. Vol. I, II.* New York: Chelsea Publishing Company. XIX, 302; XIX, 340 p. (1960).
109. Magliveras, S.S. and Leavitt, D.W. Simple 6-(33,8,36) designs from $P\Gamma L_2(32)$. Computational group theory, Proc. Symp., Durham/Engl. 1982, 337-352 (1984).
110. Neumann, P.M. A lemma that is not Burnside's. Math. Scientist 4, 133-141 (1979).
111. Oberschelp, W. Die Anzahl nicht-isomorpher m-Graphen. Monatsber. Math. 72, 220-223 (1968).
112. Oberschelp, W. Kombinatorische Anzahlbestimmungen in Relationen. Math. Ann. 174, 53-78 (1967).
113. Oberschelp, W. Strukturzahlen in endlichen Relationssystemen. Contrib. Math. Logic, Proc. Logic Colloq., Hannover 1966, 199-213 (1968).

114. O'Hara, K. M. Unimodality of Gaussian coefficients: A constructive proof. J. Comb. Theory, Ser. A 53, No.1, 29-52 (1990).
115. Paule, P. A remark on a lemma of Ingleton and Piff and the construction of bijections. Bayreuther Math. Schriften 25, 123-127 (1987).
116. Plesken, Wilhelm. Counting with groups and rings. J. Reine Angew. Math. 334, 40-68 (1982).
117. Pólya, G. Kombinatorische Anzahlbestimmungen fúr Gruppen, Graphen und chemische Verbindungen. Acta Math. 68, 145-254 (1937).
118. Pólya, G., Read, R. C. *Combinatorial enumeration of groups, graphs, and chemical compounds.* Springer Verlag, 1987.
119. Read, R. C. Every One a Winner. Ann. Discrete Math. 2, 107-120 (1978).
120. Read, R. C., Wormald, N. C. Catalogues of graphs and digraphs. Discrete Math., 224 (1980).
121. Read, R.C. Atlas of Graphs. Clarendon Press, Oxford 1998. [ISBN 0-19-853289-X].
122. Read, R.C. The use of S-functions in combinatorial analysis. Can. J. Math. 20, 808-841 (1968).
123. Read, R.C., Wormald, N.C. Counting the 10-point graphs by partition. J. Graph Theory 5, 183-196 (1981).
124. Redfield, J.H. Enumeration by frame group and range groups. J. Graph Theory 8, 205-224 (1984).
125. Redfield, J.H. The theory of group–reduced distributions. Amer. J. Math. 49, 433-455 (1927).
126. Rota, G.-C., Sagan, B. Congruences derived from group action. Eur. J. Comb. 1, 67-76 (1980).
127. Rota, G.-C., Smith, D. A. Enumeration under group action. Ann. Sc. Norm. Super. Pisa, Cl. Sci., IV. Ser. 4, 637-646 (1977).
128. Ruch, E. The diagram lattice as structural principle. Theoretica Chimica Acta 38, 167-183 (1975).
129. Ruch, E., Gutman, I. The branching extent of graphs. J. Comb. Inf. Syst. Sci. 4, 285-295 (1980).
130. Ruch, E., Hässelbarth, W., Richter, B. Doppelnebenklassen als Klassenbegriff und Nomenklaturprinzip für Isomere und ihre Abzählung. Theoretica Chimica Acta 19, 288-300 (1970).
131. Ruch, E., Klein, D.J. Double cosets in chemistry and physics. Theoretica Chimica Acta 63, 447-472 (1983).
132. Ruch, E., Schönhofer, A. Theorie der Chiralitätsfunktionen. Theoretica Chimica Acta 19, 225-287 (1970).
133. Sagan, B. E. *The symmetric group. Representations, combinatorical algorithms, and symmetric functions.* Wadsworth & Brooks/Cole Mathematics Series. Pacific Grove, xviii, 197 p. (1991). [ISBN 0-534-15540-5].
134. Scharf, Th. Die Wurzelanzahlfunktion in symmetrischen Gruppen. J. Algebra 139, 446-457 (1991).
135. Scharf, Th. Über Wurzelanzahlfunktionen voller monomialer Gruppen. Bayreuther Math. Schr. 38, 99-207 (1991).
136. Schensted, C. Longest increasing and decreasing subsequences. Canadian J. Math. 13, 179-191 (1961).
137. Schmalz, B. Verwendung von Untergruppenleitern zur Bestimmung von Doppelnebenklassen. Bayreuther Math. Schr. 31, 109-143 (1990).
138. Schützenberger, M.-P. La correspondance de Robinson. Comb. Represent. Groupe symetr., Actes Table Ronde C. N. R. S. Strasbourg 1976, Lect. Notes Math. 579, 59-113 (1977).
139. Schützenberger, Marcel P. Sur une construction de Gilbert de B. Robinson. Semin. P. Dubreil, 25e annee 1971/72, Algebre, Fasc. 1, 2, Expose 8, 4 p. (1973).

140. Schützenberger, Marcel Paul. Sur un theoreme de G. de B. Robinson. C. r. Acad. Sci., Paris, Ser. A 272, 420-421 (1971).
141. Sims, C. C. *Abstract algebra. A computational approach.* New York: J. Wiley & Sons, Inc. XV, 491 p. [ISBN 0-471-09846-9] (1984).
142. Sims, C. C. *Computation with finitely presented groups.* Encyclopedia of Mathematics, Its Applications. 48. Cambridge University Press. xiii, 604 p. (1994). [ISBN 0-521-43213-8].
143. Slepian, D. On the number of symmetry types of Boolean functions of n variables. Canadian J. Math. 5, 185-193 (1953).
144. Sloane, N.J.A. *A handbook of integer sequences.* New York-London: Academic Press, a subsidiary of Harcourt Brace Jovanovich, Publishers. XIII, 206 p. (1973).
145. Stanley, R. P. *Enumerative combinatorics. Vol. 1. 2nd ed.* Cambridge Studies in Advanced Mathematics. 49. Cambridge: Cambridge University Press. xi, 325 p., (1997). [ISBN 0-521-55309-1].
146. Stanley, R. P. *Enumerative combinatorics. Vol. 2.* Cambridge Studies in Advanced Mathematics. 62. Cambridge: Cambridge University Press (1998).
147. Stockhofe, D. Bijektive Abbildungen auf der Menge der Partitionen einer natürlichen Zahl. Bayreuther Math. Schr. 10, 1-59 (1981).
148. Stockmeyer, P. K. Enumeration of graphs with prescribed automorphism group. Ann Arbor, 1971.
149. Strehl, V. *Zykel–Enumeration bei lokal–strukturierten Funktionen.* Habilitationsschrift, Erlangen 1989.
150. Tamaschke, O. *Permutationsstrukturen. Vorlesungen an der Universität Tübingen im Wintersemester 1968/69.* B.I.-Hochschulskripten. 710/710a. Mannheim-Wien-Zürich: Bibliographisches Institut. XII, 276 S. (1969).
151. Tamaschke, O. *Schur-Ringe. Vorlesungen an der Universität Tübingen im Sommersemester 1969.* Mannheim-Wien-Zürich: Bibliographisches Institut AG XVI, 240 S. (1970).
152. L. Teirlinck. Non-trivial t-designs without repeated blocks exist for all t. Discrete Math. 65, 301-311 (1987).
153. Thomas, G. Baxter algebras, Schur–functions. Doctoral thesis, University of Wales 1974.
154. Thürlings, K.-J. Eine Verallgemeinerung des Lemmas von Cauchy-Frobenius, kombinatorische und algebraische Zusammenhänge. Bayreuther Math. Schr. 8, 39-131 (1981).
155. Tsuzuku, T. *Finite groups, finite geometries. Transl. by A. Sevenster and T. Okuyama.* Cambridge Tracts in Mathematics, 78. Cambridge University Press. XI, 328 p. (1982).
156. Turnbull, H.W. *The theory of determinants, matrices, and invariants. 3rd ed.* New York: Dover Publications, Inc., XVIII, 374 p. (1960).
157. Wagner, B. A permutation representation theoretical version of a theorem of Frobenius. Bayreuther Math. Schriften 6, 23-32 (1980).
158. Wagner, B. Symmetrische Polynome und Darstellungen der symmetrischen Gruppen. Diploma thesis, Aachen 1975.
159. White, D. E. Classifying patterns by automorphism group: an operator theoretic approach. Discrete Math. 13, 277-295 (1975).
160. White, D. E. Counting patterns with a given automorphism group. Proc. Amer. Math. Soc. 47, 41-44 (1975).
161. White, D.E. Monotonicity and Unimodality of the Pattern Inventory. Advances in Mathematics 38, 101-108 (1980).
162. Wielandt, H. Ein Beweis für die Existenz der Sylowgruppen. Arch. Math. 10, 401-402 (1959).
163. Wielandt, H. *Finite permutation groups.* New York, London: Academic Press. X, 114 p. (1964).

164. Wielandt, H. *Mathematische Werke. Vol. 2: Linear algebra, analysis. Ed. by Bertram Huppert, Hans Schneider.* Berlin: Walter de Gruyter. xx, 632 p. (1996). [ISBN 3-11-012452-1].
165. Williamson, S. G. *Combinatorics for computer science.* Computer Science Press, 1985.
166. Winkel, R. Diagram rules for the generation of Schubert polynomials. J. Comb. A (to appear).
167. Winkel, R. On Algebraic and Combinatorial Properties of Schur and Schubert Polynomials. Habilitationsschrift, Aachen 1998.
168. Wright, E.M. Burnside's lemma: A historical note. J. Comb. Theory, Ser. B 30, 89-90 (1981).
169. Zeilberger, D. Identities. Proc. Workshop, Minneapolis/MN (USA) 1988, IMA Vol. Math. Appl. 18, 35-44 (1989).
170. Zeilberger, D. Identities in search of identity. Theor. Comput. Sci. 117, 23-38 (1993).

Index

Printing (computer to film): Mercedes-Druck, Berlin
Binding: Buchbinderei Lüderitz & Bauer, Berlin